Thrombopoiesis and Thrombopoietins

Molecular, Cellular, Preclinical, and Clinical Biology

Thrombopoiesis and Thrombopoietins

Molecular, Cellular, Preclinical, and Clinical Biology

Edited by

David J. Kuter
The MGH Cancer Center,
Massachusetts General Hospital, Boston, MA

Pamela Hunt
Amgen Inc., Thousand Oaks, CA

William Sheridan
Amgen Inc., Thousand Oaks, CA
UCLA School of Medicine, Los Angeles, CA

Dorothea Zucker-Franklin
Department of Medicine, New York University Medical Center,
New York, NY

Humana Press  **Totowa, New Jersey**

For additional copies, pricing for bulk purchases, and/or information about other Humana titles, contact Humana at the above address or at any of the following numbers: Tel.: 201-256-1699; Fax: 201-256-8341; E-mail: humana@interramp.com

This publication is printed on acid-free paper. ∞
ANSI Z39.48-1984 (American Standards Institute) Permanence of Paper for Printed Library Materials.

Cover illustrations: Include Fig. 4 from Chapter 1, "Megakaryocyte Biology," by Carl W. Jackson, Julie T. Arnold, Tamara I. Pestina, and Paula E. Stenberg and Fig. 17 from Chapter 9, "The Purification of Thrombopoietin from Thrombocytopenic Plasma," by David J. Kuter, Hiroshi Miyazaki, and Takashi Kato.

Photocopy Authorization Policy:

Printed in the United States of America. 10 9 8 7 6 5 4 3 2 1

Library of Congress Cataloging-in-Publication Data

Thrombopoiesis and thrombopoietins : molecular, cellular, preclinical,
   and clinical biology / edited by David J. Kuter . . . [et al.].
      p. cm.
   Includes bibliographical references and index.
   ISBN 0-89603-379-1 (alk. paper)
   1. Thrombopoietin. I. Juter, David J.
      [DNLM: 1. Thrombopoietin—physiology. 2. Blood platelets—
cytology. 3. Hematopoiesis—physiology. QU 300 T531 1997]
QP92.1'15—dc21
DNLM/DLC
   for Library of Congress   96-44461
                                    CIP

Foreword

Shirley Ebbe, MD

The long-standing search for thrombopoietin (TPO) culminated about two years ago when a newly discovered hemopoietic growth factor was identified as TPO. In the subsequent two years the literature about TPO has increased at a meteoric rate. An understanding of cellular and subcellular events associated with megakaryocytopoiesis and its regulator that would have been unimaginable a few years ago is now being achieved. *Thrombopoiesis and Thrombopoietins: Molecular, Cellular, Preclinical, and Clinical Biology* provides a compilation of pertinent historical and recent research and projections about the potential clinical utility of TPO.

Research extending over 40–50 years indicated that there was feedback regulation of megakaryocytopoiesis. Therefore, there was an expectation that TPO existed, but conventional approaches to its purification were unsuccessful. Its identification finally hinged on a seemingly unrelated event when one mouse (of 238) injected with Friend murine leukemia virus developed a peculiar myeloproliferative disorder. The fascinating odyssey from that mouse to the identification of the *c-mpl* proto-oncogene to the realization that *c-mpl* was responsible for producing an orphan hemopoietic cytokine receptor (MPL) is presented by Wendling and Gisselbrecht in Chapter 6. The pivotal studies that identified MPL as a receptor for a growth factor that might specifically control megakaryocytopoiesis are summarized in Chapter 7 by Wendling, Debili, and Vainchenker. As testimony to the importance that was attached to discovering TPO, within a year or so after this work was published several groups had identified the ligand for MPL (Mpl-L) as the elusive TPO.

Since about 1950 there has been a sustained interest in megakaryocytopoiesis and platelet production, with scientific progress sometimes following the development and application of new technology. With the advent of accurate methods to count platelets *(1)* and transfuse them *(2)*, the importance of platelets for hemostasis in irradiated animals was appreciated. Recovery from severe depletion or overtransfusion of platelets in normal animals was characterized by transient overcorrection of the platelet count, and megakaryocyte morphology was sometimes observed to be altered in response to thrombocytopenia *(3,4)*. Therefore, it appeared that platelet production was subject to homeostatic regulation and, by analogy to erythropoietin, the concept of a TPO evolved. The capability to measure platelet survival with radiolabeled tracers permitted classification of thrombocytopenias and allowed for indirect quantification of platelet production. The pattern of labeling of megakaryocyte DNA with tritiated thymidine dispelled the preexisting concept that megakaryocyte cytoplasmic maturation and increasing nuclear DNA content proceeded hand-in-hand *(5)*. From the DNA content of individual megakaryocytes, it became clear that nuclear replication occurred in an orderly fashion and that it ceased at different ploidy levels (6). Chromosome analysis of bone marrow cells in Philadelphia–chromosome positive leukemia (7) and the morphology of splenic colonies in irradiated murine recipients of transplanted stem cells (8) demonstrated that megakaryocytes were derived from the same multipotential stem cells as granulocytes and erythroid cells, a finding confirmed by patterns of blood cell enzymes in myeloproliferative disorders.

Jackson, Arnold, Pestina, and Stenberg (Chapter 1) have integrated the results of many early and recent studies to provide a monumental description of the biology of megakaryocytes. They explain a hierarchy of megakaryocyte progenitors, development of polyploidy through mitotic events, kinetics of megakaryocyte development, effects of perturbations of the platelet count and of cytotoxic agents, instructive animal models, and clinical abnormalities of megakaryocytopoiesis. Characteristic features of the megakaryocytic response to thrombocytopenia, such as increases in size and ploidy, are identified, thus establishing some criteria for the action of TPO. The sequential acquisition of biochemical markers, development of ultrastructural organelles, and fragmentation into platelets are analyzed. The plethora of constituents, not all of which have an obvious role in hemostasis, and their meticulous subcompartmentalization are impressive. Levin presents a comparison of mammalian platelets and multifunctional amebocytes of invertebrates in Chapter 3. In addition to their well-known hemostatic potential, platelets also have some of the nonhemostatic capabilities of amebocytes, which he proposes may be evolutionary remnants. However, new insights may demonstrate the importance, or its lack, of the multiple platelet components that seem to have no apparent role in the platelet's primary function of hemostasis.

At the same time that studies of the kinetics of megakaryocytopoiesis were underway, i.e., from the 1960s through the 1980s, there were efforts to purify TPO from biological sources by concentrating procedures similar to those that were ultimately successful for erythropoietin. Mazur, in Chapter 5, has written an inclusive presentation of searches for a specific regulatory substance. As he describes, a serious problem was the lack of sensitive assays for TPO. A vagary of research is illustrated by the fact that two groups (Chapter 9 by Kuter, Miyazaki, and Kato) finally succeeded in purifying TPO from thrombocytopenic plasma by fractionation techniques at exactly the same time that Mpl-L was produced with the technology of molecular biology. Some of the reasons for their success include innovative bioassays for TPO, removal of inhibitors, and the realization that plasma TPO levels were greater in animals with hypoplastic thrombocytopenia than in those with an intact marrow. Eaton (Chapter 8) describes how Mpl-L was more rapidly characterized by purification from large quantities of plasma from animals with radiation-induced thrombocytopenia, using MPL for receptor–affinity chromatography. Another successful approach depended on the production of Mpl-L by transformation of a cell line engineered to express the receptor. In both cases, the ligand was then used to obtain the DNA encoding it, thus permitting development of recombinant technology to produce and characterize Mpl-L.

When it became possible to demonstrate colonial growth of megakaryocyte progenitor cells in cultures (9), such cultures were applied to the search for specific growth factors, an approach that had been fruitful for growth factors for leukocytes. Specific factors were not identified, but multifunctional factors were found to stimulate megakaryocyte colony formation as well as colonies of other types. Mazur (Chapter 5) and

Hoffman (Chapter 10) provide in-depth discussions of these. Some of these factors affect proliferation of progenitors, whereas others influence maturation, with little apparent overlap, thus supporting a two-level theory of the regulation of megakaryo-cytopoiesis in which cellular proliferation of early progenitors in vitro is regulated by nonspecific cytokines differing from those, including putative TPO, that regulate such later events as the development of polyploidy and cytoplasmic maturation. Mazur, however, indicates that Mpl-L supports the cellular proliferation phase as well as the later events, thus perhaps questioning the physiological relevance of the two-level model. As described by Debili, Cramer, Wendling, and Vainchenker in Chapter 14, Mpl-L affects proliferation, differentiation, and maturation when applied to progenitor cells in vitro, but the primary target cell for Mpl-L is a relatively mature megakaryocyte progenitor. Though the major effects of Mpl-L are on megakaryocytopoiesis, other lineages are also affected, especially in combination with other cytokines, thus suggesting that Mpl-L may not be totally lineage specific. The potential complexity of regulation is apparent from Hoffman's discussion of inhibitory cytokines, autocrine regulation, the microenvironment, the secondary production of one cytokine by another, and other confounding elements. Therefore, it would probably be premature to conclude that TPO is the only regulator of platelet production in all situations.

Often cited throughout this book are the findings in mice in which the gene for MPL or TPO was disabled or "knocked-out," and these studies are summarized in Chapter 21 by deSauvage and Moore. In both models, thrombocytopenia and megakaryo-cytopenia with reduced ploidy were seen. The residual platelets were macrocytic, indicating that this feature, known to occur with thrombocytopenia, may not be mediated by the TPO/MPL interaction. The fact that thrombocytopenia and megakaryocytopenia were not absolute is consistent with older observations that platelet hypertransfusion never completely eradicated megakaryocytes in rodents, in contrast to the elimination of erythropoiesis by red cell hypertransfusion. However the reasons for residual platelet production in either case are not apparent. That the TPO/MPL system may affect other hemopoietic cells was again shown by the reduction in knockout mice of colony-forming cells, but not blood cells, of other lineages. Finally, de Sauvage and Moore interpret the findings in these mice to indicate that blood levels of TPO are regulated by direct interaction with platelets rather than by adjustment of TPO production.

TPO is a structurally unique hemopoietic cytokine consisting of two domains, the biologically active domain with partial homology to erythropoietin (EPO) and a long carboxyl domain with many glycosylation sites. Gurney and de Sauvage (Chapter 11) describe the protein structure, forms of mRNA, gene structure, chromosomal location, and similarities of TPO structure in different species. In spite of homology with EPO and resemblance between the receptors for the two factors, there is no cross activation of receptors, thus eliminating the thought that some of the curious interrelationships between erythropoiesis and thrombocytopoiesis might arise from stimulation of one receptor by the cytokine for the other. Foster and Hunt (Chapter 13) expand further on the unique structure of TPO and what it means in terms of biological activity and survival of TPO in the circulation. Full-length TPO and a variety of truncated forms are detectable in human serum. The glycosylated portion appears to be important for synthesis, secretion, and stability in the circulation. It is not clear whether the longer half-life arising from preferential glycosylation or conjugation with poly[ethylene glycol] over that of truncated forms indicates lower reactivity with platelets, or whether other clearance mechanisms exist.

When cells respond to TPO, a complex of reactions occurs as a result of receptor activation. These are considered by Kaushansky, Broudy, and Drachman (Chapter 16) and Shivdasani (Chapter 12). The organization of the unique MPL receptor and how it may transmit signals for proliferation or differentiation are described. Reasons why lineage-specific patterns of gene expression have not yet been identified are given. However, the importance of nuclear transcription factors in megakaryocytopoiesis is emphasized by the maturation arrest and thrombocytopenia that occur with the elimination of one or the overexpression of another (Chapter 12). Kaushansky et al. also allude to the presence of MPL on leukemic cells, pluripotential hemopoietic cells, and erythroid progenitors. Some of these may account for the nonspecific response to TPO they and others (Farese and MacVittie, Chapter 20) observed in myelosuppressed animals and which was found during the induction of acute thrombocytopenia at the time of exposure to radiation *(10)*.

Megakaryocytes are curious cells. They mature by getting bigger and bigger, and then their cytoplasm fragments into platelets. Levin (Chapter 3) discusses, but remains puzzled by, the possible advantages that this mode of production may have over the cell-by-cell production of amebocytes or nucleated thrombocytes of so-called lower species. The question of how megakaryocyte cytoplasm actually fragments into platelets is surrounded by controversy. In Chapter 17, Choi describes the morphogenesis of megakaryocytes into proplatelets and platelets in vitro and reviews the sometimes confusing and contradictory reports of the roles of various factors on proplatelet formation. She then presents the unexpected finding from her laboratory that TPO inhibits proplatelet formation. It would be interesting to know whether the same inhibition occurs in vivo, i.e., if formation of cytoplasmic projections through the sinusoidal endothelium is affected in TPO-treated animals, since these projections may be the *in situ* counterparts of proplatelets. TPO also seems to inhibit apoptosis in mature megakaryocytes. Does this mean that fragmentation into platelets is actually a part of the process of programmed cell death? This is clearly a work in progress, and the author points out issues that are under investigation. However, it may be relevant to recall the earlier finding that partially hepatectomized rats promptly develop thrombocytopenia owing to deficient platelet production *(11)*. It appeared that the liver produced both TPO, as now thought from the localization of its mRNA, and an unidentified factor that affected release of platelets from the marrow, possibly a factor that might influence proplatelet formation.

Until now, platelet transfusions have been the treatment of choice for thrombocytopenic hemorrhage in patients with marrow aplasia. In Chapter 4, Schiffer explains why they are sometimes ineffective or even hazardous to patients. An exciting prospect is that TPO will prove to be so effective in increasing platelet counts that platelet transfusions can be avoided. Recombinant technology now appears capable of producing seemingly unlimited quantities of active Mpl-Ls, and preclinical analyses of them are promising, so this possibility can be tested. Schiffer identifies categories of patients who will be most likely to benefit from TPO therapy, and he outlines criteria for clinical trials of TPO. Results of clinical trials are not included in *Thrombopoiesis and Thrombopoietins: Molecular, Cellular, Preclinical, and Clinical Biology*, but preclinical data are.

In anticipation of administering TPO to humans, it is necessary to know whether the function of platelets might be affected. The importance of ultrastructural analysis to evaluate megakaryocytes and platelets produced under the influence of cytokines is underscored by Zucker-Franklin (Chapters 2 and 15). Some cytokines are associated with megakaryocytic or platelet cytoplasmic immaturity or ultrastructural aberrations that are not apparent with other tests. Platelets produced under the influence of cytokines could be hypo-, hyper-, or normally reactive, and ultrastructural analysis is an important tool for predicting which condition may be dominant. Harker, Marzec, and Toombs (Chapter 18) describe the effects of Mpl-L on platelet function when tested in vitro; and in Chapter 19, Harker, Toombs, and Stead disclose its effects in animal models of thrombosis. Mpl-L enhanced the aggregation of platelets by agonists, but there was no increase in thrombus formation in vivo. Thus the enhanced aggregation appeared to be a laboratory phenomenon that they propose may be caused by an increased proportion of young platelets or a sharing of signaling sequences, since both Mpl-L and platelet agonists induce protein tyrosine phosphorylation. It has been demonstrated that platelets and megakaryocytes are heterogeneous for a tyrosine phosphatase and that platelets lacking the enzyme are the more reactive *(12)*. Therefore, it could also be suggested that the ratio of negative to positive platelets might be altered by administration of Mpl-L.

In Chapter 19 the effects of megakaryocyte growth and development factor (MGDF) (a truncated Mpl-L) on platelet production and turnover in normal animals are presented. There was a close correlation between an increase in megakaryocyte mass and the increase in platelet mass turnover. In conjunction with the increasing platelet count, mean platelet volume (MPV) decreased, suggesting again that other factors are responsible for the increased MPV in thrombocytopenic individuals. In these normal animals changes in leukocyte, neutrophil, or red cell counts did not occur. Some human candidates for treatment with TPO will have marrow suppression. Chapter 20 by Farese and MacVittie is a comprehensive summary of the effects of various cytokines, alone and in combinations, in myelosuppressed animal models. Mpl-L was much superior to other cytokines in improving platelet counts, indicating that an excess may be effective even in the presence of presumably high levels of endogenous TPO (Nichol, Chapter 22) and low numbers of progenitors. In contrast to normal animals, Mpl-L also improved leukocyte counts and hematocrits suggesting that marrow cellularity may affect the proliferation and/or differentiation of multilineage or nonmegakaryocytic progenitors. The authors also describe cytokine combinations that may maximize hemopoietic recovery and preview some engineered agonists of cytokine receptors.

The cumbersome TPO bioassays of the past can now be replaced with immunoassays. These and assays for other cytokines that affect thrombocytopoiesis are reviewed by Nichol in Chapter 22, together with blood levels of TPO in different clinical conditions. Patients with hypo- or aplastic marrows show an inverse correlation between platelet count and TPO level in contrast to idiopathic thrombocytopenic purpura (ITP) in which TPO levels are normal or only slightly elevated. She speculates that the difference may be caused by increased platelet production in ITP with absorption of TPO by newly produced, but short-lived, platelets or to an effect of the megakaryocyte mass on TPO levels. Others have attributed the low levels of TPO in thrombocytolytic thrombocytopenias to the megakaryocyte mass *(13)*, a conclusion that may be favored by the knowledge that platelet production is not always increased in ITP *(14)*. That TPO levels could be affected by megakaryocyte mass might have relevance to an understanding of some compensated hypomegakaryocytic states that have been observed in animals (Chapter 1) and may have an analogy to the observation that EPO levels are influenced by the amount of erythroid marrow *(15)*. It would be instructive to know comparative survivals of TPO in the circulation of normal animals and in those with antibody-mediated or aplastic thrombocytopenia. Other interesting, and as yet unresolved, issues raised by Nichol are the elevated TPO levels in patients with primary thrombocytoses and the exceedingly high levels in some patients with chronic liver disease and persistent thrombocytopenia.

A unified concept about the regulation of platelet production is presented by Kuter in Chapter 23, based on work in his and other laboratories. He envisions that the only physiologically relevant regulator of platelet production is TPO that is constitutively produced by the liver. The amount of TPO available to influence megakaryocytopoiesis would be that amount that is not bound and removed from the circulation by platelets. The amount bound would be determined by the total body platelet mass and the ability of platelets to metabolize TPO. Elevated TPO levels in myeloproliferative disorders might be caused by dysfunctional metabolism of TPO. With the capability now to measure several facets of TPO metabolism, such as the ability of platelets to bind TPO, the number of Mpl receptors on platelets, and mRNA for TPO in different tissues, some of these postulates can now be tested experimentally.

This book contains a wealth of information about megakaryocytopoiesis and platelet production, their regulation, and areas for ongoing or future research. All authors are recognized authorities on their subjects. I have previewed the chapters, and the emphases and commentaries reflect my personal perspective. The new knowledge about TPO and its receptor opens the door to the possibility for addition of a valuable therapeutic agent to the physician's armamentarium, and it is furthering the understanding of megakaryocyte and platelet physiology. This book will serve as a valuable reference for the uninitiated as well as veteran megakaryocyte researchers, but we must continue to watch the latest publications for information in this rapidly expanding field of research.

Finally I would like to remember two men whose works contributed mightily to the scientific progress summarized in this book and who died recently: Dr. Olav Behnke and Dr. N. Raphael Shulman. They will be missed.

References

1. Brecher G, Cronkite EP. Morphology and enumeration of human blood platelets. *J Appl Physiol.* 1950;3:365–377.
2. Dillard GHL, Brecher G, Cronkite EP. Separation, concentration and transfusion of platelets. *Proc Soc Exp Biol Med.* 1951;78:796–799.
3. Craddock CG Jr, Adams WS, Perry S, Lawrence JS. The dynamics of platelet production as studied by a depletion technique in normal and irradiated dogs. *J Lab Clin Med.* 1955;45:906–919.
4. Witte S. Megakaryocyten und Thrombocytopoese bei der experimentellen thrombocytopenischen Purpura. *Acta Haematol.* 1955;14:215–230.
5. Feinendegen LE, Odartchenko N, Cottier H, Bond VP. Kinetics of megakaryocyte proliferation. *Proc Soc Exp Biol Med.* 1962;111:177–182.
6. Odell TT Jr, Jackson CW, Gosslee DG. Maturation of rat megakaryocytes studied by mi-

crospectrophotometric measurement of DNA. *Proc Soc Exp Biol Med.* 1965;119:1194–1199.

7. Whang J, Frei E III, Tjio JH, Carbone PP, Brecher G. The distribution of the Philadelphia chromosome in patients with chronic myelogenous leukemia. *Blood.* 1963;22:664–673.

8. Till JE, McCulloch EA. A direct measurement of the radiation sensitivity of normal mouse bone marrow cells. *Radiat Res.* 1961;14:213–222.

9. Metcalf D, MacDonald HR, Odartchenko, Sordat B. Growth of mouse megakaryocyte colonies in vitro. *Proc Natl Acad Sci.* 1975;72:1744–1748.

10. Ebbe S, Phalen E, Threatte G, Londe H. Modulation of radiation-induced hemopoietic suppression by acute thrombocytopenia. *Ann NY Acad Sci.* 1985;459:179–189.

11. Siemensma NP, Bathal PS, Penington DG. The effect of massive liver resection on platelet kinetics in the rat. *J Lab Clin Med.* 1975;86:817–833.

12. Behnke O. Blood platelet heterogeneity: a functional hierarchy in the platelet population. *Br J Haematol.* 1995;91:991–999.

13. Emmons RVB, Reid DM, Cohen RL, Meng G, Young NS, Dunbar CE, Shulman NR. Human thrombopoietin levels are high when thrombocytopenia is due to megakaryocyte deficiency and low when due to increased platelet destruction. *Blood.* 1996;87:4068–4071.

14. Ballem PJ, Segal GM, Stratton JR, Gernsheimer, Adamson JW, Slichter SJ. Mechanisms of thrombocytopenia in chronic autoimmune thrombocytopenic purpura. Evidence of both impaired platelet production and increased platelet clearance. *J Clin Invest.* 1987;80:33–40.

15. Stohlman F Jr. Observations on the physiology of erythropoietin and its role in the regulation of red cell production. *Ann NY Acad Sci.* 1959;77:710–724.

Preface

Our principal aims in bringing together a book on thrombopoiesis at this point after the discovery of its long-sought physiologic regulator, the Mpl ligand, are to educate investigators and to stimulate further research. We hope that we have succeeded on both counts and would like to thank the many outstanding scientists who have contributed directly and indirectly to this volume.

Since different perspectives often help one to arrive at a closer approximation to the truth, we have not attempted to remove areas of either controversy or overlap between the various chapters.

We would like to especially thank Paul Dolgert and Fran Lipton of Humana Press for their persistence and responsiveness; Lawrence Transue for technical typing; Dr. MaryAnn Foote for her tireless and patient attention to detail; and Dr. Shirley Ebbe for reading the book in its entirety and preparing the Foreword.

David J. Kuter
Pamela Hunt
William Sheridan
Dorothea Zucker-Franklin

Contents

Contributors

JULIE T. ARNOLD, PhD • *St Jude Children's Research Hospital, Memphis, Tennessee*
VIRGINIA C. BROUDY, MD • *University of Washington School of Medicine, Seattle, Washington*
ESTHER CHOI, PhD • *Amgen Inc., Thousand Oaks, California*
ELISABETH CRAMER, MD • *INSERM, Hôpital Henri Mondor, Creteil, France*
NAJET DEBILI, PhD • *INSERM, Institut Gustave Roussy, Villejuif, France*
FREDERIC J. DE SAUVAGE, PhD • *Genentech, Inc., South San Francisco, California*
JONATHAN G. DRACHMAN • *University of Washington School of Medicine, Seattle, Washington*
DAN EATON, PhD • *Genentech, Inc., South San Francisco, California*
SHIRLEY EBBE, MD • *Lawrence Berkeley National Laboratory, Berkeley, California*
ANN M. FARESE, MS, MT (ASCP) • *School of Medicine, University of Maryland Cancer Center, Baltimore, Maryland*
DONALD FOSTER, PhD • *Zymogenetics Inc., Seattle, Washington*
SYLVIE GISSELBRECHT, MD • *INSERM, Institut Cochin de Génétique Moléculaire, Paris, France*
AUSTIN L. GURNEY, PhD • *Genentech, Inc., South San Francisco, California*
LAURENCE A. HARKER, MD • *Blomeyer Professor of Medicine, Emory University School of Medicine, Atlanta, Georgia*
RONALD HOFFMAN, MD • *Professor of Medicine, The University of Illinois at Chicago, Illinois*
PAMELA HUNT, PhD • *Amgen Inc., Thousand Oaks, California*
CARL W. JACKSON, PhD • *St Jude Children's Research Hospital, Memphis, Tennessee*
TAKASHI KATO, PhD • *Kirin Brewery Co, Ltd, Gunma, Japan*
KENNETH KAUSHANSKY, MD • *University of Washington School of Medicine, Seattle, Washington*
DAVID J. KUTER, MD, DPHIL • *Massachusetts General Hospital, Boston, Massachusetts*
JACK LEVIN, MD • *Department of Laboratory Medicine and Medicine, University of California School of Medicine, San Francisco, California*
THOMAS J. MACVITTIE, PhD • *School of Medicine, University of Maryland Cancer Center, Baltimore, Maryland*
ULLA M. MARZEC • *Emory University School of Medicine, Atlanta, Georgia*
ERIC M. MAZUR, MD • *Norwalk Hospital, Norwalk, Connecticut and, Yale University School of Medicine, New Haven, Connecticut*
HIROSHI MIYAZAKI, PhD • *Kirin Brewery Co, Ltd, Gunma, Japan*
MARK W. MOORE • *Genentech, Inc., South San Francisco, California*
JANET LEE NICHOL, MS • *Amgen Inc., Thousand Oaks, California*

TAMARA I. PESTINA, PhD • *St Jude Children's Research Hospital, Memphis, Tennessee*

CHARLES A. SCHIFFER, MD • *University of Maryland Cancer Center and Department of Medicine, University of Maryland School of Medicine, Baltimore, Maryland*

WILLIAM SHERIDAN, MD • *Amgen Inc., Thousand Oaks, California; and Division of Hematology–Oncology, Department of Medicine, UCLA School of Medicine, Los Angeles, California*

RAMESH A. SHIVDASANI, MD, PhD • *Dana-Farber Cancer Institute, Boston, Massachusetts*

RICHARD B. STEAD, MD • *Amgen Inc., Thousand Oaks, Califormia*

PAULA E. STENBERG, PhD • *Oregon Health Sciences University, Portland, Oregon*

CHRISTOPHER F. TOOMBS, PhD • *Amgen Inc., Thousand Oaks, California*

WILLIAM VAINCHENKER, MD, PhD • *INSERM, Institut Gustave Roussy, Villejuif, France*

FRANÇOISE WENDLING, PhD • *INSERM, Institut Gustave Roussy, Villejuif, France*

DOROTHEA ZUCKER-FRANKLIN, MD • *Department of Medicine, New York University Medical Center, New York, New York*

I

INTRODUCTION

1

Megakaryocyte Biology

Carl W. Jackson, Julie T. Arnold, Tamara I. Pestina, and Paula E. Stenberg

1. Introduction

Megakaryocytes are giant, polyploid cells of the hemopoietic tissues, whose final differentiation step culminates in the subdivision and release of their cytoplasm into the circulation as platelets. The earliest recognizable megakaryocyte in Romanovsky-stained marrow smears is a large basophilic cell with a high nuclear-to-cytoplasmic ratio and plasma membrane blebbing. As the cells mature, the nuclear-to-cytoplasmic ratio decreases as the amount of cytoplasm dramatically increases and becomes acidophilic, with abundant cytoplasmic granules, while the nucleus becomes lobulated and the chromatin condenses. Megakaryocytes are readily distinguished from osteoclasts, the other large cells in the marrow, by their nuclear morphology; megakaryocytes usually have only one large, lobulated nucleus, whereas osteoclasts contain several small nuclei. Megakaryocytes differentiate from a committed progenitor, which, by definition, has restricted differentiation capabilities. This committed progenitor is derived from a pluripotential hemopoietic precursor. A bipotential progenitor intermediate between the pluripotential and committed precursor with capacity to differentiate along either the megakaryocyte or erythroid pathways is suggested by some studies *(1,2)*. A scheme for megakaryocyte differentiation is presented in Fig. 1.

2. Unique and Identifying Features of Megakaryocytes and Megakaryocyte Development

Megakaryocytes have several unique and identifying features, including polyploid nuclei; subdivision of their cytoplasm into membrane-bound packages, the platelets; formation of cytoplasmic dense granules containing adenine nucleotides, neuropeptides, and divalent cations; α-granules containing plasma proteins and endogenously synthesized proteins; and proteins, such as membrane glycoprotein IIb, uniquely expressed by megakaryocytes and platelets.

3. Developmental Sequence of Megakaryocytes

3.1. Pluripotent Hemopoietic Stem Cells

The cell from which megakaryocytes and the other hemopoietic cell lineages are derived is a pluripotential hemopoietic progenitor that by definition, can provide long-term hemopoietic repopulation of the marrow of a lethally irradiated animal *(3)*.

From: *Thrombopoiesis and Thrombopoietins: Molecular, Cellular, Preclinical, and Clinical Biology*
Edited by: D. J. Kuter, P. Hunt, W. Sheridan, and D. Zucker-Franklin Humana Press Inc., Totowa, NJ

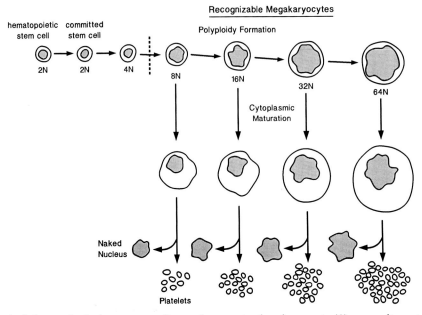

Fig. 1. Scheme depicting stages of megakaryocyte development: (1) commitment of he-mopoietic precursors to the megakaryocyte lineage; (2) differentiation of megakaryocyte pro-genitors to recognizable megakaryocytes; (3) polyploid formation; (4) cytoplasmic maturation; and (5) platelet shedding. The scheme illustrates that polyploidization precedes and is com-pleted before the rapid expansion and maturation of megakaryocyte cytoplasm; megakaryo-cytes can cease polyploidization, undergo cytoplasmic maturation, and platelet shedding at any level of polyploidy; and megakaryocyte size is related both to the degree of polyploidization and stage of maturation.

3.2. Committed Precursors: MK-BFC and MK-CFC

The megakaryocyte colony-forming cells (MK-CFC) were the first population of committed megakaryocyte progenitor cells to be identified, and the in vitro assays used for their recognition were developed in the 1970s *(4–6)*. The MK-CFC represent a heterogeneous population of cells that vary in their proliferation potential and buoyant density *(7–10)*. Murine MK-CFC undergo 1–8 cell divisions (average colony size 16–32 cells), and are more than 5 days removed from platelet formation *(4,11)*. Maximal megakaryocyte colony formation in vitro is dependent on two factors: a colony-stimu-lating factor (MK-CSF) required for proliferation of clonable progenitor cells and a maturation factor, megakaryocyte potentiator (MK-Pot), which enhances the DNA con-tent and cell size of individual megakaryocytes within a colony *(11–14)*. While interleukin (IL)-3 *(15,16)* and Mpl ligand *(17,18)* independently have both MK-CSF and MK-Pot activity, maximal numbers of megakaryocyte colonies are produced when both IL-3 and thrombopoietin (TPO) are added to the culture media *(19)*.

Development of reliable assays for human MK-CFC *(20–23)* initially proved diffi-cult because of the lack of a suitable marker that allowed detection of megakaryocytes. Identification of human MK-CFC is now routinely achieved by immunoperoxidase labeling of platelet-specific glycoproteins (GP), such as GPIIb/IIIa *(24,25)*. The growth requirements of human MK-CFC are similar to those of the murine system, being de-pendent on factors that promote proliferation and maturation *(26,27)*, but again it is apparent that the addition of Mpl ligand alone to cultures of human CD34$^+$ cells can promote growth of MK-CFC *(28,29)*.

A population of megakaryocyte progenitors more primitive than MK-CFC was first described in the mouse *(30,31)*. Originally termed a megakaryocyte burst-forming cell, MK-BFC, with high proliferative potential, this population of cells also has been denoted as Meg-HPP-CFC *(32)* or HPP-CFC-Mk *(33,34)*. In addition, a mixed high-proliferative-potential megakaryocyte (HPP-Meg-Mix) cell, a trilineage murine hemapoietic progenitor *(35)*, which is presumably even more primitive than the unilineage Meg-HPP-CFC, has been reported. Detection of murine MK-BFC in culture requires phorbol ester stimulation *(30)* or multiple early acting cytokines *(33,34)*, and incubation for 12–14 days, as opposed to 7 days for MK-CFC. Whether TPO is active on the primitive MK-BFC progenitor remains to be determined, although it is clear that TPO is able to synergize in culture with other early acting growth factors to promote megakaryocyte development from murine hemopoietic stem cells *(36,37)*.

The MK-BFC have a high proliferative capacity giving rise to large colonies of megakaryocytes (40–500 cells/colony) comprised of single *(33)* or multiple foci *(30)*. The morphologic criteria defining murine MK-BFC also apply to human MK-BFC *(38)*. Both human MK-BFC and MK-CFC express the CD34 antigen. However, only MK-CFC express detectable quantities of the HLA-DR antigen *(38,39)*. The MK-BFC and MK-CFC progenitor cell populations also differ in their elutriation and velocity sedimentation profiles and the growth-factor requirements needed for in vitro detection *(38,40–42)*. Human MK-BFC have been detected in fetal cord blood *(43,44)*, and increased numbers are reported in peripheral blood of patients after chemotherapy *(45)* and growth factor-induced mobilization *(46,47)*.

In vivo assays for identification of megakaryocyte progenitors cells have been developed only in the murine system *(48,49)*. The progenitor cells, termed megakaryocyte colony-forming cell-spleen (MK-CFC-S), were detected within the marrow of mice receiving high doses of 5-fluorouracil (5-FU) *(48)*. When the marrow from 5-FU-treated mice was transplanted into lethally irradiated mice, the MK-CFC-S cells were capable of forming large megakaryocyte colonies in the spleens of the recipients. The exact relationship of this cell to either the MK-BFC or MK-CFC is unclear. However, the MK-CFC-S most likely corresponds to the MK-BFC population based on its relative insensitivity to 5-FU treatment and its ability to generate a colony of megakaryocytes composed of several hundred cells *(49)*.

3.3. Detection and Characteristics of Immediate Megakaryocyte Precursors

Cells of the megakaryocyte lineage smaller than recognizable megakaryocytes were first detected by histochemical reactions. Zajicek found that in blood and hemopoietic tissues, the enzyme acetylcholinesterase (AChE), although primarily restricted to the erythroid lineage in humans, was found predominantly in megakaryocytes and platelets in rodents and the cat *(50)*. He observed that in addition to morphologically recognizable megakaryocytes, some smaller marrow cells expressed AChE activity. Unfortunately, the reaction product was diffusible, so that further characterization of these cells could not be done at that time. Breton-Gorius and Guichard then developed an ultrastructural histochemical technique for the demonstration of platelet peroxidase

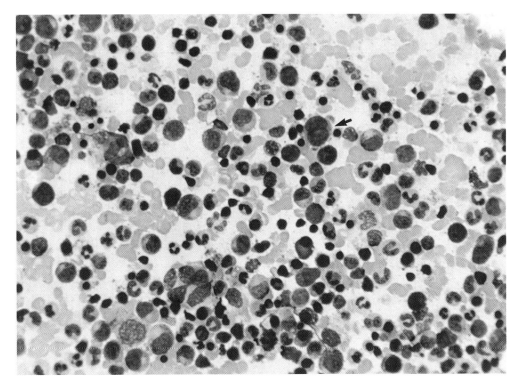

Fig. 2. Photomicrograph of a small, immature human megakaryocyte showing a high nuclear-to-cytoplasmic ratio, an even distribution of nuclear chromatin, basophilic staining of the cytoplasm, and characteristic blebbing of the plasma membrane.

and detected cells smaller than megakaryocytes that contained this enzyme activity *(51)*. However, they later found that this enzyme activity was not strictly megakaryocyte-lineage-specific, since it also was present in a population of early erythroid precursors *(51a)*. Subsequently, Karnovsky and Roots *(52)* developed an improved histochemical technique for demonstration of AChE activity that Jackson applied to analysis of the megakaryocyte lineage in rats *(53)*. Jackson confirmed the presence of AChE$^+$ cells smaller than megakaryocytes and demonstrated that the proportion of AChE$^+$ cells increased early after induction of acute thrombocytopenia. Long and Henry then reported that these cells decreased when platelet number was elevated by platelet hypertransfusion *(54)*. These data strongly suggested that the small AChE$^+$ cells were early cells of the megakaryocyte lineage. At the ultrastructural level, these cells resembled small lymphocytes with a large nucleus usually containing one nucleolus, an occasional profile of rough endoplasmic reticulum (RER), and abundant free ribosomes, all features of undifferentiated cells *(55)*. Small AChE$^+$ cells in the late stages of mitosis were occasionally observed, indicating that at least a portion of these cells are capable of cell division *(55,56)*. Jackson subsequently showed that these small cells also were labeled by platelet-specific antibody, providing more conclusive evidence that they were megakaryocyte precursors *(57)*. This led to the use of platelet antibodies directed against platelet-specific membrane proteins to detect megakaryocyte-lineage cells in human marrow *(24)*.

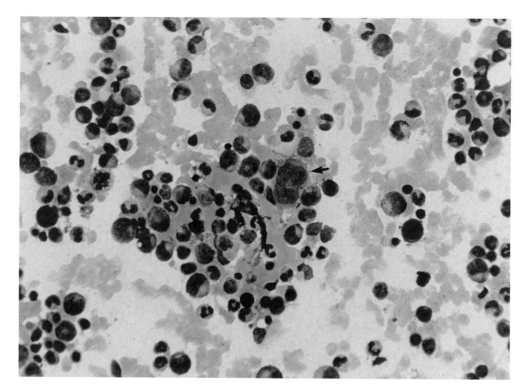

Fig. 3. Photomicrograph of an immature human megakaryocyte of intermediate size showing a high nuclear-to-cytoplasmic ratio, an even distribution of nuclear chromatin, and basophilic staining of the cytoplasm. The larger size of this immature cell as compared with the one shown in Fig. 2 is indicative of an increased level of polyploidy.

3.4. Stages of Megakaryocyte Maturation

Various classification schemes have been devised for staging megakaryocytes according to their maturity based on cytoplasmic staining characteristics and nuclear morphology *(58–60)*.

The most definitive are those in which ³H-thymidine incorporation is used as a maturation time marker *(58,59)*. The ³H-thymidine acts as a pulse label because it is available in the circulation for <20 minutes. The megakaryocytes undergoing polyploidization are the only ones initially labeled. With time, >90% of the megakaryocyte population becomes labeled as maturing and unlabeled megakaryocytes produce platelets, disappear, and are replaced by cells from a labeled precursor pool in which most of the cells are cycling. By examining the labeled cells at increasing times post-³H-thymidine injection, the morphological changes that occur during maturation can be delineated. The fact that some of the label, especially that derived from erythroid precursors undergoing enucleation, may be reused is not a complication for such classification studies *(61)*, because only the most immature megakaryocytes are actively synthesizing DNA.

The megakaryocytes initially labeled by ³H-thymidine are basophilic with only a very small rim of cytoplasm, with few if any granules *(58,59)*. They can vary in size from small (Fig. 2) to moderately large (Fig. 3), because cells undergoing polyploid-

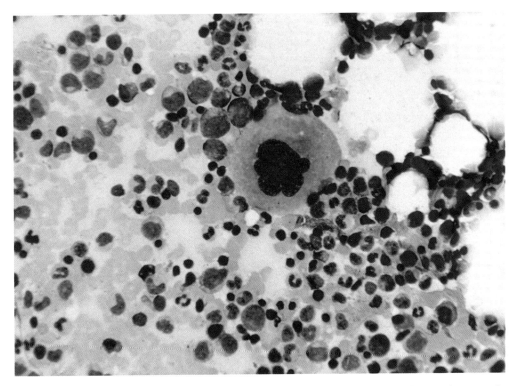

Fig. 4. Photomicrograph of a mature human megakaryocyte illustrating the vast increase in cytoplasmic volume that occurs during maturation. The cytoplasm is highly granular and acidophilic. The nuclear chromatin is condensed, and the polyploid nucleus is highly lobulated.

ization in a normal marrow vary in DNA content from 4–64N *(62–67)*, and their size is proportional to their DNA content *(59,66)*. Megakaryocytes become morphologically recognizable with Romanovsky-type stains when their DNA content exceeds 4N *(68)*. After polyploidization, both the amount of cytoplasm and granularity increases until in the most mature megakaryocytes, the cytoplasm is highly granular and acidophilic (Fig. 4) *(58,59)*. The nuclear chromatin changes from loosely packed to very condensed during cell maturation *(58,59,63)*.

3.5. Biochemical Markers for Different Stages of Megakaryocyte Maturation

Specific macromolecules are expressed at different times during megakaryocyte differentiation and should provide useful markers for analyses of megakaryocyte maturation. For instance, GPIIb is perhaps the earliest megakaryocyte marker, and is expressed on both a bipotential erythroid-megakaryocyte progenitor *(69)* and a committed megakaryocyte progenitor *(70,71)*. Platelet factor (PF)-4 is expressed later than GPIIb *(72,73)*, but considerably earlier than P-selectin, a granule membrane protein whose mRNA is expressed predominantly in morphologically mature megakaryocytes *(74)*. Another study suggests that the two forms of calpain are expressed at different stages of megakaryocyte development with *m*-calpain in the immature megakaryocytes and μ-calpain in mature forms *(75)*.

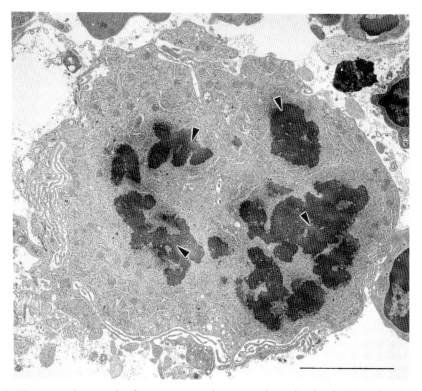

Fig. 5. Electron micrograph of a mouse megakaryocyte in endomitosis. Note the absence of a nuclear membrane. The demarcation membrane system (DMS) is beginning to develop at the cell periphery. Arrowheads indicate the condensed chromosomes. Magnification, ×5700; bar = 5 μm.

4. Specific Aspects of Megakaryocyte Development and Differentiation
4.1. Fetal Megakaryocytopoiesis

Megakaryocytes appear sequentially in the yolk sac, the fetal liver, spleen, and finally the marrow during normal embryogenesis and fetal development *(76)*.

Several reports indicate that fetal megakaryocytes are smaller than those of adults *(77–82)*. The smaller size of fetal megakaryocytes correlates with their lower ploidy, both for primary megakaryocytes *(78)* and those derived in culture *(83)*.

4.2. Polyploidy Formation

Megakaryocytes become polyploid through a process termed endomitosis, defined as DNA replication and a mitotic event with sister chromatid separation *(84)*, but no cytokinesis (Fig. 5) *(85)*. Polyploidization by megakaryocytes is unique among hemopoietic cells (hepatocytes also show a small degree of polyploidization). Polyploidization begins before megakaryocytes become morphologically recognizable and is completed in the immature basophilic stage *(68)*. Approximately 10% of recognizable rodent megakaryocytes are in the process of polyploidization *(58,59)*. Each megakaryocyte endomitotic cycle results in a doubling of the DNA content and is similar in length to that of a generation cycle in diploid cells *(86)*. The total length of one endomitotic cycle in rat megakaryocytes was calculated to be 9.3 hours, with DNA synthesis comprising 7.6 hours *(86)*. The minimum time for G_2 was 30–60 minutes and endomitosis required <1 hour *(86)*. Based on these time estimates, G_1 of the megakaryocyte endomitotic cycle is quite short *(86)*. The majority of megakaryocytes undergo three endomitotic cycles to achieve a DNA content equivalent to a $16N$ complement of chromosomes, but the number of endomitotic cycles can range from 2–5, since megakaryocytes in normal humans *(66)* and animals have polyploid megakaryocyte DNA contents of 8–$64N$ *(64,65,87)*.

At least two lines of evidence suggest that megakaryocytes undergo a mitotic event after each round of DNA synthesis *(86,88)*. First, cytogenetic analysis of polyploid metaphases in normal human marrow reveals that the chromsomes are the same size for all levels of polyploidy *(88)*. Each chromosome set consists of two chromatids. If a mitotic event did not occur after each round of DNA replication, then chromosomes of higher ploidy cells should be larger than those of lower ploidy cells or the higher ploidy cells should have multiples of sister chromatids. No such chromosomal arrangements have been reported. Second, ^{3}H-thymidine labeling indices of megakaryocyte endomitotic figures yielded a rectilinear curve, suggesting a similar cell cycle time for all megakaryocytes undergoing polyploidization *(86)*.

The level of megakaryocyte polyploidization is inversely related to platelet count *(64,89–94)*. Experimental thrombocytopenia results in higher megakaryocyte DNA contents *(64,89,91–94)*, whereas the reverse is true for thrombocytosis induced by platelet hypertransfusion *(64,89)*. Megakaryocyte ploidy is slightly greater in male than female mice *(95–97)*. Castration of male mice reduces ploidy to the level observed in female mice, suggesting that testosterone enhances megakaryocyte polyploidization *(95)*. Megakaryocyte polyploidy is also increased in the pregnant rat *(98)* in parallel with changes in plasma progesterone levels. Whether the same is true for humans remains to be determined. In contrast, megakaryocyte ploidy was reported to be decreased in diabetic BB rats *(99)*; the etiology is unknown.

The differences in gene expression in megakaryocytes that dictate polyploidization are not understood. A major obstacle here has been the lack of a model system in which a large number of polyploid cells in endomitosis can be isolated for biochemical or molecular comparisons. However, several groups have recently tried to address changes at the molecular level that may explain the process of endomitosis in megakaryocytes *(100–102)*. The assumption is that expression or function of the primary proteins regulating cell-cycle progression *(103,104)* is altered in the process of commitment of a megakaryocyte precursor to an endomitotic cycle. Data on the roles of specific cyclins and cyclin-dependent kinases in regulating the megakaryocyte cell cycle have been difficult to interpret. Several studies have been performed on megakaryoblastic cell lines that require phorbol ester stimulation for induction of polyploidization *(102,105–108)*. These cells have a limited ability to undergo ploidization in comparison to normal megakaryocyte bone marrow cells. In studies where primary megakaryocytes have been used *(101)*, cells have been unsynchronized, and therefore, determination of cyclin levels or kinase activity in a mixed cycling population may provide misinformation and overlook important phase-specific events. To date, alterations in cyclin D1 *(102)*, D3 *(101,105)*, E *(105)*, B1 *(101,107)*, and cdc-2 levels or function have been reported in the megakaryocyte endomitotic cycle. If these studies are confirmed, they infer that more than one event is altered in the megakaryocyte cell cycle, with both G_1 (cyclins D1, D3, E) and G_2/M (cyclins B, cdc-2) phases affected.

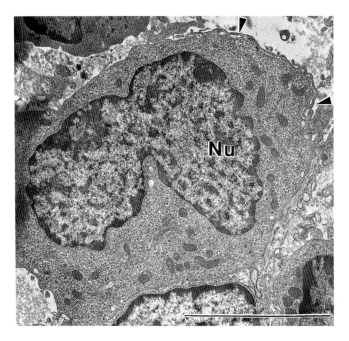

Fig. 6. Electron micrograph of an immature mouse megakaryocyte. Arrowheads identify nascent demarcation membranes. Nu, nucleus. Magnification, ×7900; bar = 5 μm.

4.3. Ultrastructural Characteristics

The earliest megakaryocyte recognizable by electron microscopy measures 10–12 μm in its long axis and contains a large, bilobed nucleus with prominent nucleoli. At this stage, the cytoplasm contains a small Golgi complex, variable amounts of rough endoplasmic reticulum (RER), abundant free ribosomes, and some mitochondria. In addition, lysosomes, a few α-granules, and some profiles of the demarcation membrane system (DMS), a membrane system in continuity with the extracellular space, are present (Fig. 6).

Megakaryocytes of intermediate maturity vary in size (approximately 15–30 μm in diameter) and polyploidy (8–32N) (Fig. 7). Nuclear changes include lobulation and chromatin condensation. The RER increases, the Golgi complex enlarges, and the DMS and α-granules become more prominent.

Mature megakaryocytes measure 30–50 μm in diameter (Fig. 8). The nucleus is usually at one pole of the cell, and nucleoli are less prominent than in immature cells. The DMS divides the entire megakaryocyte cytoplasm into "platelet fields." The RER and Golgi complex are significantly reduced, whereas the DMS and α-granules are present throughout the entire cytoplasm. At the point of platelet production, the megakaryocyte becomes irregular; the final stage in the life of the megakaryocyte is a "naked nucleus" (Fig. 9), with the cytoplasm shed as platelets by some as yet unresolved mechanism.

4.4. Cytoplasmic Maturation and the Development of Organelles

4.4.1. Demarcation Membrane System

The megakaryocyte contains a smooth membrane system, termed DMS, which at late stages of maturation is an extensive system of narrow channels homogenously

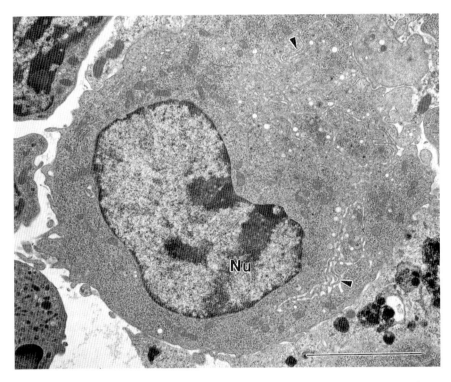

Fig. 7. Electron micrograph of a mouse megakaryocyte of intermediate immaturity. A few granules are apparent within the cytoplasm. Arrowheads indicate the developing DMS. Nu, nucleus. Magnification, ×7900; bar = 5 μm.

distributed in the cytoplasm and in contact with the external millieu *(109)*. Evidence for the patency of these channels was provided by studies using extracellular tracers, such as ruthenium red, lanthanum, horseradish peroxidase (HRP), and tannic acid *(109,110)*. The DMS was first described by Behnke *(109)* as originating from the plasma membrane in the form of tubular invaginations at multiple sites. Although the DMS is morphologically obvious in the mature megakaryocyte, elements are present as early as the promegakaryoblast stage when DNA replication is occurring. Based on Behnke's classic ultrastructural studies, the DMS was suggested to compartmentalize the platelet cytoplasm into platelet territories, which were released into the circulation as platelets at the completion of megakaryocyte maturation. Results of an ultrastructural membrane freeze-fracture study of megakaryocytes and platelets support that model *(111)*. An alternative model of platelet formation is the flow model, which envisions the role of the DMS as a membrane reservoir for evagination of cytoplasmic processes, which then subdivide into platelets *(112,113)*. This model will be discussed in detail in a later chapter (Chapter 17). Ultrastructural evidence of megakaryocyte process formation extending through the marrow sinus endothelium is also supportive of this model *(112,114)*.

The DMS was also thought to form the surface-connected canalicular system (SCCS) of platelets, which serves as a conduit for secretion of platelet granule contents *(115)*.

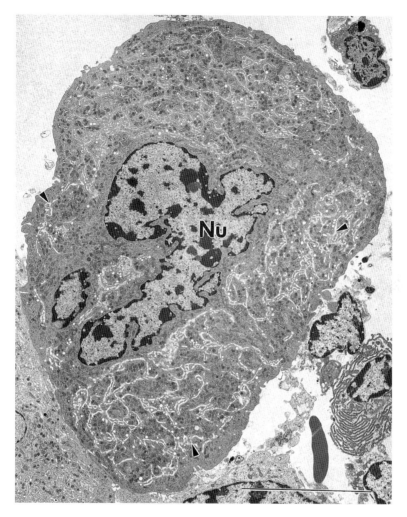

Fig. 8. Electron micrograph of a mature mouse megakaryocyte. Arrowheads mark the extensive DMS. Nu, nucleus. Magnification, ×7000; bar = 10 μm.

However, the observation that bovine platelets lack SCCS *(116,117)*, but bovine megakaryocytes have a DMS *(118,119)*, indicates that the SCCS is not derived from the megakaryocyte DMS.

4.4.2. Dense Tubular System

Histochemically, the platelet dense tubular system (DTS) and the megakaryocyte RER and nuclear envelope are identical: they all contain platelet peroxidase *(51,120)*, glucose-6-phosphatase *(121,122)*, and AChE (in rodent and cat cells) *(123–125)*. The DTS is thought to be the site of prostaglandin synthesis in platelets *(126)*. None of these structures stain with markers for surface membranes or the Golgi complex. Although neither the platelet SCCS nor the megakaryocyte DMS is connected to the DTS, both form close associations designated membrane complexes *(127,128)*.

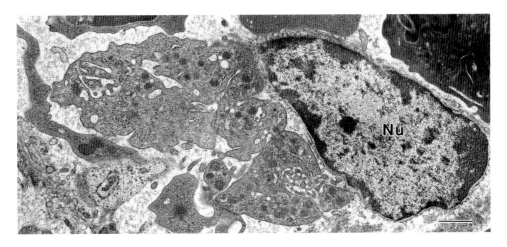

Fig. 9. Electron micrograph of a naked mouse megakaryocyte nucleus remaining after near-complete shedding of the cytoplasm to form platelets. Nu, nucleus. Magnification, ×11,600; bar = 1 μm.

4.4.3. Granule Formation and Packaging

The dynamic nature of megakaryocyte and platelet α-granules has recently been revealed. Platelet α-granule contents were thought to be synthesized solely in the megakaryocyte. A number of studies now demonstrate that both megakaryocytes and platelets are capable of endocytosing exogenous soluble proteins and packaging them within their secretory granules until they are required to function during normal hemostasis *(129–133)*.

4.4.3.1. α-GRANULES

α-Granules are present in the earliest recognizable cell of the megakaryocytic series, before the development of extensive demarcation membranes. α-Granules are thought to be derived from the megakaryocyte Golgi complex during cell maturation *(134)*. These spherical to oval granules measure 200–500 nm in diameter and contain a nucleoid within a finely granular matrix. They contain numerous protein constituents. Both integral membrane glycoproteins and granule content proteins have been identified (Table 1); some storage products have been localized to specific granule subcompartments by immunoelectron microscopy. At the time of cytoplasmic maturity, the Golgi complex and the RER are reduced, and morphologically mature α-granules are distributed homogeneously throughout the megakaryocyte cytoplasm between profiles of demarcation membranes. Platelet formation includes partitioning of α-granules into megakaryocyte cytoplasmic processes that are believed to represent the first step in the platelet birth process.

α-Granule constituents are derived by three different mechanisms: endogenous synthesis via the classic protein synthetic pathway (megakaryocytes only), receptor-mediated endocytosis (megakaryocytes and platelets), and pinocytosis (megakaryocytes and platelets). The identification of these distinct pathways provided a possible explanation for the variability of specific protein levels in granules of platelets from patients with the Gray Platelet syndrome *(135–137)* and the Wistar Furth (WF) rat *(138)* by suggesting that a defect may exist in one, but not all of these pathways *(139)*.

Table 1
Location of Megakaryocyte/Platelet α-Granule Proteins

Protein	Location	Reference
Albumin	Granule	*130,136*
β-amyloid precursor protein	Granule	*297*
β-thromboglobulin	Granule-nucleoid compartment	*140,298,299*
Clusterin	Granule	*300*
Factor V	Granule	*301*
Fibrinogen	Granule-matrix compartment[a]	*140,298,302*
Fibronectin	Granule	*303*
IgG	Granule	*130,136,139*
PF 4	Granule-nucleoid compartment	*298,304,305*
Thrombospondin	Granule	*303,306*
Vitronectin	Granule	*307*
von Willebrand factor	Granule-tubule compartment	*140,306,308*
GMP-33	Membrane	*309*
GPIb, IX	Membrane	*310*
GPIIb/IIIa	Membrane	*140,311,312*
GPV	Membrane	*310*
GPIV (CD 36)	Membrane	*313*
P-selectin	Membrane	*136,147,314*
Osteonectin	Membrane	*315*
CD9	Membrane	*316*
PECAM-1	Membrane	*316*
Rap Ib	Membrane	*317*

[a]The content of the α-granule can be subdivided into compartments. These include the relatively electron-dense nucleoid, surrounded by a fine, granular matrix. In addition, an electron-lucent area within the matrix contains tubular elements.

Endogenously synthesized platelet α-granule proteins include PF-4 and β-thrombo-globulin. These are synthesized in the megakaryocyte by the classic pathway of protein synthesis, and are therefore probably shuttled to the α-granule by transport vesicles budding from the Golgi complex. Handagama et al. *(129)* reported the first demonstration of endocytosis and storage of a soluble protein, horseradish peroxidase (HRP), into megakaryocyte and platelet α-granules. Subsequently, the uptake of albumin and IgG by pinocytosis, and fibrinogen by receptor-mediated endocytosis, and their incorporation into α-granules were shown *(130,133)*. Acquisition of fibrinogen from plasma was supported by studies showing that patients with congenital afibrinogenemia lack platelet fibrinogen *(133)*, as do patients with Glanzmann's thrombasthenia who lack the fibrinogen receptor GPIIb/IIIa on their platelet membranes *(140)*. Specific inhibition of GPIIb/IIIa was shown to prevent uptake of exogenous platelet fibrinogen into platelets and megakaryocytes *(131,141,142)*. Thus, it is now generally accepted that α-granule fibrinogen is derived from plasma via receptor-mediated uptake, and not by synthesis in megakaryocytes. Freeze-fracture analysis of platelet membranes also supports two different pathways for platelet endocytosis *(143)*.

4.4.3.2. DENSE GRANULES

By electron microscopy, dense granules are approximately 200–300 nm in diameter and are distinguished by a bull's-eye appearance consisting of a clear halo encircling a dark, central area. They are less numerous in the platelet cytoplasm than α-granules. The principal components of dense granules are adenosine triphosphate (ATP), adenosine diphosphate (ADP), pyrophosphate, calcium, and other cations *(144,145)*. The relative amounts of these are species specific *(145,146)*. Dense granule membranes contain granulophysin and P-selectin *(147)*. In addition, novel membrane proteins have been detected in bovine dense granules, which are deficient in platelets of cattle with Chediak-Higashi syndrome *(148)*.

Serotonin is not synthesized by megakaryocytes, which, like platelets, take up serotonin from the plasma *(149)*. Electron-dense granules are rarely observed in the megakaryocyte, although studies have shown that the granule body is present *(150,151)*. The inability to detect dense granules is due to the absence of serotonin and calcium storage within the megakaryocyte granule. However, megakaryocytes have the ability to incorporate serotonin into the granule, since rabbits injected with serotonin have increased numbers of dense bodies in the megakaryocyte cytoplasm *(151)*. These studies suggest that under normal conditions, the intramedullary plasma level of serotonin is too low for adequate uptake by the megakaryocyte.

4.4.4. Role of the Cytoskeleton

The role of the cytoskeleton in megakaryocytopoiesis has received little attention. Leven and Nachmias demonstrated that megakaryocytes contain actin, α-actinin, filamin, myosin, and tubulin *(152,153)*. Platelets contain many cytoskeletal proteins, including actin, actin-binding protein, α-actinin, myosin, spectrin, talin, tensin, tropomyosin, tubulin, and vinculin *(154,155)*, which must be synthesized by megakaryocytes. Stenberg et al. showed that injection of the microtubule-disrupting agent, vincristine, into rats caused the formation of large cytoplasmic membrane complexes, plasma-membrane blebbing, and dilation of the DMS, which then contained large accumulations of α-granule proteins *(156)*.

At least two lines of evidence suggest that the cytoskeleton plays an important role in the process of platelet formation. First, Leven and Tablin showed that cytochalasins stimulated cytoplasmic-process formation by isolated megakaryocytes, whereas microtubule-disrupting agents inhibited this process *(157–159)*. Second, megakaryocytes of the WF rat have defects in platelet formation and show several signs of a cytoskeletal functional defect including plasma-membrane blebbing *(160)*, large cytoplasmic membrane complexes *(161)*, and disorganized cytoplasmic organelle arrangement *(161)*. Their platelets tend to be spherical rather than disk-shaped *(156)*, and have abnormal subcellular distributions of myosin and talin *(155)*. More detailed knowledge of the role of the cytoskeleton will be required to understand fully the process of platelet formation.

4.4.5. Tyrosine Phosphorylation

Studies of the role of tyrosine phosphorylation in megakaryocyte differentiation are only now beginning. Labeling of mouse marrow by a phosphotyrosine antibody revealed heterogenous labeling among megakaryocytes *(162)*. The staining was in a granular pattern within the megakaryocyte cytoplasm, whereas the marginal zone, plasma membrane, and nucleus were devoid of label. Ultrastructural immunogold analysis revealed phosphotyrosine labeling of vesicles and granules, with the vesicular membranes strongly labeled, but the DMS not labeled *(162)*. These tyrosine-phosphorylated proteins remain to be identified.

A tyrosine kinase (MATK) localized to megakaryocytes and neural tissue has been cloned and sequenced *(163)*. This kinase has approximately 50% homology with *Csk*, and phosphorylates the carboxy-terminal tyrosine of *Src*, resulting in the downregulation of *Src* tyrosine kinase activity *(164)*. MATK can also associate with *c-kit* after stimulation with stem cell factor (SCF) *(165)*. Another recently cloned tyrosine kinase (RAFTK) also seems to be restricted to megakaryocytes and neural tissue *(166)*.

The role of tyrosine phosphorylation in TPO receptor signaling will be discussed in a later chapter (*see* Chapter 16).

4.5. Role of the Microenvironment in Megakaryocytopoiesis

In addition to the marrow, the spleen is an active hemopoietic organ in mice; however, the ratio of myeloid to erythroid production in these two organs is different. Granulopoiesis, erythropoiesis, and megakaryocytopoiesis all occur in the marrow. In contrast, splenic hemopoiesis is usually restricted to erythropoiesis and megakaryocytopoiesis. Wolf and Trentin took advantage of this fact, and showed that when a portion of bone marrow was surgically inserted into the mouse spleen, a clear distinction in the hemopoietic cell types produced in the spleen versus the marrow insert was observed *(167)*. They concluded that a difference in the microenvironment was responsible for this result and coined the term hemopoietic-inducing microenvironment *(167)*. The molecular basis for these interesting findings has not been determined.

Microenvironmental influences on megakaryocyte differentiation and platelet formation are even less well defined. The location of mature megakaryocytes adjacent to the bone marrow-sinus endothelium may be important for platelet production *(168,169)*. Some in vitro studies suggest that stromal cells can influence megakaryocyte progenitor proliferation *(170,171)* and also interact with megakaryocytes *(172)*.

4.6. Platelet Formation

Platelet formation will be discussed in detail in a later chapter of this book (*see* Chapter 17). However, two questions that have not been resolved relate to the relationship between megakaryocyte ploidy and platelet production. First, is the number of platelets produced by a given megakaryocyte related to its ploidy? Second, does platelet size vary with the ploidy of the megakaryocytes from which they are produced? Megakaryocyte size is related to the ploidy at which a megakaryocyte matures *(67,68)*, indicating that higher ploidy megakaryocytes have the capacity to produce more platelet mass, but that observation does not specify whether platelet number or size is related to ploidy.

A few other observations are worthy of note. The first is that platelet size varies considerably among different mammalian species, even though megakaryocyte size for most of the species studied so far is the same *(173)*. For instance, platelet size of normal laboratory rats is approximately 4 femtoliters (fL) *(160,174)*, whereas that of normal humans is approximately 8 fL *(175–177)*, yet the average megakaryocyte diameter of both is 21 μm *(178)*. This indicates that the same volume of megakaryocyte cytoplasm is subdivided into different numbers of platelets in different species, and suggests that these variations in platelet size must reflect differences in the proportions or activities of some of the key proteins involved in platelet formation.

Second, platelet size rapidly increases (within 8 hours) in response to acute thrombocytopenia induced by platelet destruction in rodents *(94)*. The peak increase in mean platelet volume occurs at 18–24 hours *(94,179)*. Platelet size is also increased in humans in response to thrombocytopenia due to increased platelet destruction *(180)* or blood loss. The increase in platelet size was attributed to elevated TPO levels *(94,179)*; however, animal studies using recombinant (r) TPO have not detected an early increase in mean platelet volume *(181–183a)*. Furthermore, mice with thrombocytopenia resulting from genetically engineered deletion of the TPO gene have increased mean platelet size as discussed in a later chapter (*see* Chapter 11). These observations suggest that not TPO, but another as yet undefined circulating factor, acts late in megakaryocyte differentiation to induce platelet release.

Third, thyroid function status appears to affect platelet size. Humans and dogs with hypothyroidism have smaller platelet size and increased platelet number compared with normal platelets *(184,185)*. Conversely, humans and cats with hyperthroidism have larger platelet size, but normal platelet counts *(185,186)*.

5. Kinetics of Megakaryocytopoiesis and Platelet Production

The most extensive studies of megakaryocyte kinetics have been done in rodents using ^{3}H-thymidine labeling *(58,59,187)*. This method reveals that the time required for the earliest recognizable rodent megakaryocyte to complete polyploidization, mature, and produce platelets is 2–3 days *(58,59,187)*. Human megakaryocytes are estimated to require 5 days for this process *(188)*. This marrow transit time is presumed to be dependent on the level of polyploidy achieved by a given megakaryocyte, since higher ploidy cells would have a polyploidization process longer than that of lower ploidy cells. The minimal time required from the end of the last round of endomitotic DNA synthesis to the release of rat megakaryocyte cytoplasm as platelets is estimated at 46 hours *(59)*. Thus, the time for differentiation, polyploidization, and cytoplasmic maturation for a $16N$ rat megakaryocyte from a $2N$ progenitor is estimated to be three endomitotic generation cycles at 9.3 hours each *(86)* plus the 46-hour maturation time, or a total of 74 hours. A megakaryocyte maturing at $8N$ would require 65 hours, although a $32N$ cell would need 83 hours for the same process.

Platelet survival in rodents is 4–4.5 days *(189,190)* versus 7–10 days in humans *(191)*.

6. Effects of Acute Thrombocytopenia, Elevated Platelet Count, Drugs, and Radiation on Megakaryocytopoiesis

Much of our knowledge concerning the response of megakaryocytopoiesis to altered platelet demand has been derived from animal studies in which platelet count was experimentally increased or decreased. The most studied model has been the response of megakaryocytopoiesis to acute thrombocytopenia. Acute thrombocytopenia was induced by injection of platelet antiserum *(89,192,193)*, by phlebotomy *(194)*, by exchange transfusion with platelet-poor blood *(193,195)*, or by injection of neuraminidase *(196)*. In all of these models, thrombocytopenia was achieved by removing platelets from the circulation without significantly damaging megakaryocytes and their progenitors. The earliest megakaryocytic response to acute thrombocytopenia in rodents is an increase in the proportion of small AChE cells. This occurs as early as 1 hour, with a peak at 2–6 hours after platelet antiserum injection *(53,197)*. Next is an increase in platelet size that can be detected as early as 8 hours and is maximal at 18–24 hours *(94,179,198)*. An increase in megakaryocyte ploidy is first noticeable at 12 hours and peaks at 60–72 hours *(64,89,94,179)*. Megakaryocyte size increases in concert with megakaryocyte ploidy *(89,199,200)*. Megakaryocyte number begins to increase at 24 hours and is maximal at 48–72 hours *(179,192,200)*. The proportion of recognizable megakaryocytes in the process of mitosis (endomitosis) is maximal at 32–36 hours *(179,192,201)*. The turnover time of megakaryocytes as measured by ^{3}H-thymidine labeling kinetics is also shortened *(192,193)*. Conclusions from these studies were that in response to acute thrombocytopenia: megakaryocytes can shed their cytoplasm earlier than normal, resulting in the production of larger platelets; megakaryocyte size, number, and ploidy are increased; and increased platelet production is achieved initially by increasing ploidy and with time by increasing megakaryocyte number.

Decreased platelet demand was achieved experimentally by hypertransfusing animals with platelets *(54,64,89,195,202)*. Surprisingly, to suppress megakaryocytopoiesis, platelets needed to be sustained at a very high level for several days *(89,195,202)*. Platelet-hypertransfused rodents showed decreases in small AChE$^+$ marrow cells *(54)*, megakaryocyte ploidy *(64,89)*, megakaryocyte size *(89,195)*, and megakaryocyte number *(89,195,202)*.

The effects of drugs and radiation on platelets and megakaryocytopoiesis have also provided important information relevant to basic biology and to clinical effects of these agents.

The vinca alkaloids, vincristine and vinblastine, stimulate megakaryocytopoiesis and cause thrombocytosis at relatively low dosages (0.1–0.2 mg/kg in rodents) without a preceding decrease in megakaryocytes or platelets *(203–208)*. The basis for this effect is not known, but some data suggest that damage to platelets may result in their inability to participate in the regulatory feedback loop *(208)* (*see* Chapter 23).

Persistent thrombocytosis is also seen in offspring of women abusing multiple drugs. Infants born to drug-abusing women were observed to have thrombocytosis persisting up to 30 weeks of age *(209)*. Administration of methadone (2.5 mg/kg/days sc) to female mice beginning 2 weeks before mating and continued through weaning produced a significant increase in megakaryocytes and platelets in the offspring, but not in the female mice *(210)*. The basis for the thrombocytosis in the infants is unknown.

Of all chemotherapeutic agents studied in rodents, 5-FU has a unique effect on megakaryocytopoiesis. A moderate dose (150 mg/kg in rodents) of 5-FU causes a pronounced and sustained thrombocytosis *(211–213)* seemingly inappropriate to the degree of thrombocytopenia resulting from bone marrow damage. Analysis of megakaryocytopoiesis revealed that the prolonged thrombocytosis was a consequence of a failure to downregulate megakaryocyte frequency, which, unlike megakaryocyte poidy, continued to increase even though platelet counts had rebounded above normal *(212)*.

Hydroxyurea, which kills cells in DNA synthesis, causes a transient decrease in megakaryocyte numbers, but does not produce thrombocytopenia *(214,215)*. Other chemotherapeutic agents at moderate dosage produce a more predictable megakaryocyte response to marrow damage, with a decrease in megakaryocytes and platelet numbers followed by recovery within 2 weeks in rodents *(216,217)*.

Radiation also induces marrow suppression by killing proliferating hemopoietic progenitors, but a nonlethal damaging effect on recognizable megakaryocytes may also occur, since an increase in megakaryocytes and decrease in platelets are observed during the first day after irradiation *(218)*. One interpretation of this result is that radiation causes a delay in platelet formation by mature megakaryocytes *(218)*. The pattern of thrombopoietic recovery after sublethal irradiation also reveals an unexpected finding, in that the platelet count recovers to normal levels within 20 days, but megakaryocyte number is below normal for at least 70 days *(214)*. The earlier recovery of platelet number is achieved by increased megakaryocyte size *(214)*. This suggests that megakaryocyte progenitor proliferation is reduced for an extended period after irradiation, which may help to explain why platelets are usually the last hemopoietic lineage to recover after clinical allogeneic marrow transplants *(219–221)*. Sublethally irradiated mice transplanted with a defined primitive marrow cell population also show delayed platelet recovery *(222)*. At 90 days, when both platelet and megakaryocyte numbers are normalized, megakaryocyte size is increased, whereas megakaryocyte progenitors are reduced to 50% of control *(223)*. The modified thrombopoietic response in these transplanted mice when treated with 5-FU, suggests that the ability of megakaryocyte precursors to proliferate is impaired *(223)*.

7. Models to Study Various Aspects of Megakaryocyte Differentiation

7.1. Animal Models with Hereditary Defects of Megakaryocytopoiesis and Platelet Formation

7.1.1. Sl/Sl^d and W/W^v Mice

Anemia was the first abnormality detected in these mice *(224)*. Ebbe et al. subsequently found that megakaryocytes were reduced in number, but larger in size, whereas platelet number was normal *(225,226)*. The molecular bases for these defects were subsequently found to be a mutation in SCF receptor (*c-kit*) in the W/W^v mouse, and a mutation in SCF in the Sl/Sl^d mouse *(227,228)*. The decrease in megakaryocyte number is compensated for by an increase in megakaryocyte size, so as to maintain platelet number *(229,230)*.

Ebbe et al. prepared parabiotic mouse pairs between normal and Sl/Sl^d mice to examine whether a circulating factor was responsible for the increased megakaryocyte size of Sl/Sl^d mice *(231)*. Megakaryocyte size of the normal half of the parabiotic pair was increased almost as much as that of the Sl/Sl^d partner, indicating that a soluble circulating factor was responsible for the increased megakaryocyte size of Sl/Sl^d mice *(231)*. Is this factor TPO? This question should be addressed, since a current model of thrombopoiesis proposes constitutive hepatic production of TPO, with regulation of TPO plasma levels dependent on platelet binding and metabolism *(93,232)*. Since Sl/Sl^d mice have a normal platelet number, one would expect the level of TPO in these mice to be normal, if this hypothesis is correct.

Ebbe subsequently showed that the large megakaryocyte size, but reduced megakaryocyte number of Sl/Sl^d mice could be reproduced in long-term Sl/Sl^d marrow cultures *(229)*. If TPO is the factor responsible for these in vitro observations, then the bone marrow is a likely site of TPO production. In this case then, the reduced megakaryocyte number in Sl/Sl^d marrow cultures may result in less total binding of TPO, and cause higher TPO levels that increase megakaryocyte size. These studies suggest that the marrow may be an alternative site for TPO production and both platelet and megakaryocyte mass play important roles in regulating endogenous levels.

7.1.2. WF Rat

The Wistar Furth (WF) rat has macrothrombocytopenia (large platelet size with reduced platelet number) *(160,233)* with platelet α-granule deficiency *(138)*. Both the number and size of α-granules are decreased, as is the α-granule protein content *(138)*. Bleeding time is prolonged (Stenberg, unpublished results). This phenotype resembles Gray Platelet syndrome of humans, except that the α-granule deficiency is not as severe as that reported for patients with the clinical syndrome *(138)*. The molecular basis of the WF platelet defect has not been defined, but several observations suggest a defect in cytoskeletal function. WF megakaryocytes and platelets contain large cytoplasmic membrane complexes, and display a haphazard cytoplasmic organelle arrangement *(160)*. Megakaryocytes also show an abnormal DMS *(160)*. Interestingly, platelet size increases, whereas platelet count decreases with the age of the WF rat *(160)*.

Two other Gray Platelet syndrome-like traits of the WF rat are the presence of α-granule protein accumulation in the lumena of the megakaryocyte DMS and platelet SCCS, and an increased frequency of megakaryocyte emperipolesis *(233)*, the phenomenon in which other hemopoietic cell types appear inside megakaryocytes *(234)*. The DMS and SCCS channels containing the α-granule proteins are dilated. White proposed an interesting hypothesis that the abnormal α-granule protein accumulations in the DMS act as chemoattractants to other hemopoietic cells, and that the other marrow cells are in dilated DMS channels *(235)*. The DMS *(160)* and SCCS *(138)* channels containing the α-granule proteins are dilated. The frequency of megakaryocyte emperipolesis is also increased in other clinical entities with altered hemopoiesis, such as certain malignancies and hemolytic anemias.

The macrothrombocytopenia and platelet α-granule deficiency cosegregate in an F_2 generation analysis (Jackson, unpublished results), strongly suggesting that these two abnormalities result from the same mutation. Thus, a mutation in a single gene seems to affect both subdivision of megakaryocyte cytoplasm into platelets, and α-granule formation or stability.

7.1.3. Gunmetal Mouse

The gunmetal mouse has a coat color mutation along with an abnormal platelet phenotype, including macrothrombocytopenia with α-granule deficiency, that somewhat resembles that of the WF rat, except that this mouse also has platelet-dense granule deficiency, and the increase in platelet size is not as large *(236,237)*. The molecular basis of this defect is not known, but the profile of small G-proteins in gunmetal mouse platelets is abnormal *(236)*. This mutation maps to mouse chromosome 14 *(236)*.

7.1.4. Animal Models with Heritable Hyperlipidemia

Platelet formation is abnormal in both the hyperlipidemic rat *(238)* and the Watanabe rabbit *(239)*. Platelets from hypercholesterolemic rats are reduced in size *(238)*. The hyperlipidemia in the Watanabe rabbit is due to the absence of low-density lipoprotein receptors *(240)*. In the Watanabe rabbit, both megakaryocyte and platelet size are smaller than normal, and platelet number is increased *(239)*. Watanabe rabbit platelets have normal ultrastucture, but are more lentiform than normal rabbit platelets *(239)*. Interestingly, the modal megakaryocyte ploidy is 32*N* in both Watanabe and normal New Zealand rabbits *(239)*.

7.1.5. Rodent Coat Color Mutations with Platelet-Dense Granule Deficiency

At least 11 mouse pigment-dilution mutations have been described with platelet-dense granule deficiency: beige, cocoa, light ear, maroon, mocha, muted, pale ear, pallid, pearl, ruby eye, and sandy *(241–244)*. The fawn-hooded rat also has platelet-dense granule deficiency *(245,246)*. These animals all have prolonged bleeding times and most also have lysosomal abnormalities. The mutation for pallid has been mapped to mouse chromosome 2 where it colocalizes with the gene for red cell band 4.2, a membrane skeletal protein *(247)*. Another, mocha, was reported to colocalize with the gene for ankyrin-3, whose gene product is another membrane-skeletal protein *(248)*. This suggests that many of these dense-granule deficiencies may be the result of membrane-skeletal protein mutations. Of particular interest is that platelet α-granule function appears normal in all these mice, suggesting that the proteins involved in formation of dense granules and α-granules are different.

7.1.6. Belgrade Rat

The Belgrade (b/b) rat has severe hypochromic, microcytic anemia (packed cell volume [PCV] of 14%) with decreased megakaryocytes, and moderately reduced platelet number *(249)*. The megakaryocytes are larger than normal, and this compensates in part for the decreased megakaryocyte number *(249)*. Correction of the anemia by serial red blood cell (RBC) transfusions normalized both megakaryocytes and platelets, whereas iron infusion only partially corrected the megakaryocyte profile *(250)*. These observations led to the conclusion that the megakaryocyte-platelet abnormalities of the b/b rat are the consequence of the severe hypoxia resulting from the severe anemia present in this animal *(250)*.

7.1.7. C3H Mouse

The C3H mouse presents an interesting model for the study of regulation of polyploidization, in that its modal megakaryocyte DNA content is 32*N*, rather than the usual 16*N* *(97)*. Backcross analyses suggest that multiple genes contribute to this phenotype *(251)*. The ultrastructure of C3H megakaryocytes appears normal as do their number and size *(97)*. The fact that megakaryocyte size is normal despite higher modal DNA content suggests C3H mice actually have a smaller volume of cytoplasm per unit of DNA *(97)*. This would explain why the platelet number and mean platelet volume in the C3H mouse are similar to those of other mouse strains with the normal DNA complement.

7.1.8. Canine Cyclic Hemopoiesis

The gray collie dog has an autosomal recessive hemopoietic disorder characterized by cyclic fluctuations in neutrophils, monocytes, reticulocytes, and platelets *(252–257)*. Cycles for the three hemopoietic lineages are asynchronous and occur at 11- to 14-day intervals, with platelets fluctuating between 125 and $900 \times 10^9/L$ *(257)*. Platelet size varies inversely with platelet count *(257)* and platelet function is also defective in this disorder *(257)*.

7.2. Genetically Engineered Defects in Megakaryocytopoiesis

Genetically engineered deletions (knockouts) *(258–260)* and transgenic-induced overexpression *(261,262)* of normal genes in normal mice provide models to study the role of specific genes in megakaryocytopoiesis and platelet formation. Disruption of TPO *(259)* or *Mpl (258)* genes have profound effects on megakaryocytopoiesis and platelet count as discussed in a later chapter of this book (*see* Chapter 21). Another knockout, that of the transcription factor NF-E2 *(260)*, a gene first thought to function primarily in erythroid differentiation, produced a severe, usually lethal, neonatal thrombocytopenia as discussed in a later chapter (*see* Chapter 12). The thrombocytopenia appears to result from a defect in cytoplasmic maturation, since the NF-E2$^{-/-}$ megakaryocytes, although more abundant and larger than normal with a higher modal DNA content, show a disorganized DMS with large membrane complexes and decreased granules, and are apparently incapable of platelet formation *(260)*.

7.3. The Platelet as a Model to Study Late Events in Megakaryocyte Maturation

The most mature megakaryocytes are very fragile and are difficult to obtain intact at high purity for biochemical or molecular studies. Platelets are small compartments of megakaryocyte cytoplasm expressing megakaryocytic macromolecules, which makes them suitable models for study of late events in megakaryocyte maturation. For example, protein synthesis by mature megakaryocytes may be reflected in platelets of patients with accelerated megakaryocytopoiesis, whose platelets retain significant biosynthetic capability *(263)*.

8. Clinical Abnormalities of Megakaryocytopoiesis

8.1. Chronic Myelogenous Leukemia

Thrombocytosis is a frequent occurrence in patients presenting with chronic myelogenous leukemia (CML) *(264)*, and megakaryocytes are increased in number *(265)*. However, rather than having larger megakaryocytes, the megakaryocytes in CML are usually smaller *(266,267)* and of lower-than-normal ploidy *(268,269)*, which seems discordant with the thrombocytosis. Since, in thrombocytotic states, megakaryocyte number, size, and ploidy are all usually increased, why is this not the case in CML? One explanation is that the smaller megakaryocyte size and lower ploidy are a direct consequence of the myeloproliferative defect. Alternatively, another intriguing possibility is that regulation of progenitor proliferation is abnormal, but that regulation of megakaryocyte poyploidization is normal. In that case, TPO levels would be reduced as a consequence of feedback regulation in response to elevated platelet number, and lower TPO levels would result in decreased megakaryocyte size and ploidy. Recent development of immunoassays for TPO levels allow this possibility to be examined.

Another profound thrombopoietic abnormality develops in CML. Pestina et al. found that platelet-storage pool levels of adenosine triphosphate (ATP) and adenosine diphosphate (ADP) showed a small, but significant, decrease during the chronic phase, but levels dramatically decreased as CML progressed to the accelerated phase *(270)*. This precipitous decrease in ATP and ADP represents the development of a severe abnormality in dense granule formation in megakaryocytes, and its onset may be predictive of imminent blast crisis *(270)*.

8.2. Thrombocythemia and Polycythemia Vera

Thrombocythemia and polycythemia vera represent myeloproliferative diseases in which increased platelet count is the result of both increased progenitor proliferation *(271–273)* and increased megakaryocyte polyploidization *(274,275)*. Marrow from these patients will form megakaryocyte colonies without the addition of exogenous recombinant growth factors *(271,276)*, suggesting that the signal transduction pathways that regulate progenitor proliferation and megakaryocyte polyploidization are constitutively upregulated in megakaryocyte precursors of these patients. Megakaryocytes of thrombocythemic patients have ultrastructural features indicative of stimulated thrombopoiesis, namely large cells with highly lobulated nuclei, abundant Golgi, DMS, and granules throughout the cytoplasm *(272)*. The recent cloning of TPO and understanding of signaling by Mpl *(277–282)* should allow the dissection of the molecular defects that cause abnormal proliferation and polyploidization of megakaryocytes in these diseases.

8.3. Acute Leukemia

Acute megakaryoblastic leukemia, M7 leukemia by the French-American-British (FAB) classification, is characterized by a predominance of morphologically undifferentiated blasts that can be identified as megakaryoblastic using megakaryocyte markers *(283)*. The patients usually present with thrombocytopenia, and their marrows often show extensive myelofibrosis. These leukemias respond poorly to chemotherapy. Megakaryocytes are usually reduced in number, and also can be morphologically abnormal with micromegakaryocytic forms, as in other types of acute myeloblastic leukemia (AML). Children *(284)* and adults *(285)* with non-M7 AML and small megakaryocyte size respond much more poorly to chemotherapy than those with normal megakaryocyte size. A high frequency of AML has been reported in a subset of families with autosomal dominant platelet-dense granule deficiency *(286)*. Megakaryocytes as well as other hemopoietic lineages are also reduced in number in acute lymphocytic leukemia (ALL).

8.4. TAR Syndrome

Thrombocytopenia with absent radius (TAR) syndrome is characterized by hypomegakaryocytic thrombocytopenia ($<100 \times 10^9$/L) and bilateral absence of radii inherited in an autosomal recessive pattern *(287)*. Other skeletal abnormalities especially of the arms and legs also may be present. Thrombocytopenia may be so severe at birth that hemorrhage is a complication; however, thrombocytopenia usually becomes less severe with age. Transplantation of a TAR syndrome patient with marrow from a normal sibling resulted in normal platelet production *(288)*. Correction of the platelet production defect by marrow transplantation supports the idea that the thrombopoietic defect in these patients is due to an intrinsic hemopoietic stem cell defect, rather than a deficiency of some humoral factor, such as TPO. This interpretation is consistent with a report that megakaryocyte colony-stimulating activity was high in the serum of an infant with TAR syndrome, but that megakaryocyte colony-forming cells were reduced, suggesting that the megakaryocyte progenitors were unable to respond to appropriate growth factor stimulation *(289)*.

8.5. Hereditary Macrothrombocytopenias

The hereditary macrothrombocytopenias or giant platelet disorders comprise a group of platelet formation abnormalities in which platelet number is reduced, but mean platelet volume is increased. Hemorrhage is usually not associated with the macrothrombocytopenias. These macrothrombocytopenias are usually detected by presurgery blood cell counts, and are often misdiagnosed as idiopathic thrombocytopenia (ITP). For an extensive review of macrothrombocytopenias, *see* White *(235)*. Since the size of the megakaryocytes is not reported as abnormal, the larger platelets must result from less extensive subdivision of megakaryocyte cytoplasm. May-Hegglin anomaly is one of the more frequent macrothrombocytopenias and is differentially diagnosed by the presence of leukocyte inclusions termed Döhle bodies *(235)*. Epstein's syndrome presents with nephritis and nerve deafness, as well as macrothrombocytopenia. Gray Platelet syndrome is of interest for its accompanying platelet α-granule deficency *(135,235,290–293)*. Bernard-Soulier syndrome is the only macrothrombocytopenia for which a molecular defect has been defined. This syndrome is characterized by deficiency of the platelet membrane glycoprotein Ib-IX-V complex *(294)*. Of particular interest is that GPIb serves as a linkage between the plasma membrane and the actin filament-based membrane skeleton *(295)*. These findings indicate that abnormalities in the interactions between the plasma membrane and the underlying cytoskeleton can result in abnormal platelet formation. Small numbers of patients have been described with macrothrombocytopenia with other distinguishing features *(235)*, the most recent of which is the Sebastian syndrome *(296)*.

Acknowledgments

This work was supported in part by RO1 Grant HL 51546 (C.W.J., P.E.S.) from the National Heart, Lung, and Blood Institute, P30 CA21765 Cancer Center Support Grant, and PO1 CA20180 from the National Cancer Institute, Public Health Service, Department of Health and Human Services, and by American Lebanese Syrian Associated Charities.

References

1. McDonald TP, Sullivan PS. Megakaryocytic and erythrocytic cell lines share a common precursor cell. *Exp Hematol.* 1993; 21: 1316–1320.
2. Hunt P. A bipotential megakaryocyte/erythrocyte progenitor cell: the link between erythropoiesis and megakaryopoiesis becomes stronger. *J Lab Clin Med.* 1995; 125: 303–304.
3. Golde DW. The stem cell. *Sci Am.* 1991; 265: 86–93.
4. Metcalf D, MacDonald HR, Odartchenko N, Sordat B. Growth of mouse megakaryocyte colonies in vitro. *Proc Natl Acad Sci USA.* 1975; 72: 1744–1748.
5. McLeod DL, Shreve MM, Axelrad AA. Induction of megakaryocyte colonies with platelet formation in vitro. *Nature.* 1976; 261: 492–494.
6. Nakeff A, Daniels-McQueen S. In vitro colony assay for a new class of megakaryocyte precursor: colony-forming unit megakaryocyte (CFU-M). *Proc Soc Exp Biol Med.* 1976; 151: 587–590.
7. Levin J, Levin FC, Penington DG, Metcalf D. Measurement of ploidy distribution in megakaryocyte colonies obtained from culture with studies of the effects of thrombocytopenia. *Blood.* 1981; 57: 287–297.
8. Levin J. Murine megakaryocytopoiesis in vitro: an analysis of culture systems used for the study of megakaryocyte colony–forming cells and of the characteristics of megakaryocyte colonies. *Blood.* 1983; 61: 617–623.

9. Chatelain C, de Bast M, Symann M. Identification of a light density murine megakaryocyte progenitor (LD-CFU-M). *Blood*. 1988; 72: 1187–1192.
10. Kuriya S, Ogata K, Yamada T, Gomi S, Nomura T. Three stages of differentiation in mouse megakaryocyte progenitor cells (CFU-Meg). *Exp Hematol*. 1990; 18: 416–420.
11. Williams N, Jackson H. Regulation of the proliferation of murine megakaryocyte progenitor cells by cell cycle. *Blood*. 1978; 52: 163–170.
12. Williams NT, Jackson HM, Eger RR, Long MW. The separate roles of factors in murine megakaryocyte colony formation. In: Evatt BL, Levine RF, Williams NT (eds). *Megakaryocyte Biology and Precursors: In Vitro Cloning and Cellular Properties*. New York: Elsevier/North Holland; 1981: 59–73.
13. Williams NT, Jackson HM. Kinetic analysis of megakaryocyte numbers and ploidy levels in developing colonies from mouse bone marrow cells. *Cell Tissue Kinet*. 1982; 15: 483–494.
14. Williams NT, Eger RR, Jackson HM, Nelson DJ. Two-factor requirement for murine megakaryocyte colony formation. *J Cell Physiol*. 1982; 110: 101–104.
15. Ishibashi T, Burstein SA. Interleukin-3 promotes the differentiation of isolated single megakaryocytes. *Blood*. 1986; 67: 1512–1514.
16. Kavnoudias H, Jackson H, Ettlinger K, Bertoncello I, McNiece I, Williams N. Interleukin 3 directly stimulates both megakaryocyte progenitor cells and immature megakaryocytes. *Exp Hematol*. 1992; 20: 43–46.
17. Kaushansky K, Lok S, Holly RD, et al. Promotion of megakaryocyte progenitor expansion and differentiation by c-Mpl ligand thrombopoietin. *Nature*. 1994; 369: 568–571.
18. Wendling F, Maraskovsky E, Debili N, et al. c-mpl ligand is a humoral regulator of megakaryocytopoiesis. *Nature*. 1994; 369: 571–574.
19. Broudy VC, Lin NL, Kaushansky K. Thrombopoietin (c-mpl ligand) acts synergistically with erythropoietin, stem cell factor, and interleukin-11 to enhance murine megakaryocyte colony growth and increases megakaryocyte ploidy in vitro. *Blood*. 1995; 85: 1719–1726.
20. Fauser AA, Messner HA. Identification of megakaryocytes, macrophages and eosinophils in colonies of human bone marrow containing neutrophillic granulocytes and erythroblasts. *Blood*. 1979; 53: 1023–1027.
21. Vainchenker W, Bouget J, Guichard J, Breton-Gorius J. Megakaryocyte colony formation from human bone marrow precursors. *Blood*. 1979; 54: 940–945.
22. Messner HA, Jamal N, Izaguirre C. The growth of large megakaryocyte colonies from human bone marrow. *J Cell Physiol Suppl*. 1982; 1: 45–51.
23. Messner HA, Jamal N, Yamasaki K, Solberg L, Jenkins RB. In vitro examination of human megakaryocyte precursors. In: Levine RF, Williams NT, Levin J, Evatt BL (eds). *Megakaryocyte Development and Function*. New York: Alan R Liss; 1986: 319–327.
24. Mazur EM, Hoffman R, Chasis J, Marchesi S, Bruno E. Immunofluorescent identification of human megakaryocyte colonies using an antiplatelet glycoprotein antiserum. *Blood*. 1981; 57: 277.
25. Mazur EM, Hoffman R, Bruno E, Marchesi S, Chasis J. Two classes of human megakaryocyte progenitor cells. In: Evatt BL, Levine RF, Williams NT (eds). *Megakaryocyte Biology and Precursors: In Vitro Cloning and Cellular Properties*. New York: Elsevier/North Holland; 1981: 281–288.
26. Mazur EM, Hoffman R, Bruno E. Regulation of human megakaryocytopoiesis. *J Clin Invest*. 1981; 68: 733–741.
27. Hoffman R, Stravena J, Yang HH, Bruno E, Brandt J. New insights into the regulation of human megakaryocytopoiesis. *Blood Cells*. 1987; 13: 75–86.
28. Debili N, Wendling F, Katz A, et al. The Mpl-ligand or thrombopoietin or megakaryocyte growth and differentiative factor has both direct proliferative and differentiative activities on human megakaryocyte progenitors. *Blood*. 1995; 86: 2516–2525.
29. Nichol JL, Hokom MM, Hornkohl A, et al. Megakaryocyte growth and development factor. Analyses of in vitro effects on human megakaryopoiesis and endogenous serum levels during chemotherapy-induced thrombocytopenia. *J Clin Invest*. 1995; 95: 2973–2978.

30. Long MW, Gragowski LL, Heffner CH, Boxer LA. Phorbol diesters stimulate the development of an early murine progenitor cell. The burst forming unit-megakaryocyte. *J Clin Invest.* 1985; 76: 431–438.

31. Long MW, Heffner CH, Gragowski LL. In vitro differences in responsiveness of early (BFU-Mk) and late (CFU-Mk) murine megakaryocyte progenitor cells. *Prog Clin Biol Res.* 1986; 215: 179–186.

32. Long MW. Population heterogeneity among cells of the megakaryocyte lineage. *Stem Cells.* 1993; 11: 33–40.

33. Jackson H, Williams N, Bertoncello I, Green R. Classes of primitive murine megakaryocytic progenitor cells. *Exp Hematol.* 1994; 22: 954–958.

34. Jackson H, Williams N, Westcott KR, Green R. Differential effects of transforming growth factor-$\beta1$ on distinct developmental stages of murine megakaryocytopoiesis. *J Cell Physiol.* 1994; 161: 312–318.

35. Lowry PA, Deacon DM, Whitefield P, Rao S, Quesenberry M, Quesenberry PJ. The high-proliferative-potential megakaryocyte mixed (HPP-Meg-Mix) cell: a trilineage murine hemato-poietic progenitor with multiple growth factor responsiveness. *Exp Hematol.* 1995; 23: 1135–1140.

36. Sitnicka E, Lin N, Fox N, et al. The effect of thrombopoietin on the proliferation and differentiation of murine hematopoietic stem cells. *Blood.* 1996; 87: 4998–5005.

37. Zeigler FC, de Sauvage F, Widmer HR, et al. In vitro megakaryocytopoietic and thrombopoietic activity of c-mpl ligand (TPO) on purified murine hematopoietic stem cells. *Blood.* 1994; 84: 4045–4052.

38. Briddell RA, Brandt JE, Stravena JE, Srour EF, Hoffman R. Characterization of the human burst-forming unit-megakaryocyte. *Blood.* 1989; 74: 145–151.

39. Srour EF, Brandt JE, Briddell RA, Leemhuis T, van Besien K, Hoffman R. Human CD34$^+$ HLA-DR$^-$ bone marrow cells contain progenitor cells capable of self-renewal, multilineage differentiation, and long-term in vitro hematopoiesis. *Blood Cells.* 1991; 17: 287–295.

40. Briddell RA, Brandt JE, Hoffman R. The most primitive human megakaryocyte progenitor cell does not express major histocompatibility class II antigens. *Exp Hematol.* 1988; 16: 365.

41. Briddell RA, Hoffman R. Cytokine regulation of the human burst-forming unit-megakaryocyte. *Blood.* 1990; 76: 516–522.

42. Bruno E, Cooper RJ, Briddell RA, Hoffman R. Further examination of the effects of recombinant cytokines on the proliferation of human megakaryocyte progenitor cells. *Blood.* 1991; 77: 2339–2346.

43. Zauli G, Valvassori L, Capitani S. Presence and characteristics of circulating megakaryocyte progenitor cells in human fetal blood. *Blood.* 1993; 81: 385–390.

44. Deutsch VR, Olson TA, Nagler A, Slavin S, Levine RF, Eldor A. The response of cord blood megakaryocyte progenitors to IL-3, IL-6 and aplastic canine serum varies with gestational age. *Br J Haematol.* 1995; 89: 8–16.

45. Siena S, Bregni M, Bonsi L, et al. Increase in peripheral blood megakaryocyte progenitors following cancer therapy with high-dose cyclophosphamide and hematopoietic growth factors. *Exp Hematol.* 1993; 21: 1583–1590.

46. Briddell R, Glaspy J, Shpall EJ, LeMaistre F, Menchaca D, McNiece IK. Mobilization of myeloid, erythroid and megakaryocyte progenitors by recombinant human stem cell factor (rhSCF) plus filgrastim (rhG-CSF) in patients with breast cancer. *Proc ASCO.* 1994; 13: 77 (abstract no 109).

47. Tong J, Gordon MS, Srour EF, et al. In vivo administration of recombinant methionyl human stem cell factor expands the number of human marrow hematopoietic stem cells. *Blood.* 1993; 82: 784–791.

48. Jones BC, Radley JM, Bradley TR, Hodgson GS. Enhanced megakaryocyte repopulating ability of stem cells surviving 5-fluorouracil treatment. *Exp Hematol.* 1980; 8: 61–64.

49. Thean LE, Hodgson GS, Bertoncello I, Radley JM. Characterization of megakaryocyte spleen

colony-forming units by response to 5-fluorouracil and by unit gravity sedimentation. *Blood.* 1983; 62: 896–901.

50. Zajicek J. Studies on the histogenesis of blood platelets. I. Histochemistry of acetylcholinesterase activity of megakaryocytes and platelets in different species. *Acta Haematol.* 1954; 121: 238–244.

51. Breton-Gorius J, Guichard J. Ultrastructural localization of peroxidase activity in human platelets and megakaryocytes. *Am J Pathol.* 1972; 66: 277–293.

51a. Breton-Gorius J, Villeval JL, Mitjavila MT, Vinci G, Guichard J, Rochandt H, Flandrin G, Vainchenker W. Ultrastructural and cytochemical characterization of blasts from early erythroblastic leukemias. *Leukemia.* 1987; 1: 173–178.

52. Karnovsky MJ, Roots L. A "direct-coloring" thiocholine method for cholinesterases. *J Histochem Cytochem.* 1964; 12: 219–221.

53. Jackson CW. Cholinesterase as a possible marker for early cells of the megakaryocytic series. *Blood.* 1973; 42: 413-421.

54. Long MW, Henry RL. Thrombocytosis-induced suppression of small acetylcholinesterase-positive cell in bone marrow of rats. *Blood.* 1979; 54: 1338–1346.

55. Tranum-Jensen J, Behnke O. Electron microscopical identification of the committed precursor cell of the megakaryocyte compartment of rat bone marrow. *Cell Biol Int Rep.* 1977; 1: 445–452.

56. Long MW, Williams NT, Ebbe S. Immature megakaryocytes in the mouse: physical characteristics, cell cycle status, and in vitro responsiveness to thrombopoietic stimulatory factor. *Blood.* 1982; 59: 569–575.

57. Jackson CW. Some characteristics of rat megakaryocyte precursors identified using cholinesterase as a marker. In: Baldini MG, Ebbe S (eds). *Platelets: Production, Function, Transfusion and Storage.* New York: Grune & Stratton; 1974: 33–40.

58. Ebbe S, Stohlman FJ. Megakaryocytopoiesis in the rat. *Blood.* 1965; 26: 20–34.

59. Odell TT, Jackson CW. Polyploidy and maturation of rat megakaryocytes. *Blood.* 1968; 32: 102–110.

60. Levine RF, Hazzard KC, Lamberg JD. The significance of megakaryocyte size. *Blood.* 1982; 60: 1122–1131.

61. Ebbe S, Stohlman FJ. Effects of hypertransfusion and erythropoietin on labeling of rat megakaryocytes by tritiated thymidine. *Proc Soc Exp Biol Med.* 1964; 116: 971–974.

62. De Leval M. Etude cytochimique quantitative des acides desoxyribonucleiques au cours de la maturation megacaryocytaire. *Nouv Rev Fr Hematol.* 1968; 8: 392–394.

63. Levine RF. Isolation and characterization of normal human megakaryocytes. *Br J Haematol.* 1980; 45: 487–497.

64. Jackson CW, Brown LK, Somerville BC, Lyles SA, Look AT. Two-color flow cytometric measurement of DNA distributions of rat megakaryocytes in unfixed, unfractionated marrow cell suspensions. *Blood.* 1984; 63: 768–778.

65. Corash L, Levin J, Mok Y, Baker G, McDowell J. Measurement of megakaryocyte frequency and ploidy distribution in unfractionated murine bone marrow. *Exp Hematol.* 1989; 17: 278–286.

66. Tomer A, Harker LA, Burstein SA. Flow cytometric analysis of normal human megakaryocytes. *Blood.* 1988; 71: 1244–1252.

67. Paulus JM. DNA metabolism and development of organelles in guinea-pig megakaryocytes: a combined ultrastructural, autoradiographic and cytophotometric study. *Blood.* 1970; 35: 298–311.

68. Odell TT Jr, Jackson CW, Friday TJ. Megakaryocytopoiesis in rats with special reference to polyploidy. *Blood.* 1970; 35: 775–782.

69. Prandini MH, Uzan G, Martin F, Thevenon D, Marguerie G. Characterization of a specific erythromegakaryocytic enhancer within the glycoprotein IIb promoter. *J Biol Chem.* 1995; 267: 10,370–10,374.

70. Levene RB, Lamaziere JD, Broxmeyer HE, Lu L, Rabellino EM. Human megakaryocytes. V. Changes in the phenotypic profile of differentiating megakaryocytes. *J Exp Med*. 1985; 161: 457–474.

71. Berridge MV, Ralph SJ, Tan AS. Cell-lineage antigens of the stem cell-megakaryocyte-platelet lineage are associated with the platelet IIb-IIIa glycoprotein complex. *Blood*. 1985; 66: 76–85.

72. Vinci G, Tabilio A, Deschamps JF, et al. Immunological study of in vitro maturation of human megakaryocytes. *Br J Haematol*. 1984; 56: 589–605.

73. Ryo R, Yasunaga M, Saigo K, Yamaguchi N. Megakaryocytic leukemia and platelet factor 4. *Leukemia Lymphoma*. 1992; 8: 327–336.

74. Schick PK, Konkle BA, He X, Thornton RD. P-selectin mRNA is expressed at a later phase of megakaryocyte maturation than mRNAs for von Willebrand factor and glycoprotein Ib-α. *J Lab Clin Med*. 1993; 121: 714–721.

75. Nakamura M, Mori M, Nakazawa S, et al. Replacement of *m*-calpain by μ-calpain during maturation of megakaryocytes and possible involvement in platelet formation. *Thromb Res*. 1992; 66: 757–764.

76. Gilman JR. Normal hemopoiesis in intrauterine and neonatal life. *J Pathol*. 1942; 52: 25.

77. Allen Graeve JL, de Alarcon PA. Megakaryocytopoiesis in the human fetus. *Arch Dis Child*. 1989; 64: 481–484.

78. de Alarcon PA, Graeve JLA. Analysis of megakaryocyte ploidy in fetal bone marrow biopsies using a new adaptation of the Feulgen technique to measure DNA content and estimate megakaryocyte ploidy from biopsy specimens. *Pediatr Res*. 1996; 39: 166–170.

79. Izumi T, Kawakami M, Enzan H, Ohkita T. The size of megakaryocytes in human fetal, infantile and adult hematopoiesis. *Hiroshima J Med Sci*. 1983; 32: 257–260.

80. Enzan H, Takahashi H, Kawakami M, Yamashita S, Ohkita T, Yamamoto M. Light and electron microscopic observations of hepatic hematopoiesis of human fetuses. II. Megakaryocytopoiesis. *Acta Pathol Jpn*. 1980; 30: 937–954.

81. Emura I, Sekiya M, Ohnishi Y. Two types of immature megakaryocytic series in the human fetal liver. *Arch Histol Jpn*. 1983; 46: 631–643.

82. Daimon T, David H. An automatic image analysis of megakaryocytes in fetal liver and adult bone marrow. *Z Mikrosk Anat Forsch*. 1982; 3: 454–460.

83. Hegyi E, Nakazawa M, Debili N, et al. Developmental changes in human megakaryocyte ploidy. *Exp Hematol*. 1991; 19: 87–94.

84. Radley JM, Green SL. Ultrastructure of endomitosis in megakaryocytes. *Nouv Rev Fr Hematol*. 1989; 31: 232a (abstract).

85. Japa J. A study of the morphology and development of megakaryocytes. *Br J Exp Pathol*. 1943; 24: 73–80.

86. Odell TT, Jackson CW, Reiter RS. Generation cycle of rat megakaryocytes. *Exp Cell Res*. 1968; 53: 321–328.

87. Odell TT Jr, Jackson CW, Gosslee DG. Maturation of rat megakaryocytes studied by microspectrophotometric measurement of DNA. *Proc Soc Exp Biol Med*. 1965; 119: 1194–1199.

88. Rolovic Z. Ploidy value of endoreduplicating megakaryocytes in immune and "hypersplenic" thrombocytopenia. In: Baldini MG, Ebbe S (eds). *Platelets: Production, Function, Transfusion and Storage*. New York: Grune & Stratton; 1974: 143–153.

89. Penington DG, Olsen TE. Megakaryocytes in states of altered platelet production: cell numbers, sizes and DNA content. *Br J Haematol*. 1970; 18: 447–463.

90. Mazur EM, Lindquist DL, de Alarcon PA. Evaluation of bone marrow megakaryocyte ploidy in persons with normal and abnormal platelet counts. *J Lab Clin Med*. 1988; 111: 194–202.

91. Ebbe S. Regulation of murine megakaryocyte size and ploidy by non-platelet dependent mechanisms in radiation-induced megakaryocytopenia. *Radiat Res*. 1991; 127: 278–284.

92. Kuter DJ, Rosenberg RD. Appearance of a megakaryocyte growth-promoting activity, megapoietin, during acute thrombocytopenia in the rabbit. *Blood*. 1994; 84: 1464–1472.

93. Kuter DJ, Rosenberg RD. The reciprocal relationship of thrombopoietin (c-Mpl ligand) to changes

in the platelet mass during busulfan-induced thrombocytopenia in the rabbit. *Blood*. 1995; 85: 2720–2730.

94. Corash L, Chen HY, Levin J, Baker G, Lu H, Mok Y. Regulation of thrombopoiesis: effects of the degree of thrombocytopenia on megakaryocyte ploidy and platelet volume. *Blood*. 1987; 70: 177–185.

95. Sullivan PS, Jackson CW, McDonald TP. Castration decreases thrombocytopoeisis and testosterone restores platelet production in castrated Balb/c mice: evidence that testosterone acts on a bipotential hematopoietic precursor cell. *J Lab Clin Med*. 1995; 125: 326–333.

96. McDonald TP, Swearingen CJ, Cottrell MB, Clift RE, Bryant SE, Jackson CW. Sex- and strain-related differences in megakaryocytopoiesis and platelet production in C3H and Balb/c mice. *J Lab Clin Med*. 1992; 120: 168–173.

97. Jackson CW, Steward SA, Chenaille PJ, Ashmun RA, McDonald TP. An analysis of megakaryocytopoiesis in the C3H mouse: an animal model whose megakaryocytes have 32N as the modal DNA class. *Blood*. 1990; 76: 690–696.

98. Jackson CW, Steward SA, Ashmun RA, McDonald TP. Megakaryocytopoiesis and platelet production are stimulated during late pregnancy and early postpartum in the rat. *Blood*. 1992; 79: 1672–1678.

99. Tschope D, Schwippert B, Schettler B, et al. Increased GP IIb/IIIa expression and altered DNA-ploidy pattern in megakaryocytes of diabetic BB-rats. *Eur J Clin Inv*. 1992; 22: 591–598.

100. Gewirtz AM, Calabretta B. Molecular regulation of human megakaryocyte development. *Int J Cell Cloning*. 1990; 8: 267–276.

101. Wang Z, Zhang Y, Kamen D, Lee E, Ravid K. Cyclin D3 is essential for megakaryocytopoiesis. *Blood*. 1995; 86: 3783–3788.

102. Wilhide CC, Van Dang C, Dispersio J, Kenedy AA, Bray PF. Overexpression of cyclin D1 in the Dami megakaryocytic cell line causes growth arrest. *Blood*. 1995; 86: 294–304.

103. Graña X, Reddy EP. Cell cycle control in mammalian cells: role of cyclins, cyclin dependent kinases (CDKs), growth suppressor genes and cyclin-dependent kinase inhibitors (CKIs). *Oncogene*. 1995; 11: 211–219.

104. Pines J. Cyclins and cyclin-dependent kinases: a biochemical view. *Biochem J*. 1995; 308: 697–711.

105. Datta NS, Williams JL, Long MW. The role of cyclin D3- and cyclin E-cell division kinase complexes in megakaryocyte endomitosis. *Blood*. 1993; 82: 209a (abstract no 822).

106. Chang CL, Long MW. Inactivation of CDC2 kinase during endomitotic DNA synthesis in human erythroleukemia (HEL) cells. *Exp Hematol*. 1992; 20: 822 (abstract).

107. Zhang Y, Wang Z, Kamen D, Ravid K. Analysis of a unique cell cycle during endomitosis in megakaryocytes. *Blood*. 1994; 84: 389a (abstract no 1540).

108. Datta NS, Williams JL, Long MW. Alterations in cyclin B and CDC2 protein kinase play a role in megakaryocytic endomitosis. *Blood*. 1994; 84: 389a (abstract no 1542).

109. Behnke O. An electron microscope study of the megakaryocyte of the rat bone marrow. I. The development of the demarcation membrane system and the platelet surface coat. *J Ultra Res*. 1968; 24: 412–433.

110. Bentfeld-Barker ME, Bainton DF. Ultrastructure of rat megakaryocytes after prolonged thrombocytopenia. *J Ultra Res*. 1977; 61: 201–214.

111. Zucker-Franklin D, Petursson S. Thrombocytopoiesis—analysis by membrane tracer and freeze-fracture studies on fresh human and cultured mouse megakaryocytes. *J Cell Biol*. 1984; 99: 390–402.

112. Radley JM, Scurfield G. The mechanisms of platelet release. *Blood*. 1980; 56: 996–999.

113. Radley JM, Haller CJ. The demarcation membrane system of the megakaryocyte: a misnomer? *Blood*. 1982; 60: 213–219.

114. Thiele J, Galle R, Sander, C Fischer R. Interactions between megakaryocytes and sinus wall. An ultrastructural study on bone marrow tissue in primary (essential) thrombocythemia. *J Submicrosc Cytol Pathol*. 1991; 23: 595–603.

115. White JG. A search for the platelet secretory pathway using electron dense tracers. *Am J Pathol.* 1970; 58: 31–49.
116. Zucker-Franklin D, Benson KA, Meyers KM. Absence of a surface-connected canalicular system in bovine platelets. *Blood.* 1985; 65: 241–244.
117. White JG. The secretory pathway of bovine platelets. *Blood.* 1987; 69: 878–885.
118. White JG. Mechanisms of platelet production. *Blood Cells.* 1989; 15: 48–57.
119. Menard M, Meyers KM. Storage pool deficiency in cattle with the Chediak Higashi syndrome results from an absence of dense granule precursors in their megakaryocytes. *Blood.* 1988; 72: 1726–1734.
120. White JG. Interaction of membrane systems in blood platelets. *Am J Pathol.* 1972; 66: 295–312.
121. Nichols BA, Setzer PY, Bainton DF. Glucose-6-phosphatase as a cytochemical marker of endoplasmic reticulum in human leukocytes and platelets. *J Histochem Cytochem.* 1984; 32: 165–171.
122. Daimon T, Gotoh Y. Cytochemical evidence of the origin of the dense tubular system in the mouse platelet. *Histochemistry.* 1982; 76: 189–196.
123. Paulus JM, Maigne J, Keyhani E. Mouse megakaryocytes secrete acetylcholinesterase. *Blood.* 1981; 58: 1100–1106.
124. Tranum-Jensen J, Behnke O. Acetylcholinesterase in the platelet-megakaryocyte system. I. Structural localization in platelets of the rat, mouse, and cat. *Eur J Cell Biol.* 1981; 24: 275–280.
125. Tranum-Jensen J, Behnke O. Acetylcholinesterase in the platelet-megakaryocyte system. II. Structural localization in the megakaryocytes of the rat, mouse and cat. *Eur J Cell Biol.* 1981; 24: 281–286.
126. Gerrard JM, White JG, Rao GHR, Townsend D. Localization of platelet prostaglandin production in the platelet dense tubular system. *Am J Pathol.* 1976; 83: 283–298.
127. White JG, Clawson CC. The surface-connected canalicular system of blood platelets—a fenestrated membrane system. *Am J Pathol.* 1980; 101: 353–364.
128. Breton-Gorius J. Development of two distinct membrane systems associated in giant complexes in pathological megakaryocytes. *Ser Haematol.* 1975; 8: 49–67.
129. Handagama PJ, George JN, Shuman MA, McEver RP, Bainton DF. Incorporation of circulating protein into megakaryocyte and platelet granules. *Proc Natl Acad Sci USA.* 1987; 84: 861–865.
130. Handagama PJ, Shuman MA, Bainton DF. Incorporation of intravenously injected albumin, immunoglobulin G, and fibrinogen in guinea pig megakaryocyte granules. *J Clin Invest.* 1989; 84: 73–82.
131. Handagama P, Bainton DF, Jacques Y, Conn MT, Lazarus RA, Shuman MA. Kistrin, an integrin antagonist, blocks endocytosis of fibrinogen into guinea pig megakaryocyte and platelet α-granules. *J Clin Invest.* 1993; 91: 193–200.
132. Handagama PJ, Amrani DL, Shuman MA. Endocytosis of fibrinogen into hamster megakaryocyte α-granules is dependent on a dimeric gamma A configuration. *Blood.* 1995; 85: 1790–1795.
133. Harrison P, Wilbourn B, Debili N, et al. Uptake of plasma fibrinogen into alpha granules of human megakaryocytes and platelets. *J Clin Invest.* 1989; 84: 1320–1324.
134. Jones OP. Origin of megakaryocyte granules from Golgi vesicles. *Anat Rec.* 1960; 138: 105–114.
135. Nurden AT, Kunicki TJ, Dupuis D, Soria C, Caen JP. Specific protein and glycoprotein deficiencies in platelets isolated from two patients with the gray platelet syndrome. *Blood.* 1982; 59: 709–718.
136. Rosa JP, George JN, Bainton DF, Nurden AT, Caen JP, McEver RP. Gray platelet syndrome. Demonstration of alpha granule membranes that can fuse with the cell surface. *J Clin Invest.* 1987; 80: 1138–1146.
137. Gerrard JM, Phillips DR, Rao GHR, et al. Biochemical studies of two patients with the gray platelet syndrome. Selective deficiency of platelet alpha granules. *J Clin Invest.* 1980; 66: 102–109.
138. Jackson CW, Hutson NK, Steward SA, Nagahito S, Cramer EM. Platelets of the Wistar-Furth rat have reduced levels of α-granule proteins. An animal model resembling gray platelet syndrome. *J Clin Invest.* 1991; 87: 1985–1991.

139. George JN. Platelet immunoglobin G: its significance for the evalution of thrombocytopenia and for understanding the origin of α-granule proteins. *Blood*. 1990; 76: 859–870.

140. Harrison P. The origin and physiological relevance of α-granule adhesive proteins. *Br J Haematol*. 1990; 74: 125–130.

141. Handagama P, Scarborough RM, Shuman MA, Bainton DF. Endocytosis of fibrinogen into megakaryocytes and platelet α-granules is mediated by αIIβ3 (Glycoprotein IIb-IIIa). *Blood*. 1993; 82: 135–138.

142. Harrison P, Wilbourn B, Cramer E, et al. The influence of therapeutic blocking of Gp IIb/IIIa on platelet α-granular fibrinogen. *Br J Haematol*. 1992; 82: 721–728.

143. Zucker-Franklin D. Endocytosis by human platelets: metabolic and freeze-fracture studies. *J Cell Biol*. 1981; 91: 706–715.

144. Holmsen H, Weiss HJ. Secretable storage pools in platelets. *Annu Rev Med*. 1979; 30: 119–134.

145. Holmsen M. Platelet secretion. In: Colman RW, Hirsch J, Marder VJ, Salzman EW (eds). *Hemostasis and Thrombosis: Basic Principles and Clinical Practice*. Philadelphia: Lippincott; 1982: 390–403.

146. Daimon T, David H. Precursors of monoamine-storage organelles in developing megakaryocytes of the rat. *Histochemistry*. 1983; 77: 353–363.

147. Israels SJ, Gerrard JM, Jacques YV, et al. Platelet dense granule membranes contain both granulophysin and P-selection (GMP-140). *Blood*. 1992; 80: 143–152.

148. Meyers K, Seachord C. Identification of dense granules specific membrane proteins in bovine platelets that are absent in the Chediak-Higashi syndrome. *Thromb Haemost*. 1990; 64: 319–325.

149. Fedorko ME. The functional capacity of guinea pig megakaryocytes. I. Uptake of ^{3}H-serotonin by megakaryocytes and their physiologic and morphologic response to stimuli for the platelet release reaction. *Lab Invest*. 1977; 36: 310–320.

150. White JG. Serotonin storage organelles in human megakaryocytes. *Am J Pathol*. 1971; 63: 403–410.

151. Tranzer JP, DaPrada M, Pletscher A. Storage of 5-hydroxytryptamine in megakaryocytes. *J Cell Biol*. 1972; 52: 191–197.

152. Leven RM, Nachmias VT. Cultured megakaryocytes: changes in the cytoskeleton after ADP-induced spreading. *J Cell Biol*. 1982; 92: 313–323.

153. Leven RM, Nachmias VT. Alpha-actinin arcs in megakaryocyte spreading. *Exp Cell Res*. 1984; 152: 476–485.

154. Fox JEB. The platelet cytoskeleton. *Thromb Haemost*. 1993; 70: 884–893.

155. Pestina TI, Jackson CW, Stenberg PE. Abnormal subcellular distribution of myosin and talin in Wistar Furth rat platelets. *Blood*. 1995; 85: 2436–2446.

156. Stenberg PE, McDonald TP, Jackson CW. Disruption of microtubules in vivo by vincristine induces large membrane complexes and other cytoplasmic abnormalities in megakaryocytes and platelets of normal rats like those in human and Wistar-Furth rat hereditary macrothrombocytopenias. *J Cell Physiol*. 1995; 162: 86–102.

157. Leven RM, Yee MK. Megakaryocyte morphogenesis stimulated in vitro by whole and partially fractionated thrombocytopenic plasma: a model system for the study of platelet formation. *Blood*. 1987; 69: 1046–1059.

158. Leven RM. Megakaryocyte motility and platelet formation. *Scanning Microsc*. 1987; 1: 1701–1709.

159. Tablin F, Castro M, Leven RM. Blood platelet formation in vitro. The role of the cytoskeleton in megakaryocyte fragmentation. *J Cell Sci*. 1990; 97: 59–70.

160. Jackson CW, Hutson NK, Steward SA, et al. The Wistar-Furth rat: an animal model of hereditary macrothrombocytopenia. *Blood*. 1988; 71: 1676–1686.

161. Jackson CW, Hutson NK, Steward SA, Stenberg PE. A unique talin antigenic determinant and anomalous megakaryocyte talin distribution associated with abnormal platelet formation in the Wistar Furth rat. *Blood*. 1992; 79: 1929–1737.

162. Takata K, Singer SJ. Localization of high concentrations of phosphotyrosine-modified proteins in mouse megakaryocytes. *Blood.* 1988; 73: 818–821.

163. Bennett BD, Cowley S, Jiang S, et al. Identification and characterization of a novel tyrosine kinase from megakaryocytes. *J Biol Chem.* 1994; 269: 1068–1074.

164. Avraham S, Jiang S, Ota S, et al. Structural and functional studies of the intracellular tyrosine kinase MATK gene and its translated product. *J Biol Chem.* 1995; 270: 1833–1842.

165. Jhun BH, Rivnay B, Price D, Avraham H. The MATK tyrosine kinase interacts in a specific and SH2-dependent manner with c–kit. *J Biol Chem.* 1995; 270: 9661–9666.

166. Avraham S, London R, Fu YG, et al. Identification and characterization of a novel related adhesion focal tyrosine kinase (RAFTK) from megakaryocytes and brain. *J Biol Chem.* 1995; 270: 27,742–27,751.

167. Wolf NS, Trentin JJ. Hemopoietic colony studies. V. Effect of hemopoietic organ stroma on differentiation of pluripotent stem cells. *J Exp Med.* 1968; 127: 205–214.

168. Tavassoli M, Aoki M. Localization of megakaryocytes in the bone marrow. *Blood Cells.* 1989; 15: 3–14.

169. Lichtman MA, Chamberlain JK, Simon W, Santillo PA. Parasinusoidal location of megakaryocytes in marrow: a determinant of platelet release. *Am J Hematol.* 1978; 4: 303–312.

170. Avraham H, Cowley S, Chi SY, Jiang S, Groopman JE. Characterization of adhesive interactions between human endothelial cells and megakaryocytes. *J Clin Invest.* 1993; 91: 2378–2384.

171. Rafii S, Sharpiro F, Petengell R, et al. Human bone marrow microvascular endothelial cells support long-term proliferation and differentiation of myeloid and megakaryocytopoietic progenitors. *Blood.* 1995; 86: 3353–3363.

172. Avraham H, Scadden DT, Chi S, Broudy VC, Zsebo KM, Groopman JE. Interaction of human bone marrow fibroblasts with megakaryocytes: role of the c-kit ligand. *Blood.* 1992; 80: 1679–1684.

173. von Behrens WE. Evidence of phylogenetic canalisation of the circulating platelet mass of man. *Thrombosis Diathesis Haemorrhagi.* 1972; 27: 159–172.

174. Trowbridge EA, Martin JF, Slater DN, et al. The origin of platelet count and volume. *Clin Phys Physiol Meas.* 1984; 5: 145–170.

175. Paulus JM. Platelet size in man. *Blood.* 1975; 46: 321–336.

176. Corash L. Platelet sizing techniques: biologic significance and clinical implications. *Curr Top Hematol.* 1983; 4: 99–122.

177. von Behrens WE. Mediterranean macrothrombocytopenia. *Blood.* 1975; 46: 199–208.

178. Harker LA. Megakaryocyte quantitation. *J Clin Invest.* 1968; 47: 452–457.

179. Odell TT, Murphy JR, Jackson CW. Stimulation of megakaryocytopoiesis by acute thrombocytopenia in rats. *Blood.* 1976; 48: 765–775.

180. Illes I, Pfueller S, Hussein S, Chesterman CN, Martin JF. Platelets in idiopathic thrombocytopenic purpura are increased in size but are of normal density. *Br J Haematol.* 1987; 67: 173–176.

181. Daw NC, Arnold JT, White MM, Stenberg PE, Jackson CW. Regulation of thrombopoiesis: in vivo response to a single injection of murine, PEG-megakaryocyte growth and development factor. *Blood.* 1995; 86: 897a (abstract no 3575).

182. Ulich TR, Del Castillo J, Yin S, et al. Megakaryocyte growth and development factor ameliorates carboplatin-induced thrombocytopenia in mice. *Blood.* 1995; 86: 971–976.

183. Harker LA, Hunt P, Marzed UM, et al. Dose–response effects of pegylated human megakaryocyte growth and development factor (PEG-rHuMGDF) on platelet production and function in non-human primates. *Blood.* 1995; 86: 256a (abstract no 1012).

183a. Ulich TR, Del Castillo J, Senaldi G, Kinstler O, Yin S, Kaufman S, Tarpley J, Choi E, Kirley T, Hunt P, Sheridan WP. Systemic hematologic effects of PEG-rHu-MGDF-induced megakaryocyte hyperplasia in mice. *Blood.* 1996; 87: 5006–5015.

184. van Doormaal JJ, van der Meer J, Oosten HR, Halie MR, Doorenbos H. Hypothyroidism leads to more small-sized platelets in circulation. *Thromb Haemost.* 1987; 58: 964–965.

185. Sullivan P, Gompf R, Schmeitzel L, Clift R, Cottrell M, McDonald TP. Altered platelet indices in dogs with hypothroidism and cats with hyperthyroidism. *Am J Vet Res.* 1993; 54: 2004–2009.

186. Ford HC, Toomath RJ, Carter JM, Delahunt JW, Fagerstrom JN. Mean platelet volume is increased in hyperthyroidism. *Am J Hematol.* 1988; 27: 190–193.

187. Odell TT, Burch EA, Jackson CW, Friday TJ. Megakaryocytopoiesis in mice. *Cell Tissue Kinet.* 1969; 254: 363–367.

188. Cronkite EP, Bond VP, Fliedner TM, Paglia DA, Adamik ER. Studies on the origin, production and destruction of platelets. In: Johnson SA, Monto RW, Rebuck JW, Horn RC (eds). *Blood Platelets.* Boston: Little, Brown; 1961: 595–609.

189. Jackson CW, Edwards CC. Biphasic thrombopoietic response to severe hypobaric hypoxia. *Br J Haematol.* 1977; 35: 233–244.

190. Aster RH. Studies of the mechanism of "hypersplenic" thrombocytopenia in rats. *J Lab Clin Med.* 1967; 70: 736–751.

191. Harker LA, Finch CA. Thrombokinetics in man. *J Clin Invest.* 1969; 48: 963–974.

192. Odell TT Jr, Jackson CW, Friday TJ, Charsha DE. Effects of thrombocytopenia on megakaryocytopoiesis. *Br J Haematol.* 1969; 17: 91–101.

193. Ebbe S, Stohlman F Jr, Donovan J, Overcash J. Megakaryocyte maturation rate in thrombocytopenic rats. *Blood.* 1968; 32: 787–795.

194. Odell TT, McDonald TP, Asano M. Response of rat megakaryocytes and platelets to bleeding. *Acta Haematol.* 1962; 27: 171–179.

195. Harker LA. Kinetics of thrombopoiesis. *J Clin Invest.* 1968; 47: 458–465.

196. Stenberg PE, Levin J, Baker G, Mok Y, Corash L. Neuraminidase-induced thrombocytopenia in mice: effects on thrombopoiesis. *J Cell Physiol.* 1991; 147: 7–16.

197. Kalmaz GD, McDonald TP. Effects of antiplatelet serum and thrombopoietin on the percentage of small acetylcholinesterase-positive cells in bone marrow of mice. *Exp Hematol.* 1981; 9: 1002–1010.

198. Kraytman M. Platelet size in thrombocytopenias and thrombocytosis of various origin. *Blood.* 1973; 41: 587–598.

199. Ebbe S, Stohlman F Jr, Overcash J, Donovan J, Howard D. Megakaryocyte size in thrombocytopenic and normal rats. *Blood.* 1968; 32: 383–392.

200. Burstein SA, Adamson JW, Erb SK, Harker LA. Megakaryocytopoiesis in the mouse: response according to varying platelet demand. *J Cell Physiol.* 1981; 109: 333–341.

201. Odell TT, Boran DA. The mitotic index of megakaryocytes after acute thrombocytopenia. *Proc Soc Exp Biol Med.* 1977; 155: 149–151.

202. Odell TT Jr, Jackson CW, Reiter RS. Depression of the megakaryocyte-platelet system in rats by transfusion of platelets. *Acta Haematol.* 1967; 38: 34–42.

203. Krizsa F, Kovacs Z, Dobay E. Effects of vincristine on the megakaryocyte system in mice. *J Med.* 1973; 4: 12–18.

204. Rak K. Effect of vincristine on platelet production in mice. *Br J Haematol.* 1972; 22: 617–624.

205. Robertson JH, Crozier EH, Woodend BE. The effect of vincristine on the platelet count in rats. *Br J Haematol.* 1970; 19: 331–337.

206. Klener P, Donner L, Houskova J. Thrombocytosis in rats induced by vincristine. *Haemostasis.* 1972; 1: 73–78.

207. Choi SI, Simone JV, Edwards CC. Effects of vincristine on platelet production. In: Baldini MG, Ebbe S (eds). *Platelets: Production, Function, Transfusion and Storage.* New York: Grune & Stratton; 1974: 51–61.

208. Jackson CW, Edwards CC. Evidence that stimulation of megakaryocytopoiesis by low dose vincristine results from an effect on platelets. *Br J Haematol.* 1977; 36: 97–105.

209. Burstein Y, Giardina PJV, Rausen AR, Kandall SR, Siljestrom K, Peterson CM. Thrombocytosis and increased circulating platelet aggregates in newborn infants of polydrug users. *J Pediatr.* 1979; 94: 895–899.

210. Burstein Y, Grady RW, Kreek MJ, Rausen AR, Peterson CM. Thrombocytosis in the offspring of female mice receiving dl-Methadone. *Proc Soc Exp Biol Med.* 1980; 164: 275–279.

211. Ebbe S, Yee T, Phalen E. 5-Fluorouracil-induced thrombocytosis in mice is independent of the

spleen and can be partially reproduced by repeated doses of cytosine arabinoside. *Exp Hematol.* 1989; 17: 822-826.

212. Chenaille PJ, Steward SA, Ashmun RA, Jackson CW. Prolonged thrombocytosis in mice after 5-fluorouracil results from failure to down-regulate megakaryocyte concentration. An experimental model that dissociates regulation of megakaryocyte size and DNA content from megakaryocyte concentration. *Blood.* 1990; 76: 508-515.

213. Radley JM, Hodgson GS, Levin J. Platelet production after administration of antiplatelet serum and 5-fluorouracil. *Blood.* 1980; 55: 164–165.

214. Ebbe S, Phalen E. Does autoregulation of megakaryocytopoiesis occur? *Blood Cells.* 1979; 5: 123–138.

215. Ebbe S, Phalen E. Macromegakaryocytosis after hydroxyurea. *Proc Soc Exp Biol Med.* 1982; 171: 151–157.

216. Petursson SR, Chervenick PA. Megakaryocytopoiesis and granulopoiesis following cyclophosphamide. *J Lab Clin Med.* 1982; 100: 682–694.

217. Tanum G. Megakaryocyte DNA content and platelet formation in rats after a sublethal dose of thio-TEPA. *Exp Hematol.* 1986; 14: 202–206.

218. Odell TT Jr, Jackson CW, Friday TJ. Effects of radiation on the thrombopoietic system of mice. *Radiat Res.* 1971; 48: 107–115.

219. Pendry K, Alcorn MJ, Burnett AK. Factors influencing haematological recovery in 53 patients with acute myeloid leukaemia in first remission after autologous bone marrow transplantation. *Br J Haematol.* 1993; 83: 45–52.

220. First LR, Smith BR, Lipton J, Nathan DG, Parkman R, Rappeport JM. Isolated thrombocytopenia after allogeneic bone marrow transplantation: existence of transient and chronic thrombocytopenic syndromes. *Blood.* 1985; 65: 368–374.

221. Adams JA, Gordon AA, Jiang YZ, et al. Thrombocytopenia after bone marrow transplantation for leukaemia: changes in megakaryocyte growth and growth-promoting activity. *Br J Haematol.* 1990; 75: 195–201.

222. Bradford G, Williams N, Barber L, Bertoncello I. Temporal thrombocytopenia after engraftment with defined stem cells with long-term marrow reconstituting activity. *Exp Hematol.* 1993; 21: 1615–1620.

223. Arnold JT, Barber L, Bertoncello I, Williams NT. Modified thrombopoietic response to 5-FU in mice following transplantation of Lin⁻Sca-1⁺ bone marrow cells. *Exp Hematol.* 1995; 23: 161–167.

224. Russell ES, Bernstein SE. Blood and blood formation. In: Green EL (ed). *The Biology of the Laboratory Mouse.* New York: McGraw-Hill; 1966: 351–372.

225. Ebbe S, Phalen E, Stohlman F Jr. Abnormalities of megakaryocytes in W/Wᵛ mice. *Blood.* 1973; 42: 857–864.

226. Ebbe S, Phalen E, Stohlman F Jr. Abnormalities of megakaryocytes in Sl/Slᵈ mice. *Blood.* 1973; 42: 865–871.

227. Chabot B, Stephenson DA, Chapman RM, Besmer P, Bernstein A. The proto-oncogene c-kit encoding a transmembrane tyrosine kinase receptor maps to the mouse W locus. *Nature.* 1988; 335: 88–89.

228. Huang E, Nocka K, Beier DR, et al. The hematopoietic growth factor KL is encoded by the Sl locus and is the ligand of the c-kit receptor, the gene product of the W locus. *Cell.* 1990; 63: 225–233.

229. Ebbe S, Bentfeld-Barker ME, Adrados C, et al. Functionally abnormal stromal cells and megakaryocyte size, ploidy and ultrastructure in Sl/Slᵈ mice. *Blood Cells.* 1986; 12: 217–232.

230. Ebbe S, Carpenter D, Yee T. Megakaryocytopenia in W/Wᵛ mice is accompanied by an increase in size within ploidy groups and acceleration of maturation. *Blood.* 1989; 74: 94–98.

231. Ebbe S, Phalen E, Howard D. Parabiotic demonstration of a humoral factor affecting megakaryocyte size in Sl/Slᵈ mice. *Proc Soc Exp Biol Med.* 1978; 158: 637–642.

232. Kuter DJ, Beeler DL, Rosenberg RD. The purification of megapoietin: a physiological regula-

tor of megakaryocyte growth and platelet production. *Proc Natl Acad Sci USA*. 1994; 91: 11,104–11,108.

233. Leven RM, Tablin F. Megakaryocyte and platelet ultrastructure in the Wistar Furth rat. *Am J Pathol*. 1988; 132: 417–426.

234. Lee KP. Emperipolesis of hematopoietic cells within megakaryocytes in bone marrow of the rat. *Vet Pathol*. 1989; 26: 473–478.

235. White JG. Structural defects in inherited and giant platelet disorders. In: Harris H, Hirschhorn K (eds). *Advances in Human Genetics*. New York: Plenum; 1990: 133–234.

236. Swank RT, Jiang SY, Reddington M, et al. Inherited abnormalities in platelet organelles and platelet formation and associated altered expression of low molecular weight guanosine triphosphate-binding proteins in the mouse pigment mutant gunmetal. *Blood*. 1993; 81: 2626–2635.

237. Novak EK, Reddington M, Zhen L, et al. Inherited thrombocytopenia caused by reduced platelet production in mice with the gunmetal pigment gene mutation. *Blood*. 1995; 85: 1781–1789.

238. Winocour PD, Rand ML, Kinlough-Rathbone RL, Richardson M, Mustard JF. Platelet function and survival in rats with genetically determined hypercholesterolaemia. *Atheroclerosis*. 1989; 76: 63–70.

239. Ebbe S, Dalal K, Forte T, Tablin F. Microcytic thrombocytosis, small megakaryocytes, platelet lipids and hyperreactivity to collagen, lymphocytopenia, eosinophilia and low blood volume in genetically hyperlipidemic rabbits. *Exp Hematol*. 1992; 20: 486–493.

240. Goldstein JL, Kita T, Brown MS. Defective lipoprotein receptors and atherosclerosis. Lessons from an animal counterpart of familial hypercholesterolemia. *N Engl J Med*. 1983; 309:

241. Novak EK, Hui S, Swank RT. Platelet storage pool deficiency in mouse pigment mutations associated with seven distinct genetic loci. *Blood*. 1984; 63: 536–544.

242. Swank RT, Reddington M, Howlett O, Novak EK. Platelet storage pool deficiency associated with inherited abnormalities of the inner ear in the mouse pigment mutants muted and mocha. *Blood*. 1991; 78: 2036–2044.

243. Novak EK, Sweet HO, Prochazka M, et al. Cocoa: a new mouse model for platelet storage pool deficiency. *Br J Haematol*. 1988; 69: 371–378.

244. Swank RT, Sweet HO, Davisson MT, Reddington M, Novak EK. Sandy: a new mouse model for platelet storage pool deficiency. *Genet Res*. 1991; 58: 51–62.

245. Raymond SL, Dodds WJ. Characterisitics of the Fawn-hooded rat as a model for hemostatic studies. *Thrombosis Diathesis Haemorrhagica*. 1975; 33: 361–369.

246. Tschopp TB, Zucker MB. Hereditary defect in platelet function in rats. *Blood*. 1972; 40: 217–226.

247. White RA, Peters LL, Adkinson LR, Korsgren C, Cohen CM, Lux SE. The murine pallid mutation is a platelet storage pool disease associated with the protein 4.2 (pallidin) gene. *Nat Genet*. 1992; 2: 80–83.

248. Peters LL, Eicher EM, Hoock TC, John KM, Yialamas M, Lux SE. Evidence that the mouse platelet storage pool deficiency mutation mocha is a defect in a new member of the ankyrin gene family, ANK-3. *Blood*. 1993; 82: 340a (abstract no 1344).

249. Rolovic Z, Jovanovic T, Stankovic Z, Marinkovic N. Abnormal megakaryocytopoiesis in the Belgrade laboratory rat. *Blood*. 1985; 65: 60–64.

250. Rolovic Z, Basara N, Stojanovic N, Suvajdzic N, Pavlovic-Kentera V. Abnormal megakaryocytopoiesis in the Belgrade laboratory rat. *Blood*. 1991; 77: 456–460.

251. McDonald TP, Jackson CW. Mode of inheritance of the higher degree of megakaryocyte polyploidization in C3H mice. I. Evidence for a role of genomic imprinting in megakaryocyte polyploidy determination. *Blood*. 1994; 83: 1493–1498.

252. Dale DC, Alling DW, Wolff SM. Cyclic hematopoiesis: the mechanism of cyclic neutropenia in grey collie dogs. *J Clin Invest*. 1972; 51: 2197–2204.

253. Weiden PL, Robinett B, Graham TC, Adamson J, Storb R. Canine cyclic neutropenia: a stem cell defect. *J Clin Invest*. 1974; 53: 950–953.

254. Jones JB, Lange RD, Jones ES. Cyclic hematopoiesis in a colony of dogs. *J Am Vet Med Assoc*. 1975; 166: 365–367.

255. Jones JB, Langes RD, Yang TJ, Vodopick H, Jones ES. Canine cyclic neutropenia: erythropoiesis and platelet cycles after bone marrow transplantation. *Blood.* 1975; 45: 213–219.
256. Dale DC, Graw RG Jr. Transplantation of allogeneic bone marrow in canine cyclic neutropenia. *Science.* 1974; 183: 83–84.
257. McDonald TP, Clift R, Jones JB. Canine cyclic hematopoiesis: platelet size and thrombopoietin level in relation to platelet count. *Proc Soc Exp Biol Med.* 1976; 153: 424–428.
258. Gurney AL, Carver-Moore K, de Sauvage FJ, Moore MW. Thrombocytopenia in *c-mpl*-deficient mice. *Science.* 1994; 265: 1445–1447.
259. de Sauvage FJ, Carver-Moore K, Luoh S, et al. Physiological regulation of early and late stages of megakaryocytopoeisis by thrombopoietin. *J Exp Med.* 1996; 183: 651–656.
260. Shivdasani RA, Rosenblatt MF, Zucker-Franklin D, et al. Transcription factor NF-E2 is required for platelet formation independent of the actions of thrombopoietin/MGDF in megakaryocyte development. *Cell.* 1995; 81: 695–704.
261. Guy CT, Zhou W, Kaufman S, Robinson MO. E2F-1 blocks terminal differentiation and causes proliferation in transgenic megakaryocytes. *Mol Cell Biol.* 1996; 16: 685–693.
262. Yan XQ, Lacey D, Fletcher F, et al. Chronic exposure to retroviral vector encoded MGDF (mpl-ligand) induces lineage-specific growth and differentiation of megakaryocytes in mice. *Blood.* 1995; 86: 4025–4033.
263. Kieffer N, Guichard J, Farcet J, Vainchenker W, Breton-Gorius J. Biosynthesis of major platelet proteins in human blood platelets. *Eur J Biochem.* 1987; 164: 189–195.
264. Inokuchi K, Nomura T. The relationship between the type of bcr-abl hybrid messenger RNA and thrombopoiesis in Philadelphia-positive chronic myelogenous leukemia. *Leukemia Lymphoma.* 1993; 10: *9–15.*
265. Thiele J, Wagner S, Weuste R, et al. An immunomorphometric study on megakaryocyte precursor cells in bone marrow tissue from patients with chronic myeloid leukemia (CML). *Eur J Haematol.* 1990; 44: 63–70.
266. Thiele J, Titius BR, Kopsidis C, Fischer R. Atypical micromegakaryocytes, promegakaryoblasts and megakaryoblasts: a critical evaluation by immunohistochemistry, cytochemistry and morphometry of bone marrow trephines in chronic myeloid leukemia and myelodysplastic syndromes. *Virchows Arch B Cell Pathol Incl Mol Pathol.* 1992; 62: 275–282.
267. Franzen S, Strenger G, Zaijicek J. Microplanimetric studies on megakaryocytes in chronic granulocytic leukaemia and polycythaemia vera. *Acta Hematol.* 1961; 26: 182–193.
268. Lagerlof B. Cytophotometric study of megakaryocyte ploidy in polcythemia vera and chronic granulocytic leukemia. *Acta Cytol.* 1972; 16: 240–244.
269. Renner D, Queisser W. Megakaryocyte polyploidy and maturation in chronic granulocytic leukemia. *Acta Haematol.* 1988; 80: 74–78.
270. Pestina TI, Mareeva TO, Morozov NG, Sokovinina YM, Votrin II. The level of adenine nucleotides in platelets at different stages of chronic myelogenous leukemia. *Vopr Med Khim.* 1991; 5: 89–100.
271. Kimura H, Ohkoshi T, Matsuda S, Uchida T, Kariyone S. Megakaryocytopoiesis in polycythemia vera: characterization by megakaryocytic progenitors (CFU-Meg) in vitro and quantitation of marrow megakaryocytes. *Acta Haematol.* 1988; 79: 1–6.
272. Thiele J, Jensen B, Orth KH, Orth H, Moedder B, Fischer R. Ultrastructure of megakaryocytes in the human bone marrow of patients with primary (essential)-thrombocythemia. *J Submicro Cytol S Pathol.* 1988; 20: 671–681.
273. Thiele J, Wagner S, Degel C, et al. Megakaryocyte precursors (pro- and megakaryoblasts) in bone marrow tissue from patients with reactive thrombocytosis, polycythemia vera and primary (essential) thrombocythemia. An immunomorphometric study. *Virchows Arch B Cell Pathol Incl Mol Pathol.* 1990; 58: 295–302.
274. Tomer A, Friese P, Conklin R, et al. Flow cytometric analysis of megakaryocytes from patients with abnormal platelet counts. *Blood.* 1989; 74: 594–601.

275. Queisser W, Weidenauer G, Queisser U, Kempgens U, Muller U. Megakaryocyte ploidization in myeloproliferative disorders. *Blut.* 1976; 32: 13–20.

276. Han ZC, Briere J, Abgrall JF, et al. Spontaneous formation of megakaryocyte progenitors (CFU-Mk) in primary thrombocythaemia. Acta Haematol. 1987; 78: 51–53.

277. Miyakawa Y, Oda A, Druker BJ, et al. Thrombopoietin induces tyrosine phosphorylation of Stat3 and Stat5 in human blood platelets. *Blood.* 1996; 87: 439–446.

278. Alexander WS, Metcalf D, Dunn AR. Point mutations within a dimer interface homology domain of c-Mpl induce constitutive receptor activity and tumorigenicity. *EMBO J.* 1995; 14: 5569–5578.

279. Drachman JG, Griffin JD, Kaushansky K. The c-Mpl ligand (thrombopoietin) stimulates tyrosine phosphorylation of Jak2, Shc, and c-Mpl. *J Biol Chem.* 1995; 270: 4979–4982.

280. Ezumi Y, Takayama H, Okuma M. Thrombopoietin, c-Mpl ligand, induces tyrosine phosphorylation of Tyk2, JAK2, and STAT3, and enhances agonists-induced aggregation in platelets in vitro. *FEBS Lett.* 1995; 374: 48–52.

281. Miyakama Y, Oda A, Druker BJ, et al. Recombinant thrombopoietin induces rapid protein tyrosine phosphorylation of Janus kinase 2 and Shc in human blood platelets. *Blood.* 1995; 86: 23–27.

282. Morella KK, Bruno E, Kumaki S, et al. Signal transduction by the receptors for thrombopoietin (c-mpl) and interleukin-3 in hematopoietic and nonhematopoietic cells. *Blood.* 1995; 86: 557–571.

283. Bennett JM, Catovsky D, Daniel MT, et al. Criteria for the diagnosis of acute leukemia of megakaryocyte lineage (M7). A report of the French-American-British cooperative group. *Ann Intern Med.* 1985; 103: 460–462.

284. Jackson CW, Dahl GV. Relationship of megakaryocyte size at diagnosis to chemotherapeutic response in children with acute nonlymphocytic leukemia. *Blood.* 1983; 61: 867–870.

285. Brody JP, Krause JR. Morphometric study of megakaryocyte size and prognosis in adults with acute non-lymphocytic leukemia. *Leukemia Res.* 1986; 10: 475–480.

286. Gerrard JM, McNicol A. Platelet storage pool deficiency, leukemia, and myleodysplastic syndromes. *Leukemia Lymphoma.* 1992; 8: 277–281.

287. Hall JG. Thrombocytopenia and absent radius (TAR) syndrome. *J Med Genet.* 1987; 24: 79–83.

288. Brochstein JA, Shank B, Kernan NA, Terwilliger JW, O'Reilly RJ. Marrow transplantation for thrombocytopenia-absent radii syndrome. *J Pediatr.* 1992; 121: 587–589.

289. Homans AC, Cohen JL, Mazur EM. Defective megakaryocytopoiesis in the syndrome of thrombocytopenia with absent radii. *Br J Haemol.* 1988; 70: 205–210.

290. Breton-Gorius J, Vainchenker W, Nurden A, Levy-Toledano S, Caen J. Defective α-granule production in megakaryocytes from gray platelet syndrome: ultrastructural studies of bone marrow cells and megakaryocytes growing in culture from blood precursors. *Am J Pathol.* 1981; 102: 10–19.

291. Levy-Toledano S, Caen JP, Breton-Gorius J, et al. Gray platelet syndrome: α-granule deficiency. Its influence on platelet function. *J Lab Clin Med.* 1981; 96: 831–847.

292. Raccuglia G. Gray platelet syndrome: a variety of qualitative platelet disorder. *Am J Med.* 1971; *51: 818–828.*

293. White JG. Ultrastructural studies of the gray platelet syndrome. *Am J Pathol.* 1979; 95: 445–462.

294. Nurden AT, Dupuis D, Kunicki TJ, Caen JP. Analysis of the glycoprotein and protein composition of Bernard-Soulier platelets by single and two-dimensional sodium dodecyl sulfate-polyacrylamide gel electrophoresis. *J Clin Invest.* 1981; 67: 1431–1440.

295. Fox JEB. Identification of actin-binding protein as the protein linking the membrane skeleton to glycoproteins on platelet plasma membranes. *J Biol Chem.* 1985; 260: 11,970–11,977.

296. Greinacher A, Nieuwenhuis HK, White JG. Sebastian platelet syndrome: a new variant of hereditary macrothrombocytopenia with leukocyte inclusions. *Blut.* 1990; 61: 282–288.

297. van Nostrand WE, Schmaier AH, Farrow JS, Cines DB, Cunningham DD. Protease nexin-2/amyloid β-protein precursor in blood is a platelet-specific protein. *Biochem Biophys Res Commun.* 1991; 175: 15–21.

298. Stenberg PE, Shuman MA, Levine SP, Bainton DF. Optimal techniques for immunocytochemical demonstration of β-thromboglobulin, platelet factor 4, and fibrinogen in the alpha granules of unstimulated platelets. *Histochem J.* 1984; 16: 983–1001.

299. Stenberg PE, Shuman MA, Levine SP, Bainton DF. Redistribution of alpha-granules and their contents in thrombin-stimulated platelets. *J. Cell Biol.* 1984; 98: 748–760.

300. Tschopp J, Jenne DE, Hertig S, et al. Human megakaryocytes express clusterin and package it without apolipoprotein A-1 into α-granules. *Blood.* 1993; 82: 118–125.

301. Wencel-Drake JD, Dahlback B, White JG, Ginsberg MH. Ultrastructural localization of coagulation Factor V in human platelets. *Blood.* 1986; 68: 244–249.

302. Suzuki H, Kinlough-Rathbone RL, Packham MA, Tanoue K, Yamazaki H, Mustard JF. Immunocytochemical localization of fibrinogen during thrombin-induced aggregation of washed human platelets. *Blood.* 1988; 71: 1310–1320.

303. Wencel-Drake JD, Painter RG, Zimmerman TS, Ginsberg MH. Ultrastructural localization of human platelet thrombospondin, fibrinogen, fibronectin and von Willebrand factor in frozen thin section. *Blood.* 1985; 65: 929–938.

304. Ginsberg MH, Taylor L, Painter RG. The mechanism of thrombin-induced platelet factor 4 secretion. *Blood.* 1980; 661–668.

305. Pham TD, Kaplan KL, Butler VP. Immunoelectron microscopic localization of platelet factor 4 and fibrinogen in the granules of human platelets. *J Histochem Cytochem.* 1983; 31: 905–910.

306. Cramer EM, Debili N, Martin JF, et al. Uncoordinated expression of fibrinogen compared with thrombospondin and von Willebrand Factor in maturing human megakaryocytes. *Blood.* 1989; 73: 1123–1129.

307. Gachet C, Hanau D, Spehner D, et al. Alpha II beta-III integrin dissociation induced by EDTA results in morphological changes of the platelet surface-connected canalicular system with differential location of the two separate subunits. *J Cell Biol.* 1993; 120: 1021–1030.

308. Cramer EM, Meyer D, le Menn R, Breton-Gorius J. Eccentric localization of von Willebrand factor in an internal structure of platelet α-granule resembling that of Weibel Palade bodies. *Blood.* 1985; 66: 710–713.

309. Metzelaar MJ, Heijnen HFG, Sixma JJ, Nieuwenhuis HK. Identification of a 33-Kd protein associated with the α-granule membrane (GMP-33) that is expressed on the surface of activated platelets. *Blood.* 1992; 79: 372–379.

310. Berger G, Massé J, Cramer EM. Alpha-granule membrane mirrors the platelet plasma membrane and contains glycoprotein Ib, IX and V. *Blood.* 1996; 87: 1385–1395.

311. Cramer EM, Savidge GF, Vainchenker W, et al. Alpha-granule pool of glycoprotein IIb-IIIa in normal and pathologic platelets and megakaryocytes. *Blood.* 1990; 75: 1220–1227.

312. Hardisty R, Pidard D, Cox A, et al. A defect of platelet aggregation associated with an abnormal distribution of glycoprotein IIb-IIIa complexes within the platelet: the cause of a lifelong bleeding disorder. *Blood.* 1992; 80: 696–708.

313. Berger G, Caen JP, Berndt MC, Cramer EM. Ultrastructural demonstration of CD36 in the α-granule membrane of human platelets and megakaryocytes. *Blood.* 1993; 82: 3034–3044.

314. Stenberg PE, McEver RP, Shuman MA, Jacques YV, Bainton DF. A platelet alpha granule membrane protein (GMP-140) is expressed on the plasma membrane after activation. *J Cell Biol.* 1985; 101: 880–886.

315. Breton-Gorius J, Clezardin P, Guichard J, et al. Localization of platelet osteonectin at the internal face of the α-granule membranes in platelets and megakaryocytes. *Blood.* 1992; 79: 936–941.

316. Cramer EM, Berger G, Berndt MC. Platelet α-granule and plasma membrane share two new components: CD-9 and PECAM-1. *Blood.* 1994; 84: 1722–1730.

317. Berger G, Quarck R, Tenza D, Levy-Toledano S, de Gunzburg J, Cramer EM. Ultrastructural localization of the small GTP-binding protein Rap1 in human platelets and megakaryocytes. *Br J Haematol.* 1994; 88: 372–382.

2

Platelet Structure and Function

Dorothea Zucker-Franklin

1. Introduction

Since this is a publication primarily devoted to thrombopoietin (TPO), the ligand for Mpl, and its effect on platelet production so soon after the discovery of this hormone, it is likely that many observations recorded in this book will need to be modified in the future. This will not hold true for the present chapter, which is meant as an anchor to those who have recently entered the field or who may have forgotten what is considered to be normal platelet structure and function, whether in health, in disease, or during spontaneously accelerated thrombocytopoiesis. Only the most relevant information will be recorded here, namely, the properties of platelets that may be affected by TPO when platelets are exposed to its recombinant form in vivo or in vitro. The reader is referred to an exceptionally large body of literature accumulated over many years to be apprised of further details *(1–4)*. Since many of the studies on Mpl carried out to date have, of necessity, been conducted in vitro or in animals, pitfalls in the interpretation of the results arising from differences among species will also be pointed out.

2. Normal Structure of Platelets, Subserving Normal Function

The platelet circulates as a disk-shaped piece of membrane-bound cytoplasm endowed with all organelles found in other mammalian cells except for a nucleus, a Golgi zone, and a mitotic apparatus. Platelets have a diameter of 2–3 μm and an average volume of 6–8 femtoliters (fL). Their number in circulation ranges from 150–400 × 10^9/L. Figure 1A and B depicts an equatorial and a cross-section, respectively, of pristine "resting" platelets that shall be considered "platelet units," as opposed to "compound" or giant platelets that consist of several platelet territories (*see* Section 2.3.). It is likely that in vivo, small processes form reversibly as platelets tumble through the blood vessels. These are not seen on light microscopy and should not be considered indicative of activation. At the same time, it cannot be overemphasized that some platelets will undergo irreversible shape changes due to activation in vitro unless blood is drawn directly into a syringe containing a fixative. Once platelet processes have become visible by light or scanning electron microscopy, the redistribution of membranes has become such that the body of the platelet of necessity assumes a spherical shape (Fig. 2A and B). This process is probably irreversible, if for no other reason than that in

From: *Thrombopoiesis and Thrombopoietins: Molecular, Cellular, Preclinical, and Clinical Biology*
Edited by: D. J. Kuter, P. Hunt, W. Sheridan, and D. Zucker-Franklin Humana Press Inc., Totowa, NJ

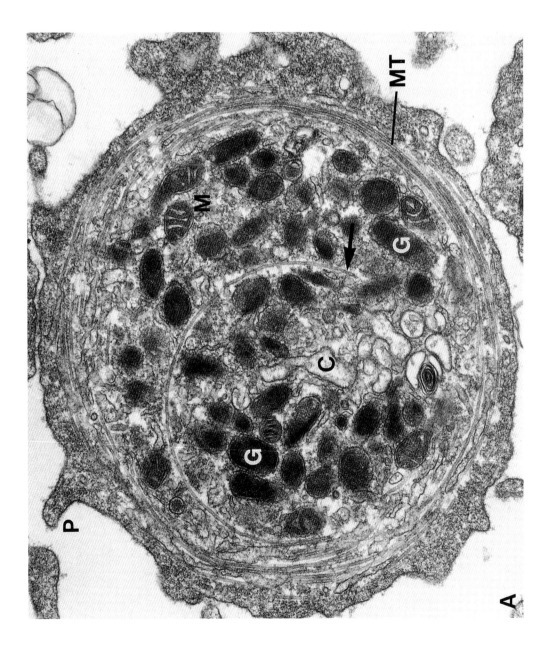

42

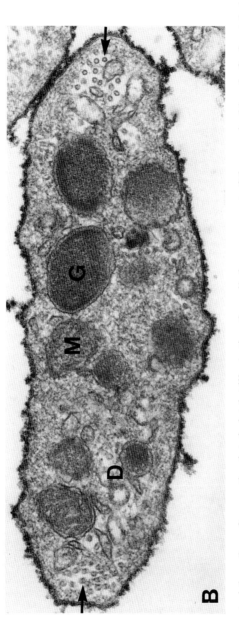

Fig. 1. (A) Equatorial section of an unstimulated platelet illustrating the coiled arrangement of the marginal band of microtubules (MT) with a "loose" end (*see* arrow). α-Granules (G) are randomly distributed. M indicates mitochondrion, and P, processes or filopodia. (Magnification, ×38,000.) (Reprinted with permission from ref. *29*.) **(B)** Unstimulated platelet in cross-section. Arrows indicate microtubules. G indicates granule; C, canaliculi; M, mitochondrion; and D, dense granule. The surface coat is delineated with ruthenium red. (Magnification, ×57,000.)

43

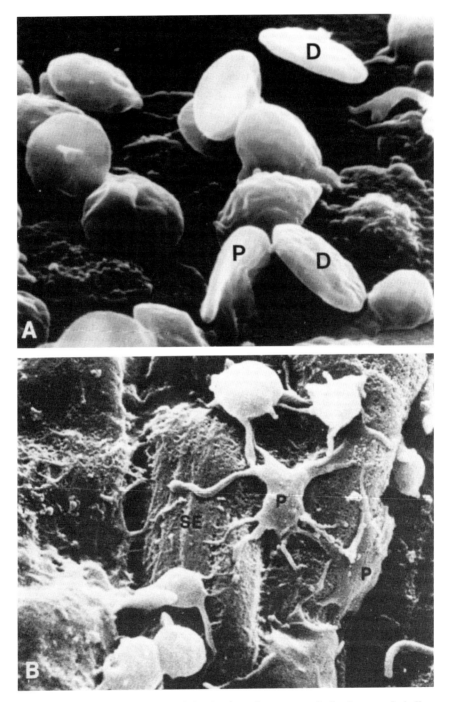

Fig. 2. Scanning electron micrographs of platelets in various stages of adhesion to endothelium treated with fibronectin **(A)** and hyaluronidase **(B)**. Some platelets in Panel A have apparently made only initial contact and are still disk-shaped (D), whereas others have begun to form processes (P). In Panel B, platelets (P) have formed long processes adhering to subendothelium (SE). (Fig. 2A courtesy of June Wencel-Drake, University of Chicago. Reprinted with permission from ref. *45*. Figure 2B courtesy of Marion Barnhart.)

vivo such filopodia must have become adherent *(5)*. Similar processes do not form in suspension unless agitation causes platelets to collide. The properties of single unstimulated platelet units will be described first.

2.1. Platelet Surface

Morphologically, three features distinguish the platelet surface from that of other cells. First, the platelet surface has a thick proteoglycan coat also referred to as the glycocalyx. Second, the surface has the ability to form fibrillar bridges joining other platelets or collagen fibrils, a process concomitant with irreversible aggregation. Third, there is a distribution of intramembranous particles within the biomolecular leaflet of the plasma membrane that is the reverse of that seen in any other mammalian cell, including megakaryocytes *(6)*. The unique characteristics of the platelet surface are probably relevant to the observation that many of the biochemically and/or immuno-logically identified epitopes that play a role in platelet function are not exposed on the surface of the resting platelet. The platelet "coat" is 150–200 nm thick and is best delineated with ruthenium red *(7,8)* (Fig. 1). Gel filtration or treatment with proteolytic enzymes removes adsorbed plasma proteins, but delineates the coat material to even better advantage, particularly since it participates in bridge formation between coherent platelets or platelets and collagen (Fig. 3). A very comprehensive review of the interaction of platelets with the vessel wall is found in ref. *5*.

It is well recognized that platelets bear an unusually large number of integrins, many of which are shared by other cells, particularly endothelial cells. It is probably correct to state that, among the integrins, the GPIIb/IIIa complex, specific for the megakaryo-cyte/platelet lineage, is the prime receptor for fibrinogen and most significant for platelet function. This receptor is latent, i.e., fibrinogen does not bind to it in resting, circulating cells. The mechanism whereby GPIIb/IIIa becomes available for fibrinogen binding is conjectural. It is, however, not remote to postulate that rearrangement of the platelet coat and components of the glycocalyx play a role in this process *(7,8)*. Fibrinogen/fibrin has been found at the site of the "bridges" by immunologic and biochemical means *(9,10)*, but whether signal transduction occurs between or across the bridges is not clear. Interestingly, it has been reported that platelets of patients with thrombasthenia or afibrinogenemia are able to form bridges when aggregated with collagen or ristocetin, although the thrombasthenic platelets bind little or no fibrinogen *(11)*. Presumably, receptors for many adhesive proteins known to play a role in hemostasis must be exposed on the surface of unactivated platelets for platelet/vessel wall interaction to occur. To date, the following receptor/ligand interactions have been recognized: the receptor for collagen, GPIa/IIa ($a_2 b_1$); the receptor for laminin, GPIc/IIa ($a_6 b_1$); the receptor for fibronectin, GPIc/IIa ($a_5 b_1$); the receptor for fibrinogen, GPIIb/IIIa (aIIb/b3); the receptor for von Willebrand factor (vWF) (GPIb/IX); the receptor for vitronectin (VN) (av/GPIIIa); and a receptor for thrombin (for a more detailed review, *see* ref. *12*).

In addition, platelets bear receptors for other ligands that may influence their function apart from their role in hemostasis. Among these, Fcγ-RII, a receptor for IgG, HLA-Class I, and P-selectin (GMP 140, PADGEM) are of particular significance (*see below*). Most germane to the thrust of this volume is the observation that platelets also appear to have receptors for TPO *(13)*.

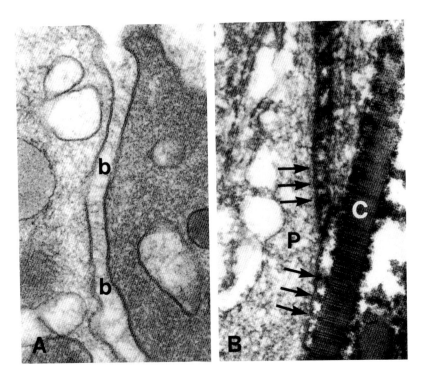

Fig. 3. Bridges between aggregated platelets and collagen. **(A)** Bridges (b) between aggregated, contiguous platelets span the extracellular space. The trilaminar structure of the plasma membrane does not appear to be altered. (Magnification, ×81,000.) **(B)** Detail of platelets (P) aggregated with fibrillar collagen fixed in the presence of ruthenium red. The periodicity of the bridges (see arrows) matches the dark bands of the collagen fibrils (C). (Magnification, ×80,000.) (Reprinted with permission from ref. *3*.)

2.2. Cytoplasmic Structures

2.2.1. Granules

Granules are the most conspicuous organelles in platelets, although they are difficult to distinguish from other organelles, such as mitochondria or dense bodies, on light microscopy. An attempt by early investigators to classify these structures has led to Greek letter designations of which only the term α-granule is still in use. In a resting platelet, the α-granules are evenly distributed (Figs. 1 and 4), are approximately 200 nm in diameter, and are located a minimum of 500 nm interior to the surface membrane. The latter point cannot be overemphasized, since fusion with the plasma membrane, even after activation, has never been observed (*see* Section 3.). Of considerable interest is that α-granules contain proteins synthesized and packaged in megakaryocytes, such as platelet factor 4 (PF-4), β-thromboglobulin, vWF, fibronectin, platelet-derived growth factor (PDGF), and P-selectin, as well as proteins taken up from plasma either by megakaryocytes or by the circulating platelets themselves *(14)*. These include fibrinogen, albumin, immunoglobulin, and amyloid β protein precursor (protease nexin II). Moreover, it has been shown that the concentration of many of these proteins in platelets reflects their level in circulating plasma *(15)*. Of further interest is the differential release of these substances on platelet activation without complete degranulation. This

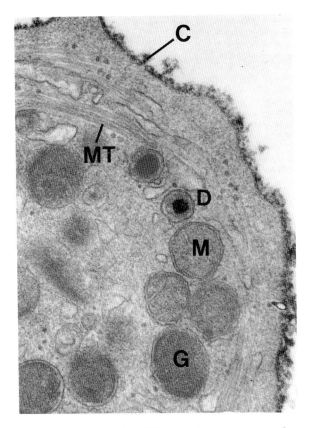

Fig. 4. Detail of a platelet showing the difference in appearance of α-granules (G), dense granules (D), and mitochondria (M). Peripheral band of microtubules, MT; ruthenium red stained surface coat, C. (Magnification, ×55,000.) (Reprinted with permission from ref. *2*.)

suggests that there is compartmentalization of granule contents, which, to some extent, has been shown to hold true for vWF and P-selectin *(16,17)*.

In addition to the α-granules, there are a few granules with extremely electron-opaque nucleoids (Figs. 1B and 4). These are often referred to as dense bodies. On the basis of studies mostly done with rabbit platelets, the numbers of dense granules are believed to be directly proportional to the serotonin content of the platelet *(18)*. Although in human platelets there are few dense granules, it has been well established that these structures contain the nonmetabolic pool of adenine nucleotides, serotonin, pyrophosphate, and calcium ions *(18–20)*. Absence of dense granules is associated with defective aggregation in vitro and a bleeding diathesis, the Hermansky-Pudlak syndrome *(21)*.

Lysosomal enzymes, such as acid phosphatase, arylsulfatase, cathepsin, and α-glucuronidase, are not contained in the α-granules, but are found in small vesicles or other components of the endoplasmic reticulum *(22,23)*. The presence of glycogen particles, as well as mitochondria, provides morphologic evidence that the platelet's energy metabolism is served by both glycolysis and oxidative phosphorylation *(24)* (Figs. 1 and 4).

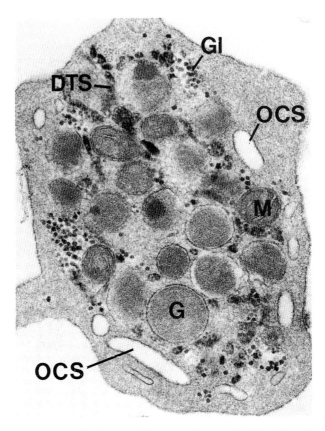

Fig. 5. Platelet stained for platelet-specific peroxidase, which has been localized within the DTS to contrast with the OCS. M, mitochondrion; Gl, glycogen; and G, granule. (Magnification, ×42,000.)

The granule-free peripheral zone of the platelet reveals a conspicuous band of microtubules believed to consist of a single or several continuous coils, since loose ends can be seen in fortuitous sections and lysed platelets *(1,25)* (Fig. 1). These structures appear to function primarily to preserve the platelet's lenticular shape (*see* Section 3.). The rich endowment of other cytoskeletal proteins is not resolved in resting platelets and will be described later.

It may be important to ascertain the presence, number, and characteristics of both α-granules and dense granules in platelets released after the administration of TPO, since some investigators have reported subtle activation of platelets after exposure to the hormone.

2.2.2. Membrane Systems

The platelet has two membrane systems: the dense tubular system (DTS) and the so-called open canalicular system (OCS), both of which are discernible in the unaltered state (Figs. 1, 4, and 5). Whether there is any connection between these systems in resting platelets is still speculative. The DTS was discovered rather late in studies of platelet ultrastructure, since it was first delineated because of its content of platelet and megakaryocyte-specific peroxidase *(26)* (Fig. 5). The presence of glucose-6-phos-

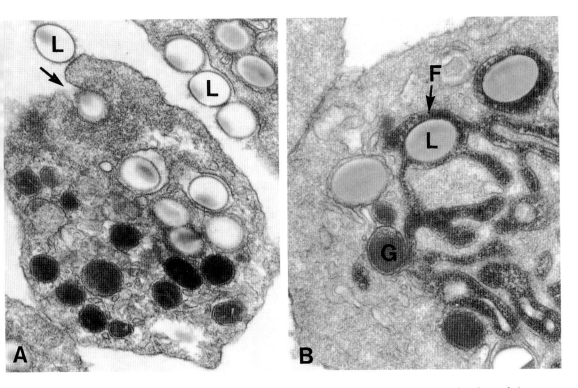

Fig. 6. Platelets fixed in the process of taking up latex particles. **(A)** Invagination of the surface membrane is apparent (*see* arrow). Fusion of α-granules and vacuoles containing latex is not seen. L indicates latex particle. (Magnification, ×36,000.) **(B)** Detail of a platelet, which was incubated first with latex particles for 10 minutes and subsequently with cationized ferritin for an additional 10 minutes. Uptake of ferritin is not energy dependent. Coalescence of canaliculi containing ferritin with vacuoles containing latex particles is evident, but fusion with α-granules is not observed. L, latex particle; F, ferritin; G, granule. (Magnification, ×48,000.) (For details of methods, *see* ref. *24.*)

phatase can also be shown within the DTS. It is possible the DTS is a major site of prostaglandin synthesis *(27)*, although this assumption has been made mostly by inference derived from studies on muscle cells *(28)*. As stated earlier, a connection between the DTS and the OCS has not been demonstrated, and the surface-connected canalicular system should not be considered "open." Fixed platelets or platelets kept at 4°C will not permit entry of tracers into the OCS even after several days of incubation when such tracers are seen adherent to the surface membrane *(24)*. It is also unlikely that the canalicular system is formed by invagination of the plasma membrane as has often been stated. The OCS should probably be considered a remnant of the megakaryocyte demarcation membrane system derived from many different sources *(5,29)*. On the other hand, large particles, such as latex beads and bacteria, are taken up by invagination of the plasma membrane. This is an energy-dependent, not a constitutive, process and resembles phagocytosis, since fusion of the invaginated plasma membrane and the canalicular membranes can be demonstrated *(24)* (Fig. 6).

2.2.3. Freeze-Fracture Technique

The freeze-fracture technique is a procedure requiring special equipment and expertise in interpretation (for a review, *see* ref. *30*). Nevertheless, it has provided information that has yielded greater understanding of platelet properties. After cells are fixed and cryoprotected with 25% glycerol, they are quick-frozen in liquid nitrogen. A mechanical impact administered at $-100°C$ in a vacuum of 10^{-6} torr causes the bilayer to fracture along the hydrophobic plane, exposing the inner surface of the outer leaflet (E face) and the outer surface of the cytoplasmic leaflet (P face) with their associated intramembranous particles (IMP). The IMP have been shown to represent transmembrane proteins. In all mammalian cells studied to date, including megakaryocytes, at least twice as many IMP are seen on the P face as on the E face of the membrane *(31,32)*. The exceptionally large number of IMP and their marked variation in size on the E face of the platelet membrane are astounding *(5,24)* (Fig. 7A and B). The other features relevant to platelet function are the pits seen on the P face, which are continuous with the canaliculi (Fig. 7C and D). These are represented by equally sized protusions on the E face. Such pits are found also in freeze-fracture replicas of the internal membranes constituting the canalicular and granule membranes (Fig. 7D). When large particulates, such as latex particles or bacteria, are interiorized, large segments of plasma membrane invaginate irrespective of the location of the pits, which remain intact *(5,24)*. It is possible that solutes and very small particulates may enter and exit through the pits.

2.3. Giant Platelet, "Proplatelet" vs "Young Platelet"

Since platelets are produced by fragmentation of megakaryocyte cytoplasm, fragments consisting of more than one platelet territory may be seen within marrow sinusoids (also in the spleen sinusoids of rodents) during normal homeostasis *(1,33,34)*. Such large fragments may be elongated and then are frequently referred to as proplatelets (Fig. 8), or they may round up, when they are usually referred to as giant platelets (Fig. 9). The term "compound platelet" might be better to describe a large megakaryocyte fragment consisting of more than one platelet territory. The question regarding where megakaryocytes shed their compendium of platelets has not been fully settled. Unquestionably, a few denuded megakaryocyte nuclei can be found in the marrow. However, there is convincing evidence—although still indirect—that, in humans, most megakaryocyte fragmentation takes place in the pulmonary vasculature *(35–38)*. The conspicuous increase in the number of denuded megakaryocyte nuclei in the marrow of human immunodeficiency virus (HIV)-infected individuals has even led to the use of this phenomenon in the diagnosis of HIV infection *(39)*. Whether TPO is involved in further fragmentation of these megakaryocyte fragments into single platelet territories or whether this occurs by mechanical or other means is not known. It has, however, been reported repeatedly that such large platelets subserve hemostasis more efficiently than small platelets *(40–45)*. Although it may be assumed that giant platelets are young platelets, small platelets are not necessarily old. They may merely represent small megakaryocyte fragments at the time they entered the circulation. However, young platelets are functionally more active as measured by in vitro variables; they are also active in shortening bleeding time *(41)*. A preponderance of giant platelets may yield some test results indicative of activation. This consideration is important in the interpretation of reports suggesting

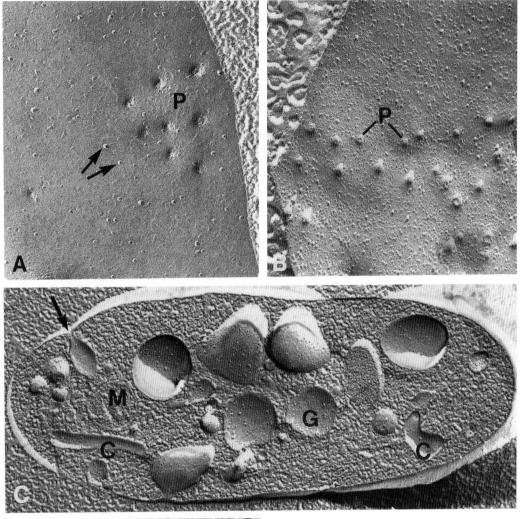

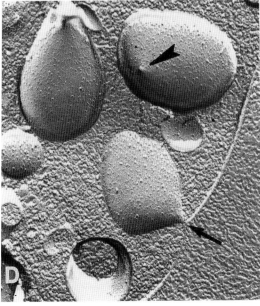

Fig. 7. Details of freeze-fracture replicas illustrating the major features relating to platelet function. **A** and **B** represent, respectively, the P and E face of the plasma membrane (*see text* for explanation). The sparsity of intramembranous particles (IMP) in A (arrows) and their abundance in B are readily apparent. These areas have been chosen because they show a concentration of "pits" (P) in A and protrusions (P) in B. These are in continuity with the OCS. In **C** and **D**, almost the entire platelet has been cleaved showing little of the plasma membrane. However, this affords a better illustration of organelle membranes, which can also be seen to have IMP, and pits (P). Arrowhead in D indicates protrusion, i.e., inside-out "pit." The arrows in C and D indicate sites of continuity between the OCS and the extracellular space. These resemble the "pits" in size. G, granule; C, canalicular system; M, mitochondrion. (Magnifications: A, ×86,000; B, ×65,000; C, ×53,000; D, ×44,000.) (A, B, and D reprinted with permission from ref. *24.*)

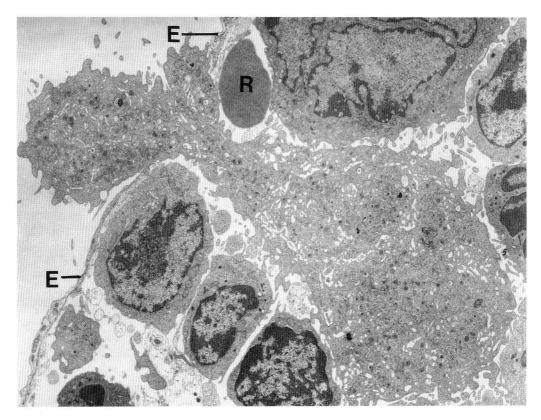

Fig. 8. Large area of megakaryocyte cytoplasm is seen to protrude into the sinusoidal space via a gap in the endothelium (E). (The megakaryocyte nucleus is not within the plane of section.) Platelet territories are present, but difficult to discern at this low magnification. Other hemopoietic cells are still within the medullary space. R, erythrocyte. (Magnification, ×5500.)

that there is subtle platelet activation after the administration of recombinant human thrombopoietins (rHuTPOs) *(46,47)*. If so, this may be attributable merely to the presence of a greater percentage of young platelets released into the circulation on stimulation with the hormone, rather than a direct effect on platelets themselves. Perhaps a better indication of a platelet's young age is the presence of ribosomes and remnants of rough endoplasmic reticulum (RER). This is observed during compensatory thrombocytosis in humans (Fig. 9) and thrombocytosis induced in mice by administration of rHuTPO (unpublished observation). The percentage of such "reticulated" platelets can be established by flow cytometry using the dye thiazole orange that reacts with RNA *(48)*. By these means, an increase in the number of young platelets has also been shown in mice 24 h after a single injection of rHuTPO *(47)*.

3. Structural Changes Associated with Activation and Degranulation
3.1. Subtle Activation

It has been recognized for some time that subtle manifestations of platelet activation, such as shape change and increased adhesiveness, precede complete degranula-

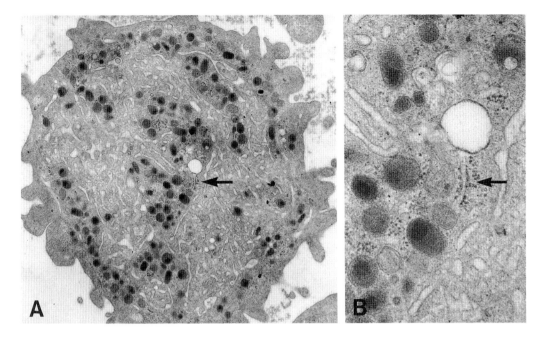

Fig. 9. (A) giant platelet from the circulating blood of a patient with immune thrombocy-topenia (ITP). Many demarcated platelet territories are seen. There is no peripheral band of microtubules. The area indicated by the arrow in **A** is illustrated at higher magnification in **B** to show ribosomes and rough endoplasmic reticulum to better advantage. (Magnification, ×10,000.) B, higher magnification of area indicated by arrow in A. (Magnification, ×33,000.)

tion. Because even subtle activation may cause increased adhesion to vessel walls, numerous tests have been devised to detect this kind of activation in clinical situations associated with a prethrombotic state. Most of these tests examine platelet adhesive-ness on glass-bead columns or cultured endothelial cells. Assays designed to detect plasma levels of β-thromboglobulin and PF-4 have also been helpful in determining whether platelet activation has taken place in vivo (reviewed in ref. *49*). Some of these tests have already been used to examine whether rHu megakaryocyte growth and development factor (MGDF) or rHuTPO administration to baboons or humans is asso-ciated with platelet activation *(50,51)*. At the time of this writing, this question is still debated. Fluorescence flow-cytometric methods may be the most practical to detect subtle changes in platelet-membrane proteins. The most promising of these may be for the detection of GMP 140, also called P-selectin, PADGEM, and CD62 antigen, an α-granule membrane protein that appears on the platelet surface before degranulation becomes grossly apparent *(52,53)* (Fig. 10). With the use of this method, P-selectin has been found on the platelet surface when rHuTPO was added to whole blood in vitro, suggesting that some activation had occurred *(46,47)*.

3.2. Changes in the Cytoskeletal Proteins Associated with Aggregation

3.2.1. Microtubules

Microtubules depolymerize into tubulin units early during activation, and when platelets are chilled or treated with sulfhydril reagents, colchicine, or vinca alkaloids

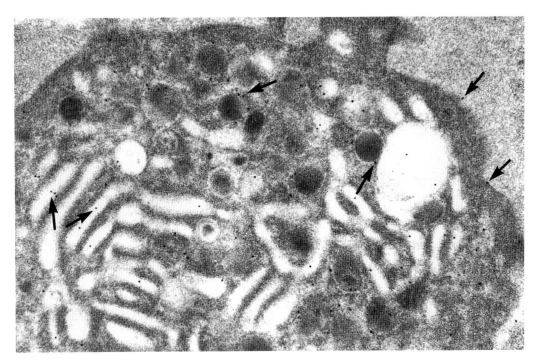

Fig. 10. Detail of an activated platelet embedded in Lowicryl K4M. Thin sections were treated with antibody to P-selectin and labeled with colloidal gold to resolve the surface as well as the intracellular antigen. Gold particles are seen in association with granule membranes, within the OCS and on the surface (*see* arrows). (Magnification, ×31,000.)

(54) (Fig. 11). Rewarming of chilled platelets is accompanied by repolymerization of microtubules. In aggregated platelets or thrombi, microtubules are often seen in pseudo-pods *(1,54)*. It may be of interest to mention that the peripheral band of microtubules is a characteristic feature of what has been referred to as the "unit" platelet and that this feature is usually not seen in "compound" platelets. Depolymerization of microtubules may not be a requirement for platelet function. When platelets are exposed to paclitaxel, an agent that stabilizes microtubules, their response to adenosine diphosphate (ADP), collagen, thrombin, sodium arachidonate, and the calcium ionophore A 23187 is not impaired *(55)*. Concomitant with microtubule depolymerization, i.e., within a minute of fixation after activation in vitro, the granules are seen in the center of the platelet (Fig. 11) before they disappear. Early investigators were able to observe this phenomenon by light microscopy. After aggregation, the granule proteins can be measured in the surrounding medium. It is difficult to understand that the literature is still replete with statements expressing the contention that the granules fuse with the plasma membrane of the platelet when discharge occurs. Membrane fusion in leukocytes during endocytosis and exocytosis was reported in 1964 *(56)*, and the study of platelet de-granulation has continued in the author's laboratory to date. To date, fusion between granule membranes and those of the platelet-surface or canalicular membranes has not been seen. Even in bovine platelets, where activation causes a movement of the gran-ules to the periphery rather than to the center, membrane fusion is not observed *(57)*. It

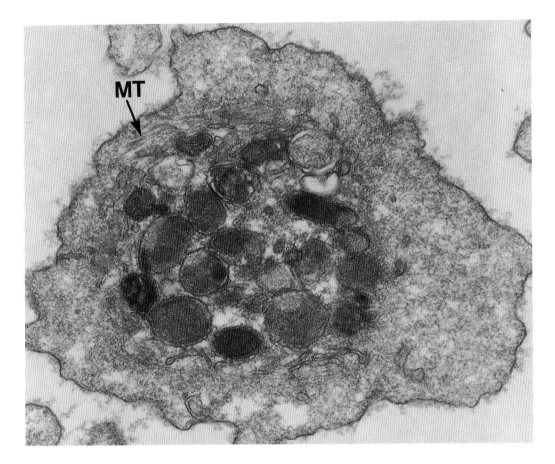

Fig. 11. Platelet briefly incubated with ADP. Microfilaments have become apparent through-out the hyaloplasm. The peripheral band of microtubules (MT) has almost completely depoly-merized. The granules are centralized and appear to be in close apposition. No granules are seen near the plasma membrane. (Magnification, ×41,000.) (Reprinted with permission from ref. *2*.)

is postulated that secretion takes place through the "pits" revealed on freeze-fracture replicas of the plasma membrane, as well as on the membranes of the canalicular system. This concept is supported by studies on bovine platelets (Fig. 12).

3.2.2. Cytoplasmic Filaments

The cytoplasm of activated platelets is replete with microfilaments that can be resolved particularly well in hypotonic media. Approximately 15% of the filamentous protein is known to consist of smooth muscle-type actin and myosin as proved by morphological and biochemical means *(54,58–60)*. Filaments are attached to the cytoplasmic aspect of the plasma and canalicular membranes, where they can be seen in longitudinal as well as in cross-section *(60)*. It is not difficult to conceive that the orientation of the filaments could be translated into a peristaltic movement along the canaliculi in a manner analogous to the mechanism afforded by the longitudinal and circular layer of smooth muscles constituting the wall of the mammalian gastrointestinal tract. This has been discussed in more detail elsewhere *(3,60)*. Other cytoskeletal proteins,

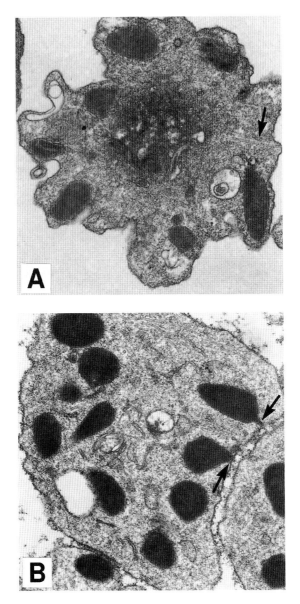

Fig. 12. Sections of bovine platelets fixed after treatment with 0.05 µg/mL thrombin for 3 minutes. The granules have moved toward the periphery. **(A)** The central condensation of microfilaments is similar to what is seen in human platelets. Note the microtubule (arrow) oriented at right angles to the plasma membrane, a phenomenon seen in many cells at the time of granule secretion. (Magnification, ×30,000.) **(B)** Bovine platelet treated with 10 m*M* ADP shows granule "neck" meeting the plasma membrane and in continuity with the extracellular space (arrows). No membrane fusion is seen. (Magnification, ×31,000.) (Reprinted with permission from ref. *2*.)

such as vimentin, talin, vinculin, and actin-binding proteins, have also been identified. Since many of these proteins are submembranous, it is assumed that they play a role in moving receptors in the plane of the membrane, as well as subserving global platelet contractility during clot retraction.

4. Platelet Pathology

It would not be commensurate with the intent of this chapter to review the entire spectrum of congenital and acquired anomalies of platelet structure and function that may be encountered in clinical practice (*see* ref. *2*). It seems, however, appropriate to draw attention to a few observations that relate to accelerated production or destruction of circulating platelets, since this may be relevant to the administration of recombinant thrombopoietic factors in a clinical setting.

4.1. "Immature" Platelets

As has already been mentioned, compensatory thrombopoiesis is usually associated with the appearance of "compound" platelets or a higher percentage of young platelets that are hemostatically more active. This alone could create a prethrombotic state in individuals who have underlying conditions, such as cardiovascular or peripheral vascular diseases; or neoplastic, inflammatory, or metabolic illnesses. The converse question is also relevant. May an immature cohort of platelets be released by the administration of rHuTPOs to thrombocytopenic patients whose plasma concentration of this hormone may already be elevated? Would this cause the release of platelets prematurely? It is conceivable that such "immature" megakaryocyte fragments would not be fully equipped with the compendium of granules, membrane receptors, and other properties necessary for normal function. It has, for instance, been shown that P-selectin appears only late during megakaryocyte maturation (*61*). Whether P-selectin is present in small megakaryocyte fragments that are prematurely released and whether this would constitute a significant functional defect is not yet known. However, the possibility that "immature" platelets are released after administration of rHuTPOs to individuals with normal platelet counts has been given some support by analysis of platelets from normal mice that developed thrombocytosis after being treated with the hormone. As illustrated in Fig. 13, such platelets not only have ribosomes and an abundance of RER, but also remnants of a Golgi complex. They appear to have very few α-granules and seem to lack the peripheral band of microtubules. The functional capacity of platelets produced in vivo after rHu thrombopoietic factor administration is discussed in Chapter 19.

4.2. Platelet Microparticles or "Dust"

The word "particle" implies that the platelet fragment is membrane-bound, whereas the word "dust" is jargon for cytoplasmic material that is not delimited by a membrane. One way to learn whether platelets elicited with rHuTPOs are more sensitive to intravascular destruction than platelets that enter the circulation spontaneously is to determine whether analysis of plasma will reveal the presence of platelet microparticles or platelet "dust." This can be done by Coulter counter analysis (*62,63*) or by electron microscopy (*63,64*). Although the latter method is more demanding, it is also more reliable, since it can distinguish between platelet and other cellular debris. To this end, the supernate of platelet-rich plasma (PRP) obtained by centrifugation for 10 minutes at 2500g is recentrifuged for 30 minutes at 27,000g. The sediment thus obtained is examined ultrastructurally (Fig. 14). Immunoelectron microscopy using an antibody to GPIIb/GPIIIa can identify the membrane fragments as being platelet derived.

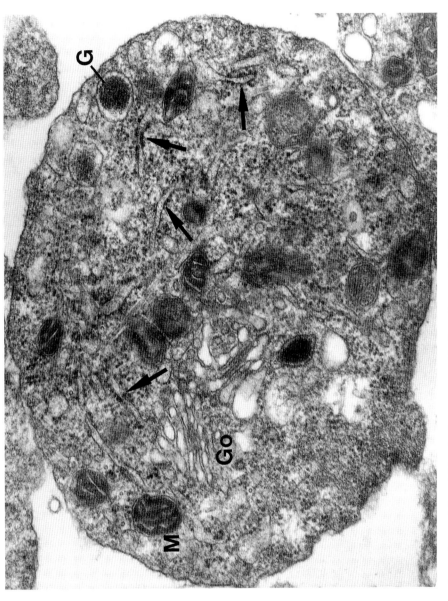

Fig. 13. Platelet from the peripheral blood of a mouse with thrombocytosis after treatment with rHuTPO. This platelet could be considered "immature," since it shows part of a Golgi complex (Go), an abundance of rough endoplasmic reticulum (arrows), and ribosomes. Note that there are more mitochondria (M) than granules (G), the latter being quite sparse. (Magnification, ×38,000.)

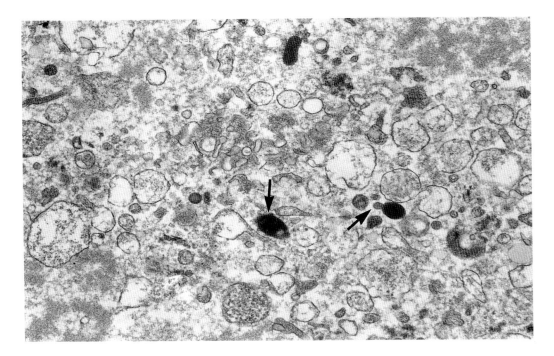

Fig. 14. Plasma sediment from the blood of a patient with ITP. Although there are some erythrocyte fragments (arrows) in the specimens from such patients, the membrane fragments and other cellular debris are platelet-derived. These are often referred to as platelet microparticles or "dust" and are not found in plasma of healthy individuals. (Magnification, ×31,000.) (Reprinted with permission from ref. *63*.)

5. Conclusion

Knowledge of platelet structure and function accumulated over the past 25 years is remarkable, and the literature on this subject is vast. Within the constraints of this chapter, it has been possible to cover only the salient features that could conceivably be affected by Mpl ligand. With the exception of the mechanism(s) involved in degranulation and the pathway that plasma proteins may take to enter into α-granules, the discussion has focused on structure/function relationships that have been confirmed by many investigators. It is regrettable that it was not possible to include many of these important studies. However, it is hoped that the reader has been given essential information in the light of which even those not engaged in platelet research can evaluate the beneficial as well as adverse impact resulting from the administration of Mpl ligand.

Acknowledgment

The author acknowledges George Grusky's expert help in preparing the illustrations for this chapter.

References

1. Zucker-Franklin D. The ultrastructure of megakaryocytes and platelets. In: Gordon AS (ed). *Regulation of Hematopoiesis*. New York: Appleton-Century-Crofts; 1970: 1553–1586.

2. Zucker-Franklin D. Megakaryocytes and platelets. In: Zucker-Franklin D, Greaves MF, Grossi CE, Marmont AM (eds). *Atlas of Blood Cells, Function and Pathology*. Philadelphia: Edi Ermes Milan, Lea & Febiger; 1989: 623–693.

3. Zucker-Franklin D. Platelet morphology and function. In: Williams WJ, Beutler E, Erslev AJ, Lichtman MA (eds). *Hematology*. New York: McGraw Hill; 1990: 1172–1181.

4. White JG. Anatomy and structural organization of the platelet. In: Colman RW, Hirsh J, Marder VJ, Salzman EW (eds). *Hemostasis and Thrombosis: Basic principles and clinical practice*. Philadelphia: Lippencott; 1994: 397–413.

5. Turitto VT, Baumgartner HR. Initial deposition of platelets and fibrin on vascular surfaces in flowing blood. In: Colman RW, Hirsh J, Marder VJ, Salzman EW (eds). *Hemostasis and Thrombosis: Basic Principles and Clinical Practice*. Philadelphia: JB Lippincott; 1994: 805–822.

6. Zucker-Franklin D, Petursson S. Thrombocytopoiesis—analysis by membrane tracer and freeze-fracture studies on fresh human and cultured mouse megakaryocytes. *J Cell Biol*. 1984; 99: 390–402.

7. Zucker-Franklin D, Rosenberg L. Platelet interaction with modified articular cartilage. Its possible relevance to joint repair. *J Clin Invest*. 1977; 59: 641–651.

8. Zucker-Franklin D, Rosenberg L. Platelet interaction with cartilage: the role of proteoglycans in vitro and in vivo. *Suppl Thromb Haemost (Germany, West)*. 1978; 68: 321–333.

9. Peerschke EI. Stabilization of platelet-fibrinogen interactions is an integral property of the glycoprotein IIb-IIIa complex. *J Lab Clin Med*. 1994; 124: 439–446.

10. Morgenstern E, Edelmann L, Reimers HJ, Miyashita C, Haurand M. Fibrinogen distribution on surfaces and in organelles of ADP stimulated human blood platelets. *Eur J Cell Biol*. 1985; 38: 292–300.

11. Mustard JF, Kinlough-Rathbone RL, Packham MA, Perry DW, Harfenist EJ, Pai KR. Comparison of fibrinogen association with normal and thrombasthenic platelets on exposure to ADP or chymotrypsin. *Blood*. 1979; 54: 987–993.

12. Ware JA, Coller BS. Platelet morphology, biochemistry and function. In: Beutler E, Williams WJ (eds). *Hematology*. New York: McGraw Hill; 1995: 1161–1201.

13. Debili N, Wendling F, Cosman D, et al. The Mpl receptor is expressed in the megakaryocytic lineage from late progenitors to platelets. *Blood*. 1995; 85: 391–401.

14. Handagama P, Scarborough RM, Shuman MA, Bainton DF. Endocytosis of fibrinogen into megakaryocyte and platelet α-granules is mediated by $a_{IIb}b_3$ (Glycoprotein IIb-IIIa). *Blood*. 1993; 82: 135–138.

15. George JN. Platelet immunoglobulin G: Its significance for the evaluation of thrombocytopenia and for understanding the origin of α-granule proteins. *Blood*. 1990; 76: 859–870.

16. Cramer EM, Meyer D, le Menn R, Breton-Gorius J. Eccentric localization of von Willebrand Factor in an internal structure of platelet α-granule resembling that of Weibel-Palade bodies. *Blood*. 1985; 66: 710–713.

17. Stenberg PE, McEver RP, Shuman MA, Jacques YV, Bainton DF. A platelet alpha-granule membrane protein (GMP-140) is expressed on the plasma membrane after activation. *J Cell Biol*. 1985; 101: 880–886.

18. Da Prada M, Pletscher A, Tranzer JP, Knuchel H. Subcellular localization of 5-hydroxytryptamine and histamine in blood platelets. *Nature*. 1967; 216: 1315–1317.

19. Davis RB, White JG. Localization of 5-hydroxytryptamine in blood platelets: an autoradiographic and ultrastructural study. *Br J Haematol*. 1968; 15: 93–99.

20. Martin JH, Carson FL, Race GJ. Calcium-containing platelet granules. *J Cell Biol*. 1974; 60: 775–777.

21. White JG, Gerrard JM. Ultrastructural features of abnormal blood platelets. A review. *Am J Pathol*. 1976; 83: 589–632.

22. Marcus AJ, Zucker-Franklin D, Safier LB, Ullman HL. Studies on human platelet granules and membranes. *J Clin Invest*. 1966; 45: 14–28.

23. Bentfeld-Barker ME, Bainton DF. Identification of primary lysosomes in human megakaryocytes and platelets. *Blood*. 1982; 59: 472–481.

24. Zucker-Franklin D. Endocytosis by human platelets: Metabolic and freeze-fracture studies. *J Cell Biol.* 1981; 91: 706–715.

25. Nachmias V, Sullender J, Asch A. Shape and cytoplasmic filaments in control and lidocaine-treated human platelets. *Blood.* 1977; 50: 39–53.

26. Breton-Gorius J, Guichard J. Ultrastructural localization of peroxidase activity in human platelets and megakaryocytes. *Am J Pathol.* 1972; 66: 277–293.

27. Gerrard JM, White JG, Rao GH, Townsend D. Localization of platelet prostaglandin production in the platelet dense tubular system. *Am J Pathol.* 1976; 83: 283–298.

28. White JG. Is the canalicular system the equivalent of the muscle sarcoplasmic reticulum? *Hemostasis.* 1975; 4: 185.

29. Zucker-Franklin D. The relationship of alpha granules to the membrane systems of platelets and megakaryocytes. *Blood Cells.* 1989; 15: 73–79.

30. Rash JE, Hudson C (eds). *Freeze-Fracture: Methods, Artifacts and Interpretations.* New York: Raven; 1979.

31. Pinto da Silva P, Branton D. Membrane splitting in freeze-etching. Covalently bound ferritin as a membrane marker. *J Cell Biol.* 1970; 45: 598–605.

32. Pinto da Silva P, Douglas SD, Branton D. Localization of A antigen sites on human erythrocyte ghosts. *Nature.* 1971; 232: 194–196.

33. Lichtman MA, Chamberlain JK, Simon W, Santillo PA. Parasinusoidal location of megakaryocytes in marrow: a determinant of platelet release. *Am J Hematol.* 1978; 4: 303–312.

34. Tavassoli M, Aoki M. Migration of entire megakaryocytes through the marrow–blood barrier. *Br J Haematol.* 1981; 48: 25–29.

35. Melamed MR, Cliffton EE, Mercer C, Koss LG. The megakaryocyte blood count. *Am J Med Sci.* 1966; 252: 301–309.

36. Kaufman RM, Airo R, Pollack S, Crosby WH. Circulating megakaryocytes and platelets released in the lung. *Blood.* 1965; 26: 720–731.

37. Trowbridge EA, Martin JF, Slater DN. Evidence for a theory of physical fragmentation of megakaryocytes implying that all platelets are produced in the pulmonary circulation. *Thromb Res.* 1982; 28: 461–475.

38. Levine RF, Eldor A, Shoff PK, Kirwin S, Tenza D, Cramer EM. Circulating megakaryocytes: Delivery of large numbers of intact, mature megakaryocytes to the lungs. *Eur J Haematol.* 1993; 51: 233–246.

39. Zucker-Franklin D, Termin CS, Cooper MC. Structural changes in the megakaryocytes of patients infected with the human immune deficiency virus (HIV-I). *Am J Pathol.* 1989; 134: 1295–1303.

40. Detwiler TC, Odell TT, McDonald TP. Platelet size, ATP content, and clot retraction in relation to platelet age. *Am J Physiol.* 1962; 203: 107–110.

41. Johnson CA, Abildgaard CF, Schulman I. Functional studies of young versus old platelets in a patient with chronic thrombocytopenia. *Blood.* 1971; 37: 163–171.

42. Murphy S, Oski FA, Naiman JL, Lusch CJ, Goldberg S, Gardner FH. Platelet size and kinetics in hereditary and acquired thrombocytopenia. *N Engl J Med.* 1972; 286: 499–504.

43. Thompson CB, Jakubowski JA, Quinn PG, Deykin D, Valeri CR. Platelet size and age determine platelet function independently. *Blood.* 1984; 63: 1372–1375.

44. Corash L, Chen HY, Levin J, Baker G, Lu H, Mok Y. Regulation of thrombopoiesis: effects of the degree of thrombocytopenia on megakaryocyte ploidy and platelet volume. *Blood.* 1987; 70: 177–185.

45. Kunicki TJ. Role of platelets in hemostasis. In: Rossi EC, Simon TL, Moss GS (eds). *Principles of Transfusion Medicine.* Baltimore, MD: Williams & Wilkins; 1991: 181–192.

46. Herceg-Harjacek L, Groopman JE, Grabarek J. Thrombopoietin induces fibrinogen-mediated platelet-endothelial cell interaction. *Blood.* 1995; 86: 84a (abstract no 323).

47. Ault KA, Mitchell J, Knowles C. Recombinant human thrombopoietin augments spontaneous and ADP induced platelet activation both in vitro and in vivo. *Blood.* 1995; 86: 367a (abstract no 1456).

48. Ault KA. Flow cytometric measurement of platelet function and reticulated platelets. *Ann NY Acad Sci.* 1993; 677: 293–308.

49. Kaplan KL. Laboratory markers of platelet activation. In: Colman RW, Hirsh S, Marder VJ, Salzman EW (eds). *Hemostasis and Thrombosis: Basic Principles and Clinical Practice.* Philadelphia: Lippincott; 1994: 1180–1196.

50. Metzelaar MJ, Korteweg J, Sixma JJ, Nieuwenhuis HK. Comparison of platelet membrane markers for the detection of platelet activation in vitro and during platelet storage and cardiopulmonary bypass surgery. *J Lab Clin Med.* 1993; 121: 579–587.

51. Harker LA, Hunt P, Marzec UM et al. Dose response effect of pegylated human megakaryocyte growth and development factor (PEG-rHuMGDF) on platelet production and function in non-human primates. *Blood.* 1995; 86: 256a (abstract no 1012).

52. Berman CL, Yeo EL, Wencel-Drake JD, Furie BC, Ginsberg MH, Furie B. A platelet alpha-granule membrane protein that is associated with the plasma membrane after activation. Characterization and subcellular localization of platelet activation-dependent granule-external membrane protein. *J Clin Invest.* 1986; 78: 130–137.

53. Isenberg WM, McEver RP, Shuman MA, Bainton DF. Topographic distribution of a granule membrane protein (GMP-140) that is expressed on the platelet surface after activation: An immunogold-surface replica study. *Blood Cells.* 1986; 12: 191–204.

54. Zucker-Franklin D. Microfibrils of blood platelets: their relationship to microtubules and the contractile protein. *J Clin Invest.* 1969; 48: 165–175.

55. White JG, Rao GH. Influence of a microtubule stabilizing agent on platelet structural physiology. *Am J Pathol.* 1983; 112: 207–217.

56. Zucker-Franklin D, Hirsch JG. Electron microscope studies on the degranulation of polymorphonuclear leukocytes during phagocytosis. *J Exp Med.* 1964; 120: 569–576.

57. Zucker-Franklin D, Benson KA, Myers KM. Absence of surface-connected canalicular system in bovine platelets. *Blood.* 1985; 65: 241–244.

58. Zucker-Franklin D, Grusky G. The actin and myosin filaments of human and bovine blood platelets. *J Clin Invest.* 1972; 51: 419–430.

59. Zucker-Frankln D, Nachman RL, Marcus AJ. Ultrastructure of thrombosthenin, the contractile protein of human blood platelets. *Science.* 1967; 157: 945–946.

60. Zucker-Franklin D. The submembranous fibrils of human blood platelets. *J Cell Biol.* 1970; 47: 293–299.

61. Schick PK, Konkle BA, He X, Thornton RD. P-selectin mRNA is expressed at a later phase of megakaryocyte maturation than mRNA for von Willebrand factor and glycoprotein Ib-alpha. *J Lab Clin Med.* 1993; 121: 714–721.

62. Khan I, Zucker-Franklin D, Karpatkin S. Microthrombocytosis and platelet fragmentation associated with idiopathic/autoimmune thrombocytopenic purpura. *Br J Haematol.* 1975; 31: 449–460.

63. Zucker-Franklin D, Karpatkin S. Red-cell and platelet fragmentation in idiopathic autoimmune thrombocytopenia purpura. *N Engl J Med.* 1977; 297: 517–523.

64. Zucker-Franklin D. Clinical significance of platelet microparticles. *J Lab Clin Med.* 1992; 119: 321–322.

3

The Evolution of Mammalian Platelets

Jack Levin

1. Introduction

As described in detail in other chapters of this book, the mammalian platelet is derived from the cytoplasm of megakaryocytes, the only polyploid hemopoietic cell. Polyploid megakaryocytes and their progeny, nonnucleated platelets, are found only in mammals. In all other animal forms, cells involved in hemostasis and blood coagulation are nucleated. The nucleated cells primarily involved in nonmammalian hemostasis are designated thrombocytes to distinguish them from nonnucleated platelets.

In many marine invertebrates, only one type of cell circulates in the blood or is present in the coelomic fluid. This single cell type plays multiple roles in the defense mechanisms of the animal, including hemostasis. Such cells are capable of aggregating and sealing wounds. Although the biochemical basis for adhesion of these cells is not understood, the participation of cell aggregation in invertebrate hemostasis suggests that this process is perhaps the earliest cell-based hemostatic function (Figs. 1 and 2) *(1–4)*. The hemocytes of the ascidian *Halocynthia roretzi*, for example, aggregate following removal from the hemolymph (i.e., the circulating body fluid, equivalent to blood, in animals with open circulatory systems) *(3)*. Aggregation depends on divalent cations and is inhibited by ethylenediaminetetraacetic acid (EDTA). Similar to the response of mammalian platelets to vascular trauma, repeated sampling of hemolymph from the same area of an individual *H. roretzi* resulted in hemocytes that were increasingly activated, as measured by aggregometry (Fig. 3). Furthermore, the aggregated hemocytes released a factor into the plasma that induced additional aggregation. The parallels with mammalian platelet function are evident.

The enormous range of types of cell-based coagulation in insects has been described by Gregoire *(5)*. The hemolymph of many insects contains coagulocytes, which on contact with a foreign surface extrude long, straight, thread-like processes that may contain cytoplasmic granules. These cytoplasmic expansions mesh with similar processes from other coagulocytes, creating a hemostatic plug. Examples of coagulocyte aggregation and cytoplasmic expansions are shown in Figs. 4 and 5.

In some invertebrates, hemostasis is provided entirely by cell aggregation at the site of a wound. In others, the hemocytes contain coagulation factors or clottable protein that are released following activation and/or aggregation of the cells *(6–8)*. In the Arthro-

From: *Thrombopoiesis and Thrombopoietins: Molecular, Cellular, Preclinical, and Clinical Biology*
Edited by: D. J. Kuter, P. Hunt, W. Sheridan, and D. Zucker-Franklin Humana Press Inc., Totowa, NJ

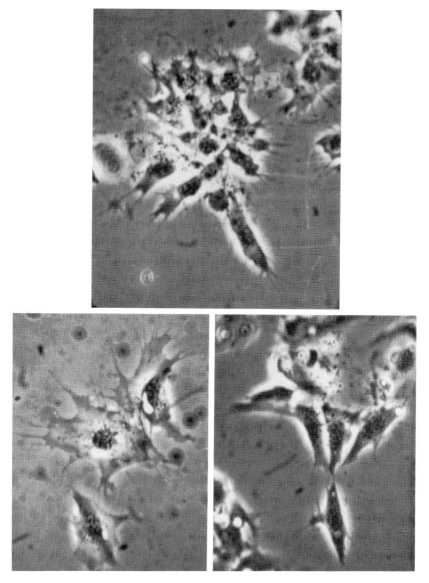

Fig. 1. Aggregation and alteration in the shape of *Limulus* amebocytes during cellular clotting on a glass surface. Note the prominent nucleus and the variety of shapes that occur following activation. The cells have also become degranulated (*see* Figs. 7, 9, and 10). Magnification, ×200 (upper) and ×320 (lower). (Reprinted with permission from ref. *1*.)

poda, as in other invertebrate classes, mechanisms of hemostasis vary widely. In some arthropods, circulating amebocytes release a coagulase that activates a clottable protein already in the plasma *(9)*. Plasma from the hemolymph of *Limulus polyphemus*, the American horseshoe crab, normally contains no coagulation factors *(7,8)*. However, following activation, amebocytes, the only type of circulating blood cell in *Limulus*, release a cascade of coagulation factors that result in coagulation of the plasma *(10,11)*.

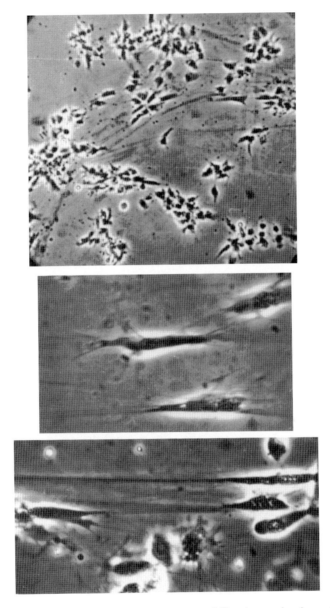

Fig. 2. Appearance of long filamentous processes following activation of *Limulus* amebocytes. These processes are often connected with those of other amebocytes. Magnification, ×128 (upper) and ×320 (middle and lower). (Reprinted with permission from ref. *1*.)

Nonmammalian vertebrates have nucleated, often spindle-shaped thrombocytes, the first cells to evolve that specialize in hemostasis *(9)*. Thrombocytes are found in fish, and in some species, multiple types of thrombocytes have been described *(12,13)*. However, some reports of multiple types of thrombocytes should be interpreted conservatively because they may be technical artifacts of the method of blood collection. Some species of estuarine cyprinodontiform fish have been described as having a seasonal variation in the ratio of mature to immature circulating thrombocytes *(14)*. Immature

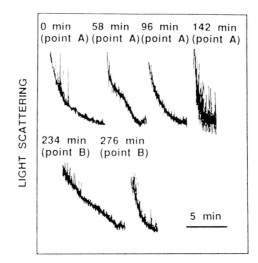

Fig. 3. Activation of hemocyte aggregation in *H. roretzi*. Point A represents the point on the tunic of the animal through which the hemolymph was repeatedly taken; point B represents a different point on the tunic. The time between the initial collection of hemolymph (0 min) and subsequent collections is indicated above each tracing. The bar represents 5 min. The extent of light scattering is shown on the ordinate in arbitrary units. (Reprinted with permission from ref. *3*.)

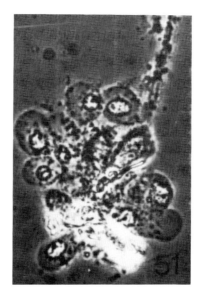

Fig. 4. *Dytiscus marginalis* L. (Coleoptera). Clustering of coagulocytes around a fragment of cuticle stimulates the reaction of hemolymph at wound sites. Magnification, ×800. (Reprinted with permission from ref. *5*.)

thrombocytes reached their highest levels during July and August, which the authors interpreted as indicating an increased rate of thrombopoiesis.

The thrombocytes of at least some birds, amphibians, and fish have a membrane system referred to as the surface-connected canalicular system (SCCS) *(15)*. This system is also a feature of mammalian platelets, which therefore has been linked to their

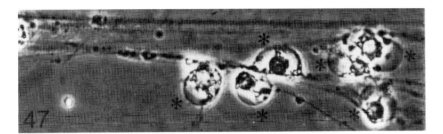

Fig. 5. *Erodius tibialis* (Coleoptera). The thread-like cytoplasmic processes are shown, carrying along the granules produced by six coagulocytes (small asterisks). Magnification, ×960. (Reprinted with permission from ref. *5*.)

derivation from the cytoplasm of megakaryocytes *(16)*. However, the presence of the SCCS in nonmammalian thrombocytes indicates that this system reflects an important function of blood cells that play a major role in hemostasis, and that the SCCS need not be derived from the demarcation membrane system (DMS) of megakaryocyte cytoplasm. Bovine platelets are apparently an exception in that they do not contain a SCCS, although bovine megakaryocytes have a DMS *(16)*.

An overview of comparative hemostasis is provided in Table 1. Because of the great heterogeneity in types of blood cells and coagulation mechanisms in invertebrates, any attempt to summarize the characteristics of hemostasis in these animals cannot avoid oversimplification and inaccuracy. Overall, however, it is clear that when cells are present in the invertebrate circulation or coelomic fluid, they always play a role in hemostasis *(17)*.

Among the extensive original studies of the blood by William Hewson (1739–1774) is a highly instructive plate that illustrates the "red particles of the blood" in a wide variety of animals (Fig. 6) *(18)*. The following statements in his text accompany this plate:

> In the blood of some insects the vesicles [blood cells] are not red, but white, as may easily be observed in a lobster (which Linnaeus calls an insect), one of whose legs being cut off, a quantity of a clear sanies flows from it; this after being some time exposed to the air jellies, but less firmly than the blood of more perfect animals. When it is jellied it is found to have several white filaments; these are principally the vesicles concreted, as I am persuaded from the following experiment. … There is a curious change produced in their shape by being exposed to the air, for soon after they are received on the glass they are corrugated, or from a flat shape are changed into irregular spheres. This change takes place so rapidly, that it requires great expedition to apply them to the microscope soon enough to observe it *(18)*.

Hewson (and his colleague Magnus Falconar) were actually observing the hemocytes of lobsters becoming activated after removal from the animal, changing shape, and then aggregating. Note the obviously altered appearance of the hemocytes, in contrast to the multiple examples of genuine "red particles of the blood" (in Fig. 6, an original Hewson plate, compare Figs. 11 and 12).

2. *Limulus polyphemus*, the Horseshoe Crab

The horseshoe crab is the last surviving member of Class Merostomata, which included marine spiders. The *Limulus* amebocyte, which is the only type of circulating

Table 1
Summary of Hemostatic Mechanisms so Far Identified
in Various Groups of Animals[a]

	Inverte-brate	Cyclo-stome	Elasmo-branch	Bony fish	Amphi-bian	Reptile	Bird	Mammal
Thrombocyte/platelet	"+"	+	+	+	+	+	+	+
Adhesion	+				+			+
Aggregation	+	+			+	+	+	+
Retraction	+	+	+	+	+		+	+
Viscous metamorphosis[b]	"+"				+		+	+
Coagulation factor	+	+	+	+	+			+
Vessel Contraction	+				+		+	+
Plasma coagulation	+	+	+	+	+	+	+	+
Fibrinogen → fibrin[c]	"+"	+	+	+	+	+	+	+
Prothrombin → thrombin		+	+	+	+	+	+	+
Spontaneous fibrinolysis	0	0	+	0	+	0	+	+

[a]Adapted from ref. *59*.
[b]For definition, *see* Table 2, footnote a.
[c]In some examples, the term "clottable protein" would be more appropriate.

blood cell in the hemolymph, has probably been the most intensely studied of the invertebrate blood cells involved in hemostasis and blood coagulation. The concentration of amebocytes in the blood of adult limuli is approximately 15×10^9/L. These are nucleated cells approximately the size of mammalian monocytes; the cytoplasm is packed with granules (Fig. 7). After activation on a foreign surface (or by bacterial endotoxin), amebocytes spread and degranulate (Fig. 7, right panel). Degranulation is associated with exocytosis *(7,19–21)*. The amebocytes, which are normally discoid (like platelets), develop pseudopodia and microspikes *(22)*. These cells are also capable of retraction (Fig. 8) apparently similar to the process of clot retraction induced by mammalian platelets.

Figure 9 shows the ultrastructure of a *Limulus* amebocyte. The cytoplasm contains at least two types of granules (major and minor) (Fig. 10A and B) that are biochemically different and contain all of the components of the blood coagulation mechanism *(11,23)*. Marginal microtubule bands are present (Fig. 11).

3. A Comparison of Human Platelets and *Limulus* Amebocytes

Mammalian platelets are appropriately considered as functioning primarily to support a range of hemostatic mechanisms, both by maintaining the integrity of small blood vessels and by contributing to the process of blood coagulation. However, platelets have many other capabilities, although often rudimentary and apparently physiologically insignificant, that appear to be unrelated to their hemostatic function. Despite approximately 400 million years of evolution, mammalian platelets retain many of the functions of *Limulus* amebocytes (and of many other comparable invertebrate blood cells). This phenomenon may have resulted from the retention of multiple functions

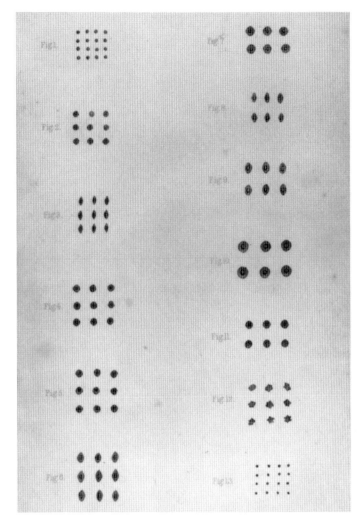

Fig. 6. "A comparative view of the flat vesicles of the blood in different animals, exhibiting their size and shape" *(18)*. The size of the red blood cell in humans (Fig. 2) is compared with that in 23 animals grouped according to the similar size of their red blood cells, thus yielding only 12 figures (Figs. 1–12) in this plate. Figures 11 and 12 show the hemocytes of the lobster before and after shape change produced by a foreign surface. Note that these are the only cells observed among the 24 animals studied that changed shape and aggregated following removal from the animal (see text). Figure 13 demonstrates "milk globules." Reprinted from ref. *18.* Plate V from Section III (originally published in 1777); from the author's collection.

previously found in a single, "all-purpose" circulating cell type, only one of whose functions was hemostasis, such as the *Limulus* amebocyte.

The multiple characteristics shared by mammalian platelets and *Limulus* amebocytes are summarized in Table 2. Platelets have rudimentary bactericidal and some phagocytic activity *(24–26)*. They contain endotoxin-binding substances *(27)*, and have been shown to interact with bacteria, endotoxin, and viruses *(28–31)*. *Staphylococcus* can

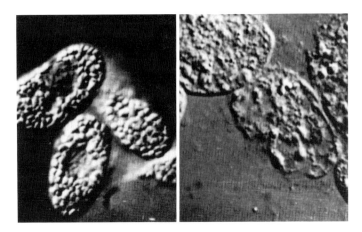

Fig. 7. *Limulus* amebocytes. Three intact normal cells are shown on the left. The cytoplasm is packed with granules. Flattened, spread, degranulated amebocytes, after exposure to a foreign surface or a bacterial endotoxin, are shown on the right. Differential interference-phase microscopy does not reveal the single large nucleus, which is located in the apparently depressed area in the middle of two of the cells in the left panel. Original magnification, ×1000. (Reprinted with permission from ref. *60*.)

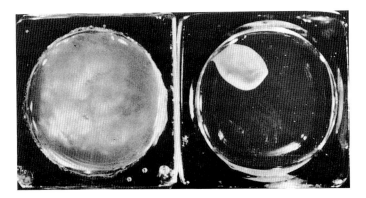

Fig. 8. Amebocyte tissue can be prepared for study by collecting blood under aseptic and endotoxin-free conditions in embryo watch glasses. After removal from the *Limulus*, the blood cells settle and aggregate into a tissue-like mass that, after an extended period in vitro, undergoes contraction. In the right-hand watch glass, the mass is 1 day old. In the upper left of this watch glass, the amebocyte tissue mass has contracted into the compact, white, button-like mass. The fluid medium was *Limulus* plasma. (Reprinted with permission from ref. *61*.)

stimulate human platelets to undergo the release reaction *(32)*. It has been demonstrated that staphylococci have a receptor for the Fc fragment of IgG that provides a mechanism for aggregation of human platelets by the formation of a complex composed of bacteria, IgG, and platelets *(33)*. Pneumococcus (*Streptococcus pneumoniae*) and vaccinia virus can induce the release of serotonin from human platelets *(34,35)*. Other investigations have demonstrated that two endocarditis-producing strains of *Streptococcus viridans* can activate human platelets in vitro *(36)*, apparently by direct

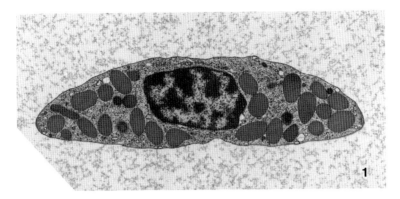

Fig. 9. Longitudinal section of a *Limulus* amebocyte. In this section, the cell is spindle-shaped. A longitudinal section cut at right angles to this one would reveal a more oval shape. Large, homogeneous secretory granules are present. Magnification, ×7000. (Reprinted with permission from ref. *62*.)

platelet–bacterial interaction. Fibrinogen in conjunction with other unidentified plasma factors was required for platelet aggregation and secretion. The aggregation of human platelets by *S. viridans*, *Streptococcus pyogenes*, and *Streptococcus sanguis* has been demonstrated *(37–39)*.

Human platelets have been reported to be cytotoxic for parasites by a mechanism involving an IgE receptor on the platelet surface *(40,41)*. Also, platelets have been shown to play a role in the excretion of *Schistosoma mansoni* in the stool of mice *(42)*. Other studies have demonstrated that platelets enhance the adherence of schistosome eggs to endothelial cells and that interleukin (IL)-6, produced by activated monocytes, markedly increases the cytotoxicity of platelets against schistosomula *(43)*. Under certain circumstances, platelets demonstrate a chemotactic response and migrate *(44,45)*, as do amebocytes *(46)*. Collectively, these and other observations indicate that bacteria, viruses, fungi, and parasites are capable of interacting with platelets. Depending on the nature of the infectious particles and other poorly understood variables, this interaction can result in platelet aggregation, release of platelet constituents, phagocytosis of the infectious agent, and ultimately, a shortened platelet life-span.

None of the above-described functions seem necessary for the current role of platelets in mammalian hemostasis. Therefore, it is likely that some of the capabilities of mammalian platelets are residua of functions originally present in the more "primitive," yet multicompetent cells from which mammalian blood cells have evolved. The roles of the amebocyte in providing hemostasis and controlling infection, and the amebocyte's reaction to endotoxin suggest that in various mammals, the response of platelets and the blood coagulation system to gram-negative infections or endotoxin is an evolutionary remnant of an ancient mechanism. The limited ability of mammalian platelets to phagocytose particles and kill bacteria may be another remnant of functions that are more important in amebocytes (and the hemostatic cells of other invertebrates). Thus, these two cell types, amebocytes from an ancient invertebrate and platelets from mammals, have remarkably similar characteristics. The relative importance of their

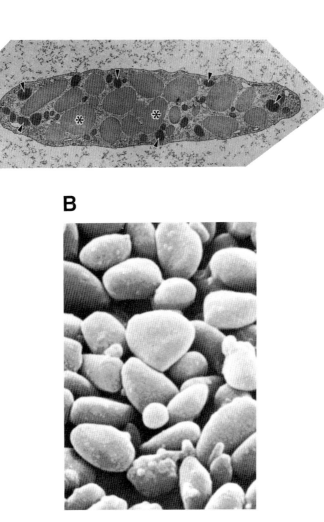

Fig. 10. A. *Limulus* amebocyte showing both major (asterisks) and minor (arrows) granules. Note the marked density of the smaller class of granules. In this section, the large nucleus is not present. Magnification, ×6140. (Reprinted with permission from ref. *62*.) **B.** Scanning electron micrograph of a preparation of intact cytoplasmic granules obtained from *Limulus* amebocytes. Magnification, ×10,000. (Reprinted with permission from ref. *23*.)

functions has changed with evolution, but after millions of years, coagulation and antibacterial mechanisms remain at least partially linked.

Although the similarities in functions of these cells are consistent with the evolution of plasmatic blood coagulation in mammals from initially cell-based mechanisms in invertebrates, there is no proof of such an evolutionary trail, despite the parallel cellular functions and the similarity of the enzymatic components, e.g., the serine proteases on which *Limulus* blood coagulation is based and some mammalian blood coagulation factors that are also serine proteases. Some evolutionary aspects of blood coagulation have been reviewed *(9,10)*. In addition, the apparently sudden evolutionary appearance of nonnucleated platelets and megakaryocytes in mammals must be considered. This

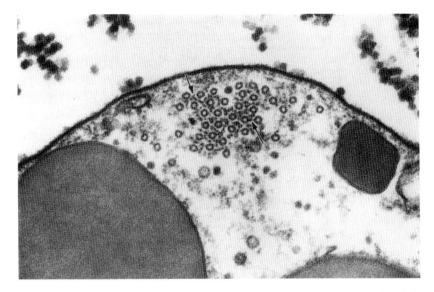

Fig. 11. A marginal microtubule band (MB) is demonstrated in a cross-section of a *Limulus* amebocyte. Projections can be seen leading from one microtubule to another (arrows). These projections may serve to stabilize the MB and are likely related to microtubule-associated proteins present in other systems. Magnification, ×104,000. (Reprinted with permission from ref. *22*.)

seems to be a marked departure from all other groups of animals, even taking into account the evolutionary concept of punctuated equilibria *(47,48)*.

4. Megakaryocytes and Mammals

Nonnucleated platelets, and presumably their polyploid megakaryocyte progenitors in the bone marrow, are present only in mammals. This suggests that some important feature of mammalian physiology benefits from this unique mechanism of producing an unprecedented anucleate cell from the cytoplasm of a larger cell, for the apparently major purpose of supporting hemostasis. However, because platelets have the rudimentary capacity to perform some of the functions that are carried out by other blood cell types in mammals, and because they also play a role in nonhemostatic defense mechanisms, we must be cautious about assuming that hemostasis is the only major platelet function. Also, all nonmammalian animal forms require a mechanism to prevent hemorrhage or uncontrolled loss of body fluids, and most (nonmammalian vertebrates being particularly relevant) have effective hemostatic mechanisms that do not involve the megakaryocyte/platelet axis. Therefore, we must ask: what is the biological advantage that led to the establishment and persistence of this cell lineage in mammals?

Members of the vertebrate class Mammalia are characterized by body hair, mammary glands, and viviparous birth, except in the egg-laying monotremes. The presence of a placenta is also characteristic of most, but not all, pregnant female mammals. The two variations in birth mechanism, oviparous birth in monotremes and viviparous, but nonplacental pregnancy in marsupials, present an opportunity to explore the potential association between placental pregnancy and platelet formation.

Table 2
Comparison of *Limulus* Amebocytes and Mammalian Platelets

Characteristics or function	*Limulus* amebocyte	Mammalian platelet
Hemostasis	Essential	Essential
Blood coagulation	Essential	Ancillary
Clottable protein	Yes	Yes
Nucleus	Yes	No
Viscous metamorphosis[a]	Yes (?)	Yes
Cellular processes	Yes	Yes
Granules	Yes	Yes
Response to endotoxin	Yes	Yes
Phagocytosis	Yes	Yes
Antibacterial function	Yes	Yes
Motility	Yes	Yes

[a]This term has been used to describe the alterations in platelet structure and morphology (including shape change) produced by platelet aggregation and contraction, and associated with platelet secretion. Previously discrete platelets become a mass of tightly packed, adherent cells, with a "fused" appearance. Apparently similar changes occur during and following the activation and aggregation of *Limulus* amebocytes.

4.1. Order Monotremata

Monotremes, considered the most primitive form of mammals, have bird-like and reptilian features. The females lay eggs. Monotremes are represented by the aquatic platypus (*Orinthorynchus anatinus*) and insectivorous echidna (spiny anteater, *Zaglossus* and *Tachyglossus*). One report on echidna platelets described two possible types, "elongated, spindle-shaped structures with a tendency to intertwine together and normal platelets with spreading and aggregating activity" *(49)*. The author suggested that the presence of two types of hemostatic cells in the echidna might indicate a link between the spindle-shaped thrombocytes of nonmammalian vertebrates and typical mammalian platelets. Platelet counts in echidna were $200–250 \times 10^9/L$. Another study of the echidna reported platelet levels of approximately $500–650 \times 10^9/L$, and that "giant multinucleated cells with a few platelet-like bodies within the cytoplasm were present in the bone marrow" *(50)*. These giant cells were believed to be megakaryocytes.

Platelet levels of approximately $400–450 \times 10^9/L$ were reported for the platypus *(51)*. The same investigators described platypus platelets as "anucleate, circular and 2–5 μm in diameter, with occasional large platelets (up to 8 μm) seen" *(52)*. Their electron microscopy studies demonstrated a homogeneous population of cells with typical platelet organelles and ultrastructure, including parallel bundles of microtubules *(52)*. This detailed report emphasized that platypus platelets were similar in appearance and size to those of other mammals, including marsupials. In this study, the two types of platelets that Hawkey *(49)* suggested were present in the echidna were not observed. Attempts by this author (J. L.) to obtain specimens of platypus bone marrow to study megakaryocytes have failed because of Australian and US regulations restricting export and import of specimens from endangered wildlife species.

4.2. Order Marsupialia

Marsupials give birth to live, but very immature young following a nonplacental pregnancy. This order is represented by the opossum, kangaroo, wombat, and bandicoot. An extensive histochemical study of the blood cells of the marsupial *Trichosurus vulpecula*, the Australian brush-tailed opossum, provided multiple photographs of typical mammalian platelets with essentially the same histochemical characteristics as human platelets *(53)*. Overall platelet size was approximately the same as that of human platelets, but a greater proportion of large platelets was reported to be present. No quantitative data were provided. In other investigations, platelet levels in five different species of marsupials ranged from approximately 200–500 × 10^9/L *(49,50,54)*, and in a sixth species (*Setonix brachyurus*, the quokka), from 425–1180 × 10^9/L. In the two marsupial species whose bone marrow has been studied, megakaryocytes were described *(50)*. Release of serotonin, a characteristic of mammalian platelets, was also demonstrated for marsupial platelets *(54)*.

Interestingly, in most of the marsupials and echidnas studied, the previously described inverse relationship between mammalian body size and platelet level was not evident *(55)*. On the basis of the previously published nomogram and platelet levels in many additional mammals, significantly higher platelet counts would have been expected in the smaller monotremes and marsupials. Perhaps the failure of this relationship to hold reflects the primitive nature of these two mammalian orders.

Blood coagulation has been described in both monotremes and marsupials as resembling human blood coagulation; clot retraction, which is consistent with normal platelet function, also has been observed in both orders *(50,54)*. Based on the above comparisons of the hemostatic systems of the aplacental and placental mammals, it appears that there is no specific association between either the development of a placenta during pregnancy (or the occurrence of live birth), or the appearance of a markedly different blood coagulation mechanism in mammals, with the presence of megakaryocytes and platelets.

5. Conclusions

There is, presumably, a biologic advantage from the presence of polyploid megakaryocytes as the source for nonnucleated platelet progeny, but this advantage has not yet been identified. An obvious benefit is the ability of a single megakaryocyte to produce many hundreds or thousands of platelets. However, augmented production can be achieved by other means: the bone marrow can markedly and adequately increase the production of red and white blood cells without resorting to a mechanism based on polyploid progenitors and cytoplasmic fragmentation. Furthermore, the many animal forms with nucleated thrombocytes have seemingly adequate hemostasis, and, as yet, platelets have not been found to be mandatory for any special feature of mammalian blood coagulation that is not present in nonmammals. However, the mechanism by which platelets are produced by megakaryocytes does allow for the rapid release of larger than normal platelets *(56,57)*. These cells are more biologically active than are the average platelets produced under steady-state conditions, and therefore may constitute an attempt to provide a maximally effective response to a pathophysiological emergency. Another unexplained element is the presence of high concentrations of acetylcholinesterase (AChE) in the megakaryocytes of only some mammalian species, such as the mouse, rat, and cat, but not in humans *(58)*. Why should only selected species have megakaryocytes that produce high concentrations of AChE? Mechanisms for the regulation of megakaryocytopoiesis and of platelet function appear identical regardless of the presence or absence of AChE in megakaryocytes. The elucidation of the evolutionary event or events that resulted in the appearance of megakaryocytes and platelets, as well as the potential evolutionary advantage of this system, remain elusive.

Acknowledgments

I wish to express my personal and professional gratitude to Frederik B. Bang for introducing me to *Limulus* over 30 years ago, and for providing the intellectual guidance that made it possible for me to begin the studies that constitute a significant portion of this chapter. I also wish to acknowledge the superb and supportive environment of the Marine Biological Laboratory, Woods Hole, MA, which nurtured my interests in comparative hemostasis, and the many staff members and scientists who provided me with the information and techniques that I required to pursue my experimental goals.

References

1. Levin J, Bang FB. A description of cellular coagulation in the Limulus. *Bull Johns Hopkins Hosp*. 1964; 115: 337–345.
2. Kenny DM, Belamarich FA, Shepro D. Aggregation of horseshoe crab (*Limulus polyphemus*) amebocytes and reversible inhibition of aggregation by EDTA. *Biol Bull*. 1972; 143: 548–567.
3. Takahashi H, Azumi K, Yokosawa H. Hemocyte aggregation in the solitary ascidian *Halocynthia roretzi*: plasma factors, magnesium ion, and met-lys-bradykinin induce the aggregation. *Biol Bull*. 1994; 186: 247–253.
4. Goffinet G, Grégoire C. Coagulocyte alterations in clotting hemolymph of *Carausius morosus* L. *Arch Int Physiol Biochim*. 1975; 83: 707–722.
5. Gregoire C. Haemolymph coagulation in insects and taxonomy. *Bull K Belg Inst Nat Wet*. 1984; 55: 3–48.
6. Ravindranath MH. Haemocytes in haemolymph coagulation of arthropods. *Biol Rev*. 1980; 55: 139–170.
7. Levin J, Bang FB. The role of endotoxin in the extracellular coagulation of Limulus blood. *Bull Johns Hopkins Hosp*. 1964; 115: 265–274.
8. Levin J, Bang FB. Clottable protein in Limulus: its localization and kinetics of its coagulation by endotoxin. *Thrombosis Diathesis Haemorrhagia* 1968; 19: 186–197.
9. Spurling NW. Comparative physiology of blood clotting. *Comp Biochem Physiol*. 1981; 68A: 541–548.
10. Levin J. The role of amebocytes in the blood coagulation mechanism of the horseshoe crab *Limulus polyphemus*. In: Cohen W (ed). *Blood Cells of Marine Invertebrates: Experimental Systems in Cell Biology and Comparative Physiology*. New York: Alan R Liss; 1985: 145–163.
11. Young NS, Levin J, Prendergast RA. An invertebrate coagulation system activated by endotoxin: evidence for enzymatic mediation. *J Clin Invest*. 1972; 51: 1790–1797.
12. Ellis AE. Leukocytes and related cells in the plaice. *Pleuronectes platessa*. *J Fish Biol*. 1976; 8: 143–156.
13. Ellis AE. The leukocytes of fish: a review. *J Fish Biol*. 1977; 11: 453–491.
14. Gardner GR, Yevich PP. Studies on the blood morphology of three estuarine cyprinodontiform fishes. *J Fish Res Board Can*. 1969; 26: 433–447.
15. Daimon T, Mizuhira V, Takahashi I, Uchida K. The surface connected canalicular system of carp (*Cyprinus carpio*) thrombocytes: its fine structure and three-dimensional architecture. *Cell Tissue Res*. 1979; 203: 355–365.

16. Stenberg PE, Levin J. Mechanisms of platelet production. *Blood Cells.* 1989; 15: 23–47.

17. Belamarich FA. Hemostasis in animals other than mammals: the role of cells. In: Spaet TH (ed). *Progress in Hemostasis and Thrombosis*, vol 3. New York: Grune and Stratton; 1976: 191–209.

18. Hewson W. On the figure and composition of the red particles of the blood, commonly called the red globules. In: Gulliver G (ed). *The Works of William Hewson, F.R.S.* London: C and J Adlard, Printers; 1846: 211–244.

19. Armstrong PB, Rickles FR. Endotoxin-induced degranulation of the *Limulus* amebocyte. *Exp Cell Res.* 1982; 140: 15–24.

20. Dumont JN, Anderson E, Winmer G. Some cytologic characteristics of the hemocytes of *Limulus* during clotting. *J Morphol.* 1966; 119: 181–208.

21. Ornberg RL, Reese TS. Beginning of exocytosis captured by rapid-freezing of *Limulus* amebocytes. *J Cell Biol.* 1981; 909: 40–54.

22. Tablin F, Levin J. The fine structure of the amebocyte in the blood of *Limulus* polyphemus. II. The amebocyte cytoskeleton: a morphological analysis of native, activated, and endotoxin-stimulated amebocytes. *Biol Bull.* 1988; 175: 417–429.

23. Mürer EH, Levin J, Holme R. Isolation and studies of the granules of the amebocytes of *Limulus polyphemus*, the horseshoe crab. *J Cell Physiol.* 1975; 86: 533–542.

24. Glynn MF, Movat HZ, Murphy EA, Mustard JF. Study of platelet adhesiveness and aggregation, with latex particles. *J Lab Clin Med.* 1965; 65: 179–201.

25. Levin J. Blood coagulation in the horseshoe crab (Limulus polyphemus): a model for mammalian coagulation and hemostasis. In: US Department of Health, Education and Welfare. *Animal Models of Thrombosis and Hemorrhagic Diseases.* Proceedings of a symposium of the National Academy of Sciences. Washington, DC: DHEW Publication No. (NIH) 76-982; 1976: 87–96.

26. Lewis JC, Maldonado JE, Mann KG. Phagocytosis in human platelets: localization of acid phosphatase-positive phagosomes following latex uptake. *Blood.* 1976; 47: 833–840.

27. Springer GF, Adye JC. Endotoxin-binding substances from human leukocytes and platelets. *Infect Immun.* 1975; 12: 978–986.

28. Clawson CC. Platelet interaction with bacteria. III. Ultrastructure. *Am J Pathol.* 1973; 70: 449–472.

29. MacIntyre DE, Allen AP, Thorne KJI, Glauert AM, Gordon JL. Endotoxin-induced platelet aggregation and secretion. I. Morphological changes and pharmacological effects. *J Cell Sci.* 1977; 28: 211–223.

30. Levin J. Bleeding with infectious diseases. In: Ratnoff OD, Forbes CD (eds). *Disorders of Hemostasis*, 3rd ed. Philadelphia: WB Saunders; 1996: 339–356.

31. Clawson CC. Platelets in bacterial infections. In: Joseph M (ed). *Immunopharmacology of Platelets.* London: Harcourt Brace; 1995: 83–124.

32. Clawson CC, Rao GHR, White JG. Platelet interaction with bacteria. IV. Stimulation of the release reaction. *Am J Pathol.* 1975; 81: 411–419.

33. Hawiger J, Steckley S, Hammond D, et al. Staphylococci-induced human platelet injury mediated by protein A and immunoglobulin G Fc fragment receptor. *J Clin Invest.* 1979; 64: 931–937.

34. Zimmerman TS, Spiegelberg HL. Pneumococcus-induced serotonin release from human platelets. Identification of the participating plasma/serum factor as immunoglobulin. *J Clin Invest.* 1975; 56: 828–834.

35. Bik T, Sarov I, Livne A. Interaction between vaccinia virus and human blood platelets. *Blood.* 1982; 59: 482–487.

36. Sullam PM, Valone FH, Mills J. Mechanisms of platelet aggregation by viridans group streptococci. *Infect Immun.* 1987; 55: 1743–1750.

37. Sullam PM, Jarvis GA, Valone FH. Role of immunoglobulin G in platelet aggregation by viridans group streptococci. *Infect Immun.* 1988; 56: 2907–2911.

38. Herzberg MC, Brintzenhofe KL, Clawson CC. Aggregation of human platelets and adhesion of *Streptococcus sanguis. Infect Immun.* 1983; 39: 1457–1469.

39. Kurpiewski GE, Forrester LJ, Campbell BJ, Barrett JT. Platelet aggregation by *Streptococcus pyogenes*. *Infect Immun*. 1983; 39: 704–708.

40. Capron A, Ameisen JC, Joseph M, Auriault C, Tonnel AB, Caen J. New functions for platelets and their pathological implications. *Int Arch Allergy Appl Immunol*. 1985; 77: 107–114.

41. Pancré V, Auriault C. Platelets in parasitic diseases. In: Joseph M (ed). *Immunopharmacology of Platelets*. London: Harcourt Brace and Co; 1995: 125–135.

42. Ngaiza JR, Doenhoff MJ. Blood platelets and schistosome egg excretion. *Proc Soc Exp Biol Med*. 1990; 193: 73–79.

43. Pancré V, Monté D, Delanoye A, Capron A, Auriault C. Interleukin-6 is the main mediator of the interaction between monocytes and platelets in the killing of *Schistosoma mansoni*. *Eur Cytokine Net*. 1990; 1: 15–19.

44. Lowenhaupt RW, Miller MA, Glueck HI. Platelet migration and chemotaxis demonstrated *in vitro*. *Thromb Res*. 1973; 3: 477–487.

45. Lowenhaupt RW, Glueck HI, Miller MA, Kline DL. Factors which influence blood platelet migration. *J Lab Clin Med*. 1977; 90: 37–45.

46. Armstrong PB. Motility of the *Limulus* blood cell. *J Cell Sci*. 1979; 37: 169–180.

47. Gould SJ. Bushes and ladders in human evolution. In: *Ever Since Darwin. Reflections in Natural History*. New York: WW Norton and Co; 1977: 56–62.

48. Gould SJ. The episodic nature of evolutionary change. In: *The Panda's Thumb. More Reflections in Natural History*. New York: WW Norton and Co; 1980: 179–185.

49. Hawkey CM. *Comparative Mammalian Haematology. Cellular Components and Blood Coagulation of Captive Wild Animals*. London: William Heinemann Medical Books; 1975: 218–227.

50. Lewis JH, Phillips LL, Hann C. Coagulation and hematological studies in primitive Australian mammals. *Comp Biochem Physiol*. 1968; 25: 1129–1135.

51. Whittington RJ, Grant TR. Haematology and blood chemistry of the free-living platypus, *Ornithorhynchus anatinus* (Shaw) (Monotremata: Ornithorhynchidae). *Aust J Zool*. 1983; 31: 475–482.

52. Canfield PJ, Whittington RJ. Morphological observations on the erythrocytes, leukocytes and platelets of free-living platypuses, *Ornithorhynchus anatinus* (Shaw) (Monotremata: Ornithorhynchidae). *Aust J Zool*. 1983; 31: 421–432.

53. Barbour RA. The leukocytes and platelets of a marsupial, *Trichosurus vulpecula*. A comparative morphological, metrical, and cytochemical study. *Arch Histol Jpn*. 1972; 34: 311–360.

54. Fantl P, Ward HA. Comparison of blood clotting in marsupials and man. *Aust J Exp Biol*. 1957; 35: 209–224.

55. Nakeff A, Ingram M. Platelet count: volume relationships in four mammalian species. *J Appl Physiol*. 1970; 28: 530–533.

56. Stenberg PE, Levin J. Ultrastructural analysis of acute immune thrombocytopenia in mice: dissociation between alterations in megakaryocytes and platelets. *J Cell Physiol*. 1989; 141: 160–169.

57. Stenberg PE, Levin J, Baker G, Mok Y, Corash L. Neuraminidase-induced thrombocytopenia in mice: effects on thrombopoiesis. *J Cell Physiol*. 1991; 147: 7–16.

58. Jackson CW. Cholinesterase as a possible marker for early cells of the megakaryocytic series. *Blood*. 1973; 42: 413–421.

59. Hawkey CM. General summary and conclusions. *Symp Zool Soc Lond*. 1970; 27: 217–229.

60. Levin J. The horseshoe crab: a model for gram-negative sepsis in marine organisms and humans. In: Levin J, Buller HR, ten Cate JW, van Deventer SJH, Sturk A (eds). *Bacterial Endotoxins. Pathophysiological Effects, Clinical Significance, and Pharmacological Control*. New York: Alan R Liss; 1988: 3–15.

61. Söderhäll K, Levin J, Armstrong PB. The effects of β1,3-glucans on blood coagulation and amebocyte release in the horseshoe crab, *Limulus polyphemus*. *Biol Bull*. 1985; 169: 661–674.

62. Copeland DE, Levin J. The fine structure of the amebocyte in the blood of *Limulus polyphemus*. I. Morphology of the normal cell. *Biol Bull*. 1985; 169: 449–457.

4

Potential Clinical Applications of Thrombopoietic Growth Factors

Charles A. Schiffer

1. Introduction

Patients receiving intensive chemotherapy or marrow-ablative regimens as treatment for a variety of cancers require repetitive transfusions of platelets. Although death from hemorrhage is a rare event in such patients, minor episodes of bleeding are common, and there are appreciable potential complications from repeated platelet transfusions (Table 1). The most clinically ominous of these side effects is the development of alloimmunization with the requirement for identification and apheresis of histocompatible donors. Often, such compatible donors cannot be readily located, and the patient remains at risk of bleeding. Another serious problem is the frequent occurrence of febrile or allergic transfusion reactions, even in patients premedicated with antipyretics and/or antihistamines; these reactions can be quite discomfiting and frightening for the patient. In neutropenic patients, such transfusion reactions often mandate hospitalization and interim coverage with broad-spectrum antibiotics until culture results are available. Although transfusion reactions are common in alloimmunized patients, the presence in the plasma of a variety of different cytokines, elaborated because of damage to contaminating leukocytes during storage, is responsible for the frequent development of transfusion reactions in nonalloimmunized recipients *(1)*.

Another problem, the frequency of which is probably underestimated, is the occurrence of bacteremia because of the administration of organisms inadvertently contaminating the platelet product at the time of collection and proliferating further during storage at room temperature *(2,3)*. This can be a serious and sometimes fatal complication, the recognition of which is probably decreased because many recipients of platelet transfusion may be "protected" by the antibiotics they are receiving for concurrent infections. Finally, the "real" cost (as distinct from hospital charges) of each platelet transfusion is considerable, varying by the type of preparation provided, with additional hidden costs, including the loss in time from employment and other tasks incurred by apheresis donations. Direct charges of between US$300 and US$700/transfusion event are representative.

Thus, any means by which the need for platelet transfusions can be eliminated or significantly reduced would be of clinical benefit, even if such maneuvers would not

From: *Thrombopoiesis and Thrombopoietins: Molecular, Cellular, Preclinical, and Clinical Biology*
Edited by: D. J. Kuter, P. Hunt, W. Sheridan, and D. Zucker-Franklin Humana Press Inc., Totowa, NJ

Table 1
Side Effects of Platelet Transfusion

Febrile transfusion reactions
    Due to alloimmunization against leukocyte antigens; may also be related to infusion of
        cytokines in stored platelets
Transfusion-related acute lung injury (TRALI)
    Hypersensitivity reactions (hives) possibly related to infused plasma proteins
Circulatory congestion
    50–60 mL plasma/U of platelet concentrate (4–10 U/transfusion)
Bacteremia
    Rare, but catastrophic
    Caused by contamination at collection and bacterial proliferation during storage
Hemolysis
    Infusion of isohemagglutinins in mismatched plasma
    Frank hemolysis is rare, but positive direct Coombs' tests are common
Alloimmunization to HLA antigens
    Refractoriness to subsequent transfusions
Sensitization to red blood cell antigens
    Small numbers of red blood cells are present in all platelet products
Graft-vs-host disease
    Caused by transfusion of potentially alloreactive lymphocytes suspended in platelet
        products
    Rare except in severely immunosuppressed recipients
    Can be prevented by γ-irradiation of platelets

alter the already very low incidence of hemorrhagic death. Because of these considerations, there has been interest in the identification of a thrombopoietic-stimulating factor for decades. The extraordinary revolution in molecular biology has resulted in the
identification and cloning of large numbers of different hemopoietic colony-stimulating factors (CSF) in the last 5–10 years. Some, such as interleukin (IL)-3, IL-6, and
IL-11, have modest to moderate thrombopoietic activity, but do not have the relatively
"clean" lineage specificity of other CSFs, such as granulocyte colony-stimulating factor (G-CSF) and erythropoietin (EPO) (4,5). In addition, there are often systemic side
effects when many of the ILs are given at clinically effective doses.

It appears that the "real thing" has been identified and synthesized in the last few
years. As detailed in other chapters in this volume and elsewhere, thrombopoietin (TPO)
cloned by a number of laboratories produces relatively specific stimulation of megakaryocyte growth and differentiation with modest, if any, effects on cells of other lineages (6–9). TPO binds specifically to a receptor expressed on megakaryocytes that
has been termed Mpl. Mice and monkeys demonstrate marked stimulation of platelet
production after repeated subcutaneous dosing with rHuTPOs without apparent side
effects (10). Importantly, experiments in both species have shown that administration
of megakaryocyte growth and development factor (MGDF), one of the recombinant
forms of the Mpl ligand, can eliminate the severe thrombocytopenia and the need for
platelet transfusion associated with the administration of either a single high dose of
radiation or single high dose of chemotherapeutic agents (10–15). Stem cell transplant
models are also being developed (16).

This is somewhat in contrast to the use of recombinant human (rHu)G-CSF in these models, which does not prevent severe neutropenia, but rather shortens the duration of neutropenia and accelerates the rate of neutrophil recovery. The difference between these results may be owing to stimulation of platelet production by megakaryocytes that had not been severely or lethally damaged by the chemotherapy or radiotherapy in the first few days. This would result in release of platelets that, with their 8–9-day survival, prevent the development of severe thrombocytopenia. In contrast, neutrophils have a 6–8-hour survival in the circulation, such that even highly effective simulators of granulopoiesis have been unable to prevent significant nadirs from occurring. Details of these findings can be found in other chapters in this volume.

Phase 1 trials in humans are in progress *(17)*. Given the observations from the preclinical models, it is reasonable to assume that thrombopoiesis with elaboration of functionally normal platelets will also be effectively stimulated in humans. On the assumption that the rHuTPOs do not produce prohibitive side effects, it is timely to consider how these compounds might be used in clinical practice and, in particular, to discuss the issues that need to be addressed to demonstrate to practicing physicians and to regulatory licensing agencies that these compounds not only elevate the platelet count, but also provide clinically meaningful benefit to thrombocytopenic patients.

2. Indications for Platelet Transfusions

There is a direct relationship between the platelet count and the bleeding time when the platelet count is $<100 \times 10^9/L$. At counts $<50 \times 10^9/L$, the bleeding time is >20 minutes, and at $<20 \times 10^9/L$, the bleeding time is prolonged indefinitely *(18)*. Despite this, major hemorrhagic events are uncommon in patients treated with intensive chemotherapy, and major surgery can be performed on patients with platelet counts in the $40–50 \times 10^9/L$ range *(19)*. Although some of this relatively benign clinical behavior may be the result of the use of platelet transfusion and close attention by clinicians to other therapeutic maneuvers that can help ameliorate hemorrhage (Table 2), mostly it is a tribute to the apparent resilience of the vascular system, even at very low platelet counts. Indeed, many patients with amegakaryocytic thrombocytopenia survive for months to years without significant bleeding at platelet counts of $<10 \times 10^9/L$, particularly if there are no other complicating medical problems, such as intervening infection and fever. Thus, it would be difficult to demonstrate a favorable impact of the use of rHuTPOs by assessments of the rate of serious hemorrhage. Furthermore, it is difficult to quantify more minor degrees of hemorrhage, such as ecchymosis, hematuria, or guaiac-positive stools. Unfortunately, even a considerable reduction in the number of platelet transfusions is unlikely to be reflected in a decrease in the rate of alloimmunization, since the development of this complication is not related to the number of transfusions received *(20,21)*. Consequently, potential benefit of an rHuTPO should be evaluated primarily by a reduction in the number of platelet transfusions, and by inference, shorter duration of severe thrombocytopenia and a decrease in the likelihood of some of the other complications enumerated in Table 1. It is, therefore, critical to identify groups of patients who reproducibly require significant numbers of platelet transfusions as part of their treatment.

There are wide variations in clinical practice concerning criteria for administering platelet transfusions *(22)*. Although most clinicians prescribe platelet transfusions pro-

Table 2
Other Approaches to Decrease the Likelihood or Severity of Hemorrhage in Thrombocytopenic Patients

---

Rapid diagnosis and treatment of infection

Avoidance of unnecessary invasive procedures

Avoidance of medications that affect platelet function (aspirin and other nonsteroidals)

Exclusion of other treatable disorders, such as idiopathic thrombocytopenic purpura (ITP), sepsis, or thrombotic thrombocytopenic purpura (TTP)

Antifibrinolytic agents in selected, usually clinically stable, patients

Prompt recognition of alloimmunization with selection of histocompatible donors

Estrogen therapy to prevent menses in younger women

Recognition and correction of coagulation disorders (e.g., vitamin K deficiency, hypofibrinogenemia owing to L-asparaginase)

Avoidance/treatment of renal dysfunction

Red blood cell transfusions to prevent severe anemia and retinal hemorrhage

Recognition of situations where hemorrhage is more likely (hyperleukocytosis in acute myeloid leukemia [AML], certain tumor types, central nervous system metastases in malignant melanoma)

Development of cytokines to stimulate thrombopoiesis, attenuate mucositis, or "downregulate" the inflammatory effects of other cytokines

---

phylactically to prevent bleeding in patients actively receiving therapy, there is considerable variation in the platelet count "trigger." This ambiguity is reinforced by an absence of rigorous, prospectively controlled data, although it is possible to summarize what appear to be recent trends in clinical practice. In the past, a platelet count of 20×10^9/L was perhaps the most quoted figure for the use of prophylactic platelet transfusion. This number was probably based on misinterpretation of data published during the "infancy" of platelet transfusion that demonstrated an increased rate of hemorrhage in children with acute lymphocytic leukemia (ALL) exhibiting platelet counts $<20 \times 10^9$/L *(23–25)*. Also notable about those early papers was, as mentioned earlier, that most children did not have serious bleeding even at these lower platelet counts. Additionally, it was common in the early 1960s to use aspirin as an antipyretic, and finally, it was more difficult to control active infection, particularly with gram-negative organisms, because of the inferior antibiotics that were available.

Recent editorials and a small randomized trial have called attention to the arbitrary nature of this 20×10^9/L "trigger" *(26–28)*, and many institutions are now much more comfortable in observing clinically stable patients at lower counts without immediately administering platelet transfusion. At the University of Maryland Cancer Center, the decision to administer platelets is individualized according to the patient's unique clinical circumstances *(29)*. Frequently, prophylactic platelet transfusions will not be given to patients who are clinically stable without active hemorrhage (e.g., patients with aplastic anemia or chronic myelodysplasia), even at platelet counts of $\leq 5 \times 10^9$/L. In contrast, platelet transfusions are often given at higher platelet counts to patients who are bleeding; have an active infection; have evidence of disseminated intravascular coagulation; require invasive procedures; or are at the early, rather than late, stages of their chemotherapy course.

Clinical trials evaluating rHuTPOs must, therefore, adopt a reasonably standard and generally accepted policy for platelet transfusions. In general, there would probably be relatively little disagreement among clinicians regarding criteria for administering transfusion early during leukemia induction. However, criteria may vary considerably in clinically stable patients in whom bone marrow recovery is imminent. Such potentially small variations in clinical practice could have a significant effect on the interpretation of clinical trials evaluating rHuTPOs, particularly if, in the clinical circumstance being studied, there is usually a requirement for a relatively small number of platelet transfusions.

Because of these variables and vagaries of clinical practice, as well as the heterogeneity inherent in any group of seriously ill patients, randomized trials will be necessary to demonstrate the clinical benefit of rHuTPOs. Similar issues were faced when myeloid growth factors were evaluated. In general, clinical trials should include patients in whom the rHuTPOs are to be most widely used clinically. There are three such general groups of patients: those receiving marrow-ablative therapy with stem cell support, those receiving induction or postremission-consolidation therapy for acute leukemia, and those undergoing intensive chemotherapy for solid tumors or lymphomas. In the first two circumstances, all patients receiving such therapy require frequent platelet transfusion, and a relatively simple trial design would be to randomize patients to receive rHuTPOs or a placebo. In contrast, a relatively small percentage of patients receiving conventional therapy for most solid tumors need platelet transfusions, and those patients who do, receive fewer repeated transfusions. There is, however, a subset of patients receiving intensive, sometimes experimental regimens, without stem cell support who require platelet transfusions. An efficient trial design in such patients could include identification of patients who have required platelet transfusions during the immediately previous course of chemotherapy, with randomization to rHuTPOs or a placebo during the next course of therapy if dose modification is not implemented. This strategy has the advantage of identifying the subgroup of patients who, for one reason or another, requires platelet transfusion with particular regimens.

It is critical that all such trials be double blinded and placebo controlled to evaluate side effects of therapy objectively, as well as to control for potential biases that could influence the decision to prescribe platelet transfusions, particularly toward the end of the aplastic period. The value of a placebo control was evident in a recent randomized study of the use of rHu granulocyte-macrophage colony-stimulating factor (GM-CSF) to attenuate the duration of neutropenia in older patients with acute myeloid leukemia (AML) *(30)*. In this blinded, placebo-controlled study, approximately one-third of patients were removed from treatment because of toxicity attributed to the growth factor. Of note is that an identical number of patients in both the growth factor and placebo groups were thought to have experienced such toxicities. Although many of the cytokines can cause a number of different symptoms, including fever, edema, occasional respiratory symptomatology, and rashes, it is hoped that the apparent lineage specificity of the rHuTPOs, analogous to that of EPO, might be associated with fewer symptoms. This is of particular importance to patients who are concurrently neutropenic. The development of fever and new symptoms in these patients often results in initiation of or changes in antibiotic therapy, and sometimes leads to the empiric administration of amphotericin-B for fevers of unknown origin. Amphotericin-B has a number of significant and well-known side effects, so it is desirable to avoid this medication if it is not necessary.

In all trials, there should be some standardization of the dose of platelets adminis-tered, and possibly, some guidelines concerning ABO compatibility and the duration of storage. It is possible (although unproved), for example, that institutions that routinely administer larger numbers of units of platelet concentrates per transfusion might use fewer overall transfusions because of the higher increments achieved. This could be of the greatest impact during the later stages of a sequence of platelet transfusions. In addition, although their use in patients with myeloid leukemias is somewhat controver-sial, rHuG-CSF or rHuGM-CSF are used relatively routinely in patients receiving very high-dose therapy. Therapeutic or toxic interactions between the myeloid growth fac-tors and rHuTPOs will, therefore, also need to be considered. In some circumstances, it may be advisable to compare three approaches (i.e., placebo and rHuTPOs with and without myeloid CSF).

There are also a number of issues that could influence the interpretation and design of clinical trials unique to each of the groups of patients that might benefit from an effective rHuTPO.

3. Adult Acute Leukemia

3.1. Induction Therapy

All patients with AML receiving initial induction therapy and most adults receiving induction therapy for ALL require large numbers of platelet transfusions. Fifty-one patients with AML treated recently at the University of Maryland Cancer Center who achieved complete remission with standard daunorubicin and cytarabine induction therapy received an average of 12.7 platelet transfusions (range, 5–36; 95% Cl, 10.9–14.4) (Fig. 1). In general, the patients requiring the largest number of transfusions included individuals who became alloimmunized, patients with acute progranulocytic leukemia with disseminated intravascular coagulation, or patients with very serious infections. Because adult AML is a relatively common disorder with subgroups of pa-tients being cured after receiving intensive therapy *(31)*, this is obviously an important group of patients in which to evaluate the benefit of thrombopoietic agents. There are a number of unique issues that must be considered however:

- Varying with patient age, between 50 and 80% of newly diagnosed patients with de novo AML attain complete remission and normal marrow function. Patients who do not achieve marrow recovery, usually because of persistence of drug-resistant leukemia, are not "in-formative" in terms of the kinetics of stimulation of thrombopoiesis because of persistent marrow dysfunction and the absence of recovery of normal precursors. Thus, there are many patients who cannot benefit from even effective thrombopoietic agents because of inadequacies of current chemotherapy. These patients cannot be identified beforehand and, as a consequence, all patients must be randomized and considered in the analysis, thereby increasing the necessary sample size.
- Most AML induction regimens include 7-day infusions of cytarabine with some recent protocols, including even longer durations of therapy *(31,32)*. There has been concern about coadministration of hemopoietic stimulatory factors and chemotherapy because of the possibility that normal precursors may be recruited into cycle with enhanced cytotox-icity from the chemotherapy. As mentioned previously, the preclinical models of rHuTPO

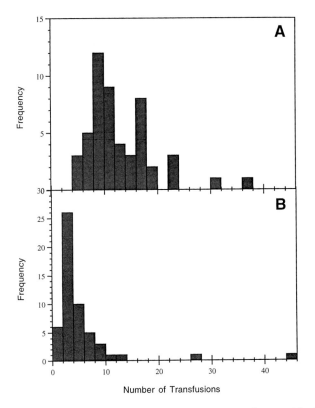

Fig. 1. Number of platelet transfusions administered to patients with AML undergoing chemotherapy. **(A)** 51 patients who achieved either complete or partial remission after induction chemotherapy with daunorubicin and cytarabine. **(B)** 54 courses of postremission intensification chemotherapy with high-dose cytarabine. The number of platelet tranfusions is significantly greater after induction compared with consolidation chemotherapy (median, 10 versus 3; $p < 10^{-7}$).

efficacy used very short durations of therapy, and it is unknown whether the dramatic benefits seen in those circumstances would be apparent when the rHuTPOs are given after the completion of longer courses of chemotherapy.

- Preliminary studies have suggested that Mpl can be expressed on myeloid blasts from patients with a variety of different French-American-British (FAB) morphologic classifications *(33–36)*. Some studies have suggested that rHuTPOs can enhance leukemia cell growth in vitro *(35)*, whereas others indicate minimal enhancement of proliferation *(36)*. It is unknown whether this presents a danger of stimulation of leukemic growth in vivo. It is reassuring that large numbers of clinical trials evaluating the use of rHuG-CSF and rHuGM-CSF after the completion of induction chemotherapy in patients with AML have failed to demonstrate an increased incidence of treatment failure owing to drug-resistant leukemia *(30,37,38)*. Nonetheless, it is critical to monitor trials in patients with AML for the emergence of either obvious leukemic growth or an unexpectedly low complete response rate. Patients with megakaryoblastic leukemia (FAB M7) should probably be excluded from at least the initial trials.

- Approximately 20–30% of patients entering remission receive two courses of chemotherapy during induction, usually separated by 10–21 days. Studies with rHuG-CSF or rHuGM-CSF have dealt with this problem in different ways. Some protocols have continued the growth factor during the second treatment, whereas others have stopped the growth factor, restarting it after completion of the second course. The latter approach probably is preferable given the theoretic concerns about leukemia stimulation and "cycling" of normal precursors.

3.2. Postremission Consolidation Therapy

Recent clinical trials have demonstrated that intensive postremission chemotherapy can be curative in a substantial fraction of adults with AML *(31)*. Most regimens incorporate one to four courses of a high-dose cytarabine regimen. In a study at University of Maryland Cancer Center incorporating four sequential courses of high-dose cytarabine *(31)*, 27 patients receiving 54 courses of treatment received an average of 4.9 platelet transfusions (range, 1–44; 95% Cl, 3.1–6.6)/ course (Fig. 1). Particularly given the curative intent of therapy, this is also a critical clinical scenario where the rHuTPOs might be of value. There are some unique issues inherent in these patients as well:

- Essentially 100% of these patients eventually recover normal marrow function because, by definition, they are starting with a remission bone marrow and normal blood counts. Because the mortality rate is extremely low, virtually all patients would be "informative" for assessing the kinetics of platelet recovery.
- It is possible that there could be some beneficial stimulation of residual normal megakaryocyte precursors even after the administration of chemotherapy, thereby attenuating both the depth and duration of thrombocytopenia.
- Many of these patients do not become infected and febrile, and some remain as outpatients for the duration of their thrombocytopenia. Because of the smaller number of transfusions required, variations in criteria for administering prophylactic transfusion could profoundly affect the results. It may also be difficult to demonstrate a statistically significant reduction in the number of transfusions given because of the small number required, unless there was a large patient sample size. There is also the possibility of debate about the clinical significance of a reduction of only one or two platelet transfusions.
- Approximately 20–30% of patients with AML receiving multiple platelet transfusions become alloimmunized because of antibody against HLA antigens and require transfusions from histocompatible donors *(20)*. Depending on the rapidity with which this complication is recognized and the ability to identify suitable donors, alloimmunization can markedly affect the number and frequency of platelet transfusions administered, which, in turn, could possibly complicate the analysis of this end point. Because antibody formation can be delayed in patients receiving induction chemotherapy, this is more often a clinical issue with patients receiving postremission therapy, although it can be a problem during induction therapy as well *(20,21)*. It is therefore advisable to monitor for the development of lymphocytotoxic antibody in patients on clinical trials of rHuTPOs and to anticipate the need for histocompatible platelets during both induction and consolidation in patients with serologic evidence of alloimmunization. There is no reason to suspect that the rHuTPOs will alter the incidence of alloimmunization, although it would be of interest to monitor for this end point as well.
- Stimulation of leukemia growth is unlikely to be a problem in this circumstance.
- Multiple courses of therapy are usually given, and analysis should be done of individual treatments, as well as the entire course of consolidation. This could increase the complexity of the analysis.

Because of these differences, patients receiving induction and consolidation therapy should probably be evaluated in separate trials.

3.3. Stem Cell Transplantation

There has been a dramatic increase in the use of stem cell transplants in the support of both marrow-ablative and intensive chemotherapy for patients with hematologic malignancies and solid tumors, all of whom experience transient thrombocytopenia. In most patients, there is a predictable return of the platelet count; occasional "outliers" with prolonged thrombocytopenia are noted. This is particularly true when bone marrow is used as the source of reconstituting stem cells. It is likely, however, that peripheral blood stem cells will replace bone marrow in the autologous setting in all but very occasional circumstances in the future. Most studies have noted more reliable and rapid reconstitution of both neutrophils and platelets using peripheral blood stem cells, such that most patients require only a few platelet transfusions *(39)*. It may, therefore, be difficult to "improve" on these results even with an effective thrombopoietic factor. Before initiating such trials, a careful appraisal should be made of the number of platelet transfusions generally required for a given disorder. It would also be interesting to evaluate the rHuTPOs in the occasional patient who has delayed platelet recovery. Benefit from myeloid growth factors has been demonstrated in analogous patients with delayed neutrophil recovery.

3.4. Solid Tumors/Lymphomas

Except for patients receiving high-dose therapy for tumors, such as sarcoma or testicular cancer, patients receiving standard regimens infrequently require platelet transfusion, and when they do, it is uncommon that more than one or two transfusions are administered per course. Nonetheless, as described earlier, clinical trials in this subset of patients could be informative, particularly if patients are identified as being "at risk" by virtue of having required transfusions on previous courses of therapy.

4. Use of Recombinant Human Thrombopoietic Factors in Myelodysplasia

In many patients with myelodysplasia, severe thrombocytopenia is a major cause of morbidity and altered quality of life. Because there is an inverse relationship between the level of platelet count and endogenous TPO levels *(40,41)*, it is predictable that most such patients will have high endogenous levels and perhaps already have maximally stimulated megakaryopoiesis *(see also* Chapters 22 and 23). This is true with regard to EPO levels in most myelodysplasia patients, although clinical trials have demonstrated a small percentage of patients who respond to pharmacologic doses of EPO *(42)*. Similarly, some patients with myelodysplasia have increased neutrophil counts when stimulated with rHuG-CSF or rHuGM-CSF, providing at least transient benefit combating infection *(43)*. The rHuTPOs should be evaluated in selected myelodysplastic patients exhibiting hemorrhagic problems. An assessment of pretreatment levels should be included to determine whether there are some patients, presumably those with lower endogenous production, who might benefit from such an approach. Similar considerations would pertain to the management of patients with aplastic anemia. Obviously, both groups of patients have the same underlying problem of markedly impaired marrow function and decreased ability to respond to growth factors.

5. Platelet Donation

5.1. Autologous Platelet Donation

Platelets can be stored in the liquid state for at least 7 days (although they are licensed for only 5 days at present) and can be cryopreserved with dimethyl sulfoxide (DMSO) virtually indefinitely (44–46). The University of Maryland Cancer Center has a program of autologous frozen platelet transfusion that focuses on patients who are alloimmunized, but who require further courses of intensive chemotherapy (46). Platelets are obtained at the time of remission and cryopreserved for subsequent use with the number frozen directly related to the predonation platelet count. An obvious consideration would be to stimulate patients to provide higher preapheresis platelet counts, thereby increasing the yield and efficiency of the procedure. It would also be appropriate to consider this maneuver in patients who are undergoing apheresis for collection of stem cells. It is conceivable that platelets could be collected either simultaneously or at a subsequent apheresis with storage either in the liquid or frozen state, depending on the interval between stem cell collection and subsequent treatment and thrombocytopenia. It is also possible that patients pretreated with rHuTPOs before intensive therapy may, because of starting at a higher platelet count, take longer to become thrombocytopenic, thereby decreasing, or perhaps even eliminating, the need for platelet transfusion.

5.2. Allogeneic Platelet Donation

It is frequently possible to identify only a small number of platelet donors for a given alloimmunized patient (47). Again, both the yield of platelets obtained from apheresis and the expected increments are directly related to the donor's precollection platelet count and the number transfused. Although there may be concerns about premedicating nonrelated donors for this purpose, family members frequently represent the major source of histocompatible platelets, and in this situation, any risk from premedication would be balanced by benefit to the patient. These donor safety issues are the same considerations that have applied in the past to the use of corticosteroids and hydroxethyl starch administered to granulocyte donors and, more recently, to the use of rHuG-CSF to stimulate the neutrophil counts in granulocyte donors (48). More information about the safety of the rHuTPOs, particularly about the potential for thrombotic episodes, would be available as further clinical experience accrues.

Allogenic peripheral blood stem cells are being used increasingly in patients undergoing allogenic transplantation (49). These donors are routinely prestimulated with rHuG-CSF, and it may be possible to add an rHuTPO to these collections, thereby providing large numbers of transfused platelets as a means of increasing the recipient's platelet count at the same time that the stem cells are being infused. All of these approaches could serve to decrease the overall transfusion requirements, particularly since the duration of thrombocytopenia after allogenic peripheral blood stem cell transplant is relatively short.

6. Conclusions

There are multiple potential clinical applications of rHuTPOs either alone or in combination with other cytokines. Given the rapid pace of development of both rHuEPO and the myeloid CSFs, it is highly likely that answers to the clinical questions posed will be forthcoming soon. Indeed, the explosive growth of recent abstracts and publications dealing with the rHuTPOs attest to the intense interest and research activity in this area. It can be expected that a major new weapon will be added to the armamentarium of those providing supportive care for patients receiving intensive forms of chemotherapy. The dimensions of its eventual clinical use will be determined by the balance between the magnitude of the clinical benefit and the clinical risks and costs of the new cytokine.

References

1. Heddle NM, Klama L, Singer J, et al. The role of plasma from platelet concentrates in transfusion reactions. *N Engl J Med.* 1994; 331: 625–628.
2. Buchholz DH, Young VM, Friedman NR, Reilly JA, Mardiney MR, Jr. Bacterial proliferation in platelet products stored at room temperature. Transfusion-induced enterobacter sepsis. *N Engl J Med.* 1971; 285: 429–433.
3. Braine HG, Kickler TS, Charache P, et al. Bacterial sepsis secondary to platelet transfusion: an adverse effect of extended storage at room temperature. *Transfusion.* 1986; 26: 391–393.
4. Ishibashi T, Kimura T, Uchida T, Kariyone S, Friese P, Burstein SA. Human interleukin 6 is a direct promoter of maturation of megakaryocytes in vitro. *Proc Natl Acad Sci USA.* 1989; 86: 5953–5957.
5. Musashi M, Yang YC, Paul SR, Clark SC, Sudo T, Ogawa M. Direct and synergistic effects of interleukin 11 on murine hemopoiesis in culture. *Proc Natl Acad Sci USA.* 1991; 88: 765–769.
6. Kaushansky K. Thrombopoietin: The primary regulator of platelet production. *Blood.* 1995; 86: 419–431.
7. Bartley TD, Bogenberger J, Hunt P, et al. Identification and cloning of a megakaryocyte growth and development factor that is a ligand for the cytokine receptor Mpl. *Cell.* 1994; 77: 1117–1124.
8. Lok S, Kaushansky K, Holly RD, et al. Cloning and expression of murine thrombopoietin cDNA and stimulation of platelet production in vivo. *Nature.* 1994; 369: 565–568.
9. de Sauvage FJ, Hass PE, Spencer SD, et al. Stimulation of megakaryocytopoiesis and thrombopoiesis by the c-Mpl ligand. *Nature.* 1994; 369: 533–538.
10. Ulich T, del Castillo J, Yin S, et al. Megakaryocyte growth and development factor ameliorates carboplatin-induced thrombocytopenia in mice. *Blood.* 1995; 86: 971–976.
11. Farese AM, Hunt P, Boone TC, MacVittie TJ. Recombinant human megakaryocyte growth and development factor stimulates thrombocytopoiesis in normal non-human primates. *Blood.* 1995; 86: 54–59.
12. Farese AM, Hunt P, Grab LB, MacVittie TJ. Combined administration of recombinant human megakaryocyte growth and development factor and granulocyte colony stimulating factor enhances multilineage hematopoietic reconstitution in nonhuman primates after radiation induced marrow aplasia. *J Clin Invest.* 1996; 97: 2145–2151.
13. Andrews RG, Winkler A, Woogerd P, et al. Recombinant human megakaryocyte growth and development factor (rHUMGDF) stimulates thrombopoiesis in normal baboons and accelerates platelet recovery after chemotherapy. *Blood.* 1995; 86: 371a (abstract no 1471).
14. Kaushansky K, Broudy VC, Grossmann A, et al. Thrombopoietin expands erythroid progenitors, increases red cell production, and enhances erythroid recovery after myelosuppressive therapy. *J Clin Invest.* 1995; 96: 1683–1687.

15. Hokom MM, Lacey D, Kinstler OB, et al. Peglayted megakaryocyte growth and development factor abrogates the lethal thrombocytopenia associated with carboplatin and irradiation in mice. *Blood*. 1995; 86: 4486–4492.

16. Molineux G, Hartley C, McElroy T, McCrea C, McNiccc I. Megakaryocyte growth and development factor (MGDF) accelerates platelet recovery in peripheral blood progenitor cell (PBPC) transplant recipients. *Blood*. 1995; 86: 461a (abstract no 1831).

17. Basser R, Clarke K, Fox R, et al. Randomized, double-blind, placebo-controlled phase I trial of pegylated megakaryocyte growth and development factor (PEG-rHuMGDF) administered to patients with advanced cancer before and after chemotherapy—early results. *Blood*. 1995; 86: 257a (abstract no 1014).

18. Harker LA, Slichter SJ. The bleeding time as a screening test for evaluation of platelet function. *N Engl J Med*. 1972; 287: 155–159.

19. Bishop JF, Schiffer CA, Aisner J, Matthews JP, Wiernik PH. Surgery in acute leukemia: A review of 167 operations in thrombocytopenic patients. *Am J Hematol* 1987; 26: 147–155.

20. Dutcher J, Schiffer CA, Aisner J, Wiernik PH. Long-term follow-up of patients with leukemia receiving platelet transfusions: Identification of a large group of patients who do not become alloimmunized. *Blood*. 1981; 58: 1007–1011.

21. Dutcher J, Schiffer CA, Aisner J, Wiernik PH. Alloimmunization following platelet transfusion: the absence of a dose response relationship. *Blood*. 1981; 57: 395–398.

22. National Institutes of Health Consensus Development Consensus Conference. Platelet transfusion therapy. *JAMA*. 1987; 257: 1777–1780.

23. Gaydos LA, Freireich EJ, Mantel N. The quantitative relation between count and hemorrhage in patients with acute leukemia. *N Engl J Med*. 1962; 266: 905–909.

24. Freireich EJ, Kliman A, Gaydos LA, et al. Response to repeated platelet transfusion from the same donor. *Ann Intern Med*. 1963; 59: 277–287.

25. Djerassi I, Farber S, Evans AE. Transfusions of fresh platelet concentrates to patients with secondary thrombocytopenia. *N Engl J Med*. 1963; 268: 221-226.

26. Schiffer CA. Prophylactic platelet transfusion. *Transfusion*. 1992; 32: 295–298.

27. Beutler E. Platelet transfusions: the 20,000/µL trigger. *Blood*. 1993; 81: 1411–1413.

28. Gmur J, Burger J, Schanz U, Fehr J, Schaffner A. Safety of stringent prophylactic platelet transfusion policy for patients with acute leukaemia. *Lancet*. 1991; 338: 1223-1226.

29. Heyman MR, Schiffer CA. Platelet transfusion therapy for the cancer patient. *Semin Oncol*. 1990; 17: 198-209.

30. Stone RM, Berg DT, George SL, et al. Granulocyte-macrophage colony stimulating factor after initial chemotherapy for elderly patients with primary acute myeloid leukemia. *N Engl J Med*. 1995; 332: 1671–1677.

31. Mayer RJ, Davis RB, Schiffer CA, et al. Intensive postremission chemotherapy in adults with acute myeloid leukemia. *N Engl J Med*. 1994; 331: 896–903.

32. Archimbaud E, Thomas X, Leblond V, et al. Timed sequential chemotherapy for previously treated patients with acute myeloid leukemia: long-term follow-up of the etoposide, mitoxantrone, and cytarabine-86 trial. *J Clin Oncol*. 1995; 13: 11–18.

33. Vigon I, Dreyfus F, Melle J, et al. Expression of the c-*mpl* proto-oncogene in human hematologic malignancies. *Blood*. 1993; 82: 877–883.

34. Matsumura I, Kanakura Y, Kato T, et al. Growth response of acute myeloblastic leukemia cells to recombinant human thrombopoietin. *Blood*. 1995; 86: 703–709.

35. Piacibello W, Sanavio F, Garetto L, et al. The effect of human megakaryocyte growth and development factor (MGDF) on human myeloid leukemia cell growth. *Blood*. 1995; 86: 45a (abstract no 168).

36. Slack JL, Baer MR, Bernstein SH, et al. Acute myeloid leukemia (AML) blast cell proliferation and differentiation in response to megakaryocyte growth and development factor (MGDF). *Blood*. 1995; 86: 520a (abstract no 2070).

37. Rowe JM, Andersen J, Mazza JJ, et al. A randomized placebo-controlled phase III of granulo-

cyte-macrophage colony stimulating factor (GM-CSF) in adult patients (55 to 70 years) with acute myelogenous leukemia (AML). A study of the Eastern Cooperative Oncology Group (ECOG). *Blood.* 1995; 86: 257–262.

38. Dombret H, Chastang C, Feraux P, et al. A controlled study of recombinant human granulocyte colony stimulating factor in elderly patients after treatment for acute myeloid leukemia (AML). *N Engl J Med.* 1995; 332: 1678–1683.

39. Weaver CH, Buckner CK, Longin K, et al. Syngeneic transplantation with peripheral blood mononuclear cells collected after the administration of recombinant human granulocyte colony-stimulating factor. *Blood.* 1993; 82: 1981–1984.

40. Nichol JL, Hornkohl A, Selesi D, Wyres M, Hunt P. TPO levels in plasma of patients with thrombocytopenia or thrombocytosis. *Blood.* 1995; 86: 371a (abstract no 1474).

41. Nichol J, Hokom MM, Hornkohl, A, et al. Megakaryocyte growth and development factor. Analyses of in vitro effects on human megakaryopoiesis and endogenous serum levels during chemotherapy-induced thrombocytopenia. *J Clin Invest.* 1995; 95: 2973–2978.

42. Stein RS, Abels RI, Krantz SB. Pharmacologic doses of recombinant human erythropoietin in the treatment of myelodysplastic syndromes. *Blood.* 1991; 78: 1658–1665.

43. Negrin RS, Haeuber DH, Nagler A, et al. Maintenance treatment of patients with myelodysplastic syndromes using recombinant human granulocyte colony stimulating factor. *Blood.* 1990; 76: 36–43.

44. Hogge DE, Thompson BW, Schiffer CA. Platelet storage for seven days in second generation blood bags. *Transfusion.* 1986; 26: 131–135.

45. Schiffer CA, Lee EJ, Ness PM, Reilly J. Clinical evaluation of platelet concentrates stored for 1–5 days. *Blood.* 1986; 67: 1591–1594.

46. Schiffer CA, Aisner J, Wiernik PH. Frozen autologous platelet transfusion for patients with leukemia. *N Engl J Med.* 1978; 299: 7–12.

47. O'Connell BA, Lee EJ, Rothko K, Hussein MA, Schiffer CA. Selection of histocompatible apheresis platelet donors by cross-matching random donor platelet concentrates. *Blood.* 1992; 79: 527–531.

48. Bensinger WI, Price TH, Dale DC, et al. The effects of daily recombinant human granulocyte colony-stimulating factor administration on normal granulocyte donors undergoing leukapheresis. *Blood.* 1993; 81: 1883–1888.

49. Korbling M, Huh YO, Durett A, et al. Allogeneic blood stem cell transplantation: peripheralization and yield of donor-derived primitive hematopoietic progenitor cells (CD34+Thy-1dim) and lymphoid subsets, and possible predictors of engraftment and graft-versus-host disease. *Blood.* 1995; 86: 2842–2848.

II

THE SEARCH FOR THE PHYSIOLOGIC REGULATOR OF PLATELET PRODUCTION

5

Historical Perspective and Overview

Eric M. Mazur

1. Introduction

Although the normal peripheral platelet count varies considerably from person to person, it is clear that platelet production is a regulated physiological process. That such a regulatory system exists is evidenced by the constancy of the normal peripheral platelet count in individuals over time, the capacity of the bone marrow to increase platelet production in the setting of accelerated platelet turnover, and the other alterations that characterize platelets and megakaryocytes produced in the clinical context of varying peripheral platelet demand (1).

It is well established that platelets, the smallest cellular element of the peripheral blood, are derived from bone marrow megakaryocytes, the "giant cells" of the bone marrow. Furthermore, megakaryocytes develop, as do all hemopoietic cells, from a common pluripotent hemopoietic stem cell. The developmental process for megakaryocytes is complex, and incorporates the commitment of the pluripotent hemopoietic progenitor cell to megakaryocytic differentiation, mitotic amplification of the megakaryocyte precursor cells, variable endomitotic nuclear division, cytoplasmic growth and maturation with the development of platelet-specific structures and organelles, and platelet formation with shedding into the circulation (1,2). Each of these developmental processes affects the net output of platelets to the circulation, and thus, each is a potential focus of regulatory control. Final platelet production is directly correlated with the number of megakaryocytes as well as megakaryocyte ploidy, size, and extent of cytoplasmic maturation (3,4). The net effect of early mitotic events on quantitative platelet production is amplified by postmitotic megakaryocyte development. Therefore, regulatory perturbations affecting megakaryocyte stem cell expansion potentially manifest the greatest effect on quantitative platelet output, but because this effect is early, exhibit the slowest response time. Conversely, regulation-induced changes affecting the postmitotic phases of megakaryocyte development may more rapidly alter platelet production, but are limited in their maximum net quantitative effects. Thrombopoietic regulatory control also appears to have the capacity to stimulate the rate of megakaryocyte maturation, which has been observed to increase in the context of accelerated peripheral platelet use (5).

From: *Thrombopoiesis and Thrombopoietins: Molecular, Cellular, Preclinical, and Clinical Biology*
Edited by: D. J. Kuter, P. Hunt, W. Sheridan, and D. Zucker-Franklin Humana Press Inc., Totowa, NJ

2. The Platelet Radioisotope-Incorporation Bioassay

It is in the context of this understanding of the processes of megakaryocytopoiesis and platelet production that the search for the physiologic regulator of platelet production must be considered. The term "thrombopoietin" was first proposed by Keleman et al. in 1958 to describe an activity in human thrombocythemic sera *(6)*. During the next few years, initial biological studies focused on thrombocytopenic animals and the capacity of their test plasma to increase the circulating platelet counts in normal, like-species recipients *(7,8)*. Although these initial studies were encouraging and suggested that a circulating thrombopoietic factor existed, it was recognized shortly thereafter that the thrombocytosis in the recipient animals was a nonspecific effect, not necessarily the result of a physiologic platelet regulator. It was reported that both powdered glass and egg albumin increased platelet counts in the recipient animals (in addition to nonthrombocytopenic sera from experimental animals injected with such foreign agents) *(9)*. Although it was correctly inferred that some types of inflammation may stimulate the endogenous production of a substance stimulating platelet production (which we now believe to be interleukin [IL]-6), the lack of specificity of the assay system precluded additional substantive investigation of a true, physiologic platelet regulator or thrombopoietin. The use of induced thrombocytosis in recipient animals as a biologic assay for the as yet to be defined thrombopoietin was all but abandoned.

A more specific bioassay system for a circulating thrombopoietic regulator was reported by a number of investigators in 1969 and 1970 *(10–12)*. This system measured the capacity of test plasma from rodents to stimulate the incorporation of exogenous, radiolabeled amino acids into the cellular protein of newly produced, circulating platelets in recipient animals of the same species. The use of thrombocytotic animals whose endogenous thrombopoiesis was almost completely suppressed enhanced the sensitivity of this bioassay system *(13)*. The standard radioisotopes employed were ^{75}Se-selenomethionine and ^{35}S-sulfate. These radioisotopes were not in any way specifically targeted toward or uniquely incorporated into platelets. Rather, they labeled all newly synthesized proteins. Platelet specificity was achieved by measuring radioisotope activity only within circulating platelets harvested from the experimental animal at a fixed interval after injection of the radioisotope.

Using this approach, a circulating plasma activity, at this time operationally termed thrombopoietin (and refered to explicitly in this chapter as TPN), was identified in acutely thrombocytopenic animals, which consistently increased platelet radioisotope incorporation two- to threefold above control levels determined using normal plasma *(10–13)*. Conversely, plasma from animals with thrombocytosis because of platelet hypertransfusion (and presumably suppressed endogenous thrombocytopoiesis) stimulated significantly less radioisotope incorporation into platelets than the normal control plasma (Fig. 1) *(11,14)*. This assay was believed to be more specific than the animal thrombocytosis assay and more sensitive, since most of the recipient test animals did not develop increases in their platelet counts. Of note is that the thrombocytopenic test plasma used in these assays was almost uniformly harvested from animals whose thrombocytopenia was induced experimentally by a consumptive process, i.e., either the injection of platelet antiserum or exchange transfusion using platelet-poor red blood cells.

A similar, if not identical activity (called thrombopoiesis-stimulating factor or TSF) was identified in growth media conditioned by human embryonic kidney cells (HEK)

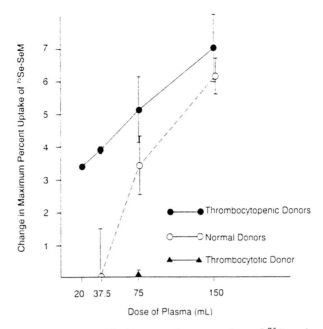

Fig. 1. Dose–response relationship between incorporation of ^{75}Se-selenomethionine into platelets and volume of plasma administered. Total volumes of 37.5, 75, or 150 mL of plasma from normal or thrombocytopenic donors were administered in three equal doses to rabbits with transfusion-induced thrombocytosis. ^{75}Se-selenomethionine was injected 6 hours after the last dose of plasma. Increase in the maximum incorporation of ^{75}Se-selenomethionine into platelets is plotted as the change from the mean maximum percent incorporation of six hypertransfused control rabbits. The mean ± 1 SE of each group of rabbits is shown. Plasma from normal donors produced a lesser effect than plasma from thrombocytopenic donors. Seventy-five milliliters of plasma from thrombocytotic donors had no demonstrable effect *(14)*.

(15). TSF exhibited activity identical to that of plasma TPN in the platelet radioisotope incorporation bioassay and, thus, was suspected of being indistinguishable from the physiologic regulator of platelet production. It is important to emphasize that TPN and TSF were both identified and defined by this platelet radioisotope-incorporation bioassay. Therefore, their physiologic relevance was only as secure as the bioassay was valid in reflecting physiologic platelet regulation. Many concerns were raised regarding the physiologic relevance of the assay. Since platelet counts were not reproducibly increased by TPN and TSF in vivo, and since only a small fraction (<10%) of the radiolabel appeared in platelets, the radioisotope-incorporation assay was criticized as only a very indirect measurement of thrombopoiesis, reflecting only minor increases in platelet size and/or protein content *(1,16)*.

Scattered early clinical reports in humans had also supported the concept that a humoral thrombopoietic activity existed. However, in all instances, such activity was inferred from the capacity of allogeneic plasma to increase the platelet count in human recipients. Two pediatric patients were described with chronic thrombocytopenia, postulated as resulting from a deficiency of endogenous thrombopoietic activity, who responded to infusions of normal human plasma with significant increases in their platelet counts *(17,18)*. In both instances, the increases in the recipients' platelet counts were accompanied by increases in plasma thrombopoietic stimulatory activity as measured by the radioisotope-incorporation assay (19,20). Unfortunately, these case reports are flawed, since in addition to thrombocytopenia, both patients had microangiopathic hemolytic anemia. In retrospect, despite allegations to the contrary (18), these two patients likely had a *forme fruste* of thrombotic thrombocytopenic purpura (TTP), a disorder resulting in consumptive thrombocytopenia and known to respond reliably to plasma infusions. A more convincing study by Adams et al. (21) demonstrated that the infusion of autologous, postplatelet-apheresis plasma into normal volunteers at baseline platelet counts induced an approximate 50% increase in the circulating count. This increase was accompanied by an increase in immature megakaryocytes in the recipients' bone marrow. The application of the rodent platelet radioisotope-incorporation bioassay to plasma from patients with thrombocytopenias of multiple etiologies failed to demonstrate a coherent pattern for increases in plasma thrombopoietic activity. For example, of 10 human donors with chronic consumptive thrombocytopenia, only 1 exhibited an increase in functional thrombopoietic activity (22).

With the development of experimental techniques to isolate and culture bone marrow megakaryocytes, more sophisticated and potentially physiologically relevant assays were applied to the study of TPN and TSF. Using isolated megakaryocyte preparations, TPN and/or TSF was shown to stimulate the synthesis of platelet factor (PF)-4 and platelet glycosaminoglycans, to promote ^{3}H-TdR DNA incorporation and to enhance nuclear endoreduplication (23–25). Furthermore, TSF was shown to induce the full maturation of low-ploidy, transitional megakaryocyte precursor cells in culture (26), and fractionated thrombocytopenic plasma stimulated megakaryocyte cytoplasmic process formation and fragmentation in vitro (27). In vivo, exogenous TSF administration to experimental animals stimulated increases in platelet number, platelet size, megakaryocyte size, and average polyploid megakaryocyte DNA content (15). Both TSF and plasma TPN were partially purified and characterized biochemically (15,28). A similar activity, called megakaryocyte-stimulating factor or MSF, was purified from media conditioned by HEK cells based on its ability to stimulate PF-4 synthesis by a rat promegakaryoblast cell line (29). Unfortunately, the cumbersome nature of the bioassays and, in the case of plasma TPN, its limited supply, precluded the successful purification of MSF, TSF, or TPN to homogeneity sufficient to permit protein sequencing. Hence, the existence and nature of unique physiologic thrombopoietic factors could not be confirmed at a molecular level.

3. Megakaryocyte Colony-Forming Assays

The development of colony-forming assays for rodent and human megakaryocyte progenitors (the megakaryocyte colony-forming cell or MK-CFC) provided new opportunities for the analysis of megakaryocytopoietic regulation. In these assays, bone marrow or peripheral blood mononuclear cells were cultured in a single-cell suspension in semisolid growth media (30–32). After a period of incubation, megakaryocyte colonies were enumerated based on either their unique morphologic appearance or the immunolabeling of colonies with platelet-specific antisera. In these systems, detection of MK-CFC-derived colonies is dependent on the following set of condtions: presence of megakaryocyte committed stem cells (MK-CFC) in the cultured cell population; sensitivity of these MK-CFC to the megakaryocyte-growth stimulator provided; adequate concentrations of a megakaryocyte-growth stimulator in culture to support colony growth; mitotic expansion of the MK-CFC (at least two generations for a colony of four cells); and sufficient maturation of the MK-CFC-derived progeny cells to render them identifiable as megakaryocytes in the culture. Perturbations of any of these factors will affect the number of megakaryocyte colonies enumerated *(2).*

In initial studies using this new bioassay system, potential thrombopoietic regulators were re-evaluated for their capacity to support megakaryocyte colony growth in vitro. Neither TPN nor TSF proved to be active in independently supporting megakaryocyte-colony development *(32,33).* The only materials that were found initially to stimulate megakaryocyte-colony growth consistently and reproducibly were media conditioned by spleen cells, lectin-stimulated lymphocytes (PHA-LCM), and WEHI-3 cells *(33,34).* Notably and in addition, crude preparations of erythropoietin (EPO) also exhibited variable megakaryocyte colony-stimulating activity (MK-CSA) depending on the reporting laboratory *(32,33,35).* Despite their inactivity alone in stimulating megakaryocyte colony growth, both TPN and TSF were found to act synergistically, in conjunction with other sources of MK-CSA, to enhance megakaryocyte colony numbers in megakaryocyte clonogenic assays *(16,36).* Materials exhibiting such synergistic activity were defined functionally as "megakaryocyte potentiators" *(37).* It was assumed that megakaryocyte potentiators served to increase the detection of megakaryocyte colonies in culture by stimulating megakaryocyte size, ploidy, and maturity, since both TPN and TSF had been previously shown to stimulate postmitotic aspects of megakaryocyte development in other experimental systems (*vide supra*). These observations resulted in a "two-level" regulatory hypothesis for the control of megakaryocyte development. The early, proliferative events of the megakaryocyte progenitor or MK-CFC were believed to be modulated separately from the later, postmitotic phases of megakaryocyte development. It was postulated that MK-CSA was responsible for the former, whereas "megakaryocyte potentiators" (including TPN and TSF) were responsible for the latter *(1,2,37).*

A new and potent source of MK-CSA was identified by Hoffman et al. *(38)* and Mazur et al. *(31)* in plasma and sera from patients with bone marrow megakaryocyte aplasia. Patients with aplastic anemia *(31,38),* isolated amegakaryocytic thrombocytopenia (syndrome of congenital thrombocytopenia absent radius or TAR syndrome) *(39),* and patients at hematologic nadirs after cytotoxic chemotherapy *(40)* or radiotherapy *(41)* exhibited high levels of circulating MK-CSA that supported megakaryocyte colony formation in vitro with a direct dose–response relationship. Not only were colony numbers increased with plasma or serum MK-CSA, but the number of constituent megakaryocytes per colony was also increased (Fig. 2). Although MK-CSA was almost uniformly present in sera from patients whose thrombocytopenia resulted from bone marrow megakaryocyte hypoplasia, MK-CSA was not detected in sera from patients with consumptive thrombocytopenia of equal severity *(38).* This observation further supported the two-level regulatory model for megakaryocytopoiesis and suggested that MK-CSA was produced in vivo only in response to the depletion of bone marrow megakaryocytes or megakaryocyte stem cells. It was proposed that thrombopoietin, the putative postmitotic regulator of megakaryocyte development, was elaborated in vivo in the clinical context of accelerated platelet turnover with an intact bone marrow megakaryocyte compartment *(1).* Attempts to purify a specific megakaryocyte colony-stimulating factor from

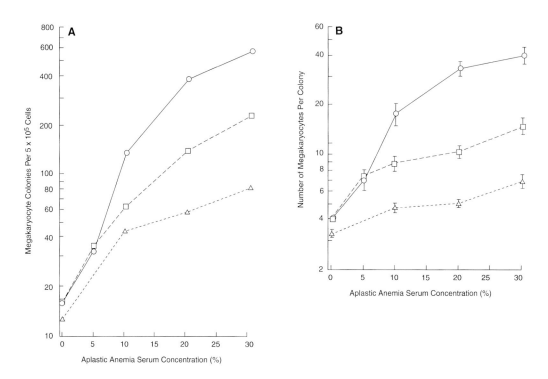

Fig. 2. (A) Effects of the concentration of sera from three patients with aplastic anemia on the number of megakaryocyte colonies cultured from human bone marrow mononuclear cells. Each line is derived from a single representative experiment performed in duplicate. A proportional dose–response effect is demonstrated in each experiment. Patient 1, ○, $r = 0.817$, $P < 0.01$; patient 2, □, $r = 0.980$, $P < 0.01$; patient 3, △, $r = 0.978$, $P < 0.01$. **(B)** Effects of the concentration of sera from three patients with aplastic anemia on the number of megakaryocytes present in individual megakaryocyte colonies. Each point represents the mean ± SEM number of mega-karyocytes/colony derived from examining 50 colonies cultured during a single experiment at different aplastic anemia serum concentrations. At 0% aplastic serum concentration, the number of colonies examined varied from 38 to 50 because of the limited colony numbers present. Data for these three experiments are derived from the same experiments presented in Fig. 2A. In each experiment, a proportional dose–response relationship is present between the mean number of megakaryocytes per colony and the concentration of aplastic anemia serum. Patient 1, ○, $r = 0.985$, $P < 0.01$; patient 2, □, $r = 0.975$, $P < 0.01$; patient 3, △, $r = 0.967$, $P < 0.05$ *(31)*.

human plasma *(42)* and irradiated, aplastic canine sera *(43)* met with only partial success. Quantities of plasma and sera were limited, and the megakaryocyte colony-forming bioassay was both cumbersome and required an excessively long turnaround time (i.e., approximately 12 days). However, these initial purification attempts did support the concept that a distinct megakaryocyte-lineage humoral regulator existed.

4. Recombinant DNA Technology and Cytokines Affecting Thrombopoiesis

The successful cloning of a number of hemopoietic cytokines occurred in the late 1980s. Several of these cytokines exhibited stimulatory effects on megakaryocytopoiesis and platelet production, both in vitro and in vivo. A comprehensive review of these cytokines and their effects on megakaryocytopoiesis was published by Hoffman *(44)*. Like the impure activities previously described (i.e., MK-CSA and "megakaryocyte potentiator"), the recombinant hemopoietic cytokines segregated themselves into two groups: colony-stimulating activities, which stimulated primarily megakaryocyte colony growth; and thrombopoietin-like agents, which stimulated primarily postmitotic megakaryocyte maturation (and acted as "megakaryocyte potentiators" in colony-forming assays). However, unlike the impure plasma and serum thrombopoietic activities, none of the recombinant cytokines that stimulated megakaryocytopoiesis exhibited any degree of megakaryocyte-lineage specificity.

4.1. Interleukin-3

The most potent of the recombinant cytokines in supporting megakaryocyte colony growth was IL-3. In vitro, recombinant (r)IL-3 stimulated increases in both megakaryocyte colony number and colony size (i.e., the number of constituent cells per colony) in a concentration-dependent manner *(45,46)*. This stimulation was focused preferentially on the early developmental stages of megakaryocytopoiesis, with rIL-3 acting to expand mitotically and differentiate partially megakaryocyte progenitor cells to low-ploidy, postmitotic precursors expressing megakaryocyte-specific phenotypic markers. Studies of colony megakaryocytes in vitro demonstrated that rIL-3 alone was insufficient to support full megakaryocyte nuclear endoreduplication *(2,45,46)*. The MK-CSA exhibited by both WEHI-3-conditioned medium and by PHA-LCM was found to be attributable to IL-3 *(34,47)*.

The stimulatory activity of IL-3 on megakaryocytopoiesis in vitro was confirmed in vivo. The administration of rIL-3 to primates and humans for 7–15 days resulted in up to an average twofold increase in the peripheral blood platelet concentration *(48,49)*. Notably, the thrombopoietic response tended to occur late in the treatment period and often peaked during the week after discontinuation of rIL-3. This clinical observation further supported the postulate that IL-3 stimulates an early, rather than late, component of megakaryocyte development and suggested that more prolonged rIL-3 administration may result in an even more profound thrombocytotic effect.

4.2. Interleukin-6

IL-6 was the other recombinant cytokine that most consistently exhibited a stimulatory effect on megakaryocytopoiesis. In contrast with IL-3, however, IL-6 appeared to act late in megakaryocytopoiesis, affecting primarily the postmitotic megakaryocyte developmental compartment. In most investigations, IL-6 alone did not support progenitor cell-derived megakaryocyte colony growth in vitro *(50,51)*. IL-6 did, however, significantly potentiate colony formation in the presence of IL-3 *(50,51)*. Furthermore, the "megakaryocyte potentiator" activity previously reported to be present in media conditioned by lung and bone marrow tissue was found to be attributable to an IL-6-like molecule *(52)*. The mechanisms by which IL-6 influenced megakaryocytopoiesis and augmented megakaryocyte colony detection were further elucidated in studies using a variety of in vitro experimental systems. IL-6 was demonstrated to stimulate increases in megakaryocyte size, ploidy, and maturation *(53)*. Ravid et al. *(54)* did not confirm the effect of IL-6 on megakaryocyte ploidy, but did observe a threefold increase in transcriptional activity of the gene for PF-4, a platelet α-granule protein gene specific for cells of the megakaryocyte lineage. IL-6 also stimulated the development of low-ploidy, transitional, murine megakaryocyte precursors into large, cytologically mature megakaryocytes (51), and induced the formation of extensive proplatelet cytoplasmic extensions by isolated guinea pig megakaryocytes (53). Since both IL-6 and its receptor are expressed by megakaryocytes and megakaryocytic cell lines, it was postulated that IL-6 participated in an autocrine regulatory loop controlling postmitotic megakaryocyte development (55).

The stimulatory effects of IL-6 on megakaryocytopoiesis in vitro were fully reproducible in vivo. Recombinant IL-6 administration to nonhuman primates for 8–14 days approximately doubled the peripheral platelet count and significantly increased megakaryocyte size and ploidy (56–58). The thrombopoietic effect of rIL-6 was further enhanced by approximately 50% (i.e., to a threefold increase in the platelet count) by administering rIL-3 and rIL-6 in a sequential fashion (59). The in vivo synergy between IL-3 and IL-6 was expected given their postulated early and late loci of activity in the megakaryocyte developmental process. Murine models confirmed the in vivo stimulatory effects of rIL-6 on platelet numbers, megakaryocyte size, and megakaryocyte endoreduplication (60). Coincident with its stimulation of thrombopoiesis in vivo, IL-6 also stimulated physiologic changes characteristic of an acute inflammatory state. Nonhuman primates receiving 11–14 day courses of rIL-6 developed dose-dependent weight loss, anemia, neutrophilia, elevated serum levels of acute phase-reacting proteins, and reductions in both serum prealbumin and albumin (56,58). In patients, elevated serum levels of IL-6 were found to correlate better with ongoing inflammation than with accelerated platelet demand (61), and in experimental animals, serum IL-6 levels did not increase after the induction of immune thrombocytopenia (62). Therefore, although IL-6 was not likely to be either the most important or even a contributing physiologic regulator of thrombopoiesis, it might be in part responsible for the reactive thrombocytosis observed in vivo in association with chronic inflammation and malignancy (2,62,63).

4.3. Other Cytokines

Several other cytokines have also been evaluated as potential physiologic regulators of platelet production. IL-11, a pleiotropic hemopoietic growth factor cloned from a primate stromal cell line (20), exhibits a spectrum of in vitro and in vivo activities similar to IL-6. Although not supporting megakaryocyte colony development alone, IL-11 does synergize with IL-3 to increase megakaryocyte colony formation in vitro (64). In vivo, rIL-11 stimulates a modest thrombocytosis (130–150% of baseline) in normal experimental animals, and accelerates both neutrophil and platelet recovery after cytotoxic chemotherapy, both with and without bone marrow transplantation (65,66). Leukemia-inhibitory factor (LIF), another cytokine with similarities to IL-6, also shares its modest stimulatory effects on thrombocytopoiesis. Recombinant human granulocyte-macrophage colony-stimulating factor (rHuGM-CSF), now in clinical use to accelerate neutrophil recovery after bone marrow transplantation, had been previously shown to have low-level, MK-CSA in colony cultures of human megakaryocyte progenitor cells (67). However, no significant stimulatory effects on circulating platelet counts or platelet recovery in vivo have been observed clinically.

4.4. Recombinant Thrombopoietins

By the early 1990s, many investigators openly doubted the existence of a distinct thrombopoietin (TPO) or any other megakaryocyte lineage-specific humoral regulator. It was proposed that megakaryocytopoiesis and platelet production were regulated by a complex interplay of the known cytokines, interacting locally in the bone marrow microenvironment to control megakaryocyte development. Megakaryocyte mitotic expansion and early development were believed to be regulated by IL-3 (and perhaps GM-CSF) and postmitotic maturation by IL-6, IL-11, LIF, and even EPO. The MK-CSA of aplastic plasma was postulated to be the result of high levels of a combination of active, known cytokines (although increased circulating levels of IL-3 were never detected) *(34)* and the activity of plasma from acutely thrombocytopenic experimental animals in the rodent radioisotope-incorporation bioassay was all but disregarded as artifact. The failure of any laboratory to purify a "megakaryocyte colony-stimulating factor" or a "TPO" from active plasma was promulgated as evidence against their existence.

However, experimental data and a rationale contradicting this emerging "conventional wisdom" were available. As noted above, rIL-3 was the only recombinant cytokine exhibiting significant MK-CSA in human bioassays, yet endogenous IL-3 was virtually never detectable in vivo, either circulating in plasma or in the bone marrow, even at the transcriptional level. The lineage-nonspecific multifactor hypothesis for thrombopoietic regulation did not explain the high levels of MK-CSA detectable in infants with congenital amegakaryocytic thrombocytopenia, a single lineage defect. Accumulating clinical observations in humans demonstrated that the therapeutic administration of rHuEPO, rHuG-CSF, and rHuGM-CSF did not significantly increase the circulating platelet counts. In addition, both rHuIL-6 and rHuIL-11, although capable of modestly raising the platelet count in vivo, could only augment platelet production by 50–200%, far below the eight- to tenfold increases in platelet output observed clinically in the context of severe consumptive thrombocytopenia *(4)*. Furthermore, IL-6 stimulated physiologic responses characteristic of acute inflammation *(56,58)*, responses incompatible with a physiologic hemopoietic regulator and not observed clinically when platelet production was maximally stimulated.

A direct assessment of the relationship of the MK-CSA in human aplastic sera to the known recombinant cytokines was performed *(34)*. Neutralizing antisera to IL-3 and GM-CSF were found to abrogate completely the MK-CSA of rIL-3 and rGM-CSF, respectively. Notably, the MK-CSA in human lymphocyte-conditioned media was also neutralized completely by a combination of IL-3 and GM-CSF antisera. In contrast, the MK-CSA in aplastic human sera was absolutely resistant to neutralization by either anti-IL-3, anti-GM-CSF, or their combination, regardless of the titer of the neutralizing antisera. Since none of the other recombinant cytokines (i.e., rHuG-CSF, rHuEPO, rHuIL-6) could support human megakaryocyte colony growth, it was concluded that the activity in aplastic plasma (MK-CSA) was a unique thrombopoietic regulator *(34)*.

Other investigators continued to believe that a second unique thrombopoietic regulator governing postmitotic megakaryocyte maturation also existed *(15,36)*. This plasma regulator of postmitotic megakaryocyte development continued to be called TPO. TPO was believed to be distinct from the MK-CSA detected in plasma from patients and experimental animals with productive or amegakaryocytic thrombocytopenia.

However, there were a few hints that MK-CSA and TPO were not necessarily distinct from one another. Experimental work indicated that the effects of aplastic plasma on megakaryocytopoiesis in colony-forming assays were in part concentration-dependent. Low concentrations preferentially stimulated megakaryocyte size and ploidy, whereas higher concentrations stimulated mitotic expansion of the megakaryocyte progenitor cell, resulting in the development of large numbers of low-ploidy megakaryocytes *(68)*. A new thrombopoietic bioassay developed by Kuter and Rosenberg *(69)* (*see* Chapter 9) demonstrated that plasma from animals with acute consumptive thrombocytopenia stimulated increases in both the number and ploidy of megakaryocytes developing in liquid cultures of megakaryocyte-depleted bone marrow. Since this substance, called megapoietin, stimulated both megakaryocyte numbers as well as ploidy, one might infer effects of megapoietin on both the mitotic and endomitotic phases of megakaryocyte development.

Resolution of these issues required the cloning and functional characterization of these putative megakaryocytopoietic regulatory factors. As has been recently reviewed *(70)*, and is outlined elsewhere in this volume, identification and cloning of the receptor Mpl and the recognition that its expression was limited to cells and progenitors of megakaryocyte lineage set this process in motion. Several laboratories recognized simultaneously that the MK-CSA in aplastic plasma (obtained from irradiated animals) could be adsorbed and/or neutralized by exposure to Mpl. Furthermore and somewhat unexpectedly, it was also demonstrated that a soluble form of Mpl abrogated the stimulatory effect of normal sera on megakaryocyte polyploidization. Contradicting the prevailing theory of two-factor thrombocytopoietic regulation, it was proposed that Mpl ligand, TPO, and MK-CSA might be the same molecule *(71)*.

Cloning of the receptor Mpl led rapidly to the identification and cloning of its ligand. Two laboratories exploited affinity chromatography using immobilized Mpl to purify the ligand from aplastic plasma of irradiated pigs *(72)* and dogs *(73)* (*see* Chapter 8). Almost simultaneously, partial amino acid sequence was obtained for megapoietin, and it too was found to be identical to Mpl ligand *(74)* (*see* Chapter 9). Furthermore, megapoietin, initially detected in animals with acute consumptive thrombocytopenia, was also detected after chemotherapy-induced megakaryocyte aplasia *(75)*. Functional characterization of the ligand confirmed its stimulatory effects on both phases of megakaryocyte development: (1) megakaryocyte colony formation and (2) megakaryocyte endoreduplication and cytoplasmic maturation *(76)*. Plasma levels of Mpl ligand were found to be increased in the clinical contexts of both consumptive and productive thrombocytopenia *(76)*. Administered in vivo, Mpl ligand stimulated up to six- to sevenfold increases in the platelet counts in nonhuman primates *(77)*. The currently available evidence is compelling that both mitotic and postmitotic megakaryocyte development are regulated by the same molecule, Mpl ligand, which is the physiologic, lineage-restricted regulator of platelet production. MK-CSA and TPO are one in the same.

Yet incompletely defined is the relationship between Mpl ligand and the late-acting plasma factors TPN and TSF (*vide supra*), first identified using the platelet radioisotope-incorporation bioassay. Both TPN and TSF exhibit stimulatory effects on megakaryocyte size, ploidy, and maturation that are similar to those observed with Mpl ligand. However, neither of these factors has been demonstrated to support megakaryocyte colony development in vitro, whereas Mpl ligand is a potent stimulator of such colony growth. In addition, the study by Nichol et al. *(76)* indicates that Mpl ligand does not synergize with IL-3 in supporting megakaryocyte colony development in human cells in vitro, although such in vitro synergy is characteristic of both TPN and TSF. In contrast, a recent report by McDonald et al. *(78)* suggests that TSF and Mpl ligand are identical based on functional equivalence, similarity in molecular masses, the stimulation of cell lines expressing Mpl ligand by TSF, and the detection of Mpl ligand mRNA by polymerase chain reaction (PCR) in HEK cells. Whether there are additional, late-acting physiologic regulators of platelet production yet to be cloned remains unknown.

5. Summary

Over 35 years of research have resulted in the identification and cloning of a humoral, hemopoietic growth factor specific for cells of the megakaryocytic lineage, Mpl ligand. This single factor exhibits concentration-dependent effects on both the proliferative and postmitotic phase of megakaryocyte development. This single observation seriously weakens the two-regulator hypothesis for megakaryocytopoiesis developed during the past 20 years. However, the availability of this cytokine presents new and exciting opportunities to dissect the true physiological control systems governing megakaryocytopoiesis and platelet production. More importantly, perhaps, is that the clinical availability of a physiologic platelet-stimulatory cytokine is likely to find application in treating a wide range of human diseases.

References

1. Mazur EM. Megakaryocytopoiesis and platelet production: a review. *Exp Hematol*. 1987; 15: 340-350.
2. Mazur EM. Megakaryocytes and megakaryocytopoiesis. In: Loscalzo J, Schafer A (eds). *Thrombosis and Hemorrhage*. Oxford: Blackwell Scientific Publications; 1994: 161-194.
3. Harker LA. Kinetics of thrombopoiesis. *J Clin Invest*. 1968; 47: 458–465.
4. Harker LA, Finch CA. Thrombokinetics in man. *J Clin Invest*. 1969; 48: 963–974.
5. Ebbe S, Stohlman F Jr, Donovan J, Overcash J. Megakaryocyte maturation rate in thrombocytopenic rats. *Blood*. 1968; 32: 383–392.
6. Keleman E, Cserhati I, Tanos B. Demonstration and some properties of human thrombopoietin in thrombocythemic sera. *Acta Haematol*. 1958; 20: 350–353.
7. Odell TT Jr, McDonald TP, Detwiler TC. Stimulation of platelet production by serum of platelet-depleted rats. *Proc Soc Exp Biol Med*. 1961; 108: 428–431.
8. Spector B. In vivo transfer of a thrombopoietic factor. *Proc Soc Exp Biol Med*. 1961; 108: 146–149.
9. Odell TT Jr, McDonald TP, Howsden FL. Native and foreign stimulators of platelet production. *J Lab Clin Med*. 1964; 64: 418–424.
10. Evatt BL, Levin J. Measurement of thrombopoiesis in rabbits using selenomethionine. *J Clin Invest*. 1969; 48: 1615–1626.
11. Harker LA. Regulation of thrombopoiesis. *Am J Physiol*. 1970; 218: 1376–1380.
12. Penington DG. Isotope bioassay for "thrombopoietin." *Br Med J*. 1970; 1: 606–608.
13. Shreiner DP, Levin J. Detection of thrombopoietic activity in plasma by stimulation of suppressed thrombopoiesis. *J Clin Invest*. 1970; 49: 1709–1713.
14. Levin J, Evatt BL, Shreiner DP. Measurement of plasma thrombopoiesis stimulating activity using selenomethionine-[75]Se: studies in rabbits and mice. In: Baldini MG, Ebbe S (eds). *Platelets: Production, Function, Transfusion, and Storage*. New York: Grune & Stratton; 1974: 63–72.

15. McDonald TP. Thrombopoietin: its biology, purification, and characterization. *Exp Hematol.* 1988; 16: 201–205.

16. Ebbe S. Biology of megakaryocytes. *Prog Hemost Thromb.* 1976; 3: 787–795.

17. Schulman I, Pierce M, Lukens A, Currimbhoy Z. Studies on thrombopoiesis. I. A factor in normal human plasma required for platelet production; chronic thrombocytopenia due to its deficiency. *Blood.* 1960; 16: 943–957.

18. Miura M, Koizumi S, Nakamura K, et al. Efficacy of several plasma components in a young boy with chronic thrombocytopenia and hemolytic anemia who responds repeatedly to normal plasma infusions. *Am J Hematol.* 1984; 17: 307–319.

19. McDonald TP, Green D. Demonstration of thrombopoietin production after plasma infusion in a patient with congenital thrombopoietin deficiency. *Thromb Haemost.* 1977; 37: 577–579.

20. McDonald TP, Miura M, Koizumi S. Thrombopoietin production in a patient with chronic thrombocytopenia after plasma infusion. *Thromb Res.* 1985; 38: 353–359.

21. Adams WH, Liu YK, Sullivan LW. Humoral regulation of thrombopoiesis in man. *J Lab Clin Med.* 1978; 91: 141–147.

22. Shreiner DP, Weinberg J, Enoch D. Plasma thrombopoietic activity in humans with normal and abnormal platelet counts. *Blood.* 1980; 56: 183–188.

23. Kellar KL, Evatt BL, McGrath CR, Ramsey RB. Stimulation of DNA synthesis in megakaryocytes by thrombopoietin *in vitro.* In: Evatt BL, Levine RF, Williams N (eds). *Megakaryocyte Biology and Precursors: In Vitro Cloning and Cellular Properties.* New York: Elsevier; 1981: 21–34.

24. Greenberg SM, Kuter DJ, Rosenberg RD. In vitro stimulation of megakaryocyte maturation by megakaryocyte stimulatory factor. *J Biol Chem.* 1987; 262: 3269–3277.

25. Hill RJ, Leven RM, Levin FC, Levin J. The effect of partially purified thrombopoietin on guinea pig megakaryocyte ploidy in vitro. *Exp Hematol.* 1989; 17: 903–907.

26. Long MW, Williams N, Ebbe S. Immature megakaryocytes in the mouse: physical characteristics, cell cycle status, and in vitro responsiveness to thrombopoietic stimulatory factor. *Blood.* 1982; 59: 569–575.

27. Leven RM, Yee MK. Megakaryocyte morphogenesis stimulated in vitro by whole and partially fractionated thrombocytopenic plasma: a model system for the study of platelet formation. *Blood.* 1987; 69: 1046–1052.

28. Hill R, Levin J. Partial purification of thrombopoietin using lectin chromatography. *Exp Hematol.* 1986; 14: 752–759.

29. Tagrien G, Rosenberg RD. Purification and properties of a megakaryocyte stimulatory factor present both in the serum-free conditioned medium of human embryonic kidney cells and in thrombocytopenic plasma. *J Biol Chem.* 1987; 262: 3262–3268.

30. Nakeff A, Daniels-McQueen S. In vitro colony assay for a new class of megakaryocyte precursor: colony-forming unit megakaryocyte (CFU-M) (39265). *Proc Soc Exp Biol Med.* 1976; 151: 587–590.

31. Mazur EM, Hoffman R, Bruno E. Regulation of human megakaryocytopoiesis. An in vitro analysis. *J Clin Invest.* 1981; 68: 733–741.

32. Mazur EM, Hoffman R, Chasis J, Marchesi S, Bruno E. Immunofluorescent identification of human megakaryocyte colonies using an antiplatelet glycoprotein antiserum. *Blood.* 1981; 57: 277–286.

33. Levin J, Levin FC, Hull DF III, Penington DG. The effects of thrombopoietin on megakaryocyte-CFC, megakaryocytes, and thrombopoiesis: with studies of ploidy and platelet size. *Blood.* 1982; 60: 989–998.

34. Mazur EM, Cohen JL, Newton J, et al. Human serum megakaryocyte colony-stimulating activity appears to be distinct from interleukin-3, granulocyte–macrophage colony-stimulating factor, and lymphocyte-conditioned medium. *Blood.* 1990; 76: 290–297.

35. Vainchenker W, Bouguet J, Guichard J, Breton-Gorius J. Megakaryocyte colony formation from human bone marrow precursors. *Blood.* 1979; 54: 940–945.

36. Hill RJ, Levin J, Levin FC. Correlation of in vitro and in vivo biological activities during the partial purification of thrombopoietin. *Exp Hematol.* 1992; 20: 354–360.
37. Williams N, Eger RR, Jackson HM, Nelson DJ. Two-factor requirement for murine mega-karyocyte colony formation. *J Cell Physiol.* 1982; 110: 101–104.
38. Hoffman R, Mazur E, Bruno E, Floyd V. Assay of an activity in the serum of patients with disorders of thrombopoiesis that stimulates formation of megakaryocytic colonies. *N Engl J Med.* 1981; 305: 533–538.
39. Homans AC, Cohen JL, Mazur EM. Defective megakaryocytopoiesis in the syndrome of throm-bocytopenia with absent radii. *Br J Haematol.* 1988; 70: 205–210.
40. Mazur EM, de Alarcon P, South K, Miceli L. Human serum megakaryocyte colony-stimulating activity increases in response to intensive cytotoxic chemotherapy. *Exp Hematol.* 1984; 12: 624–628.
41. de Alarcon P, Schmieder JA. Megakaryocyte colony stimulating activity (Mk-CSA) in serum from patients undergoing bone marrow transplantation. In: Levine RF, Williams N, Levin J, Evatt BL (eds). *Megakaryocyte Development and Function.* New York: Alan R. Liss; 1986: 335–340.
42. Hoffman R, Yang HH, Bruno E, Straneva JE. Purification and partial characterization of a megakaryocyte colony-stimulating factor from human plasma. *J Clin Invest.* 1985; 75: 1174–1182.
43. Mazur EM, South K. Human megakaryocyte colony-stimulating factor in sera from aplastic dogs: partial purification, characterization, and determination of hematopoietic cell lineage specificity. *Exp Hematol.* 1985; 13: 1164–1172.
44. Hoffman R. Regulation of megakaryocytopoiesis. *Blood.* 1989; 74: 1196–1212.
45. Segal GM, Stueve T, Adamson JW. Analysis of murine megakaryocyte colony size and ploidy: effects of interleukin-3. *J Cell Physiol.* 1988; 137: 537–544.
46. Mazur EM, Cohen JL, Bogart L, et al. Recombinant gibbon interleukin-3 stimulates mega-karyocyte colony growth in vitro from human peripheral blood progenitor cells. *J Cell Physiol.* 1988; 136: 439–446.
47. Williams N, Sparrow R, Gill K, Yasmeen D, McNiece I. Murine megakaryocyte colony stimu-lating factor: its relationship to interleukin 3. *Leukemia Res.* 1985; 9: 1487–1496.
48. Donahue RE, Seehra J, Metzger M, et al. Human IL-3 and GM-CSF act synergistically in stimulating hematopoiesis in primates. *Science.* 1988; 241: 1820–1823.
49 Lindemann A, Ganser A, Herrmann F, et al. Biologic effects of recombinant human interleukin-3 in vivo. *J Clin Oncol.* 1991; 9: 2120–2127.
50. Warren MK, Conroy LB, Rose JS. The role of interleukin 6 and interleukin 1 in megakaryocyte development. *Exp Hematol.* 1989; 17: 1095–1099.
51. Navarro S, Debili N, Le Couedic JP, et al. Interleukin-6 and its receptor are expressed by human megakaryocytes: in vitro effects on proliferation and endoreplication. *Blood.* 1991; 77: 461–471.
52. Banu N, Fawcett J, Williams N De Giorgio T, Withy R. Tissue sources of murine megakaryo-cyte potentiator: biochemical and immunological studies. *Br J Haematol.* 1990; 75: 313–318.
53. Leven RM, Rodriguez A. Immunomagnetic bead isolation of megakaryocytes from guinea-pig bone marrow: effect of recombinant interleukin-6 on size, ploidy, and cytoplasmic fragmenta-tion. *Br J Haematol.* 1991; 77: 267–273.
54. Ravid K, Kuter DJ, Rosenberg RD. rmIL-6 stimulates the transcriptional activity of the rat PF4 gene. *Exp Hematol.* 1995; 23: 397–401.
55. Williams N, De Giorgio T, Banu N, Withy R, Hirano T, Kishimoto T. Recombinant interleukin 6 stimulates immature murine megakaryocytes. *Exp Hematol.* 1990; 18: 69–72.
56. Asano S, Okano A, Ozawa K, et al. In vivo effects of recombinant human interleukin-6 in primates: stimulated production of platelets. *Blood.* 1990; 75: 1602–1605.
57. Stahl CP, Zucker-Franklin D, Evatt BL, Winton EF. Effects of human interleukin-6 on mega-karyocyte development and thrombocytopoiesis in primates. *Blood.* 1991; 78: 1467–1475.

58. Mayer P, Geissler K, Valent P, Ceska M, Bettelheim P, Liehl E. Recombinant human interleukin 6 is a potent inducer of the acute phase response and elevates the blood platelets in nonhuman primates. *Exp Hematol.* 1991; 19: 688–696.

59. Geissler K, Valent P, Bettelheim P, et al. In vivo synergism of recombinant human interleukin-3 and recombinant human interleukin-6 on thrombopoiesis in primates. *Blood.* 1992; 79: 1155–1160.

60. Ishibashi T, Shikama Y, Kimura H et al. Thrombopoietic effects of interleukin-6 in long-term administration in mice. *Exp Hematol.* 1993; 21: 640–646.

61. Straneva JE, van Besien KW, Derigs G, Hoffman R. Is interleukin 6 the physiological regulator of thrombopoiesis? *Exp Hematol.* 1992; 20: 47–50.

62. Hill RJ, Warren K, Levin J. Does interleukin 6 mediate the thrombopoietic response to acute immune thrombocytopenia? *Exp Hematol.* 1990; 18: 704 (abstract).

63. Tefferi A, Ho TC, Ahmann GJ, Katzmann JA, Greipp PR. Plasma interleukin-6 and C-reactive protein levels in reactive versus clonal thrombocytosis. *Am J Med.* 1994; 97: 374–378.

64. Bruno E, Briddell RA, Cooper RJ, Hoffman R. Effects of recombinant interleukin 11 on human megakaryocyte progenitor cells. *Exp Hematol.* 1991; 19: 378–381.

65. Neben TY, Loebelenz J, Hayes L, et al. Recombinant human interleukin-11 stimulates megakaryocytopoiesis and increases peripheral platelets in normal and splenectomized mice. *Blood.* 1993; 81: 901–908.

66. Du XX, Neben T, Goldman S, Williams DA. Effects of recombinant human interleukin-11 on hematopoietic reconstitution in transplant mice: acceleration of recovery of peripheral blood neutrophils and platelets. *Blood.* 1993; 81: 27–34.

67. Mazur EM, Cohen JL, Wong GG, Clark SC. Modest stimulatory effect of recombinant human GM-CSF on colony growth from peripheral blood human megakaryocyte progenitor cells. *Exp Hematol.* 1987; 15: 1128–1133.

68. Arriaga M, South K, Cohen JL, Mazur EM. Interrelationship between mitosis and endomitosis in cultures of human megakaryocyte progenitor cells. *Blood.* 1987; 69: 486–492.

69. Kuter DJ, Rosenberg RD. Appearance of a megakaryocyte growth-promoting activity, megapoietin, during acute thrombocytopenia in the rabbit *Blood.* 1994; 84: 1464–1472.

70. Kaushansky K. Thrombopoietin: the primary regulator of platelet production. *Blood.* 1995; 86: 419–431.

71. Wendling F, Maraskovsky E, Debill N, et al. c-Mpl ligand is a humoral regulator of megakaryocytopoiesis. *Nature.* 1994; 369: 571–574.

72. de Sauvage FJ, Hass PE, Spencer SD, et al. Stimulation of megakaryocytopoiesis and thrombopoiesis by the c-Mpl ligand. *Nature.* 1994; 369: 533–538.

73. Bartley TD, Bogenberger J, Hunt P, et al. Identification and cloning of a megakaryocyte growth and development factor that is a ligand for the cytokine receptor Mpl. *Cell.* 1994; 77: 1117–1124.

74. Kuter DJ, Beeler DL, Rosenberg RD. The purification of megapoietin: a physiological regulator of megakaryocyte growth and platelet production. *Proc Natl Acad Sci USA.* 1994; 91: 11,104–11,108.

75. Kuter DJ, Rosenberg RD. The reciprocal relationship of thrombopoietin (c-Mpl ligand) to changes in the platelet mass during busulfan-induced thrombocytopenia in the rabbit. *Blood.* 1995; 85: 2720–2730.

76. Nichol JL, Hokom MM, Hornkohl A, et al. Megakaryocyte growth and development factor. Analyses of in vitro effects on human megakaryopoiesis and endogenous serum levels during chemotherapy induced thrombocytopenia. *J Clin Invest.* 1995; 95: 2973–2978.

77. Farese AM, Hunt P, Boone T, MacVittie TJ. Recombinant human megakaryocyte growth and development factor stimulates thrombocytopoiesis in normal nonhuman primates. *Blood.* 1995; 86: 54–59.

78. McDonald TP, Wendling F, Vainchenker W, et al. Thrombopoietin from human embryonic kidney cells is the same factor as c-mpl-ligand. *Blood.* 1995; 85: 292–294.

6

The Murine Myeloproliferative Leukemia Virus MPLV, *v-mpl* Oncogene, and *c-mpl*

Françoise Wendling and Sylvie Gisselbrecht

1. Discovery

Retroviruses are widely distributed in avian and mammalian species, and cause various neoplastic diseases in their natural hosts *(1)*. The type-C retroviruses have been divided into two groups: the replication-competent viruses that induce leukemia in susceptible animals with a long latent period, but do not usually transform cells in culture; and the acute replication-defective viruses that cause leukemias or sarcomas in animals within a few days or weeks and transform cells in vitro. The replication defect of acutely transforming retroviruses is due to the loss of structural genes that are essential for replication with the concomitant acquisition of altered cellular sequences, named oncogenes, responsible for cell transformation. As a result of their replication defectiveness, acutely transforming retroviruses require a helper replication-competent virus for propagation.

The murine myeloproliferative leukemia virus is a naturally occurring leukemogenic, nonsarcomagenic retrovirus complex composed of a replication-competent Friend murine leukemia virus (F-MuLV) and a defective component myeloproliferative leukemia virus (MPLV) carrying the *v-mpl* oncogene. The original viral complex was isolated from a mouse inoculated at birth with a molecular clone of the helper virus contained in the Friend virus complex (F-MuLV 57) *(2)*. Although several clonal F-MuLV isolates can induce a rapid erythroblastosis in inoculated neonatal Swiss or Balb/c mice, the DBA/2 strain of mice was found to be resistant to this early syndrome *(3,4)*. Nevertheless, these mice developed various types of hemopoietic malignancies (myeloblastic, lymphoblastic, or erythroblastic leukemias) after a long latent interval of 7–12 months. In general, these leukemias were associated with a more or less severe anemia *(5,6)*.

Of 238 DBA/2 mice inoculated at birth with F-MuLV 57, a unique mouse developed, 7 months after infection, an hepatosplenomegaly unusually accompanied by polycythemia (hematocrit > 70%). Cell-free homogenate prepared from the spleen of this mouse, blood plasma, or the supernatant medium from an in vitro permanent erythroblastic cell line derived from the leukemic spleen was used to inoculate adult mice of various strains. Virtually 100% of the infected mice developed a fatal acute myeloproliferative syndrome within 1–3 months with similar clinical pictures no matter which strain of mice

From: *Thrombopoiesis and Thrombopoietins: Molecular, Cellular, Preclinical, and Clinical Biology*
Edited by: D. J. Kuter, P. Hunt, W. Sheridan, and D. Zucker-Franklin Humana Press Inc., Totowa, NJ

was studied. Because several hemopoietic lineages were obviously involved in this disease, the virus isolate was named myeloproliferative leukemia virus or MPLV.

2. In Vivo and In Vitro Pathological Effects of MPLV

2.1. Leukemogenic Properties

MPLV virus stocks (F-MuLV + MPLV) prepared from the original erythroblastic cell line were used to determine the in vivo biological properties of MPLV. After inoculation of adult mice, the MPLV complex provokes an acute myeloproliferative syndrome characterized by hepatosplenomegaly, polycythemia, and myelemia. There is no thymic or lymph node involvement (2). The spleen and liver are massively infiltrated with maturing precursor cells of the granulocytic, erythrocytic, and megakaryocytic lineages. The blood is also extensively invaded by morphologically normal polymorphonuclear cells, erythroblasts, and platelets (2). A characteristic property of the MPLV-transforming activity is rapid suppression of growth factor requirements for in vitro colony formation for a large spectrum of committed as well as multipotential progenitor cells. A primary manifestation of viral infection is a switch to erythropoietin (EPO) independence of the erythroid colony-forming cell (E-CFC) population, which is complete in the spleen after 6 days of infection (7). The size and the maturation level of the autonomously growing E-CFC were similar to those of control uninfected cultures stimulated with EPO. Addition of neutralizing anti-EPO antibodies to the culture medium did not impair the spontaneous colony formation, ruling out a possible stimulating effect of EPO present or secreted in the culture medium.

Methylcellulose serum-deprived cultures were used to investigate the effects of MPLV on the early erythroid progenitors, erythroid burst-forming cells (E-BFC). Well-hemoglobinized pure and mixed erythroid colonies developed without the addition of either interleukin (IL)-3 or EPO to the culture media. Again, addition of neutralizing antibodies specific for these cytokines did not reduce the number of spontaneous erythroid colonies and did not impair their terminal maturation. In contrast to uninfected control cultures, the majority of erythroid colonies developing within 7 days in MPLV-infected cultures were large and contained numerous polyploid megakaryocytes intermingled with the erythroblasts. Approximately 12% of the colonies revealed three or more lineages of differentiation (7). Further studies documented that in vivo infection of mice with MPLV also resulted in the spontaneous colony formation by a variety of myeloid progenitors, namely, granulocyte-macrophage colony-forming cells (GM-CFC), granulocyte colony-forming cells (G-CFC), megakaryocyte progenitors (MK-CFC), and multipotent progenitors (GEMM-CFC), and that this was owing to infection of these progenitors and not a consequence of paracrine secretion of soluble colony-stimulating factors by accessory cells (8). Together, these studies demonstrated that progenitor cells from various myeloid lineages infected with MPLV acquire the ability to proliferate and terminally differentiate, independently of the signals normally provided by cytokine receptor activation.

2.2. In Vitro Transformation Properties of MPLV

2.2.1. Myeloid Cell Transformation

To analyze the in vitro transforming properties of MPLV, a high-titer, helper-free MPLV stock was produced using the psi-CRE packaging cell line (9). Bone marrow cells enriched in high proliferative potential primitive progenitors by pretreatment of the donors with 5-fluorouracil (5-FU) were infected with nonreplicating MPLV. A 2-hour incubation of the bone marrow cells with infectious supernatant was sufficient to induce autonomous colony formation of approximately 30% of the colony-forming cells present in the preparation *(10)*. Cytologically, approximately half of these spontaneous colonies were composed of only either granulocytes, megakaryocytes, or erythroblasts, whereas the remainder were mixed colonies containing two (usually erythroblasts and megakaryocytes), three, or more lineages of differentiation. These results provided the formal proof that MPLV was the viral component responsive for the in vitro myeloproliferative effects of the viral complex.

Another interesting property of MPLV is its ability to generate immortalized, factor-independent, hemopoietic cell lines *(10)*. Bone marrow cells infected in vitro with helper-free MPLV and cultured in liquid medium containing fetal bovine serum (FBS), but no added cytokines, produced rapidly dividing cell populations growing in suspension. These nonadherent cell populations could be serially transferred and established as permanent cell lines. Cytological examination of cells from early passages revealed maturing and terminally differentiating erythroblasts, megakaryocytes, polymorphonuclear cells, and mast cells, in association with immature blast cells. On continuous passages, the majority of the cell lines evolved toward a more restricted phenotype, which remained stable over several months. Diverse, spontaneously differentiating, immortalized cell lines were isolated. Of note, of 24 cell lines, 5 contained only megakaryocytic cells and 5 were composed of megakaryocytes and hemoglobinized erythroblasts. Analyses of proviral-cell DNA junctions were performed to determine the clonal status of these cell lines. Although cultures were polyclonal early after infection, only one or a few major proliferating clones were detected after a few weeks. Hemopoietic cells immortalized by MPLV were not tumorigenic in syngeneic mice. However, after several months in culture, more than 50% of the cell lines produced subcutaneous tumors, indicating that additional genetic events were necessary to reach full malignancy *(10)*. Based on these data, the authors have suggested that these permanent cell lines probably arose from the outgrowth of a single or few MPLV-immortalized multipotential stem cells.

2.2.2. Lymphoid Cell Transformation

After in vivo inoculation of the MPLV complex, no gross abnormalities of the lymphoid system were described. However, it was demonstrated that helper-free MPLV could infect and transform pre-B, pro-B, and early pro-B cells in a long-term lymphoid culture system *(11)*. Although this event was quite rare, transformed B-progenitor cells and precursor cells were able to grow in liquid medium in the absence of a feeder-cell layer or exogenous growth factors, and were able to form colonies spontaneously in semisolid medium. Transformed B-cells appeared to be blocked at an early stage and did not acquire the markers of terminal B-cell differentiation. Evidence was presented that the autonomous proliferation of MPLV-infected B-cells resulted from an autocrine stimulation by an unidentified growth factor *(11)*.

Several experiments were performed to learn whether helper-free MPLV could also transform T-lymphoid progenitors. An increase in proliferation of MPLV-infected, purified thymic T-cells was observed, but isolation of continuous cell lines was unsuccessful (Ezine S, personal communication).

3. Molecular Characterization of the Oncogene *v-mpl*

3.1. Virological Studies

Initial analyses of the MPLV virus stock (or MPLV virus isolate) complex by nucleic acid hybridization revealed that it contained two distinct retroviral genomes: one was the parental F-MuLV, and the second was a smaller retroviral component now designated as MPLV. Comparison of viral RNA species expressed in F-MuLV or in F-MuLV + MPLV-producing cells by Northern blot analysis showed that MPLV was 0.8 kb shorter than F-MuLV and that a deletion had probably occurred in the MPLV *env* gene *(12)*. By comparing the restriction endonuclease cleavage maps of MPLV and F-MuLV genomes, it was confirmed that MPLV was derived from F-MuLV, had conserved the F-MuLV *gag* and *pol* regions, and was deleted and rearranged in the *env* region *(12)*. MPLV has no transforming properties on NIH 3T3 fibroblasts *(2)*; however, because the MPLV titer in the original isolate was approximately equivalent to that of F-MuLV, nonproducer cells containing MPLV were successfully derived by the technique of limiting dilution and single-cell cloning. When these cells were superinfected with various replicating helper viruses, culture supernatants reproduced the same in vivo syndrome as the original isolate *(12)*.

3.2. Molecular Cloning of MPLV and Gene Structure

The sequence responsible for MPLV pathogenicity was identified by structural and sequence analyses of a full-length molecular MPLV clone. This clone was isolated from a genomic library prepared from a nonproducer Mus Dunni epithelial clone containing a single infectious copy of the provirus *(10)*. Comparison of the MPLV and F-MuLV *env* gene nucleotide sequences confirmed the data obtained by restriction endonuclease analysis, and suggested that MPLV had acquired nonviral sequences. The MPLV *env* gene is composed of 191 nucleotides of the 5'-end of the F-MuLV *env* gene, followed by 110 nucleotides derived from a central portion of the F-MuLV *env* gene and 623 nonviral nucleotides (Fig. 1). The MPLV *env* gene has an open reading frame with a coding capacity for an *env-mpl* fusion polypeptide of 284 amino acids containing one potential *N*-glycosylation site; it has an estimated molecular size of 31 kDa. This protein contains 64 amino acids derived from the amino terminus of the F-MuLV gp70, including the 34 amino acids of the GP70 signal peptide, 36 amino acids from a central region of the F-MuLV GP70, and 184 amino acids encoded by *v-mpl* sequences *(10)*. A hydrophobicity plot of the putative protein showed that the *v-mpl* domain contained a stretch of 22 hydrophobic amino acids, suggesting that the *env-mpl* fusion gene encoded a product presenting the features of a transmembrane protein. The mature protein would comprise an extracellular domain of 109 amino acids, a single transmembrane domain of 22 amino acids, and a cytoplasmic domain of 119 amino acids without consensus sequence for kinase activity *(10)*. A computer search of the MPLV-specific sequence showed no homologies with viral genes, oncogenes of any known retroviruses, or any known genes *(10)*.

Most oncogenic sequences found in the genome of acutely transforming retroviruses are derived from cellular genes conserved through evolution *(13)*. The use of MPLV-derived specific probes revealed discrete bands in a Southern blot analysis of *Eco*RI-digested genomic DNA from mouse, rat, mink, dog, cow, and human *(10)*. It was, therefore, concluded that the nonviral sequences of MPLV proviral DNA were derived

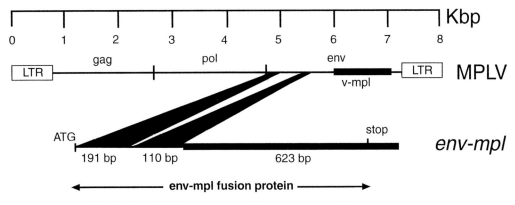

Fig. 1. The MPLV genome and the structure of the *env-mpl* gene.

from a cellular gene, and they were designated as *v-mpl*. Amino acid examination of the *v-mpl*-encoded extracellular domain showed a tyrosine-serine-alanine-tyrosine-serine sequence (or WSAWS motif) located close to the transmembrane domain. This motif is highly conserved in numerous receptors belonging to the cytokine receptor family, also called "the superfamily of hematopoietin receptors" *(14,15)*. Based on these observations, it was proposed that the *v-mpl* oncogene could correspond to the truncated and constitutively activated form of a member of this family *(10)*. The main MPLV characteristics are summarized in Table 1.

3.3. Characterization of v-mpl Domains Required for Pathogenicity

3.3.1. The env-mpl Gene Is Responsible for MPLV Pathogenicity

To formally demonstrate that the *env-mpl* fusion product was responsible for the biological properties of MPLV, a fragment encompassing the whole *env-mpl* coding sequence was inserted into a modified Harvey-*ras*-derived retroviral vector *(16)*. This vector is devoid of *gag* and *pol* genes, but contains noncoding rat-derived VL 30 sequences placed under the control of Moloney-derived long terminal repeats (LTR). The Ha-MPLV construct was cotransfected with helper F-MuLV DNA into NIH 3T3 fibroblasts. Viral supernatants injected into adult mice reproduced the same acute pathology as the wild-type isolate *(16)*. This experiment provided two important observations: the LTR, *gag*, and *pol* regions of MPLV are not involved in MPLV pathogenicity, and the MPLV *env-mpl* gene is sufficient to induce all biological characteristics of the acute myeloproliferative syndrome.

3.3.2. The env-mpl Product Functions as a Constitutively Activated Cytokine Receptor

The cytoplasmic domain of cytokine receptors is devoid of tyrosine kinase activity, but interacts with cytoplasmic tyrosine kinases, among which is the JAK family of tyrosine kinases. Once activated upon ligand binding, these kinases induce tyrosine phosphorylation and activation of the "signal transducers and activators of transcription" (STAT) family of transcription factors (reviewed in ref. *17*). It was demonstrated that STAT proteins were constitutively activated in growth factor-dependent cell lines that became growth factor-independent after expressing the *env-mpl* oncogene, dem-

Table 1
Characteristics of MPLV (F-MuLV)

It was isolated from a DBA/2 mouse inoculated at birth with F-MuLV that developed a my-
    eloproliferative syndrome
It induces an acute leukemia in adult mice of various strains
It is not sarcomatogenic in vivo and does not morphologically transform NIH 3T3 fibroblasts
    in culture
It is replication-defective; the replication-defective transforming genome can be rescued
    from nonproductively MPLV-infected cells by infection with replication-competent leuke-
    mia viruses
It is highly related to F-MuLV; it has conserved the *gag* and *pol* genes, but has deletions and
    an insertion of a nonviral sequence in the *env* gene
It transforms in vivo and in vitro a broad range of hemopoietic progenitor cells, including
    erythroid, granulocyte-macrophage, megakaryocytes, mast cells, and multipotent colony-
    forming cells, and early B-lymphoid progenitor cells
MPLV-infected hemopoietic progenitors acquire the ability to proliferate and differentiate in
    the absence of growth factors
It probably immortalizes multipotent progenitor cells in vitro and generates permanent cul-
    tures of various hemopoietic lineages containing maturing elements
It contains the oncogene, *v-mpl*

onstrating that indeed the *env-mpl* product functions as a constitutively activated
receptor *(18)* (*see* Chapter 16).

3.3.3. The WS Motif Is Not Essential for MPLV Pathogenicity

Most cytokine receptors, including the *v-mpl* oncogene, display the conserved WS
motif close to the transmembrane domain *(14)*. This motif has an important role in
receptor function by creating receptor chain interactions. To test its role in *v-mpl*, a
short fragment of 24 nucleotides encompassing the WS region was deleted from *v-mpl*
by mutagenesis (ΔWS-MPLV), and supernatants of NIH 3T3 cells cotransfected with
ΔWS-MPLV and F-MuLV DNA were inoculated into mice. All infected mice developed typical MPLV disease. This indicates that the WS motif had no role in the constitutive activation of the *v-mpl* oncogene *(19)*.

3.3.4. Conserved Box 1 and Box 2 Within the v-mpl Cytoplasmic Region Are Necessary for MPLV Pathogenicity

A series of viral mutants deleted in the transmembrane or cytoplasmic *v-mpl* sequences were constructed to delimit the regions important for MPLV oncogenic properties *(19)*. Deletions of either the transmembrane *v-mpl* domain (22 amino-acid
deletion) or the cytoplasmic domain (119 amino-acid deletion) totally abolished the
in vivo and in vitro transforming properties of MPLV. Deletions suppressing the 50
C-terminal *v-mpl* amino acids did not impair MPLV pathogenicity. In contrast, deletions
encompassing 69 juxtamembrane *v-mpl* amino acids resulted in nonpathogenic viruses
(Fig. 2). It was observed that this particular *v-mpl* subdomain contains the two motifs
named box 1 and box 2, which are highly conserved in cytokine receptor-transducing
chains and are absolutely required for signal transduction *(20–22)*. These experiments
have shown that the transmembrane region of *v-mpl* and a limited region within the
v-mpl cytoplasmic domain are required to deliver constitutive proliferation signals.

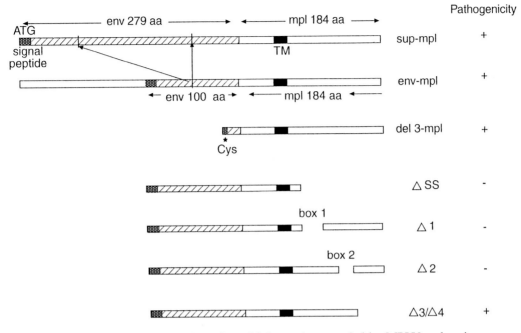

Fig. 2. Schematic representation of env-Mpl proteins encoded by MPLV and various constructs, and their pathogenicity. Hatched boxes represent GP70 *env* sequences; white boxes represent *mMpl*-specific sequence; black boxes indicate the transmembrane (TM) domain.

3.3.5. The Internal F-MuLV env Gene Deletion Is Not Involved in MPLV Pathogenicity

The *v-mpl* sequences had been transduced in frame into a partially deleted F-MuLV *env* gene. To determine whether this deletion had any role in *v-mpl* activation, the 179 amino acids deleted from the F-MuLV *env* gene were reinserted into the *env* portion of the *env-mpl* fusion gene (SUPMPLV virus in Fig. 2) *(23)*. The biological activity of the provirus construct was tested in mice inoculated with culture supernatants of NIH 3T3 cells cotransfected with SUPMPLV and F-MuLV DNAs. All mice developed the same pathology as mice infected with the MPLV original isolate. The only difference observed was a longer latency period for the appearance of the disease, most probably as a result of less efficient spread of SUPMPLV because of viral interference between SUPMPLV and the helper F-MuLV. Thus, the internal deletion of the *env* gene is not required for MPLV pathogenicity, but may favor efficient virus replication.

3.4. Isolation of a Natural MPLV Variant

In the course of these studies, a spontaneous mutant of SUPMPLV was isolated. The genomic size of the variant virus, DEL3MPLV, was even smaller than that of MPLV. Sequence analysis showed that DEL3MPLV had a deletion encompassing the 22 C-terminal amino acids of the F-MuLV-encoded peptide signal, all the F-MuLV *env* sequences, and 18 N-terminal amino acids of the *v-mpl* extracellular domain *(23)*. Thus, the extracellular domain of DEL3MPLV was composed of only 12 N-terminal residues of the *env* F-MuLV-encoded signal peptide fused in frame to 25 juxtamembrane amino acids of the *v-mpl* extracellular domain (Fig. 2). No mutations or deletions were found in the transmembrane and cytoplasmic domains. On in vivo inoculation in mice, the variant DEL3MPLV was as oncogenic as wild-type MPLV. A cysteine residue at position 3 in the uncleaved signal sequence was shown to be strictly required for pathogenicity. In addition, strong biochemical arguments demonstrated that unlike the *env-mpl* product of MPLV, the *del3-mpl* oncogene is constitutively activated by disulfide-linked homodimerization *(23)*. This observation is reminiscent of ligand-independent activation of the EPO receptor, where a point mutation in the receptor extracellular domain converts an arginine to a cysteine, leading to the formation of constitutively activated disulfide-linked homodimers *(24)*.

4. The Proto-Oncogene *c-mpl*
4.1. Human Proto-Oncogene Structure

Analyses of cellular DNA from humans and other mammals indicated that the cellular homolog of *v-mpl* is an evolutionarily conserved gene predominantly expressed in hemopoietic tissues *(10)*. To isolate a full-length *c-mpl* cDNA, mRNAs from various human leukemic cell lines were analyzed by Northern blot analysis. The proto-oncogene *c-mpl* was found to be expressed as a major 3.7-kb and a minor 2.8-kb mRNA in the erythrocytic and megakaryocytic HEL cell line. A cDNA library was prepared, and six clones hybridizing with *v-mpl* riboprobes were isolated *(25)*. Endonuclease restriction mapping allowed classification of these six clones into two groups: three clones had a *Pst*I site not present in the others (referred to as clone P), whereas three clones had a *Kpn*I site absent in clone P (referred to as clone K). Sequence analyses proved that all six clones had common sequences at their 5'-ends, but differences between clones P and K were observed at the 3'-ends. Clone P encodes a protein of 635 amino acids. Hydrophobicity plot analysis predicted two major hydrophobic regions, one corresponding to a signal peptide, and a second spanning amino acids 492–513, which was highly suggestive a transmembrane domain. From the deduced polypeptide sequences, it was demonstrated that the mature protein was composed of an extracellular domain of 463 amino acids, a transmembrane domain of 22 amino acids, and a cytoplasmic domain of 122 amino acids. Of note, polypeptides encoded by clones P and K had common extracellular and transmembrane domains, but their cytoplasmic domain differed following a stretch of nine common amino acids after the transmembrane region. Both proteins contain four potential *N*-glycosylation sites within the extracellular domain *(25)*.

Comparison of Mpl protein sequences with other members of the cytokine receptor superfamily definitively proved that the encoded product of the proto-oncogene *c-mpl* was an orphan cytokine receptor. In particular, the extracellular Mpl domain contains a duplication of the canonical hemopoietin receptor domain (HRD) characterized by a distinctive 200 amino-acid module composed of two pairs of cysteine residues and the WSXWS motif *(14,15)* (Fig. 3). Such a duplicate HRD is found in a minority of hemopoietin receptors, namely the common β chain of the receptors for IL-3, granulocyte-macrophage colony-stimulating factor (GM-CSF), and IL-5, and the leukemia-inhibitory factor (LIF) receptor α chain *(26,27)*. The first HRD of Mpl is longer than the second one, since it contains a 51 amino-acid hydrophobic subdomain not observed in other cytokine receptors. Importantly, the cytoplasmic region of human clone P shares 92% homology with the murine *v-mpl*, whereas clone K diverges completely *(25)*.

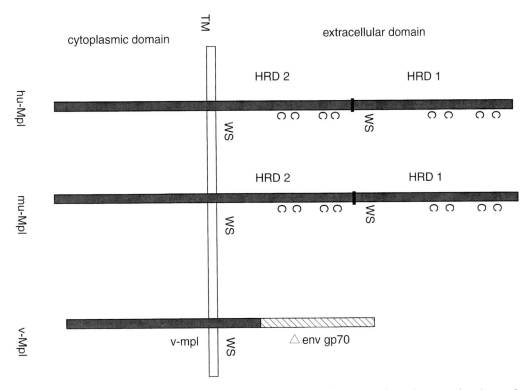

Fig. 3. Schematic structure of the proteins encoded by human and murine *c-mpl* and *v-mpl*. Abbreviations are transmembrane domain (TM); conserved cysteine residues (c); WSXWS motif (WS); hematopoietin receptor domain (HRD1 and HRD2).

4.2. Murine Proto-Oncogene Structure

Two independent groups have cloned murine *c-mpl* cDNA either from mouse spleen or fetal liver cDNA libraries *(28,29)*. The overall structure of murine MPL is similar to the human protein (Fig. 3). The open reading frame encodes a protein of 625 amino acids, including a signal peptide of 25 amino acids. The extracellular portion of the protein contains 465 amino acids arranged in two subdomains, each with the four conserved cysteines and a WSXWS motif. A hydrophobic transmembrane domain of 22 amino acids is followed by a 121-amino acid cytoplasmic domain. Clones containing a 257-nucleotide deletion were also isolated, which could correspond to a potentially soluble isoform of the receptor *(28,29)*. However, a soluble form of the receptor was not detected in the culture media of COS or Ba/F3 cells transfected with this cDNA form *(28)*. Sequence comparison of the murine and human (clone P) Mpl reveals that the two proteins share 81% peptide identity. Aligment of *v-mpl* and murine *c-mpl* coding regions showed that, of the 465 extracellular amino acids, only 43 are present in *v-mpl (29)* (Fig. 3). In addition, the only change between the two sequences is the absence of the two C-terminal amino acids in the extreme 3'-end of the oncogene *(29)*. These two missing amino acids are not involved in the transforming properties of *v-mpl (21)*.

4.3. Signal Transduction by Mpl Cytoplasmic Domain

The question of whether Mpl could deliver a proliferative signal after ligand binding was originally investigated by the construction of two types of chimeric receptors. In these experiments, extracellular and transmembrane domains of the human IL-4 receptor were fused to the cytoplasmic domain of Mpl *(28)*, and the extracellular domain of the human G-CSF receptor was fused to the transmembrane and the cytoplasmic domain of Mpl *(29)*. Chimeric constructs were electroporated into IL-3-dependent Ba/F3 cells. Transfected cells were able to proliferate in response to either IL-4 or G-CSF. These data further demonstrated that the Mpl cytoplasmic domain was competent for signaling. However, in the absence of a specific ligand, it was not clear whether Mpl corresponded to a β-transducing chain of a multimeric receptor or a receptor capable of both binding and signaling.

4.4. Structure and Transcription of the c-mpl Genomic Locus

4.4.1. Structure of the c-mpl Locus

Both the human and murine structures of the *c-mpl* genomic locus are characterized *(30,31)*. The *c-mpl* gene spans 17 kb (human) or 15 kb (murine) of genomic DNA. Both genes are organized into 12 exons, which closely conform to the pattern observed in several other cytokine receptor genes. The structure of exon–intron junctions is in agreement with the established consensus GT-AT rule *(32)*. Exon 1 contains a small untranslated region and the signal peptide sequence. The two extracellular cytokine receptor domains (HRD) are encoded by four exons of approx 150 bp each (exons 2–5 for HRD 1, and exons 6–9 for HRD 2). Unlike other cytokine receptors, Mpl contains a hydrophilic insert of 51 amino acids. As a result, exon 4 is 299 bp long. The transmembrane domain is encoded by exon 10, and the cytoplasmic domain by exons 11 and 12.

Shorter (257 bp truncated) human and murine *c-mpl* cDNAs encoding a potentially soluble Mpl protein have been described *(28,29)*. Such transcripts are generated by an alternative splicing of the primary transcript between exon 8 and exon 11, which suppresses exon 9 encoding the transmembrane domain and exon 10 *(30)*. Messenger RNA encoding detectable soluble forms of cytokine receptors have been described in several cases. Using an enzyme-linked immunoadsorbent assay (ELISA), the concentration of soluble Mpl in plasma of normal human donors was characterized *(33)*. In addition, *c-mpl* transcript heterogeneity within the first extracellular HRD of the murine receptor has been reported. These transcripts differ by the presence or absence of a 24-bp insertion at the end of exon 4 *(34)* or a 60 amino-acid deletion in exon 4 *(33)*.

Cloning analyses of human *c-mpl* cDNA described two colinear forms of Mpl (Mpl P and Mpl K), which differ in their cytoplasmic domain after a common stretch of nine amino acids *(23)*. Sequencing of intron 10, which immediately follows exon 9 encoding the transmembrane domain, showed that the Mpl K cytoplasmic region was produced by a read through in intron 10 leading to a premature termination of the transcript *(30)*. Whether the Mpl K form has a functional role is unknown.

4.4.2. c-mpl Promoter

The nucleotide sequence of the putative *c-mpl* promoter region is determined for both the human and the murine genes *(30,31)*. Promoters extend approximately 700

Table 2
Characteristics of the *c-mpl* Proto-Oncogene and Its Product

It was discovered as a truncated form transduced in the genome of MPLV

The human and murine *mpl* locus both span 17 kb; it is composed of 12 exons; the transmembrane domain is encoded by exon 9

Both murine and human *c-mpl* may encode a potential soluble form of the receptor

The *c-mpl* promoter sequence contains consensus binding sites for Sp1, GATA, and Ets transcription factors

c-mpl is located on human chromosome 1 at p34 and on the distal part of mouse chromosome 4

It encodes a transmembrane protein of 81 kDa

It is a member of the cytokine receptor superfamily

Its extracellular domain contains a duplication of the canonical HRD characterized by a 200 amino-acid module with two pairs of cysteines and the WSXWS motif

Its cytoplasmic domain is competent for signaling

Human and murine Mpl proteins have identical structure and share 86% overall homology

---

nucleotides 5' to the translation initiation codon and reveal extensive homology, especially in the proximal region. Both promoters are G + C rich and lack TATA or CAAT motifs. Several consensus motifs for Sp1, Ets, GATA, and NF-IL6 transcription factors are present. A combination of GATA and Ets consensus sequences in human and murine promoters has been demonstrated to be important for megakaryocytic-specific gene expression *(35–37)* (*see* Chapter 12). Promoter activity determined by CAT assays indicates that the *c-mpl* promoter is completely inactive in Jurkat T-lymphoid cells or carcinomatous HeLa cells, whereas high-level CAT activity was detected in the erythroid K562 and the erythrocytic and megakaryocytic HEL cells *(30)*. Interestingly, however, K562 cells, unlike HEL cells, do not express *c-mpl* mRNA *(32)*. This indicates that other sequences are important in the regulation of *c-mpl* gene expression.

4.5. Chromosomal Localization

In situ hybridization experiments located the murine *c-mpl* gene locus on the terminal region of chromosome 4 band D *(29)*. Analysis of a large panel of DNA samples prepared from interspecific mouse back-crosses confirmed this location and showed that *c-mpl* was linked to several markers of mouse chromosome 4 with the following gene order: interferon-β (IFNβ), c-*jun*, *c-mpl*, and serotonin receptor (Htr-1d).

The human *c-mpl* locus was localized on chromosome 1p34 by *in situ* hybridization *(38)*. This band contains the loci of several genes involved in the regulation of hemopoiesis, such as the lymphocyte-specific protein tyrosine kinase (LCK), and the TAL/SCL gene encoding a transcription factor involved in the erythrocytic and megakaryocytic differentiation. The main characteristics of the *c-mpl* proto-oncogene and its product are summarized in Table 2.

5. Conclusions

Some growth factor receptors have been isolated from oncogenes carried in the genome of acutely transforming, naturally occurring retroviruses. For example, *v-fms* corresponds to an altered receptor for the macrophage colony-stimulating factor (M-CSF or CSF-1), *v-kit* codes for a truncated version of the stem cell factor (SCF) receptor, and *v-erb* encodes a truncated form of the epidermal growth factor (EGF) receptor *(33–41)*. These receptors are members of the tyrosine kinase receptor family, and structural alterations of these genes constitutively activate their kinase function leading to cell transformation and tumorigenesis. In contrast, *c-mpl* belongs to the hematopoietin receptor superfamily, a receptor family with no intrinsic kinase activity. Therefore, understanding of the possible mechanism(s) responsible for the constitutive activation of the truncated *v-mpl* form remains an important question. A recurring theme within hematopoietin receptors is receptor subunit oligomerization either through the formation of heterooligomers or homodimers. Oligomerization follows ligand binding and stimulates receptor activation. A constitutively activated and oncogenic form of the EPO receptor has been isolated. It was demonstrated that changing a specific asparagine to a cysteine residue within the extracellular domain leads to disulfide-linked homodimerization and constitutive activation *(42)*. A recent report has shown that substitution of cysteines for residues in a region of homology between the extracellular domains of the Mpl and the EPO receptors causes ligand-independent Mpl activation via disulfide-linked homodimerization *(43)*. Moreover, FDC-P1 and Ba/F3 cells expressing active receptor mutants no longer require exogenous growth factors and are tumorigenic in transplanted mice *(43)*. These results imply that the normal process of ligand-induced Mpl activation occurs through receptor homodimerization. In addition, a similar conclusion was drawn from the results obtained with the MPLV variant del3MPLV *(23)*. However, the actual mechanism(s) of activation of the *env-mpl* fusion protein in MPLV remains unclear, since no cysteine residue exists within the extracellular domain of the mature protein. Thus, the activation of *v-mpl* might result from conformational changes within the N-terminus of two fusion proteins, leading to intermolecular interactions and constitutive tyrosine phosphorylation of downstream signaling pathways.

Acknowledgments

Work from our laboratories is supported by the Institut National de la Santé et de la Recherche Médicale (INSERM), the Ligue Nationale Contre le Cancer (LNCC), l'Association de Recherche contre le Cancer (ARC), and the Fondation pour la Recherche Médicale (FRM).

References

1. Reddy EP, Salka AM, Curran T (eds). *The Oncogene Handbook.* New York: Elsevier; 1988.
2. Wendling F, Varlet P, Charon M, Tambourin P. MPLV: a retrovirus complex inducing an acute myeloproliferative leukemic disorder in adult mice. *Virology.* 1986; 149: 242–246.
3. Troxler DH, Scolnick EM. Rapid leukemia induced by cloned Friend strain of replicating murine type-C virus. Association with induction of xenotropic-related RNA sequences contained in spleen focus-forming virus. *Virology.* 1978; 85: 17–27.
4. Ruscetti S, Davis L, Field J, Oliff A. Friend murine leukemia virus-induced leukemia is associated with the formation of mink cell focus-inducing viruses and is blocked in mice expressing endogenous mink cell focus-inducing xenotropic viral envelope genes. *J Exp Med.* 1981; 154: 907–920.
5. Shibuya T, Mak TW. Host control of susceptibility to erythroleukemia and to the types of leukemia induced by Friend murine leukemia virus: Initial and late stages. *Cell.* 1982; 31: 483–493.
6. Wendling F, Fichelson S, Heard JM, Gisselbrecht S, Varet B, Tambourin P. Induction of myeloid leukemias in mice by biologically cloned ecotropic F-MuLV. In: Scolnick EM, Levine AJ (eds). *Tumor Viruses and Differentiation.* New York: Liss; 1983: 357–362.

7. Wendling F, Penciolelli JF, Charon M, Tambourin P. Factor-independent erythropoietic progenitor cells in leukemia induced by the myeloproliferative leukemia virus. *Blood.* 1989; 73: 1161–1167.

8. Wendling F, Vigon I, Souyri M, Tambourin P. Myeloid progenitor cells transformed by the myeloproliferative leukemia virus proliferate and differentiate in vitro without addition of growth factors. *Leukemia.* 1989; 3: 475–480.

9. Danos O, Mulligan RC. Safe and efficient generation of recombinant retroviruses with amphotropic and ecotropic host range. *Proc Natl Acad Sci USA.* 1988; 85: 6460–6464.

10. Souyri M, Vigon I, Penciolelli JF, Heard JM, Tambourin P, Wendling F. A putative truncated cytokine receptor gene transduced by the myeloproliferative leukemia virus immortalizes hematopoietic progenitors. *Cell.* 1990; 63: 1137–1147.

11. Fichelson S, Vigon I, Dusanter I, et al. In vitro transformation of murine pro-B and pre-B cells by *v-mpl*, a truncated form of a cytokine receptor. *J Immunol.* 1995; 154: 1577–1586.

12. Penciolleli JF, Wendling F, Robert-Lezenes J, Barque JP, Tambourin P, Gisselbrecht S. Genetic analysis of myeloproliferative leukemia virus, a novel acute leukemogenic replication-defective retrovirus. *J Virol.* 1987; 61: 579–583.

13. Bishop JM. Cellular oncogenes and retroviruses. *Annu Rev Biochem.* 1983; 52: 301–354.

14. Cosman D, Lyman SD, Idzerda RL, et al. A new cytokine receptor superfamily. *Trends Biochem Sci.* 1990; 15: 265–270.

15. Bazan JF. Structural design and molecular evolution of a cytokine receptor superfamily. *Proc Natl Acad Sci USA.* 1990; 87: 6934–6938.

16. Bénit L, Charon M, Cocault L, Wendling F, Gisselbrecht S. The 'WS motif' common to *v-mpl* and members of the cytokine receptor superfamily is dispensable for myeloproliferative leukemia virus pathogenicity. *Oncogene.* 1993; 8: 787–790.

17. Schindler C, Darnell JE Jr. Transcriptional responses to polypeptide ligands: the JAK-STAT pathway. *Annu Rev Biochem.* 1995; 64: 621–651.

18. Pallard C, Gouilleux F, Bénit L, et al. Thrombopoietin activates a STAT5-like factor in hematopoietic cells. *EMBO J.* 1995; 14: 2847–2856.

19. Bénit L, Courtois G, Charon M, Varlet P, Dusanter-Fourt I, Gisselbrecht S. Characterization of *mpl* cytoplasmic domain sequences required for myeloproliferative leukemia virus pathogenicity. *J Virol.* 1994; 68: 5270–5274.

20. Fukunaga R, Ishizaka-Ikeda E, Pan CX, Seto Y, Nagata S. Functional domains of the granulocyte colony-stimulating factor receptor. *EMBO J.* 1991; 10: 2855–2865.

21. Murakami M, Narazaki M, Hibi M, et al. Critical cytoplasmic region of the interleukin 6 signal transducer gp130 is conserved in the cytokine receptor family. *Proc Natl Acad Sci USA.* 1991; 88: 11,349–11,353.

22. Ziegler SF, Bird TA, Morella KK, Mosley B, Gearing DP, Baumann H. Distinct regions of the human granulocyte-colony-stimulating factor receptor cytoplasmic domain are required for proliferation and gene induction. *Mol Cell Biol.* 1993; 13: 2384–2390.

23. Courtois G, Bénit L, Mikaeloff Y, et al. Constitutive activation of a variant of the *env-mpl* oncogene product by disulfide-linked homodimerization. *J Virol.* 1995; 69: 2794–2800.

24. Watowich SS, Yoshimura A, Longmore GD, Hilton DJ, Yoshimura Y, Lodish HF. Homodimerization and constitutive activation of the erythropoietin receptor. *Proc Natl Acad Sci USA.* 1992; 89: 2140–2144.

25. Vigon I, Mornon JP, Cocault L, et al. Molecular cloning and characterization of MPL, the human homolog of the *v-mpl* oncogene: Identification of a member of the hematopoietic growth factor receptor superfamily. *Proc Natl Acad Sci USA.* 1992; 89: 5640–5644.

26. Hayashida K, Kitamura T, Gorman DM, Arai K, Yokota T, Miyajima A. Molecular cloning of a second subunit of the receptor for human granulocyte-macrophage colony-stimulating factor (GM-CSF): Reconstitution of a high-affinity GM-CSF receptor. *Proc Natl Acad Sci USA.* 1990; 87: 9655–9659.

27. Gearing DP, Thut CJ, Vantden Bos T, et al. Leukemia inhibitory factor receptor is stucturally related to the IL-6 signal transducer, gp130. *EMBO J.* 1991; 10: 2839–2848.

28. Skoda RC, Seldin DC, Chiang MK, Peichel CL, Vogt TF, Leder P. Murine c-mpl: a member of the hematopoietic growth factor receptor superfamily that transduces a proliferative signal. *EMBO J.* 1993; 12: 2645–2653.

29. Vigon I, Florindo C, Fichelson S, et al. Characterization of the murine Mpl proto-oncogene, a member of the hematopoietic cytokine receptor family: molecular cloning, chromosomal location and evidence for a function in cell growth. *Oncogene.* 1993; 8: 2607–2615.

30. Mignotte V, Vigon I, Boucher de Crèvecoeur E, Roméo PH, Lemarchandel V, Chrétien S. Structure and transcription of the human c-mpl gene (MPL). *Genomics.* 1994; 20: 5–12.

31. Alexander WS, Dunn AR. Structure and transcription of the genomic locus encoding murine c-Mpl, a receptor for thrombopoietin. *Oncogene.* 1995; 10: 795–803.

32. Jackson IJ. A reappraisal of non-consensus mRNA splice sites. *Nucleic Acids Res.* 1991; 19: 3795–3798.

33. Hornkohl A, Selesi B, Bennett L, Hockman H, Nichol J, Hunt P. Thrombopoietin and SMPL concentrations in the serum and plasma of normal donors. *Blood.* 1995; 86: 838a (abstract no 3580).

34. Lofton-Day C, Buddle M, Berry J, Lok S. Differential binding of murine thrombopoietin to two isoforms of murine c-mpl. *Blood.* 1995; 86: 594a (abstract no 2362).

35. Roméo PH, Prandini MH, Joulin V, et al. Megakaryocytic and erythrocytic lineages share specific transcription factors. *Nature.* 1990; 344: 447–449.

36. Martin DI, Zon LI, Mutter G, Orkin SH. Expression of an erythroid transcription factor in megakaryocytic and mast cell lineages. *Nature.* 1990; 344: 444–447.

37. Lemarchandel V, Ghysdael J, Mignotte V, Rahuel C, Roméo PH. GATA and Ets cis-acting sequences mediate megakaryocyte-specific expression. *Mol Cell Biol.* 1993; 13: 668–676.

38. Le Coniat M, Souyri M, Vigon I, Wendling F, Tambourin P, Berger R. The human homolog of the myeloproliferative virus maps to chromosome band 1p34. *Hum Genet.* 1989; 83: 194–196.

39. Sherr CJ, Rettenmier CW, Sacca R, Roussel MF, Look AT, Stanley ER. The *c-fms* proto-oncogene product is related to the receptor for the mononuclear phagocyte growth factor, CSF-1. *Cell.* 1985; 41: 665–676.

40. Besmer P, Murphy JE, George PC, et al. A new acute transforming feline retrovirus and relationship of its oncogene *v-kit* with the protein kinase gene family. *Nature.* 1986; 320: 415–421.

41. Downward J, Yarden Y, Mayes E, et al. Close similarity of epidermal growth factor receptor and v-erb-B oncogene protein sequences. *Nature.* 1984; 307: 521–527.

42. Watowich SS, Yoshimura A, Longmore GD, Hilton DJ, Yoshimura Y, Lodisch HF. Homodimerization and constitutive activation of the erythropoietin receptor. *Proc Natl Acad Sci USA.* 1992; 89: 2140–2144.

43. Alexander WS, Metcalf D, Dunn AR. Point mutations within a dimer interface homology domain of c-Mpl induce constitutive receptor activity and tumorigenicity. *EMBO J.* 1995; 14: 5569–5578.

7

Mpl Expression and Functional Role in Human Megakaryocytopoiesis

Françoise Wendling, Najet Debili, and William Vainchenker

1. Introduction

The *c-mpl* gene was discovered in 1990 as the cellular homologue of *v-mpl*, the oncogene of the murine myeloproliferative leukemia virus, MPLV (*see* Chapter 6) *(1)*. In mice, *v-mpl* induces a lethal myeloproliferative disease involving multiple hemopoietic progenitors *(2–4)*. In vitro, *v-mpl* transformed multipotential and lineage-committed progenitors, leading to acquisition of growth-factor independence for both proliferation and terminal maturation *(1)*. These data showed that *v-mpl* was a potent deregulator of hemopoiesis, but they did not provide any information about the function of the normal *c-mpl* proto-oncogene.

The cellular *c-mpl* gene encodes a cell-surface receptor belonging to the hematopoietin or cytokine receptor superfamily *(5–7)*. The first evidence that *c-mpl* might play a role restricted to hemopoiesis was provided by hybridization of poly(A)$^+$ RNA prepared from different organs of mice. Northern blot analysis, using riboprobes derived from *v-mpl*, detected a single 3.0-kb mRNA in fetal liver, bone marrow, and spleen of adult mice, but not in other tissues, such as adult liver, kidney, brain, testis, or thymus *(1)*. Furthermore, analysis of various phenotypically characterized murine leukemic cell lines revealed 3.0 kb *c-mpl* mRNA in multipotential and erythrocytic-megakaryocytic cell lines, whereas no expression was seen in myeloblastic, erythroblastic, pre-B-lymphoblastic cell lines, or cultured plasmacytomas *(7)*. Of note, *c-mpl* transcripts were also detected in interleukin (IL)-3-dependent murine pro-B Ba/F3 cells *(6)*. In human leukemic cell lines, a major 3.7-kb and a minor 2.7-kb mRNA corresponding to the two forms of the human *c-mpl*-encoded receptors, Mpl-P and Mpl-K, respectively, were detected by Northern blot analysis only in the erythrocytic-megakaryocytic HEL cells *(5)*. Together, these data strongly suggested a functional role of the *c-mpl*-encoded receptor, Mpl, in hemopoiesis, but did not precisely define the hemopoietic lineage(s) in which *c-mpl* plays a regulatory role (*see* Chapter 6).

1.1. c-mpl *Expression by Human Leukemic Cell Lines*

To clarify the hemopoietic lineages in which *c-mpl* was expressed, a large panel of phenotypically well-characterized human leukemic cell lines was examined by the

From: *Thrombopoiesis and Thrombopoietins: Molecular, Cellular, Preclinical, and Clinical Biology*
Edited by: D. J. Kuter, P. Hunt, W. Sheridan, and D. Zucker-Franklin Humana Press Inc., Totowa, NJ

RNA-based polymerase chain reaction (PCR) technique *(8)*. Reverse transcription (RT) was performed with random hexanucleotides as primers on 100 ng total cellular RNA. The resulting cDNA was subjected to 35 amplification cycles with primers specific for a 220-base sequence of the *c-mpl* message *(9)* and a 116-base sequence of the β_2 microglobulin message *(10)* to compare and monitor for efficient cDNA synthesis. Primers specific for *c-mpl* transcripts were sense primer (nucleotide 843) 5'-TGGAGATGCAGTGGCACTTG-3', and antisense primer (nucleotide 1029) 5'-TGATGTCTGGGGTGTCAAGA-3'. PCR products were subjected to electrophoresis through a 2% agarose gel, and amplification products were detected by an overnight hybridization to internal synthetic ^{32}P-γ-ATP-labeled oligomer probes: *c-mpl* 5'-TTCTACCACAGCAGGGCACG-3', and β_2-microglobulin 5'-GCCCAAGATAGTTAAGTGGG-3'.

The *c-mpl* transcripts were detected in UT-7, TF-1, ELF153, CHRF, AP217, LAMA, DAMI, KU812, HEL, and Mo-7E cell lines (Fig. 1A). Importantly, these leukemic cell lines represent cells arrested during differentiation along the megakaryocytic lineage, since all of them spontaneously and unambiguously expressed the platelet-specific surface glycoproteins GPIIb (CD41) and GPIIIa (CD61). A weaker expression was also detected in the myeloid line KG1a, a cell clone that has previously been shown to express a low level of GPIIb/IIIa *(11)*. In contrast, cell lines negative for the *c-mpl* transcript have the phenotypes of cells committed to either the erythroblastic (K562), lymphoblastic (CEM), plasmacytoblastic (RPMI), or monocytic (U937) lineages (Fig. 1A) or myeloblastic (HL60), lymphoblastic B (Raji), and T (MO and Jurkat) lineages *(8)*. Among various RT-PCR-positive cell lines, the major 3.7-kb *c-mpl* transcript can be detected by Northern blot analysis of total RNA in only a few cell lines, namely CHRF, DAMI *(12)*, and HEL *(5)*. In other cell lines (UT-7, AP217, KU 812, or Mo-7E), the levels of *c-mpl* transcription were too low for detection by Northern blot analysis *(5,12)*.

1.2. Surface Expression of Mpl by Immunofluorescence

To study further the expression pattern and distribution of the Mpl protein, a monoclonal antibody (MAb) directed against the extracellular domain of the receptor was generated. Briefly, a recombinant protein was obtained by subcloning a fragment encompassing amino acids 26–284 of the extracellular domain of human *c-mpl* cDNA into a yeast expression vector under the control of a leader sequence for secretion. The cDNA was also modified at the N-terminus by the addition of a Flag epitope tag to be used for antibody purification by affinity chromatography using an anti-Flag MAb (for details, *see* ref. *12*). A Balb/c mouse was immunized by repeated injections of the purified protein. Hybridomas were generated, and one single-cell clone was used to produce the anti-Mpl MAb designated M1. Using a typing kit, M1 was identified as an IgG1. The Mpl displayed on various human leukemic cell lines was examined by indirect immunofluorescence and flow cytometry analyses (Fig. 1B). Binding of anti-Mpl M1 is clearly detected on all cell lines found positive by the RT-PCR analysis. However, the cell-surface expression of the Mpl protein did not totally correlate with the amount of *c-mpl* transcripts. The best examples are the Mo-7E and KU812 cells. KU812 showed a high level of *c-mpl* mRNA, but very few receptors on the cell surface; in contrast, Mo-7E cells displayed a high fluorescent profile, whereas the amount of *c-*

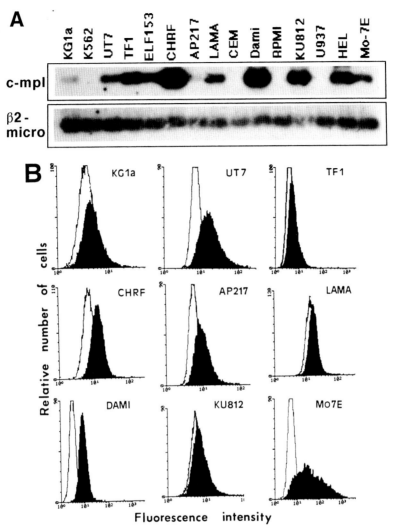

Fig. 1. Expression of *c-mpl* mRNA and cell-surface Mpl protein on human leukemic cell lines. **(A)** RT-PCR studies. RT-PCR was performed on 100 ng total RNA prepared from pluripotential (UT-7, TF-1, KU812), erythrocytic-megakaryocytic (Mo-7E, HEL, DAMI, LAMA), megakaryoblastic (ELF 153, CHFR, AP217), myeloblastic-megakaryoblastic (KG1a), erythroleukemic (K562), monocytic (U937), lymphoblastic T (CEM), and plasmacytoblastic (RPMI) cell lines. The β_2 microglobulin (β_2) transcripts served as internal controls for each sample. Autoradiographs were exposed for 6 hours at –80°C with intensifying screens. **(B)** Fluorescence profiles of representative cell lines. Cells were labeled with the anti-Mpl M1 MAb (black profile) or with a irrelevant control IgG1 (thin line) and indirectly stained with FITC-conjugated sheep antimouse IgG F(ab)' 2 fragments. Cytofluorographs were obtained on a FACsort.

mpl transcript was clearly weaker than in KU812. These discrepancies are not clearly understood, and they may be owing to a posttranscriptional processing of the mRNA encoding the protein and/or to its intracellular trafficking *(12)*.

Together, the data provided both by RT-PCR techniques, and by immunofluorescence analyses, demonstrated that the product of the *c-mpl* gene was only found in cell lines expressing an erythrocytic/megakaryocytic phenotype. This approach indicated a possible involvement of Mpl in the regulation of megakaryocytopoiesis.

2. *c-mpl* Expression in Normal Human Hemopoietic Cells

2.1. Analysis of Purified Hemopoietic Subsets by RT-PCR

Expression of *c-mpl* was further examined on hemopoietic cell populations highly purified from peripheral blood or bone marrow from normal adult donors *(8)*. Nucleated peripheral blood cells were separated into T or B lymphocytes, granulocytes, and monocytes by panning with MAb or by adherence. Acidophilic erythroblasts were obtained from pure erythroid burst-forming cell (E-BFC)-derived colonies plucked from methylcellulose after 10 days in culture. Megakaryocytes were isolated from day 12 liquid cultures *(13)* by means of the immunomagnetic bead technique using an anti-CD61 MAb (Y2-51). Platelets were obtained from platelet-rich plasma cleared of leukocytes by filtration. CD34$^+$ cells were isolated from bone marrow or leukapheresis material using magnetic beads *(14,15)*, and endothelial cells (HUVEC) were cultured from umbilical cord veins. RT was performed as described using the same primers. Southern blot analysis of amplified products revealed signals corresponding to *c-mpl* transcripts only in purified CD34$^+$ cells, highly enriched megakaryocytes, and platelets. A faint signal was seen on cultured endothelial cells, whereas the other cell populations were negative (Fig. 2).

2.2. Expression of Mpl on Normal Megakaryocytes and Platelets by Immunofluorescence Labeling

To determine whether Mpl was detectable on the cell surface of megakaryocytes, bone marrow cells were cultured in liquid cultures in the presence of 10% serum from patients with aplastic anemia. From previous studies, it was known that serum from patients with platelet counts $<20 \times 10^9$/L contained a stimulatory activity that promotes the growth and differentiation of megakaryocytic progenitors and promegakaryoblasts *(16)*. Cultures were depleted of adherent cells by successive transfers in tissue-culture-treated flasks. After 10–12 days in culture, megakaryocytes represented as much as 10% of the nonadherent cell population *(13)*. To identify megakaryocytes, cells were double-labeled with a phycoerythrin (PE)-conjugated anti-CD41 (GPIIb) or anti-CD61 (GPIIIa) MAb, and the anti-Mpl M1 MAb indirectly stained with fluorescent isothiocyanate (FITC). Flow cytometry analyses revealed that the great majority of CD41$^+$ or CD61$^+$ cells (megakaryocytes) were positively and unambiguously stained with M1. Furthermore, immunostaining of cytospin smears of the cultures revealed that large cells with multilobed nuclei presented a strong surface fluorescence with M1 and a strong cytoplasmic positivity when costained with a polyclonal anti-von Willebrand factor (vWF) antibody. In contrast, M1 labeling was not seen on vWF-negative contaminating cells *(12)*. These data suggested that the surface density of Mpl molecules must increase with megakaryocyte maturation, since the greatest staining was detected on the largest megakaryocytes. Similar immunofluorescence studies were performed with peripheral blood platelets. Platelets were positively stained with M1, but just at the threshold of detection (Fig. 3). Immunoblotting of platelet lysates with M1 revealed a band migrating at 82 kDa that corresponded to the expected size of the full-length, glycosylated Mpl protein *(12)*.

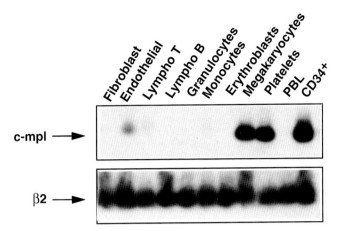

Fig. 2. Expression of *c-mpl* and β_2 microglobulin mRNA in normal purified hemopoietic cell populations. RT-PCR was performed on 100 ng total RNA, except for the CD34$^+$ population where total RNA was extracted from a pellet containing 1×10^5 cells. Autoradiographs were exposed for 6 hours at $-80°C$ with intensifying screens. (Reprinted with permission from ref. *12*.)

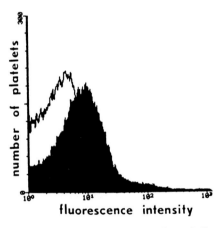

Fig. 3. Cell-surface expression of Mpl on human platelets. Cells were labeled with the anti-Mpl M1 MAb (black profile) or with an irrelevant control IgG1 (thin line), and indirectly stained with FITC-conjugated sheep antimouse IgG F(ab)' 2 fragments. Cytofluorographs were obtained on a FACsort *(12)*.

Thus, these results confirmed the RT-PCR data, and demonstrated that megakaryocytes and platelets express detectable levels of Mpl protein at their surface.

2.3. Expression of Mpl by CD34$^+$ Cells

2.3.1. Flow Cytometry Analysis

Since transcripts encoded by the *c-mpl* gene were found by RT-PCR analysis in the CD34$^+$ population, an investigation to determine which progenitor cell subsets expressed the Mpl was undertaken *(12)*. Bone marrow CD34$^+$ cells were isolated by immunomagnetic beads *(15)*, labeled with M1, and analyzed by flow cytometry. Only a very small fraction (<1%) reacted positively with the antibody. In general, the fluorescent staining was much fainter than on mature megakaryocytes, but clearly distinct from the labeling obtained with a control isotype-matched irrelevant MAb. The faint fluorescent display is likely owing to the low numbers of cytokine receptors usually expressed on immature progenitor cells (17,18). It was shown by dual labeling with M1 and CD61 (GPIIIa) or CD41 (GPIIb) that the large majority of the M1-labeled CD34$^+$ cells coexpressed the glycoproteins GPIIb/IIIa, suggesting that the CD34$^+$/Mpl$^+$ population could well represent megakaryocyte progenitor or precursor cells (19).

2.3.2. Cell Sorting and Progenitor Cell Assays

To demonstrate the nature of progenitor cells contained in the CD34$^+$/Mpl$^+$ sub-population, the CD34$^+$/Mpl$^+$ and CD34$^+$/Mpl$^-$ fractions were sorted. Cells were plated in semisolid culture medium supplemented with serum from patients with aplastic anemia as a source of megakaryocyte colony-stimulating factor (MK-CSF) (16,20), and recombinant human (rHu)IL-3, rHu granulocyte-macrophage colony-stimulating factor (GM-CSF), rHu granulocyte colony-stimulating factor (G-CSF), and rHu erythropoietin (EPO) to reveal colonies from different lineages. Cultures were grown in a cocktail of all these cytokines. The CD34$^+$/Mpl$^-$ subset contained numbers of erythroid-burst forming cells (E-BFC), granulocyte-macrophage colony-forming (GM-CFC), and megakaryocyte colony-forming cells (MK-CFC) similar to those found in the unseparated CD34$^+$ fraction. In contrast, the CD34$^+$/Mpl$^+$ subset was almost totally depleted in E-BFC and GM-CFC, but enriched in late MK-CFC and megakaryoblasts giving rise to small megakaryocyte clusters composed of two to three megakaryocytes within 7 days in culture (12). Unexpectedly, it was noted that the CD34$^+$/Mpl$^+$ fraction was always contaminated by a fair percentage of late E-BFC. This was even more evident when the CD34$^+$/Mpl$^+$-sorted fraction was grown in liquid cultures with aplastic serum, IL-3, GM-CSF, G-CSF, and EPO, since the main population observed after day 8 in culture contained almost exclusively erythroblasts (Fig. 4). At the time of this experiment, the gene encoding the Mpl ligand (thrombopoietin, TPO) was not cloned, and the authors could not ascertain whether late E-BFC expressed functional Mpl. However, recent experiments have shown that TPO synergizes with EPO to promote the growth of erythroid progenitors (21,22). Thus, it seems likely that the observation was not the result of an artifact in the cell sorting, but that Mpl might, in fact, be present on erythroid progenitor cells. By flow cytometry, however, Mpl could not be detected on primitive erythroid progenitor cells (12).

3. Demonstration of a Functional Role of Mpl in Megakaryocytopoiesis

3.1. c-mpl Antisense Oligodeoxynucleotides

In the absence of a specific known ligand for Mpl, demonstration that the receptor plays a key role in regulating megakaryocytopoiesis was provided indirectly by studies performed with c-mpl antisense oligodeoxynucleotides (8). At least two different mechanisms account for the effects of antisense oligodeoxynucleotides in cells. Oligonucleotide/RNA duplexes are either degraded in the cytoplasm by RNase H, and/or the formation of oligonucleotide/RNA duplexes prevents RNA translation by ribosomes (23,24).

Six synthetic unmodified 18-mer antisense, sense, or scrambled oligodeoxynucleotides complementary to different regions of the first c-mpl extracellular domain (5,9)

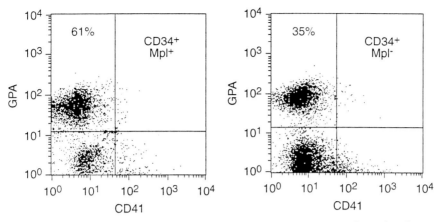

Fig. 4. FACsort analysis of CD34$^+$/Mpl$^+$ and CD34$^+$/Mpl$^-$ sorted fractions showing presence of Mpl$^+$ erythroid precursors. Cells were grown in liquid cultures with aplastic serum, IL-3, GM-CSF, G-CSF, and EPO. After 8 days in culture, cells were dual labeled with an antiglycophorin A (GPA) and an antiGPII b (CD41) MAb. Percentages of erythroid cells (GPA-positive) are given in the upper left panel.

were derived. CD34$^+$ cells were incubated at a concentration of 1×10^6 cells/mL in serum-deprived medium *(25)* supplemented with recombinant cytokines in the presence of 10 μM (70 μg/mL) oligodeoxynucleotides. After an overnight incubation at 37°C, 5 mM oligodeoxynucleotides were added, and incubation was continued for an additional 6 hours. Cells were then washed, and an equal number of cells (1×10^5 cells) from each group were used for RNA extraction and RT-PCR analysis.

Two oligodeoxynucleotides, antisense 3 (AS3, 5'-GGCCCAGGAGGGCATCTT-3') encompassing the first methionine codon of the coding sequence (nucleotides –3 to +15) and antisense 6 (AS6, 5'-TGCTGTCAGAGCTGAAGC-3') complementary to a sequence present in exon 4 (nucleotides +690 to +708) produced a significant *(70%)* decrease in *c-mpl* mRNA levels. In contrast, their respective sense (S3 and S6), a scrambled antisense to AS6 (AS7), or the other antisense oligomers (derived from nucleotides +1 to +18, +22 to +39, and +13 to +30, respectively) did not show any effect *(8)*. Controls for RNA integrity and efficient RT were given by amplification of β_2 microglobulin transcripts. In addition, the specific action of the antisense oligonucleotides was suggested by their inability to alter *c-kit* message, which is known to be expressed by primitive and committed hemopoietic progenitors *(26)*. Because no decrease in either the β_2 microglobulin or *c-kit* messages was seen in any groups, it was concluded that treatment of CD34$^+$ cells by antisense AS3 and AS6 oligodeoxynucleotides resulted in a specific degradation of *c-mpl* mRNA. Moreover, the life-span of the Mpl protein was evaluated by exposing DAMI cells to actinomycin D for varying intervals. After 8 hours of treatment, no Mpl could be detected on the cell surface by labeling with the M1 MAb, suggesting that the half-life of the Mpl protein was about 4 hours *(8)*.

3.2. Specific Inhibition of MK-CFC by Antisense Treatment of CD34$^+$ Cells

The biological significance of a decrease in *c-mpl* mRNA was examined by plating oligodeoxynucleotide-treated CD34$^+$ cells in fibrin clot cultures *(8)*. Exposure of CD34$^+$

cells to sense or antisense oligomers did not affect the development of E-BFC- or GM-CFC-derived colonies compared with untreated controls (Fig. 5). The influence of oligomer treatment on the development of MK-CFC was examined by plating CD34$^+$ oligonucleotide-treated cells in medium containing 10% serum from aplastic patients to provide a source of megakaryocyte colony-stimulating activity (presumably TPO) *(16)* and rHuIL-3 *(27)*. Furthermore, to see MK-CFC unambiguously, cultures were stained with the Y2-51 MAb that specifically recognizes a GPIIIa epitope, and antibody fixation was revealed by immunoenzymatic labeling with alkaline phosphatase/antialkaline phosphatase *(28)*. In contrast to the lack of effect seen on E-BFC or GM-CFC, antisense oligos AS3 and AS6 (which decreased *c-mpl* mRNA) significantly reduced the number of MK-CFC-derived colonies, whereas their respective sense oligomers (S3 and S6) had no detectable effect. In four repeated experiments, the decrease in MK-CFC ranged from 54–74% for oligo AS3, and 54–81% for oligo AS6. These percentages were calculated from the number of colonies obtained from cells exposed to the respective sense oligonucleotides. An important observation was the demonstration that the decrease in MK-CFC formation resulting from oligomer treatment was not owing to a possible toxic effect of the antisense oligomers since a normal number of MK-CFC were obtained when a cocktail of cytokines (stem cell factor [SCF] + IL-3 + GM-CSF + IL-6 + IL-11) was added to the culture medium in place of aplastic serum.

4. Expression of *c-mpl* in Human Hematologic Malignancies and Adenocarcinoma

Several investigators have shown that deregulated expression of growth factors and/or growth factor receptors could be involved in the pathogenesis of human leukemias and myeloproliferative syndromes *(25,29,30)*. Studies were undertaken to analyze *c-mpl* expression in a large series of patients with different hemopoietic malignances *(31,32)*. By Northern blot analysis, very low or undetectable levels of *c-mpl* transcripts were found in cells from patients with lymphoid malignancies or with myeloproliferative and chronic phase myelodysplastic syndromes. However, *c-mpl* expression was detectable in 50% of cases of acute myeloblastic leukemia (AML) and frequently in patients with secondary AML of poor prognosis developing during the progression of myeloproliferative or myelodysplastic syndromes *(31)*. These data suggest that abnormal expression of the *c-mpl* gene product may occur as part of the leukemic phenotype in myeloid blast cells. In addition, further studies performed in vitro with fresh leukemic blasts from patients with AML showed that blast cell proliferation is induced with the Mpl ligand *(32,33)*, or with a combination of Mpl ligand and SCF *(34)* in 50% of the cases studied. In contrast, among 20 human primary solid tumors of various origin, none were found to express *c-mpl* transcripts by RT-PCR *(35)*.

5. Conclusions

These results provided the first clear evidence for a functional role of Mpl in the regulation of megakaryocytopoiesis. Analyses by both RT-PCR and immunofluorescence demonstrate that Mpl, the product of the *c-mpl* proto-oncogene, is expressed by purified CD34$^+$ progenitor cells, megakaryocytes, platelets, and endothelial cells. The oligonucleotide antisense strategy revealed a direct correlation between the inhibition of *c-mpl* mRNA and a profound decrease in MK-CFC formation. Based on these data,

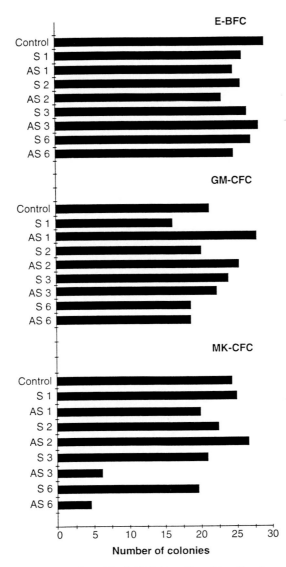

Fig. 5. Effects of treatment of purified CD34$^+$ cells with oligodeoxynucleotides, demonstrating specfic inhibition of MK-CFC. To analyze E-BFC and GM-CFC formation, cells were plated in the fibrin-clot culture system at a concentration of 200 cells/0.5 mL of medium containing 20% preselected fetal bovine serum (FBS) and rHu IL-3 (100 U/mL), rHuGM-CSF (2.5 ng/mL), rHuG-CSF (100 ng/mL), and rHuEPO (2 U/mL). Cultures were harvested after 12 days of incubation, stained with benzidine, counterstained with hematoxylin, and the number of colonies (more than 50 cells) counted. For MK-CFC formation, cells were seeded at a concentration of 500 cells/0.5 mL in fibrin-clot culture containing 10% serum from aplastic patients + rHuIL-3 (100 U/mL). Clots were harvested onto slides after 12 days of growth, fixed with neutral formalin, and reacted with the mouse Y2-51 antiGPIIIa MAb. Antibody fixation was revealed by incubation with a mouse IgG + IgM F(ab)'2 coupled to alkaline phosphatase. Data represent mean colony numbers of four different combined experiments (four plates/point/ experiments). (Reprinted with permission from ref. *12*.)

the hypothesis that Mpl could be the receptor for a cytokine specifically regulating thrombocytopoiesis was strengthened. Furthermore, the observation that stimulation of MK-CFC development occurred in medium containing human aplastic serum provided strong evidence that a ligand able to bind and activate Mpl was present in sera from thrombocytopenic animals. In fact, the successful isolation and cloning of TPO, the ligand of Mpl, stems from these results. As described in another chapter (Chapter 8), TPO was biochemically purified from sera of thrombocytopenic animals by affinity chromatography using columns on which a soluble form of the Mpl was captured.

Acknowledgments

We are very grateful to David Cosman (Immunex Corp., Seattle, WA) for his helpful collaboration. Work from our laboratory was supported by the Institut National de la Santé et de la Recherche Médicale (INSERM), and grants from the Ligue Nationale contre le Cancer, the Association pour la Recherche contre le Cancer (ARC), and the Fondation de France. We are grateful to Immunex Corp. (Seattle, WA), Amgen Inc. (Thousand Oaks, CA), and Genetics Institute (Cambridge, MA) for their generous gifts of recombinant human cytokines.

References

1. Souyri M, Vigon I, Penciolelli JF, Heard JM, Tambourin P, Wendling F. A putative truncated cytokine receptor gene transduced by the myeloproliferative leukemia virus immortalizes hematopoietic progenitors. *Cell*. 1990; 63: 1137–1147.
2. Wendling F, Varlet P, Charon M, Tambourin P. MPLV: a retrovirus complex inducing an acute myeloproliferative leukemic disorder in adult mice. *Virology*. 1986; 149: 242–246.
3. Wendling F, Penciolelli J-F, Charon M, Tambourin P. Factor-independent erythropoietic progenitor cells in leukemia induced by the myeloproliferative leukemia virus. *Blood*. 1989; 73: 1161–1167.
4. Wendling F, Vigon I, Souyri M, Tambourin P. Myeloid progenitor cells transformed by the myeloproliferative leukemia virus proliferate and differentiate in vitro without addition of growth factors. *Leukemia*. 1989; 3: 475–480.
5. Vigon I, Mornon JP, Cocault L, et al. Molecular cloning and characterization of MPL, the human homolog of the v-mpl oncogene: identification of a member of the hematopoietic growth factor receptor superfamily. *Proc Natl Acad Sci USA*. 1992; 89: 5640–5644.
6. Skoda RC, Seldin DC, Chiang MK, Peichel CL, Vogt TF, Leder P. Murine c-mpl: a member of the hematopoietic growth factor receptor superfamily that transduces a proliferative signal. *EMBO J*. 1993; 12: 2645–2653.
7. Vigon I, Florindo C, Fichelson S, et al. Characterization of the murine Mpl proto-oncogene, a member of the hematopoietic cytokine receptor family: molecular cloning, chromosomal location and evidence for a function in cell growth. *Oncogene*. 1993; 8: 2607–2615.
8. Methia N, Louache F, Vainchenker W, Wendling F. Oligodeoxynucleotides antisense to the proto-oncogene c-mpl specifically inhibit in vitro megakaryocytopoiesis. *Blood*. 1993; 82: 1395–1401.
9. Mignotte V, Vigon I, Boucher de Crevecoeur E, Roméo PH, Lemarchandel V, Chrétien S. Structure and transcription of the human c-mpl gene (MPL). *Genomics*. 1994; 20: 5–12.
10. Güssow D, Rein R, Ginjaar I, et al. The human β2-microglobulin gene. Primary structure and definition of the transcriptional unit. *J Immunol*. 1987; 139: 3132–3139.
11. Berridge MV, Ralph SJ, Tan AS. Cell-lineage antigens of the stem cell-megakaryocyte-platelet lineage are associated with the platelet IIb-IIIa glycoprotein complex. *Blood*. 1985; 66: 76–85.
12. Debili N, Wendling F, Cosman D, et al. The Mpl receptor is expressed in the megakaryocytic lineage from late progenitors to platelets. *Blood*. 1995; 85: 391–401.

13. Debili N, Hegyi E, Navarro S, et al. In vitro effects of hematopoietic growth factors on the proliferation, endoreplication, and maturation of human megakaryocytes. *Blood.* 1991; 77: 2326–2388.

14. Civin CI, Strauss LC, Brovall C, Fackler MJ, Schwartz JF, Shaper JH. Antigenic analysis of hematopoiesis. III. A hematopoietic progenitor cell surface antigen defined by a monoclonal antibody raised against KG-1a cells. *J Immunol.* 1984; 133: 157–165.

15. Sawada K, Kranz SB, Dai CH, et al. Purification of human blood burst forming units-erythroid and demonstration of the evolution of erythropoietin receptors. *J Cell Physiol.* 1990; 142: 219–230.

16. Hoffman R. Regulation of megakaryocytopoiesis. *Blood.* 1989; 74: 1196–1212.

17. Wognum AW, Krystal G, Eaves CJ, Eaves AC, Lansdorp PM. Increased erythropoietin-receptor expression on CD34-positive bone marrow cells from patients with chronic myeloid leukemia. *Blood.* 1992; 79: 642–649.

18. Wognum AW, van Gils FCJM, Wagemaker G. Flow cytometric detection of receptors for interleukin-6 on bone marrow and peripheral blood cells of humans and rhesus monkeys. *Blood.* 1993; 81: 2036–2043.

19. Debili N, Issaad C, Massé JM, et al. Expression of CD34 and platelet glycoproteins during human megakaryocytic differentiation. *Blood.* 1992; 80: 3022–3035.

20. Vainchenker W, Bouguet J, Guichard J, Breton-Gorius J. Megakaryocyte colony formation from human bone marrow precursors. *Blood.* 1979; 54: 940–945.

21. Fibbe WE, Heemskerk DPM, Laterveer L, et al. Accelerated reconstitution of platelets and erythrocytes after syngeneic transplantation of bone marrow cells derived from thrombopoietin pretreated donor mice. *Blood.* 1995; 86: 3308–3313.

22. Kaushansky K, Broudy VC, Grossmann A, et al. Thrombopoietin expands erythroid progenitors, increases red cell production, and enhances erythroid recovery after myelosuppressive therapy. *J Clin Invest.* 1995; 96: 1683–1687.

23. Dash P, Lotan I, Knapp M, Kandel ER, Goelet P. Selective elimination of mRNAs in vivo: complementary oligodeoxynucleotides promote RNA degradation by an RNase H-like activity. *Proc Natl Acad Sci USA.* 1987; 84: 7896–7900.

24. Boiziau C, Kurfurst R, Cazenave C, Roig V, Thuong NT, Toulmé JJ. Inhibition of translation initiation by antisense oligonucleotides via an RNase-H independent mechanism. *Nucleic Acids Res.* 1991; 19: 1113–1119.

25. Mitjavila M-T, Le Couedic JP, Casadevall N, et al. Autocrine stimulation by erythropoietin and autonomous growth of human erythroid leukemic cells in vitro. *J Clin Invest.* 1991; 88: 789–797.

26. Ratajczak MZ, Luger SN, Deriel K, Abraham J, Calabretta B, Gewirtz AL. Role of the KIT protooncogene in normal and malignant human hematopoiesis. *Proc Natl Acad Sci USA.* 1992; 89: 1710–1714.

27. Mazur EM, Cohen JL, Bogart L, et al. Recombinant gibbon interleukin-3 stimulates megakaryocyte colony growth in vitro from human peripheral blood progenitor cells. *J Cell Physiol.* 1988; 136: 439–446.

28. Cordell JL, Falini B, Erber WN, et al. Immunoenzymatic labeling of monoclonal antibodies using immune complexes of alkaline phosphatase and monoclonal anti-alkaline phosphatase (APAAP complexes). *J Histochem Cytochem.* 1984; 32: 219–229.

29. Rambaldi A, Wakamiya N, Vellenga E, et al. Expression of the macrophage colony-stimulating factor and c-fms genes in human acute myeloblastic leukemia cells. *J Clin Invest.* 1988; 8: 1030–1035.

30. Eaves AC, Eaves CJ. Erythropoiesis in culture. *Clin Haematol.* 1984; 13: 371–391.

31. Vigon I, Dreyfus F, Melle J, et al. Expression of the c-mpl proto-oncogene in human hematologic malignancies. *Blood.* 1993; 82: 877–883.

32. Matsumura I, Kanakura Y, Kato T, et al. Growth response of acute myeloblastic leukemia cells to recombinant human thrombopoietin. *Blood.* 1995; 86: 703–709.

33. Piacibello W, Sanavio F, Garetto L, et al. The effect of human megakaryocyte growth and dvelopment factor (MGDF) on human myeloid leukemia cell growth. *Blood.* 1995; 86: 45a (abstract no 168).
34. Fontenay-Roupie M, Picard F, Casadevall N, et al. Thrombopoietin effects on the in vitro proliferation of acute myeloblastic leukemia and myelodysplasia-deriving cells. *Blood.* 1995; 86: 11a (abstract no 34).
35. Columyoe L, Loss M, Scadden DT. Thrombopoietin receptor expression in human cancer cell lines and primary tissues. *Cancer Res.* 1995; 55: 3209–3512.

8

The Purification and Cloning of Human Thrombopoietin

Dan Eaton

1. Introduction

For more than 20 years, numerous groups have attempted purification of thrombopoietin (TPO) from the plasma of thrombocytopenic or pancytopenic animals and humans (reviewed in refs. *1–3*, and Chapter 5). These efforts were based on observations, dating from the late 1950s, of a factor in such plasma that specifically regulates platelet production *(4,5)*. The observation that nephrectomized mice become thrombocytopenic led others to attempt isolation of TPO from kidney cell-culture medium *(3)*. Because of the lack of a specific TPO assay and the extremely low level of endogenous TPO in thrombocytopenic plasma or cell-culture media, these efforts were historically of limited success. However, these experiments provided information that, when combined with the discovery of a putative TPO receptor, Mpl, aided in the relatively rapid isolation of TPO from the plasma of aplastic animals by several groups, who reported the isolation and cloning of TPO in 1994 *(6–10)*. During the same period, two different groups succeeded in purifying TPO using conventional chromatographic procedures (*see* Chapter 9).

2. The *c-mpl* Proto-oncogene Product, Mpl, Is the TPO Receptor

The discovery of the cytokine receptor Mpl (*see* Chapter 6) certainly can be described as a critical breakthrough in the search for TPO. Not only did this discovery allow for the development of essential tools for the isolation of TPO, but also strongly supported the notion that a hemopoietic growth factor that selectively regulated megakaryocytopoiesis did indeed exist (*see* Chapter 7). A portion of the *c-mpl* gene was initially discovered fused to viral sequences encoding the envelope protein of the mutant murine leukemia virus, myleoproliferative leukemia virus (MPLV) *(11)*. The ability of *v-mpl* to transform hemopoietic progenitor cells and its homology to members of the cytokine receptor superfamily prompted speculation that the cellular homolog of *v-mpl* was a novel cytokine receptor. The isolation of human *c-mpl* confirmed that it belonged to the cytokine receptor superfamily *(12)*. Methia et al. *(13)* subsequently demonstrated that the expression of *c-mpl* mRNA and Mpl protein was restricted to CD34+ bone marrow cells, megakaryocytes, and platelets, indicating that *c-mpl* is down regulated in all hemopoietic lineages except megakaryocytes. They also

From: *Thrombopoiesis and Thrombopoietins: Molecular, Cellular, Preclinical, and Clinical Biology*
Edited by: D. J. Kuter, P. Hunt, W. Sheridan, and D. Zucker-Franklin Humana Press Inc., Totowa, NJ

demonstrated that *c-mpl* antisense oligonucleotides selectively inhibited megakaryo-cyte colony formation without affecting erythroid or granulocyte-macrophage colony growth in vitro. These two observations led these investigators to predict a critical role for Mpl in the regulation of megakaryocytopoiesis and that its ligand might be the long-sought TPO.

Subsequently, two groups demonstrated that Mpl was capable of transducing a proliferative signal *(14,15)*. In each case, these groups made chimeric receptors con-sisting of the extracellular domain of either the interleukin (IL)-4 *(15)* or granulocyte colony-stimulating factor (G-CSF) *(14)* receptors and the intracellular domain of Mpl, and transfected them into the murine IL-3-dependent cell line, Ba/F3. Once trans-fected with these chimeric receptors, the Ba/F3 cells became responsive to IL-4 or G-CSF, indicating that Mpl transduces a proliferation signal. Because of this finding, a very specific and simple proliferation assay was developed by engineering the Ba/F3 cell line to express full-length Mpl(Ba/F3-Mpl$^+$), making them responsive to the putative ligand. This assay was then used to screen for Mpl and subsequently monitor its isolation.

3. Thrombocytopenic Plasma Contains TPO

The presence of thrombopoietic and megakaryopoietic activities in the plasma of thrombocytopenic animals and humans has been well documented in the last 30 years *(1–3)*. Until the discovery of *c-mpl*, however, many had speculated that these activities were owing to the synergistic activity of previously identified factors (*see* Chapter 5). Today it appears that Mpl ligand, or TPO, is likely responsible for most, if not all, of the thrombopoeitic activity present in the plasma of thrombocytopenic animals and humans. Wendling et al. *(16)* demonstrated that plasma from pancytopenic mice, rats, dogs, and pigs all contain an activity that stimulates the proliferation of a Ba/F3-Mpl$^+$ cell line. This activity was neutralized by a soluble Mpl-IgG fusion protein consisting of the extracellular domain of Mpl and the Fc domain of IgG, indicating that this plasma contained Mpl ligand. In addition, the Mpl-IgG fusion protein also completely inhib-ited the megakaryocytopoietic activity of the plasma from the pancytopenic animals. From this study, it was therefore concluded that Mpl ligand was responsible for the megakaryocytopoietic activity of aplastic plasma and might be the long-sought TPO.

4. Isolation and Cloning of TPO

TPO, the ligand for the cytokine receptor, Mpl, was isolated by two different ap-proaches. In one strategy, researchers at Genentech (South San Francisco, CA) *(6)* and Amgen (Thousand Oaks, CA) *(8)* purified the ligand present in the plasma of irradiated pigs or dogs, respectively, using purification schemes centered around a receptor-affin-ity chromatography step. The group at Genentech initially fractionated aplastic porcine plasma by hydrophobic (phenyl-Sepharose) and ligand dye (blue-Sepharose) chroma-tography followed by Mpl-affinity chromatography. Blue-Sepharose was used based on the speculation that TPO may be structurally similar to erythropoietin (EPO), which binds to this resin. The group at Amgen purified canine TPO using sequential chroma-tography on wheat germ agglutin-Sepharose, Mpl-affinity, ion-exchange, and gel-fil-tration colomns followed by reverse-phase high-pressure liquid chromatography (HPLC). In both cases, TPO was purified approximately 4 million- to 10 million-fold from the starting plasma. The proliferation of an IL-3-dependent murine cell line transfected with human *(6)* or mouse *(8) c-mpl* was used to monitor the purification.

Clearly the use of the Mpl-affinity column in these schemes allowed for relatively straightforward purifications. In the Genentech scheme, the purification from the Mpl column alone was approximately 8000-fold. This degree of purification was achieved not only because of the high affinity of the ligand for Mpl, but also because of the stability of the ligand in chaotropic agents, such as urea and sodium dodecyl sulfate (SDS). This latter characteristic allowed the Mpl-affinity column to be run in the presence of 2*M* urea, keeping nonspecific binding to a minimum. Stability in SDS allowed for gel elution experiments enabling identification of the active ligand with high resolution. Thus, from 5 L of aplastic porcine plasma, sufficient quantities of Mpl ligand were obtained to allow determination of its amino-terminal sequence. In both cases, these groups isolated what is now known to be truncated versions of the full-length ligand. From the purification, the Genentech group obtained active proteins of 30, 28, and 18 kDa, all of which had identical amino-terminal sequences (Fig. 1). From gel elution experiments, an activity of 70 kDa was also detected. This latter observation suggested that the 30-, 28-, and 18-kDa fragments were derived from a larger precursor. Similarly, the group from Amgen isolated active fragments of 31, 25, and 17 kDa.

The amino-terminal sequence obtained from the purified proteins was used to design degenerate oligonucleotides to isolate the corresponding cDNA fragments by polymerase chain reaction (PCR). Initially the group at Genentech isolated a porcine genomic DNA fragment that encoded the first 23 amino acids of the porcine ligand. Synthetic olignucleotides based on this fragment were then used to amplify human genomic DNA by PCR, and a fragment encoding a putative exon was isolated. Subsequently, a human fetal liver cDNA library was screened, using as primers oligonucleotides corresponding to the ends of the exon sequence. The group from Amgen initially amplified a canine kidney cDNA library by PCR, and the subsequent PCR product was used to screen a human fetal liver cDNA library. Both groups then isolated a full-length cDNA clone encoding the ligand.

The group at Zymogenetics (Seattle, WA) *(7)* used a unique and risky approach, based on the observation that autonomous transformation of hemopoietic cell lines is often owing to autocrine production of a growth factor for which the cells have a receptor. They speculated that mutagenesis of the Ba/F3-Mpl$^+$ cells could result in an Mpl ligand-producing line. The chemical agent, 2-ethylmethanesulfonate (EMS), was used as a mutagen to induce random gene expression in Ba/F3-Mpl$^+$ cells, and the subsequent mutants were selected for autonomous growth (IL-3-independent). Although many of the isolated clones secreted IL-3, one of the clones selected by this technique produced an activity that could be neutralized by the extracellular domain of soluble Mpl. cDNA library pools were prepared from this cell line and transfected into baby hamster kidney (BHK) cells. Conditioned media from these cells were assayed by the Ba/F3-Mpl$^+$ cell proliferation assay. A positive pool was identified, from which a full-length cDNA encoding murine Mpl ligand was isolated using standard pool-breakdown techniques.

5. Structure of TPO

The isolated cDNA for human TPO *(6,8)* has an open reading frame encoding 353 amino acids consisting of a 21 amino-acid signal peptide and a mature human TPO

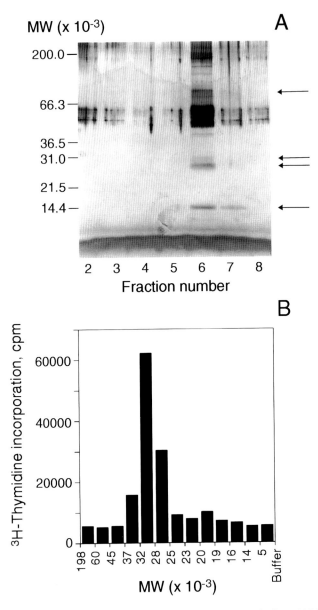

Fig. 1. Identification of Mpl ligand by SDS-PAGE and gel elution. **(A)** Mpl-affinity column fractions resolved on SDS-PAGE. Arrows indicate proteins from which the amino-terminal sequence as well as activity was obtained. **(B)** Elution of Mpl ligand activity from an SDS gel. The affinity-purified ligand was resolved by SDS-PAGE, and after electrophoresis, the gel was sliced, and proteins were electroeluted and assayed in the Ba/F3-Mpl⁺ assay.

molecule of 332 amino acids composed of two domains (Fig. 2). An amino terminal domain of 153 amino acids shows 23% identity to EPO, and 50% similarity when conservative substitutions are taken into account. This domain also shows a low level of identity to interferon (IFN)-α and IFN-β *(7)*. The EPO-like domain of TPO contains four cysteines, three of which are conserved with EPO, including the first and last,

```
h-ML    1  S P A P P A C D L R V L S K L L R D S H V L H S R L S Q C P E V H P L P T P V L L P A V D F S L G E
h-epo   1  A P P R L I C D S R V L E R Y L L E A K E A E N I T T G C A E H C S L N E N I T V P D T K V N F Y A

h-ML   51  W K T Q M E E T K A Q D I L G A V T L L L E G V M A A R G Q L G P T C L S - - S L L G Q L S G Q V R
h-epo  51  W K R M E V G Q Q A V E V W Q G L A L L S E A V L R G Q A L L V N S S Q P W E P L Q L H V D K A V S

h-ML   99  L L - - L G A L Q S L L G T Q - - - L P P Q G R T T A H K D P N A I F L S F Q H L L R G K V R F L -
h-epo 101  G L R S L T T L L R A L G A Q K E A I S P P D A A S A A P L R T I T A D T F R K L F R V Y S N F L R

h-ML  143  - - M L V G G S T L C V R R A P P T T A V P S R T S L V L T L N E L P N R T S G L L E T N F T A S A
h-epo 151  G K L K L Y T G E A C R T G D R

h-ML  191  R T T G S G L L K W Q Q G F R A K I P G L L N Q T S R S L D Q I P G Y L N R I H E L L N G T R G L F

h-ML  241  P G P S R R T L G A P D I S S G T S D T G S L P P N L Q P G Y S P S P T H P P T G Q Y T L F P L P P

h-ML  291  T L P T P V V Q L H P L L P D P S A P T P T P T S P L L N T S Y T H S Q N L S Q E G
```

Fig. 2. Comparison of human Mpl ligand (h-ML) and EPO sequences. Identical amino acids are boxed, and asparagine-linked glycosylation sites are underlined with a solid line for Mpl ligand and a dashed line for EPO. The two cysteines that are important for function are indicated by a dot.

which form an essential disulfide in EPO *(17)*. Similarly, the first and last cysteine of TPO form an essential disulfide *(18,19)*. However, unlike EPO, the disulfide formed between the two internal cysteines of TPO also appears to be essential *(18,19)*. None of the asparagine-linked glycosylation sites present in EPO are conserved in the EPO-like domain of TPO; however, the EPO-like domain of TPO likely contains two or three O-linked glycosylation sites. The 181 amino-acid carboxy-terminal domain of TPO does not bear homology to any other known protein and contains six potential *N*-glycosylation sites. It is rich in proline, serine, and threonine. The function of this domain remains to be elucidated. However, because of its high degree of glycosylation, this region may act to stabilize TPO and increase its circulating half-life. In the case of EPO, unglycosylated forms are fully active in vitro, but are relatively inactive in vivo because of rapid clearance *(20)*. Similarly, a truncated form of TPO consisting of just the EPO-like domain is fully functional in vitro, but significantly less active in vivo when compared with glycosylated full-length TPO *(21)*.

Sequence comparison between the mouse, pig, and dog TPO shows that the overall homology is approximately 70% with the EPO-like domain being highly conserved (approximately 80% identity), while the glycosylated C-terminal domain shows more species divergence (approximately 60% identity) *(22)*. In all forms, the carboxy-terminal domain contains six to seven potential asparagine-linked glycosylation sites. Interestingly, the two domains of TPO are separated by a potential dibasic proteolytic cleavage site that is conserved among the various species examined. Processing at this site could be responsible for releasing the C-terminal region from the EPO-like domain in vivo. The low-molecular-weight forms of TPO purified from the plasma of thrombocytopenic animals *(6–10)* could correspond to the EPO-like domain after cleavage at this site. The physiological relevance of this potential cleavage site is unclear at this time *(see* Chapter 13). The exact circulating form of TPO has not been elucidated.

When aplastic porcine plasma was subjected to gel filtration chromatography, TPO activity present in this plasma resolved with a relative molecular weight (M_r) of approximately 150,000 kDa (unpublished observation). Purified full-length TPO also resolves at this M_r, whereas the truncated forms resolve with M_r ranging from 18,000–30,000 kDa. Using a TPO enzyme-linked immunosorbent assay (ELISA) that selectively detects either full-length or truncated TPO, it has also been shown that full-length TPO is the predominant form in the plasma of marrow transplantation patients *(23)*. These results suggest that TPO circulates in the full-length form *(see* Chapter 13).

6. Conclusion

The isolation of TPO ended more than three decades of speculation that a specific factor regulated platelet production. Before its discovery, it was widely speculated that megakaryocytopoiesis was regulated by multiple factors acting synergistically, a notion that was then quickly dispelled. In other chapters, the profound effects of TPO on megakaryocytopoiesis and platelet production in vitro and in vivo are described. From these studies, it is clear that TPO is a physiological regulator of platelet production, has dramatic effects on platelet production in animals, and, to date, is the most potent thrombopoietic agent described. TPO not only stimulates platelet production in normal animals, but also accelerates platelet recovery in myelosuppressed animals. Because of this, recombinant forms of TPO may be of enormous benefit to cancer patients who become thrombocytopenic after myelosuppresive chemotherapy, radiation therapy, or bone marrow transplantation.

References

1. Gordon MS, Hoffman R. Growth factors affecting human thrombocytopoiesis: potential agents for the treatment of thrombocytopenia. *Blood.* 1992; 80: 302–307.
2. Avraham H. Regulation of megakaryocytopoiesis. *Stem Cells.* 1993; 11: 499–510.
3. McDonald TP. Thrombopoietin: its biology, clinical aspects and possibilities. *Am J Pediatr Hematol Oncol.* 1992; 14: 8–21
4. Yamamoto S. Mechanisms of the development of thrombocytosis due to bleeding. *Acta Haematol Jpn.* 1957; 20: 163–178.
5. Kelemen E, Cserhati I, Tanos B. Demonstration and some progenitors of human thrombopoietin. *Acta Haematol.* 1958; 20: 350–355.
6. de Sauvage FJ, Hass PE, Spencer SD, et al. Stimulation of megakaryocytopoiesis and thrombopoiesis by the c-Mpl ligand. *Nature.* 1994; 369: 533–538.
7. Lok S, Kaushansky K, Holly RD, et al. Cloning and expression of murine thrombopoietin cDNA and stimulation of platelet production in vivo. *Nature.* 1994; 369: 565–568.
8. Bartley TD, Bogenberger J, Hunt P, et al. Identification and cloning of a megakaryocyte growth and development factor that is a ligand for the cytokine receptor Mpl. *Cell.* 1994; 77: 1117–1124.
9. Sohma Y, Akahori H, Seki N, et al. Molecular cloning and chromosomal localization of the human thrombopoietin gene. *FEBS Lett.* 1994; 353: 57–61.
10. Kuter DJ, Beeler DL, Rosenberg RD. The purification of megapoietin: a physiological regulator of megakaryocyte growth and platelet production. *Proc Natl Acad Sci USA* 1994; 91: 11,104–11,108.

11. Souyri M, Vigon I, Penciolelli JF, Heard JM, Tambourin P, Wendling F. A putative truncated cytokine receptor gene transduced by the myeloproliferative leukemia virus immortalizes hematopoietic progenitors. *Cell.* 1990; 63: 1137–1147.

12. Vigon I, Mornon JP, Cocault L, et al. Molecular cloning and characterization of MPL, the human homolog of the *v-mpl* oncogene: identification of a member of the hematopoietic growth factor receptor superfamily. *Proc Natl Acad Sci USA* 1992; 89: 5640–5644.

13. Methia N, Louache F, Vainchenker W, Wendling F. Oligodeoxynucleotides antisense to the proto-oncogene *c-mpl* specifically inhibit in vitro megakaryocytopoiesis. *Blood.* 1993; 82: 1395–1401.

14. Vigon I, Florindo C, Fichelson S, et al. Characterization of the murine Mpl proto-oncogene, a member of the hematopoietic cytokine receptor family: molecular cloning, chromosomal location and evidence for a function in cell growth. *Oncogene.* 1993; 8: 2607–2615.

15. Skoda RC, Seldin DC, Chiang MK, Peichel CL, Vogt TF, Leder P. Murine *c-mpl*: a member of the hematopoietic growth factor receptor superfamily that tranduces a proliferative signal. *EMBO J.* 1993; 12: 2645–2653.

16. Wendling F, Maraskovsky E, Debili N, et al. c-Mpl ligand is a humoral regulator of megakaryocytopoiesis. *Nature.* 1994; 369: 571–574.

17. Wang FF, Kung CKH, Goldwasser E. Some chemical properties of human erythropoietin. *Endocrinology.* 1985; 116: 2286–2292.

18. Linden H, O'Rork C, Hart CE, et al. The structural and functional role of disulfide bonding in thrombopoietin. *Blood.* 1995; 86: 255a (abstract no 1008).

19. Kato T, Ozawa T, Muto T, et al. Essential structure for the biological activity of thrombopoietin. *Blood.* 1995; 86: 365a (abstract no 1448).

20. Takeuchi M, Inoue N, Strickland TW, et al. Relationship between sugar chain structure and biological activity of recombinant human erythropoietin produced in Chinese Hamster Ovary cells. *Proc Natl Acad Sci USA* 1989; 86: 7819–7822.

21. Eaton DL, Gurney A, Malloy B, et al. Biological activity of human thrombopoietin (TPO), the c-MPL ligand, and TPO variants and the chromosomal localization of TPO. *Blood.* 1994; 84: 241a (abstract no 948).

22. Gurney AL, Kuang WJ, Xie MH, Malloy BE, Eaton Dl, de Sauvage FJ. Genomic structure, chromosomal localization and conserved alternative splice forms of thrombopoietin. *Blood.* 1995; 85: 981–988.

23. Meng YG, Martin TG, Peterson ML, Shuman MA, Cohen RL, Wong WL. TPO regulation following bone marrow transplantation. *Blood.* 1995; 86: 313a (abstract no 1237).

9

The Purification of Thrombopoietin from Thrombocytopenic Plasma

David J. Kuter, Hiroshi Miyazaki, and Takashi Kato

1. Introduction

The identification and purification of the humoral regulator of platelet production has long been a major goal in the field of hematology. The initial isolation of erythropoietin (EPO) and subsequently of granulocyte colony-stimulating factor (G-CSF), granulocyte-macrophage colony-stimulating factor (GM-CSF), and monocyte colony-stimulating factor (M-CSF) suggested that a similar lineage-specific factor regulated platelet production. The commercial success of the previously identified lineage-specific factors provided further incentive.

Over the past 30 years, many efforts (reviewed in more detail in Chapter 5) have been undertaken to purify the putative humoral regulator of platelet production, "thrombopoietin" (TPO) *(1)*, from plasma, urine, or tissue-culture medium. Of these, the purifications by McDonald *(2–6)*, Levin *(7,8)*, Rosenberg *(9,10)*, and Murphy *(11,28)* have been the most extensively reported. Despite well-conceived efforts, some of epic proportions, none of these purifications yielded amounts of a single molecule adequate for amino-acid sequencing. As described below, the failure of these purifications can be attribututed to a number of factors. First, there was no rapid, reliable, biologically relevant TPO assay. Second, the appropriate source of starting material was unknown. Third, the required amount of starting material was underestimated given what is now known to be the circulating concentration of endogenous TPO. Finally, there was an inadequate understanding of the probable physiology of endogenous TPO.

Although present in the circulation, few hemopoietic growth factors have been successfully purified from plasma. EPO was purified from urine after many frustrating efforts to isolate it from plasma. G-CSF and GM-CSF were isolated from tissue-culture lines. In contrast, TPO has recently been successfully purified from the plasma by two different methods. The first used affinity chromatography with the recently isolated cytokine receptor Mpl to purify TPO from thrombocytopenic plasma *(12,13)*. In a second approach, two independent groups, one at Kirin's Pharmaceutical Research Laboratory (Tokyo, Japan) *(14,15)* and the other at Massachusetts Institute of Technology (MIT) (Boston, MA) *(16–18)*, were able to isolate TPO directly from thrombocytopenic

From: *Thrombopoiesis and Thrombopoietins: Molecular, Cellular, Preclinical, and Clinical Biology*
Edited by: D. J. Kuter, P. Hunt, W. Sheridan, and D. Zucker-Franklin Humana Press Inc., Totowa, NJ

plasma using standard purification methods. This chapter describes how the success of this latter approach was the result of a better understanding of the relevant physiology and the development of innovative bioassays, as well as an appreciation of the source and amount of starting material required.

2. Review of the Previous Major Purifications

Most previous attempts to purify TPO were modeled after the purification of EPO from anemic animal plasma or urine using erythroid-specific bioassays, such as reticulocyte counts or ^{59}Fe incorporation in vivo or in vitro. Although there is no platelet equivalent to the reticulocyte, early investigators *(6)* found that injection of thrombocytopenic plasma into rodents increased the number of circulating platelets. Unfortunately the injection of control foreign proteins also led to increased platelet counts, and this method was soon abandoned *(see* Chapter 1).

Subsequent purification attempts by McDonald *(4)* and Levin *(7)* used bioassays that measured the stimulation of platelet production in vivo using radioisotopes. McDonald and colleagues developed a thrombocytosis-stimulating factor (TSF) assay *(2)*, in which mice were initially made thrombocytopenic by the injection of antiplatelet serum and 5 days later developed rebound thrombocytosis. The thrombocytotic animals were then injected with test materials, and several days later, the incorporation of $^{35}SO_4$ into platelets was measured. Thrombocytopenic plasma *(2)* as well as human embryonic kidney (HEK)-conditioned medium *(3)* were found to stimulate this assay. Using this assay, a 30-kDa (nonreduced) protein was purified approximately 164,000-fold from HEK-conditioned medium *(4)* using Sephadex G-75, ethanol precipitation, preparative sodium dodecyl sulfate polyacrylamide gel electrophoresis (SDS-PAGE), and reverse-phase high-pressure liquid chromatography (RP-HPLC) on a C8 column (application in $0.01M$ NaH_2PO_4 and elution with CH_3CN). Although inadequate for protein sequencing, the partially purified molecule stimulated megakaryocyte colonies and small acetylcholinesterase-positive (SACHE$^+$) cells in vitro; increased megakaryocyte size, number, ploidy, and maturation in vivo; and increased platelet production *(see* ref. *5* for review).

Levin and colleagues also developed a platelet-production bioassay in which ^{35}Se-selenomethionine incorporation into platelets was determined after the injection of test materials into mice. ^{35}Se-selenomethionine incorporation increased after the injection of rabbit plasma collected 4–6 hours after the onset of thrombocytopenia *(7)*. From this thrombocytopenic rabbit plasma, a 40–47 kDa protein was partially purified. Purification steps included a 40–60% ammonium sulfate precipitation, wheat germ agglutinin (WGA)-Sepharose chromatography, concanavalin A (Con A) chromatography, and finally gel-filtration chromatography using a TSK-3000 column *(7)*. Although the partially purified molecule had little effect on megakaryocyte colony assays *(7)* or ploidy *(8)*, it did increase the growth of single megakaryocytes in vitro as well as increase acetylcholinesterase activity in murine liquid bone marrow cultures.

To overcome the deficiencies of these murine bioassays, Tayrien and Rosenberg *(9)* used the rat promegakaryoblast (RPM) cell line as a target, and showed that both HEK-conditioned medium and thrombocytopenic rabbit serum stimulated the production of platelet factor (PF)-4 by these cells. A 15-kDa protein was purified, but did not provide

reliable sequence data. The purified protein stimulated ^{14}C-PF-4 and ^{35}S-sulfate proteoglycan production in isolated megakaryocytes *(10)*, but did not increase megakaryocyte colony-forming cells (MK-CFC) or megakaryocyte ploidy in vitro, or increase the
platelet count in vivo (Kuter DJ, 1987, unpublished results).

Conventional MK-CFC assays were used by Erickson-Miller et al. *(11)* and Murphy,
Miyake, and colleagues *(28)* to identify a stimulatory activity in urine from patients
with aplastic anemia. A partial purification of this material was obtained following
ultrafiltration, desalting with a G-50 column, preparative PAGE, Mono P chromato-
focusing, and cation exchange HPLC. The 15-kDa protein stimulated megakaryocyte
colonies in serum-free cultures and did not possess activity of any then-known cytokine. After 3 days of administration of the partially purified protein to animals, the
platelet count increased 40%, and the number of splenic megakaryocytes also increased
(28). The biological activity was not affected by treatment with neuraminidase *(28)*.

3. The Development of New Bioassays for TPO

The previous TPO assays were inadequate for monitoring protein purification. The
^{35}SO$_4$ and ^{35}Se-selenomethionine murine bioassays were expensive, tedious, not always reliable, not easily quantifiable, and not amenable to assaying large numbers of
samples. Although the use of megakaryocytic cell lines was appealing, the RPM cells
were probably not very "megakaryocytic" *(19)*. Interestingly, a wide range of megakaryocyte cell lines have failed to show a mitogenic response when exposed to thrombocytopenic plasma (Kuter DJ, 1984–1992, unpublished results). Only those cell lines
that have been constructed both to contain Mpl and to be dependent on TPO for survival appear to demonstrate a mitogenic response to this cytokine. Finally, MK-CFC
assays are tedious, not easily quantified, unsuitable for large numbers of specimens,
and influenced by other cytokines, such as interleukin (IL)-3.

Given this state of affairs, newer bioassays for TPO were developed by the groups at
MIT and Kirin. While the workers at MIT created an assay dependent on late megakaryocyte maturation, those at Kirin developed an assay based on very early megakaryocyte growth.

Preliminary work at MIT had demonstrated that megakaryocyte ploidy was a unique
feature of the feedback loop between platelets and the megakaryocyte *(20)*. Ploidy was
then chosen as a marker and an in vitro ploidy assay developed *(21)*. In brief, rat bone
marrow was isolated, red blood cells lysed, and the marrow depleted of extant megakaryocytes using a series of Percoll gradients and filtration through fine meshes. This
decreased the background number of megakaryocytes, such that over 3 days in culture
(Fig. 1) newly produced megakaryocytes could be identified by flow cytometry, their
number counted, and their ploidy distribution determined. Subsequent experiments
demonstrated that this basal megakaryocyte growth could be stimulated markedly by
thrombocytopenic plasma (Fig. 2) and the extent of stimulation quantified by flow
cytometry. Because changes in the megakaryocyte ploidy distribution could be measured with exquisite precision *(22)*, megakaryocyte ploidy was the sole marker subsequently quantified (Fig. 3). The ploidy distribution for each sample tested could then
be precisely expressed by a simple value, the geometric mean ploidy *(23)*. This ploidy
assay demonstrated a linear dose–response curve for thrombocytopenic plasma (Fig. 4)

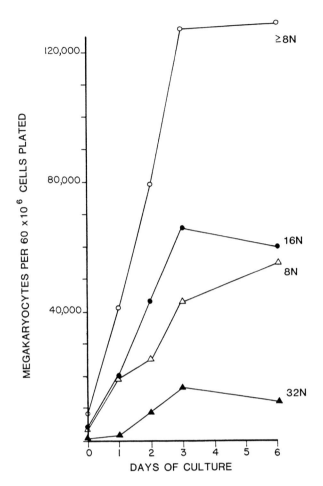

Fig. 1. Megakaryocyte growth from rat bone marrow cultures. Rat bone marrow was depleted of megakaryocytes and cultured in the presence of rat serum derived from platelet-poor plasma *(20)*. The number and ploidy of the megakaryocytes that grew from these cultures were measured daily by flow cytometry *(20)*.

and a run-to-run coefficient of variation of 5%, and was unaffected by any known cytokine except transforming growth factor-β (TGF-β), which inhibited the assay *(22)*. The major limitation of the ploidy assay for protein purification was that because the assay was based on flow cytometry, only a limited number of samples (20–30/day) could be assayed. A nonflow cytometric surrogate assay was therefore devised in which ^{14}C-serotonin was added to the culture wells 3 hours before harvesting and its uptake measured. More than 95% of the serotonin was located in megakaryocytes and this assay perfectly correlated with the ploidy assay (Fig. 5). As many as 300 samples could be easily assayed in 1 day.

Workers at Kirin had appreciated the deficiencies of the MK-CFC assays using agar or methylcellulose, and sought to develop a liquid culture system using MK-CFC cells. A GPIIb/IIIa-positive population of MK-CFC cells was purified from rat bone marrow

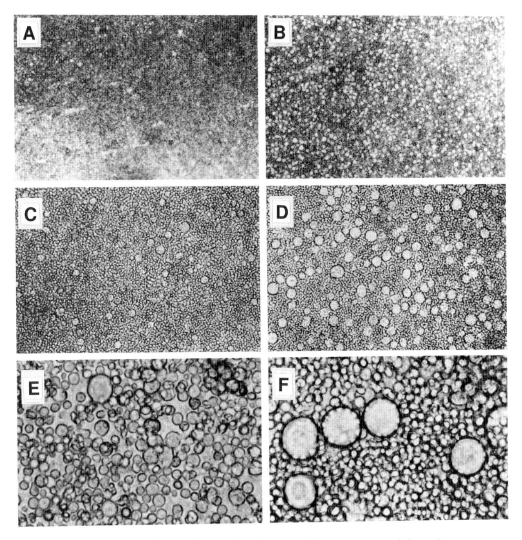

Fig. 2. Thrombocytopenic rat plasma stimulates megakaryocyte growth in rat bone marrow cultures. Photomicrographs of rat bone marrow cultures grown for 3 days in the presence of either normal **A, C, E)** or thrombocytopenic **(B, D, F)** rat plasma. All of the large cells are megakaryocytes, and the rest are early myeloid cells. (Magnification: ×5 [A, B]; ×12.5 [C, D]; ×50 [E, F].)

using an anti-GPIIb/IIIa monoclonal antibody (MAb) *(24)* after Percoll gradient separation and adherance depletion. This population contained more-differentiated MK-CFC that were more responsive to TPO than less-differentiated MK-CFC (H Miyazaki, 1995, unpublished). When these cells (containing approximately 10% MK-CFC) were placed into liquid culture, they developed into mature megakaryocytes over 4 days *(25)*. The extent of megakaryopoiesis could then be quantified after a 3-hour incubation with ^{14}C-serotonin (Fig. 6). The incorporation of serotonin was energy dependent, and was proportional to both the size and number of megakaryocytes present *(15)*. In

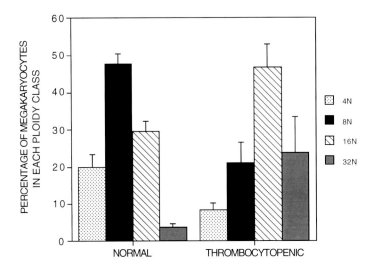

Fig. 3. Flow cytometry analysis of bone marrow cultures grown in the presence of plasma from normal or thrombocytopenic rats. The cultures shown in Fig. 2 were prepared for flow cytometry (20), and the megakaryocyte ploidy distribution analyzed. The $64N$ ploidy class was not quantified. The mean ploidy of megakaryocytes grown in normal plasma was 9.1, and the mean ploidy of megakaryocytes grown in thrombocytopenic plasma was 14.6.

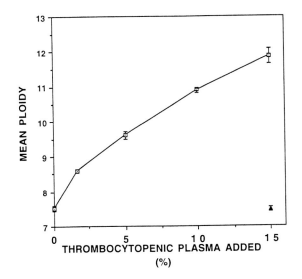

Fig. 4. Dose–response curve of thrombocytopenic sheep plasma in the ploidy assay. Rat bone marrow cultures were grown for 3 days in the presence of 15% fetal bovine serum (previously dialyzed for 24 hours versus Hank's Balanced Salt Solution without calcium or magnesium) plus increasing concentrations of normal (filled triangle) or thrombocytopenic (open squares) sheep serum. The cultures were harvested and the ploidy distribution (mean ploidy ± SD) determined by flow cytometry (22).

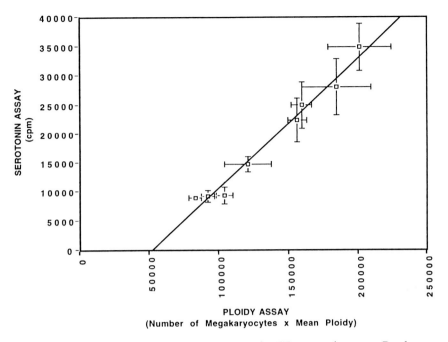

Fig. 5. Correlation between MIT ploidy assay and MIT serotonin assay. Rat bone marrow cultures were cultured in quadruplicate with increasing amounts of purified sheep TPO for 3 days *(17)*. Half of the wells were analyzed by ^{14}C-serotonin incorporation, and the other half analyzed by flow cytometry. The megakaryocyte mean ploidy and the number of megakaryocytes per well were measured, and the product (mean ploidy × number of megakaryocytes ± SD) plotted versus the ^{14}C-serotonin incorporation (cpm, ± SD). Statistical analysis: $r^2 = 0.977$.

this culture system, IL-3 promoted significant dose-dependent megakaryocyte growth, whereas IL-6, GM-CSF, EPO, and stem cell factor (SCF) promoted relatively minor growth; and IL-11 and leukemia-inhibitory factor (LIF) had no effect (Fig. 7).

Thrombocytopenic rat plasma promoted dose-dependent megakaryocyte growth, whereas normal rat plasma had no effect (Fig. 8). The stimulatory effect was easily recognized by microscopic observation and offered visual confirmation of the serotonin incorporation. This assay was rapid, quantitative, and allowed the preparation of one 96-well plate from the marrow from three rats.

4. Optimizing the Starting Material for Purification

Previous purification efforts had used as starting material either urine, HEK-conditioned medium, or plasma from thrombocytopenic animals. Although urine had been a primary source for the purification of EPO, there were few physiological data to suggest that it contained significant amounts of TPO. Early experiments *(16)* suggested that concentrated urine from thrombocytopenic rabbits did not stimulate the megakaryocyte ploidy assay. HEK-conditioned medium also lacked physiological correlates and contained numerous other hemopoietic cytokines *(26)*. Although thrombocytopenic plasma was a reasonable starting material, earlier workers had used plasma and/or se-

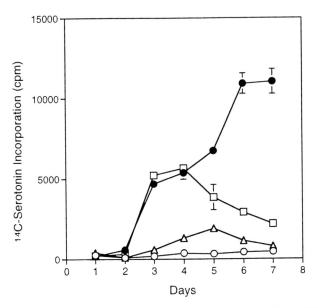

Fig. 6. Time-course for megakaryocyte growth stimulated with IL-3, EPO, and rat TPO. GPIIb/IIIa-positive cells were cultured with 1 U/mL IL-3 (filled circles), 10 ng/mL EPO (open triangles), 10 mg/mL of partially purified rat TPO obtained from the YMC-Pack Protein-RP column (open squares), and medium alone (open circles) for as long as 7 days, and the incorporation of ^{14}C-serotonin was measured daily. The data represent the mean (±SD) from one experiment. Similar data were obtained in another experiment *(15)*.

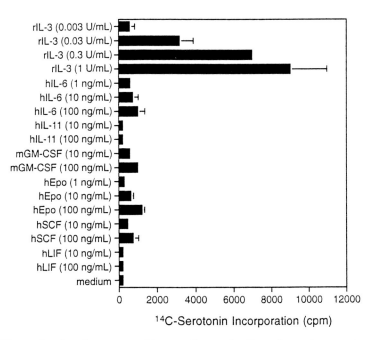

Fig. 7. Effects of various known cytokines on the production of megakaryocytes from GpIIb/IIIa-positive cells. GPIIb/IIIa-positive cells were cultured with varying concentrations of known cytokines for 4 days, and the incorporation of ^{14}C-serotonin was measured. The data represent the mean (±SD) from three separate experiments *(15)*.

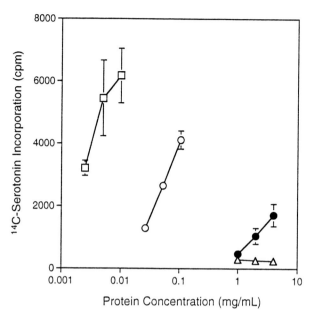

Fig. 8. The effects of normal rat plasma, thrombocytopenic rat plasma, and partially purified fractions of rat TPO on the production of megakaryocytes from GPIIb/IIIa-positive cells. GPIIb/IIIa-positive cells were cultured for 3 days with varying concentrations of normal rat plasma (open triangles), thrombocytopenic rat plasma (filled circles), TPO partially purified from thrombocytopenic rat plasma after the phenyl Sepharose column (open circles), or the YMC-Pack Protein-RP column (open squares), and the incorporation of ^{14}C-serotonin was measured. The data represent the mean (±SD) from two separate experiments *(15)*.

rum obtained 1–6 hours after inducing thrombocytopenia by the infusion of antiplatelet antibody. As described below, this immune model gave too brief a period of thrombocytopenia and resulted in the collection of plasma at times when TPO levels were barely increased in the circulation. Furthermore, by using serum rather than plasma, inhibition by TGF-β was a potential problem for the bioassays *(22)*.

To optimize the amount of TPO in plasma, scientists at both MIT and Kirin first established time-courses for the appearance of TPO in the circulation. In antiplatelet antibody models of thrombocytopenia (*see* Chapter 23) in rats and rabbits at MIT, levels of thrombopoietic activity increased half-maximally at 8 hours, peaked at 24 hours *(16)*, and then rapidly decreased. Unfortunately, small variations in this time-course between animals made it very difficult to obtain adequate amounts of plasma during the brief appearance of TPO in the circulation.

At Kirin, antibody-induced thrombocytopenic rat plasma (TRP) was examined to determine whether it was a suitable starting material for purification. An unknown thrombopoietic factor existed in the TRP that was both chromatographically and biologically distinguishable from other known cytokines. A large pool of TRP was then laboriously prepared by injecting antiplatelet antibody into 1000 rats. Although important biochemical information about TPO was obtained, a complete purification could not be accomplished. In comparison to the amount of EPO necessary for

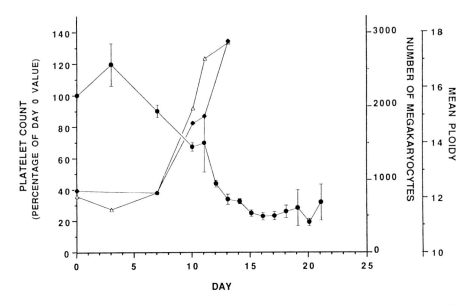

Fig. 9. The relationship of TPO concentration to the platelet count in sheep. After administration of busulfan to 106 sheep, platelet counts (filled circles, ±SD) as well as the stimulatory effect of plasma samples on the number (filled squares) and mean ploidy (open triangles) of megakaryocytes in vitro were determined at frequent intervals. All values for mean ploidy and number of megakaryocytes after day 7 differ from pretreatment values ($p < 0.05$) *(17)*.

its purification, the levels of TPO in TRP were estimated to be too low to permit successful purification.

Because of these limitations of the immune models of thrombocytopenia, both groups developed nonimmune models of prolonged thrombocytopenia in which much higher levels of thrombopoietic activity were present for a longer period of time. The MIT group used busulfan (Fig. 9), whereas the Kirin group used irradiation to generate severe, prolonged thrombocytopenia 14 days later. Both groups carefully harvested blood into citrate anticoagulant to prevent the generation of inhibitory substances, such as TGF-β. Finally, to provide adequate amounts of starting material, the MIT researchers plasmapheresed a large number of thrombocytopenic sheep, whereas the Kirin workers bled 1100 thrombocytopenic rats.

5. Purification Schemes

The previous attempts to purify TPO had used many different chromatographic methods, some of which were used in the subsequent purifications at MIT and Kirin. The purification schemes for both groups are shown in Tables 1 and 2.

There are a number of important aspects of the 11-step MIT purification. The polyethylene glycol (PEG) step eliminated high-molecular-weight serum proteins, especially fibrinogen, whereas the carboxymethyl (CM) and diethylaminoethyl (DEAE) Sepharose steps allowed a significant reduction in volume and a modest degree of purification. At both the CM and DEAE steps, multiple eluting forms were always de-

Table 1
The Purification of Thrombopoietin from Thrombocytopenic Sheep Plasma[a]

	Volume, mL	Protein, mg/mL	Total protein, mg	TPO, U/mL	Total TPO, U	Recovery, %	Specific activity	Purification
Plasma	21132	48.4	1023000	0.022	502	100	0.0005	1
PEG precipitation	49344	12.4	609393	0.011	537	100	0.0009	1.69
CM Sepharose-1	3636	7.73	24656	0.143	514	96	0.02	40
CM Sepharose-2	524	13.4	6999	1	477	102	0.069	152
DEAE Sepharose	150	6.85	875	0.725	117	26	0.133	311
Hydroxylapatite (High NaCl)	355	0.0367	12.9	0.346	124	28	9.41	23084
Hydroxylapatite (Low NaCl)	9.92	0.652	3.68	11.8	84.2	16.2	21.17	40829
Mono S	5.34	0.119	0.436	10.55	55.1	11.9	80	187374
WGA Sepharose	1.42	0.092	0.12	10.71	26.2	5.43	274	458068
Heparin Sepharose	1.25			19.2	24	4.8		
TMS PRC	0.16	0.025	0.004	57.5	7.16	1.9	1785	4973039
TSK HPLC	3	0.0006	0.0017		7.95	1.6	4676	9352941

[a]Average of 12 separate 20-L purifications with quantitation of recoveries. The CM-Sepharose product was arbitrarily chosen to have 1 U/mL of thrombopoietin activity.

Table 2
The Purification of Thrombopoietin from Irradiated Rat Plasma

Process	Total protein, mg	Relative specific activity	Total activity, $\times 10^{-5}$	Activity yield, %
Citrated plasma	493000	—	—	—
Ca-treatment & Sephadex G-25	480300	2	8.65	100
Q Sepharose FF	314400	9	27.00	313
WGA-Agarose	15030	1.32×10^2	19.90	230
TSK-gel AF-Blue 650 MH	4236	9.05×10^2	38.30	443
Phenyl Sepharose 6 FF/LS	2762	8.47×10^2	23.40	271
Sephacryl S-200HR (F3)	51	2.00×10^4	10.20	118
YMC-pack Protein-RP	2.854	1.30×10^5	3.71	43
YMC-pack CN-AP	0.373	8.00×10^5	2.98	35
Capcell Pak C1 (FA)	0.0396[a]	4.89×10^6	1.94	22
15% SDS-PAGE	0.0017[b]	1.49×10^8	2.50	29

[a]An 11-step standard procedure was devised to purify rat thrombopoietin directly from the plasma of 1100 irradiated rats. The relative specific activity was arbitrarily calculated by using an assay standard obtained from thrombocytopenic rat plasma after the administration of antibody against rat platelets *(14)*.

[b]Estimate based on the absorbance at 280 nm of the chromatogram, assuming A_{280} of 1.0 is equal to 1 mg of protein per mL.

[c]Estimate based on the intensity of silver staining of BSA after SDS-PAGE.

tected and only the major species was carried on to the next step. A crucial aspect of subsequent steps was the addition of 3-([3-cholamidoproply]-dimethyl-ammonio)-1-propanosulfonate (CHAPS) (0.25% final concentration) to minimize the extensive aggregation of TPO that was experienced during purification. A wide range of detergents were effective, but only CHAPS did not interfere with the bioassay. A high degree of intermediate purification was then attained on a newly developed ceramic hydroxylapatite column. At high sodium chloride concentrations, TPO did not bind to this matrix, but when reapplied at lower sodium chloride concentration, it bound and then eluted as a single peak. Sulfonated Sepharose (Mono S) was used to perform a high-resolution cationic exchange step, and the eluting activity peak was concentrated on WGA Sepharose. A major protein copurifying at this step was identified by sequence analysis as protein C inhibitor. Since this protein was known to bind tightly to heparin, heparin Sepharose was used to remove it. The remaining TPO activity bound irreversibly to C-18 and C-8 hydrophobic matrices, but could be eluted from trimethyl silane (TMS) as a single peak. A few remaining high-molecular-weight contaminants were removed on Tosohaas TSK G-3000 (TSK), and the final TPO molecule eluted as a single peak. From each 20-L batch of sheep plasma, approximately 1.7 µg TPO was recovered with a final purification of approximately 9 million-fold. The final TSK product consisted of three different molecular weight species (Fig. 10) of 41.6, 35.7, and 27.8 kDa each with the same specific activity. Amino acid analysis of the entire TSK product produced equimolar amounts of four nonoverlapping peptides of unambiguous sequence (Fig. 11) and no minor peptides.

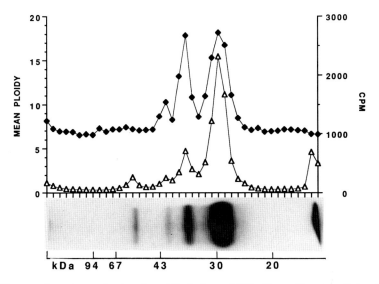

Fig. 10. Electrophoretic analysis of purified sheep TPO. Two aliquots of the final TSK-purified protein, one iodinated, were subjected to electropheresis (unreduced) on SDS gels. The iodinated lane was subjected to autoradiography and also quantified with the phosphorImager (open triangles). The other lane was eluted, and TPO activity was measured by assessing the effect on mean ploidy (filled squares) *(17)*.

At Kirin, the liquid MK-CFC assay was used to follow the 11-step purification (Table 2) of TPO from 730 mL aliquots (total of 8 L) of thrombocytopenic rat plasma. Key aspects of the purification were as follows. Fibrinogen was removed by clotting the plasma, then adding (*para*-aminodiphenyl) methananesulfonyl fluoride hydrochloride and filtering through Sepharose G-25. A threefold increase in total activity was found after elution from Q Sepharose FF suggesting the removal of an inhibitory substance. (The MIT group found that CM-Sepharose removed a similar inhibitor from the plasma of some thrombocytopenic sheep as well as from all plasmas from both normal and thrombocytopenic rabbits.) TPO activity bound to WGA-agarose, and this step gave a 14.7-fold increase in specific activity and led to the removal of most protease activity. Preliminary experiments had demonstrated strong binding of TPO to triazine dye (AF-Blue). Active fractions were eluted with NaSCN and then immediately applied to phenyl Sepharose to remove the chaotropic reagent. Sephacryl S-200HR gel filtration produced four peaks of activity, one of which *(F3)* had the highest purity and was subjected to further purification on three different types of reverse-phase matrices. Kirin researchers found that the maximum recovery and separation on each reverse-phase matrix were obtained by using a propanol solvent system. These chromatographic conditions had the added advantage of decreasing aggregation. The final peak eluting off the C1 reverse-phase column was subjected to SDS-PAGE (nonreduced) and three bands with molecular weights of 19, 14, and 11 kDa were detected on silver staining (Fig. 12). MK-CFC activity was associated only with the 19-kDa protein (Fig. 12), which was then digested and the peptides sequenced. Three of the six peptides recov-

```
                    1        10        20        30        40        50        60
Rat       MELTDLLLVAILLLTARLTLSSPVPPACDPRLLNKLLRDSYLLHSRLSQCPDVNPLSIPVLLPAVDFSLGEWKTQTEQSKA
Murine    MELTDLLLAAMLLAVARLTLSSPVAPACDPRLLNKLLRDSHLLHSRLSQCPDVDPLSIPVLLPAVDFSLGEWKTQTEQSKA
Canine    MELTELLLVVMLLLTARLDPCLPAPPACDPRLLNKMLRDSHVLHSRLSQCPDIYPLSTPVLLPAVDFSLGEWKTQKEQTKA
Human     MELTELLLVVMLLLTARLTLSSPAPPACDLRVLSKLLRDSHVLHSRLSQCPEVHPLPTPVLLPAVDFSLGEWKTQMEETKA
Ovine                     SPVPPACDPRLL                                              KA

                    70        80        90       100       110       120       130      140
Rat       QDILGAVSLLLEGVMAARGQLEPSCLSSLLGQLSGQVRLLLGALQGLLGTQLPPQGRTTAHKDPSALFLSLQQLLRGKVR
Murine    QDILGAVSLLLEGVMAARGQLEPSCLSSLLGQLSGQVRLLLGALQGLLGTQLPLQGRTTAHKDPNALFLSLQQLLRGKVR
Canine    QDVWGAVALLLDGVLAARGQLGPSCLSSLLGQLSGQVRLLLGALQGLLGTQLPPQGRTTIHKDPNAIFLSFQQLLRGKVR
Human     QDILGAVTLLLEGVMAARGQLGPTCLSSLLGQLSGQVRLLLGALQSLLGTQLPPQGRTTAHKDPNAIFLSFQHLLRGKVR
Ovine     QDVLGTTTLLLEAVMAARGQLGPTXL                                 KDPSAIFLNFQQLLRGK

                   150       160       170       180       190       200       210      220
Rat       FLLLVEGPALCVRRTLPTTAVPSRTSQL-LTLNKFPNRTSGLLEINFSVVARTAGPGLLNRLQGFRAKIIPGQLNQTSGS
Murine    FLLLVEGPTLCVRRTLPTTAVPSSTSQL-LTLNKFPNRTSGLLEINFSVTARTAGPGLLSRLQGFRVKITPGQLNQTSRS
Canine    FLLLVAGPTLCAKQSQPTTAVPTNTS-LFLTLRKLPNRTSGLLEINSSISARTTGSGLLKRLQGFRAKI-PGLLNQTSRS
Human     FLMLVGGSTLCVRRAPPTTAVPSRTS-LVLTLNELPNRTSGLLEINFTASARTTGSGLLKWQQGFRAKI-PGLLNQTSRS
Ovine            KRAPPAXAVPGSISPL-LTLNKLPXRTSGLLETXSSVSARTTGFGLP

                   230       240       250       260       270       280       290      300
Rat       LDQIPGYLNGTHEPVNGIHGLFAGTSLQT-LEAPDVVPGAFNKGSLPLNLQSGLPPIPSLAADG-YTLFPPSPTFPTP-G
Murine    PVQISGYLNRTHGPVNGIHGLFAGTSLQT-LEASDISPGAFNKGSLAFNLQGGLPPSPSLAPDGH-TPFPPSPALPTTHG
Canine    LNQTPGHLSRTHGPLNGIHGLLPGLSL-TALGAPDIPPGTSDMDALPPNLWPRYSPSPIHPPPGQYTLFSPLPTSPTPQ-
Human     LDQIPGYLNRIHELLNGTRGLFPGPSRRT-LGAPDISSGTSDTGSLPPNLQPGYSPSPTHPPTGQYTLFPLPPTLPTP--
Ovine

                   310       320       330       340
Rat       SPPQLPPVS.............................
Murine    SPPQLHPLFPDPSTTMPNSTAPHPVIMYPHPRNLSQET..
Canine    NP--LQPPPPDPSATA-NSTSPLLIAAHPHFQNLSQEE..
Human     -VVQLHPLLPDPSAPTPTPTSPLLNTSYTHSQNLSQEG..
Ovine
```

Fig. 11. Comparison of rat, mouse, dog, human, and sheep TPO amino acid sequences. Except for the sheep sequence, all are deduced from the cloned cDNA sequences *(12–14,27).*

ered contained unique sequences. The two major peptides (SPVPPACDPRLL, RRTLPTTAVPS) recovered were confirmed to be derived from rat TPO by cDNA cloning (Fig. 11).

6. Properties of the Purified "Native" TPO

The TPO purified by both groups was not the full-length molecule predicted from the DNA sequence, but consisted primarily of the "EPO domain." Similar truncated molecules were also obtained when TPO was purified from thrombocytopenic plasma using Mpl affinity columns *(12,13,27).* Gel-filtration and SDS gel analysis performed on the material before the DEAE step demonstrated heterogeneous biological activity between 70 and 80 kDa (Kuter DJ, 1990, unpublished results). Subsequent purification steps contained biologically active protein of smaller molecular weight that may represent the product of protease digestion. Although detectable proteolytic activities in plasma were almost completely removed after the WGA chromatography step (Kato T, unpublished data) used in the purification at both Kirin and MIT, trace proteolytic ac-

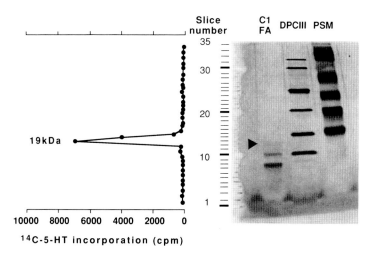

Fig. 12. SDS-PAGE analysis of purified rat TPO. The final C1-RP-HPLC fraction of rat TPO purified from irradiated rat plasma (Table 2) was subjected to 15% SDS-PAGE under nonreducing conditions. The TPO activity (filled circles) extracted from the gel was detected in the region corresponding to the diffuse 19-kDa silver-stained band. Its molecular weight was determined by the DPCIII low-molecular-weight markers (Daiichi Pure Chemicals, Japan). PSM is a mixture of prestained markers (Bio-Rad, Japan) *(14).*

tivity could still be present. Peptides obtained by both groups contained an intact Arg153-Arg154 suggesting that cleavage by dibasic peptidases at this potential site might not be physiologically important. IP-TPO was found to be sensitive to reduction and to protease activity, but stable at pH 4.0–10.0 and after boiling at 100°C. At many stages of purification, TPO behaved as a very hydrophobic protein and tended to associate with itself and other proteins, unless detergents or chaotrophic agents were present. The purified TPO is a glycoprotein as demonstrated by its binding to WGA Sepharose *(17)* and WGA agarose *(14).* Finally, based on the recoveries after purification, the TPO concentration in plasma from thrombocytopenic sheep and rats is approximately 167 and 38 p*M*, respectively.

"Native" TPO is produced and secreted by livers of normal and thrombocytopenic rats, as well as from three rat hepatoma cell lines *(29).* These studies confirmed that the liver was the major site of TPO production and identified a source of the cDNA library used to clone rat TPO *(30).* TPO was partially purified from the culture medium of rat liver and rat hepatoma cell lines, and eluted at the same retention time on reverse-phase chromatography (Fig. 13) as did TPO purified from the plasma of irradiated rats. Interestingly, the TPO partially purified from each cell line showed a variable molecular weight distribution on SDS-PAGE compared with the TPO purified from plasma (Fig. 14). A truncated form was also observed in TPO purified from serum-free conditioned medium of cell lines that had not been exposed to blood components (Kato T et al., unpublished data).

TPO purified from thrombocytopenic canine plasma has been shown to stimulate MK-CFC in vitro *(27),* and the availability of "native" TPO, purified as above, allowed

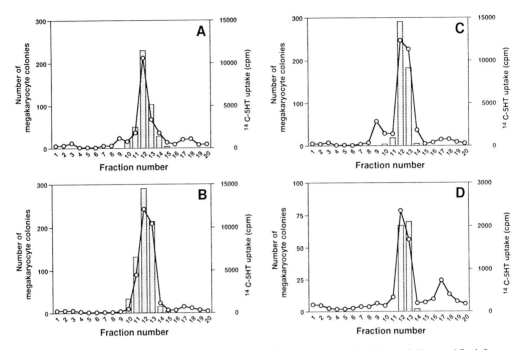

Fig. 13. The elution profile of MK-CFC and TPO activity of TPO partially purified from rat hepatoma cells lines and plasma from irradiated rats. Partial purification of TPO from three rat hepatoma cell lines (H4-II-E, McA-RH8994, HTC) and plasma from irradiated rats was performed with Sephadex G25, Q-Sepharose, WGA-agarose, TSK-gel AF-Blue, phenyl-Sepharose, and Vydac Protein C4 RP-HPLC columns. In all cases significant activity on both MK-CFC and serotonin MK-CFC assays was observed and eluted at the same retention time on RP-HPLC. **(A)** H4-II-E. **(B)** McA-RH8994. **(C)** HTC. **(D)** plasma from irradiated rats. In the same manner, TPO partially purified from culture medium from primary hepatocyte cultures obtained from normal, immune thrombocytopenic, and irradiated thrombocytopenic rats also showed significant TPO activity in the same reverse-phase fractions (not shown) *(29)*.

a range of biological experiments to be performed using nonrecombinant TPO. All showed properties identical to those of the recombinant molecule. In vitro, the plasma-derived TPO increased the number, size, and ploidy of megakaryocytes (Fig. 15), increased the number of MK-CFC both with and without IL-3 (Fig. 16), and stimulated proplatelet formation (Fig. 17). On injection into animals, it increased the number, size, and ploidy of megakaryocytes (Fig. 18) and increased the platelet count (Fig. 19). It was also neutralized by a soluble form of Mpl (Fig. 20).

7. Conclusions

As described elsewhere (Chapters 6–8), the discovery of Mpl is a dramatic example of how the identification of a previously unknown hemopoietic receptor could be used to identify and purify a new hemopoietic cytokine. Unfortunately, not all biologically relevant molecules will be amenable to this approach. The pu-

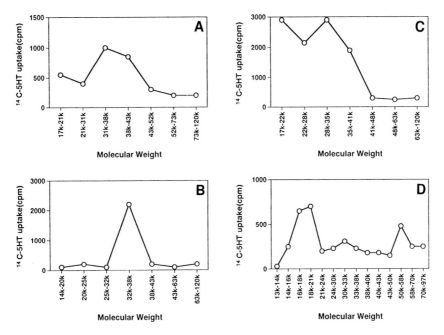

Fig. 14. The variable molecular weight of rat TPO partially purified from hepatoma cell lines and irradiated rat plasma. TPO was partially purified by Vydac C4 RP-HPLC (as described in Fig. 13) from three rat hepatoma cell lines (H4-II-E, McA-RH8994, HTC), and plasma from irradiated rats, and subjected to SDS-PAGE. Each gel was sliced into 10–20 pieces. TPO activity was extracted from each piece and assayed in the ^{14}C-serotonin assay. **(A)** H4-II-E. **(B)** McA-RH8994. **(C)** HTC. **(D)** plasma from irradiated rats *(29)*.

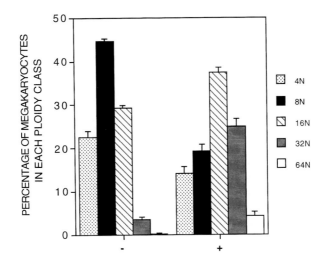

Fig. 15. Purified sheep TPO stimulates megakaryocyte ploidy in vitro. Either purified sheep TPO (+, 100 pM) or buffer (–) was added to rat bone marrow cultures *(17)* and the effect on megakaryocyte ploidy measured by flow cytometry. In the absence of TPO, the mean ploidy was 10.4, and in its presence, the mean ploidy was 18.8.

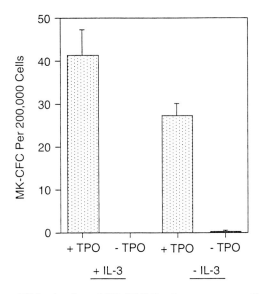

Fig. 16. Purified sheep TPO stimulates MK-CFC. Rat bone marrow cells were prepared as previously described *(22)* and MK-CFC grown in methylcellulose for 7 days in the presence or absence or IL-3 and supplemented with sheep TPO (50 pM) or buffer. Colonies consisting of three or more megakaryocytes were identified by staining for acetylcholinesterase and are presented ±SD.

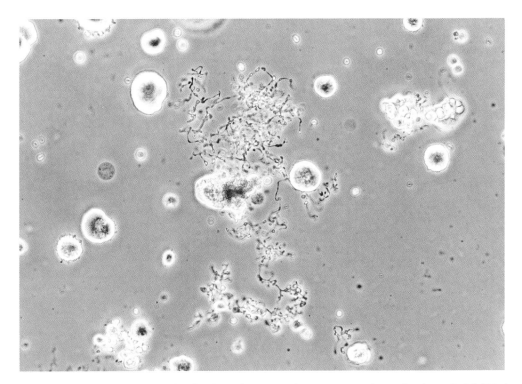

Fig. 17. A photomicrograph of cultured rat megakaryocytes and proplatelets. GpIIb/IIIa[+] cells highly enriched for rare MK-CFC were cultured in liquid medium containing partially purified rat TPO for 4 days. MK-CFC differentiated into mature megakaryocytes, some of which formed proplatelets.

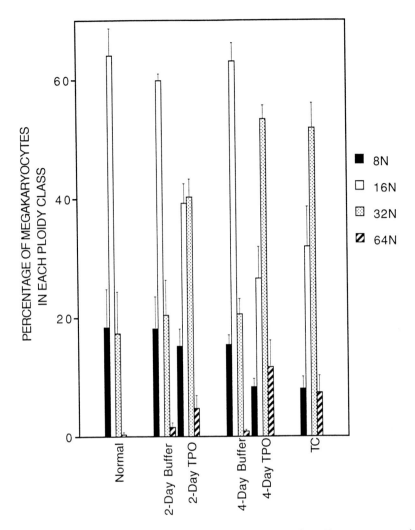

Fig. 18. Sheep TPO stimulates megakaryocyte ploidy in vivo. Four rats were injected twice daily for 2 days with either sheep TPO (240 ng/injection) or an equivalent volume of buffer, and harvested on day 3 for flow cytometry *(21)*. Five rats were injected twice daily with TPO (240 ng/injection) or buffer for 4 days, and harvested on day 5 for flow cytometry. Shown for comparison are the megakaryocyte ploidy distributions for 356 normal rats and for 11 rats made severely thrombocytopenic (platelet count 7.6% of normal) by the injection of antiplatelet antibody *(21)*. The mean ploidy values are: normal, 17.5; 2-day buffer, 18.5; 2-day TPO, 23.5; 4-day buffer, 18.5; 4-day TPO 29.5; thrombocytopenic (TC), 27.0 *(17)*.

rification of TPO by workers at MIT and Kirin illustrates a more classical approach for the isolation of new hemopoietic cytokines from plasma. By adopting a clearer understanding of the relevant biology, new bioassays were developed that were physiologically meaningful. When coupled with a better appreciation of the source and amount of starting material, successful purification of a new hemopoietic growth factor, TPO, was accomplished.

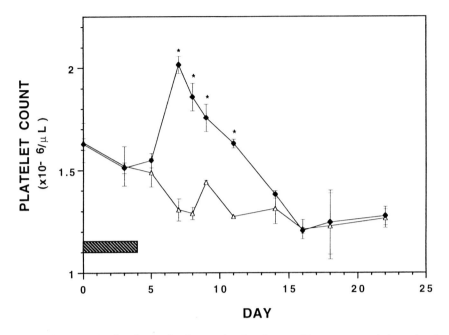

Fig. 19. Sheep TPO stimulates platelet production in rats. Two rats were injected twice daily with either sheep TPO (filled squares, 228 ng/injection) or buffer (open triangles) for 4 days (hatched area), and the platelet counts (±SD) were determined at intervals. * $p < 0.01$ compared with buffer *(17)*.

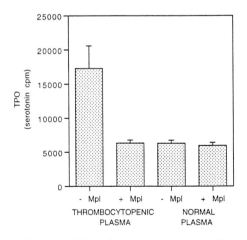

Fig. 20. All of the stimulatory activity of thrombocytopenic sheep plasma can be removed by the TPO receptor, Mpl. Plasma-derived serum from normal or thrombocytopenic sheep was incubated with soluble Mpl *(18)* for 1 hour and then added to rat bone marrow cultures *(17)*. The effect on serotonin incorporation (±SD) was measured 3 days later.

Acknowledgment

This work was supported in part NIH Grant HL54838 (D. J. K.).

References

1. Kelemen E, Cserhati I, Tanos B. Demonstration and some properties of human thrombopoietin. *Acta Haematol.* 1958; 20: 350–355.
2. McDonald TP, Cottrell M, Clift R. Hematologic changes and thrombopoietin production in mice after X-irradiation and platelet-specific antisera. *Exp Hematol.* 1977; 5: 291–298.
3. McDonald TP, Clift R, Lange RD, Nolan C, Tribby IL, Barlow GH. Thrombopoietin production by human embryonic kidney cells in culture. *J Lab Clin Med.* 1975; 85: 59–66.
4. McDonald TP, Clift RC, Cottrell MB. A four-step procedure for the purification of thrombopoietin. *Exp Hematol.* 1989; 17: 865–871.
5. McDonald TP. Thrombopoietin: its biology, purification and characterization. *Exp Hematol.* 1988; 16: 201–205.
6. McDonald TP. The hemagglutination-inhibition assay for thrombopoietin. *Blood.* 1973; 41: 219–233.
7. Hill RJ, Levin J, Levin FC. Correlation of in vitro and in vivo biological activities during the partial purification of thrombopoietin. *Exp Hematol.* 1992; 20: 354–360.
8. Hill RJ, Leven RM, Levin FC, Levin J. The effect of partially purified thrombopoietin on guinea pig megakaryocyte ploidy in vitro. *Exp Hematol.* 1989; 17: 903–907.
9. Tayrien G, Rosenberg RD. Purification and properties of a megakaryocyte stimulatory factor present both in the serum-free conditioned medium of human embryonic kidney cells and in thrombocytopenic plasma. *J Biol Chem.* 1987; 262: 3262–3268.
10. Greenberg SM, Kuter DJ, Rosenberg RD. In vitro stimulation of megakaryocyte maturation by megakaryocyte stimulatory factor. *J Biol Chem.* 1987; 262: 3269–3277.
11. Erickson-Miller CL, Ji H, Parchment RE, Murphy MJ. Megakaryocyte colony-stimulating factor (Meg-CSF) is a unique cytokine specific for the megakaryocyte lineage. *Br J Haematol.* 1993; 84: 197–203.
12. de Sauvage FJ, Hass PE, Spencer SD, et al. Stimulation of megakaryocytopoiesis and thrombopoiesis by the c-Mpl ligand. *Nature.* 1994; 369: 533–538.
13. Bartley TD, Bogenberger J, Hunt P, et al. Identification and cloning of a megakaryocyte growth and development factor that is a ligand for the cytokine receptor Mpl. *Cell.* 1994; 77: 1117–1124.
14. Kato T, Ogami K, Shimada Y, et al. Purification and characterization of thrombopoietin. *J Biochem.* 1995; 118: 229–236.
15. Miyazaki H, Horie K, Shimada Y, et al. A simple and quantitative liquid culture system to measure megakaryocyte growth using highly purified CFU-MK. *Exp Hematol.* 1995; 23: 1224–1228.
16. Kuter DJ, Rosenberg RD. Appearance of a megakaryocyte growth-promoting activity, megapoietin, during acute thrombocytopenia in the rabbit. *Blood.* 1994; 84: 1464–1472.
17. Kuter DJ, Beeler DL, Rosenberg RD. The purification of megapoietin: a physiological regulator of megakaryocyte growth and platelet production. *Proc Natl Acad Sci USA.* 1994; 91: 11,104–11,108.
18. Kuter DJ, Rosenberg RD. The reciprocal relationship of thrombopoietin (c-Mpl ligand) to changes in the platelet mass during busulfan-induced thrombocytopenia in the rabbit. *Blood.* 1995; 85: 2720–2730.
19. Myrseth LE, Gulowsen AC, Davies CL, Grandaunet J. Is the RPM cell line a useful model for the study of megakaryocyte development? *Exp Hematol.* 1990; 18: 1073–1077.
20. Kuter DJ, Greenberg SM, Rosenberg RD. Analysis of megakaryocyte ploidy in rat bone marrow cultures. *Blood.* 1989; 74: 1952–1962.
21. Kuter DJ, Rosenberg RD. The regulation of megakaryocyte ploidy in vivo in the rat. *Blood.* 1990; 75: 74–81.
22. Kuter DJ, Gminski D, Rosenberg RD. Transforming growth factor-beta inhibits megakaryocyte growth and endomitosis. *Blood.* 1992; 79: 619–626.

23. Arriaga M, South K, Cohen JL, Mazur EM. Interrelationship between mitosis and endomitosis in cultures of human megakaryocyte progenitor cells. *Blood.* 1987; 69: 486–492.
24. Miyazaki H, Inoue H, Yanagida M, et al. Purification of rat megakaryocyte colony forming cells using a monoclonal antibody against rat platelet glycoprotein IIb/IIIa. *Exp Hematol.* 1992; 20: 855–861.
25. Inoue H, Ishii H, Tsutsumi M, et al. Growth factor-induced process formation of megakaryocytes derived from CFU-MK. *Br J Haematol.* 1993; 85: 260–269.
26. Withy RM, Rafield LF, Beck AK, Hoppe H, Williams N, McPherson JM. Growth factors produced by human embryonic kidney cells that influence megakaryopoiesis include erythropoietin, interleukin 6, and transforming growth factor-beta. *J Cell Physiol.* 1992; 153: 362–372.
27. Hunt P, Li Y-S, Nichol JL, et al. Purification and biologic characterization of plasma-derived Megakaryocyte Growth and Development Factor (MGDF). *Blood.* 1995; 86: 540–547.
28. Miyake T, Kawakita M, Enomoto K, Murphy MJ. Partial purification and biological properties of thrombopoietin extracted from the urine of aplastic anemia patients. *Stem Cells.* 1982; 2: 129–144.
29. Shimada Y, Kato T, Ogami K, et al. Production of thrombopoietin (TPO) by rat hepatocytes and hepatoma cell lines. *Exp Hematol.* 1995; 23: 1388–1396.
30. Sohma Y, Akahori H, Seki N, et al. Molecular cloning and chromosomal localization of the human thrombopoietin gene. *FEBS Lett.* 1994; 353: 57–61.

10

The Role of Other Hemopoietic Growth Factors and the Marrow Microenvironment in Megakaryocytopoiesis

Ronald Hoffman

1. Introduction

The recent identification, purification, and cloning of thrombopoietin (TPO), a lineage-specific regulator of platelet production, is the culmination of several decades of research dealing with the regulation of megakaryocytopoiesis *(1–5)*. Although endogenous TPO plays a pivotal role in promoting the proliferation and maturation of megakaryocytic progenitor cells and megakaryocytes, a considerable amount of data exists that indicates that other growth factors are also capable of altering this finely regulated biological process *(6,7)*. These non-TPO regulatory factors are capable not only of promoting megakaryocytic proliferation and maturation, but also of downregulating these cellular processes *(6,7)*. These growth factors might play an important role in the physiological regulation of megakaryocytopoiesis and may be instrumental in the pathogenesis of a number of clinical syndromes *(8–12)*. In addition, recombinant forms of these growth factors have considerable potential in the treatment of clinical disorders of thrombopoiesis *(7)*.

2. Regulation of Megakaryocyte Development

Physiological control of the megakaryocyte lineage occurs at at least two levels: expansion of megakaryocyte numbers and regulation of megakaryocyte maturation. Whether the final stage of platelet biogenesis, platelet shedding, is a regulated process is an area of active investigation. Although TPO appears to influence several of the stages of platelet development, other growth factors also have important biological effects. The importance of these other growth factors in the regulation of megakaryocytopoiesis is highlighted by the studies of Gurney et al. *(13)*. In an attempt to investigate further the role of the TPO receptor, Mpl, in the regulation of megakaryocytopoiesis, Mpl-deficient mice were generated by gene targeting. These mice possessed platelet counts and megakaryocyte numbers that were 15% of that observed in normal mice. These results clearly show that Mpl specifically regulates thrombopoiesis and definitively indicates that the ligand for Mpl, TPO, plays a pivotal role in the regulation of megakaryocytopoiesis. However, it is important to emphasize that these animals were

From: *Thrombopoiesis and Thrombopoietins: Molecular, Cellular, Preclinical, and Clinical Biology*
Edited by: D. J. Kuter, P. Hunt, W. Sheridan, and D. Zucker-Franklin Humana Press Inc., Totowa, NJ

not totally deficient in megakaryocytes and platelets, suggesting a redundancy of growth factors involved in the regulation of megakaryocytopoiesis. Confirmatory proof of the physiological importance of these non-TPO growth factors was provided by deSauvage et al. *(14)*. These investigators generated TPO-deficient mice by gene targeting. TPO-deficient mice had a >80% decrease in platelets and megakaryocytes, but normal levels of other cells belonging to hemopoietic lineages *(14)* (*see* Chapter 21). This reduction in the numbers of megakaryocytes was accompanied by a predominance of lower ploidy megakaryocytes and by reduced numbers of assayable megakaryocyte progenitor cells. Again, in these animals, platelet production was not totally halted, indicating that other growth factors can promote thrombopoiesis to a limited degree in the absence of TPO. Although these studies confirm the pivotal role that TPO plays in the regulation of megakaryocytopoiesis, they also clearly suggest that additional growth factors are involved in the regulation of this process.

Megakaryocyte production can be regulated at multiple cellular levels *(15–18)*. A number of growth factors can serve as proliferation factors that amplify platelet production by expanding the progenitor cell pool *(6,7)*. A second group of hemopoietic growth factors serve as maturation factors acting on more differentiated cells to promote polyploidization, cytoplasmic and membrane development, and likely platelet release *(6,7)*. It had been previously hypothesized that two independent biofeedback loops regulate these processes *(18–20)*. Megakaryocyte numbers, size, and ploidy were thought to be related to platelet counts, whereas progenitor cell numbers were related not to platelet counts, but rather to marrow megakaryocyte numbers *(18–20)*. Both the hemopoietic growth factors that affect numerous lineages (lineage nonspecific) and the megakaryocyte lineage-specific growth factors are capable of influencing platelet production. TPO has recently been shown to effect virtually every stage of megakaryocyte development *(1–5)*. Whether the two-stage regulatory hypothesis is truly valid will, therefore, require further careful investigation (*see* Chapter 5).

Although the biological activity of TPO was originally thought to be restricted to the megakaryocyte lineage, recent studies show that TPO can affect multiple hemopoietic lineages. Several groups have demonstrated that TPO can affect the proliferative capacity of the pluripotent stem cell and synergistically augment in vitro activity *(21–24)*. It is possible that the earliest stages of megakaryocyte development are controlled primarily by either lineage nonrestricted growth factors (interleukin [IL]-3, granulocyte-macrophage colony-stimulating factor [GM-CSF], stem cell factor [SCF], IL-1, flk-2) or a lineage-specific megakaryocyte colony-stimulating factor, and that later stages of megakaryocyte development are controlled primarily, but not exclusively, by TPO, which has a limited effect on proliferation, but a major effect on maturation.

Some light has been shed on the effect of non-TPO growth factors on thrombopoiesis by studies of a rare acquired disorder, cyclical amegakaryocytic thrombocytopenia *(12,25)*. Cyclical amegakaryocytic thrombocytopenia is characterized by periods of thrombocytopenia with absent or diminished bone marrow megakaryocytes, followed by recovery of platelets with normal to increased numbers of megakaryocytes. In these patients, there is no evidence of either increased platelet destruction or antiplatelet antibodies to account for the thrombocytopenia. The platelet counts typically decrease slowly and have a rapid recovery phase. These cycles range from 30–70 days in duration and are not characterized by cyclical pertubations of white blood cells or erythrocytes. Recently, Zent et al. showed that endogenous TPO levels cycle appropriately in an inverse manner with circulating platelet numbers in a patient with cyclical amegakaryocytic thrombocytopenia, indicating that this disorder cannot be attributed to defects of TPO production *(25)*. Our group had previously studied another patient with acquired cyclical amegakaryocytic thrombocytopenia and attributed this syndrome to an antibody that selectively blocked the action of GM-CSF, but not IL-3 on megakaryocyte progenitor cells *(12)*. The blocking activity could not be overcome by a 10-fold increase in the concentration of GM-CSF. Interestingly, exposure of marrow cells to the blocking antibody for an hour abrogated the megakaryocyte colony-stimulating activity of subsequently added GM-CSF, but not IL-3. The addition of GM-CSF blocked the ability of the patient's IgG to inhibit the colony-stimulating activity of GM-CSF *(12)*. These findings indicated that the target of the patient's IgG fraction was probably a cellular component present on megakaryocyte progenitor cells that must be activated for GM-CSF to promote megakaryocyte colony formation. Thus, perturbations of non-TPO growth factors can lead to failure of thrombopoiesis, and lineage-nonspecific factors, such as GM-CSF, may play a role in the regulation of baseline megakaryocytopoiesis.

3. Lineage-Nonspecific Growth Factors That Promote Megakaryocytopoiesis

Considerable insight has been gained into the effects of lineage-nonspecific hemopoietic growth factors on thrombocytopoiesis *(6,7,26–48)*. These hemopoietic growth factors affect both progenitor cell proliferation and megakaryocyte maturation. Such growth factors either act directly at the level of megakaryocyte progenitor cells, or alternatively, synergistically amplify the proliferative effects of other growth factors.

IL-3 is the most effective lineage-nonspecific cytokine for stimulating colony formation by both megakaryocyte burst-forming cells (MK-BFC) and megakaryocyte colony-forming cells (MK-CFC) *(26,27,31–33)*. GM-CSF also possesses this same megakaryocyte colony-stimulating activity, although to a more limited degree *(26,27,31–33)*. The actions of both GM-CSF and IL-3 are additive *(31)*. The additive effect of these two growth factors has been simulated pharmacologically with the construction of a genetically engineered GM-CSF/IL-3 fusion protein (PIXY-321) *(49,50)*. PIXY-321 has a profound effect on both MK-BFC and MK-CFC *(50)*. Recombinant TPO has recently been shown to also have megakaryocyte-colony stimulating activity *(49)* *(see* Chapter 14). Optimal levels of IL-3 and rTPO were capable of stimulating similar numbers of megakaryocyte colonies and the actions of these growth factors were additive. SCF, IL-11, and erythropoietin (EPO) can also synergize with TPO to stimulate megakaryocyte colony formation *(51)*. Broudy et al. have conjectured that there are actually two populations of megakaryocyte progenitor cells: one that is solely responsive to IL-3 and GM-CSF that maintains baseline thrombopoiesis, and one that is responsive to TPO and contributes to megakaryocyte expansion in response to acute thrombocytopenia *(51)*.

IL-11 and SCF, in contrast, lack the ability to stimulate MK-BFC, but each amplifies the colony-stimulating activity of IL-3 *(52,53)*. However, only SCF can potentiate the colony-stimulating activity of GM-CSF at the level of the MK-BFC *(53)*. Furthermore, SCF, IL-6, leukemia-inhibiting factor (LIF), flk-2, oncostatin-m (OSM), and IL-11 have the ability to promote the effects of IL-3 on MK-CFC proliferation *(34,52–57)*. Such combinations of cytokines result not only in increased numbers of megakaryocyte colonies, but also larger individual colonies *(52–57)*. Many growth factors (IL-1, IL-6, SCF) also secondarily modulate megakaryocytopoiesis by inducing marrow accessory cells to release other growth factors with colony-stimulating activity *(31,33,58)*. These secondarily elaborated growth factors also may affect the ability of TPO to promote megakaryocyte colony formation.

Other than TPO, several hemopoietic growth factors exert an important effect on megakaryocyte maturation. IL-3, IL-6, and IL-11, LIF, SCF, OSM, and EPO each individually promote human megakaryocyte maturation in vitro as determined by their effects on megakaryocyte size, number, and ploidy *(36,41,59)*. The maturation effects of IL-6 and IL-11 have been reported to exceed that of LIF *(59)*. At least in vitro, IL-11 also enhances the number of megakaryocytes appearing in marrow suspension cultures *(55)*. The megakaryocyte maturational effects of LIF, IL-6, and IL-11 are either partially (LIF and IL-6) or totally (IL-11) additive to those of IL-3. IL-11, IL-6, LIF, and OSM share GP-130, the IL-6 signal transducer, as part of their cell–receptor complex *(60)*. The receptors for each of these cytokines have distinct binding properties and different intracellular signalling pathways that may affect the mechanisms by which they augment platelet production *(60)*.

Megakaryocytes and platelets are known to contain basic fibroblast growth factor (bFGF), a potent modulator of hemopoiesis *(61,62)*. It stimulates growth of adherent stromal cells in human long-term bone marrow cultures, thereby promoting hemopoietic cell development *(63)*. bFGF is deposited in a complex with proteoglycans within the extracellular matrix and is also found on the cell surface of stromal cells *(63,64)*. It appears to affect human megakaryocytopoiesis by directly promoting progenitor cell proliferation, as well as by stimulating marrow accessory cells to release other growth factors *(65)*. Thus, bFGF is probably a component of the marrow microenvironment that plays an important role in the control of human megakaryocytopoiesis.

The physiological relevance of the effects of IL-6 on thrombopoiesis has been the subject of considerable investigation. A carefully controlled study by Hill et al. indicated that IL-6 levels were essentially unchanged after the induction of acute, severe thrombocytopenia in animals *(66)*. Underscoring this experimental observation, three prospective studies of patients with platelet disorders demonstrated that IL-6 was elevated only in reactive thrombocytosis, but not in primary thrombocytosis *(8–10)*. Studies conducted by Straneva et al. and Hollen et al. also showed that 80–90% of the patients with increased IL-6 levels had ongoing inflammatory processes *(8–10)*. Furthermore, a reciprocal relationship between IL-6 levels and platelet counts was not documented. These data indicate that IL-6 is unlikely to be a primary physiological regulator of megakaryocyte maturation. Increased levels of IL-6 may account, however, for the secondary thrombocytosis observed in some patients with inflammation *(8–10)*. Beck et al. have recently presented data which clearly demonstrates the potential of IL-6 to produce secondary thrombocytosis *(11)*. Castleman's disease is a heterogeneous group of lymphoproliferative disorders associated with elevated IL-6 levels and thrombocytosis. Beck et al. infused an anti-IL-6 monoclonal antibody (MAb) into a patient with Castleman's disease and observed resolution of thrombocytosis *(11)*. It remains possible that aberrant production of other megakaryocyte maturation factors, such as LIF, IL-11, or TPO may also account for the production of secondary thromb-ocytosis in additional patients.

Presently, bone marrow fibroblasts and marrow stomal cells are thought to provide the main site for the production and release of growth factors that mediate megakaryo-cyte development. If this hypothesis is correct, the megakaryocyte would serve as a passive target cell for the action of growth factors. Recently, convincing data have been provided, however, that suggest that megakaryocytopoiesis is controlled by both autocrine and paracrine feedback loops regulated by numerous mediators *(67,68)*. The first hint of this regulatory mechanism was provided by Navarro et al., who showed that megakaryocytes express and synthesize IL-6 and the IL-6 receptor *(67)*. Further-more, Wickenhauser et al., using an elegant assay in which cytokine release by single megakaryocytes could be examined, documented the release of IL-11, IL-3, IL-6, and GM-CSF by single normal human megakaryocytes *(68)*. In addition, IL-3, and to a lesser extent IL-11, augmented the release of IL-6 by megakaryocytes *(68)*. Further-more, IL-3 led to a significant increase in the release of IL-3 and GM-CSF, but not IL-11, by megakaryocytes *(68)*. The most convincing data supporting autocrine-regulatory mechanisms were the demonstration of measurable amounts of IL-6 mRNA following stimulation of a highly enriched megakaryocyte population with IL-3 *(68)*. Cytokine production by megakaryocytes might not only affect megakaryocyte development, but could also influence other hemopoietic lineages and modulate the proliferation and function of marrow stroma. Excessive release of cytokines by megakaryocytes has been proposed as the basis for the development of the myelofibrosis associated with mega-karyocyte hyperplasia that is characteristic of many myeloproliferative disorders.

4. Inhibitory Cytokines

There is a considerable amount of data indicating that megakaryocyte colony forma-tion is affected not only by factors that promote proliferation, but also by factors that inhibit megakaryocytopoiesis *(44–48,69–71)*. Products released from platelets were first thought to inhibit in vitro megakaryocytopoiesis based on the observation that platelet-poor plasma promoted megakaryocyte colony formation to a greater degree than serum *(46,69)*. A number of cytokines, including transforming growth factor (TGF)-β, platelet factor (PF)-4, and the interferons, have been shown to inhibit in vitro megakaryocytopoiesis *(31,44,46,69,70)*. In addition, a low-molecular-weight protein, known as platelet-released glycoprotein, has also been shown to inhibit megakaryocyte development *(45)*. The identity of this molecule remains unknown.

TGF-β is a potent stimulus of the biosynthesis of matrix molecules and their recep-tors. It is a pleiotropic cytokine that has complex effects on cell proliferation, either stimulating or inhibiting this process, depending on the cellular system, and is a plate-let α-granule constituent that has been shown by a number of laboratories to have a considerable ability to inhibit in vitro megakaryocyte development *(44,46)*.

A number of groups have focused on the ability of chemokines to inhibit in vitro megakaryocytopoiesis *(47,70,72)*. Chemokines are 70–100 amino-acid peptides that contain four cysteine residues *(73)*. The α or CXC chemokine family subgroup (PF-4, β-thromboglobulin, IL-8, neutrophil activating peptide-2) is characterized by four con-served cysteine residues separated from each other by one additional amino acid resi-due *(73)*. The CC or β-chemokines (macrophage-inhibiting protein [MIP]-1 α) are characterized by the first two cysteine residues being immediately adjacent to each other (73). Both PF-4 and β-thromboglobulin are α-granule constituents. The biologically active form of β-thromboglobulin is a neutrophil cathepsin N-terminal cleavage product, neutrophil-activating product-2 (NAP-2) (74). Gewirtz et al. and Han et al. have shown that PF-4, a platelet-specific CXC chemokine, can directly inhibit in vitro megakaryocytopoiesis (47,70,72). In addition, they have demonstrated that other CXC chemokines, NAP-2, IL-8, and even the more distantly related CC chemokines, MIP-α, MIP-1β, and C10, have direct inhibitory effects on in vitro megakaryocyte colony formation (72). The role of chemokines in the regulation of megakaryocytopoiesis has been further confirmed by the expression of the α and β isoforms of the IL-8 receptor by megakaryocytes (72). Whether chemokines elaborated by ancillary cells in the marrow microenvironment play a role in the autocrine or paracrine regulation of megakaryocytopoiesis has yet to be determined. Intuitively, one would expect that increased platelet destruction would stimulate rather than inhibit platelet production. The significance of the increased elaboration of inhibitors associated with platelet destruction and release of the contents of α-granules remains unclear.

5. Extracellular Influences and Cell Interactions

Blood cells develop within the marrow microenvironment in the context of their interactions with neighboring cells and extracellular molecules. Within the past decade, a number of investigators have demonstrated that both stromal cells and the extracellular matrix are dynamic and inductive (or permissive) components of all developing cell systems (75,76). With respect to hemopoiesis, numerous studies have shown that hemopoietic progenitor cells interact with growth factors, accessory cells, such as T-cells, stromal cells, and extracellular matrix components (75,76). This developmental network is further complicated by observations that stromal cells express membrane-associated growth factors and that the extracellular matrix both binds hemopoietic growth factors, and presents these cytokines in a biologically functional manner (77–80).

Cell–cell and cell–extracellular matrix communications involving developing megakaryocytes are poorly understood. Structurally mature (platelet-shedding) megakaryocytes are located on the abluminal surface of the bone marrow sinusoid. Megakaryocytes are thought to extend pseudopods through or between sinusoidal endothelial cells, thus allowing shear forces to fragment platelets into the circulation (81). Both the location and putative mechanisms of platelet shedding imply that megakaryocyte–matrix or megakaryocyte–endothelial cell interactions are important to thrombopoiesis. Isolated megakaryocytes adhere to bovine corneal endothelial cell-derived matrix, and platelet-like structures are induced under these conditions (82,83). In addition, megakaryocytes adhere to collagen and secrete both a collagenase and a gelatinase, suggesting a possible mechanism for pseudopod infiltration of the surrounding extracellular matrix (82,83).

Studies of megakaryocyte progenitor cells show that cell–extracellular matrix relationships are important in regulating megakaryocyte proliferation. Approximately 30% of MK-CFC adhere to the extracellular matrix proteins, fibronectin, and thrombospondin (84). Interestingly, 60–80% of MK-BFC attach to thrombospondin, whereas they fail to bind to fibronectin (83). Therefore, primitive megakaryocyte progenitor cells show both altered expression of cytoadhesion molecule attachment and altered responsiveness to complex matrix-cytokine regulatory signals as compared with the characteristics of progenitor cells belonging to other lineages *(84)*.

6. Clinical Evaluation of Hemopoietic Growth Factors Other Than TPO in Promoting In Vivo Thrombopoiesis

The next step in the evaluation of the therapeutic potential of growth factors involves implementation of preclinical in vivo trials followed by a series of clinical trials in humans. This tedious process is necessary to select active agents. Some, but not all, of these trials have met with success.

EPO is capable of promoting megakaryocyte maturation, and in mice, EPO administration resulted in increased platelet numbers *(85)*. This effect of EPO on platelet production may be owing to the structural homology between EPO and TPO *(1–5)* or to synergy of EPO with endogenous TPO on proliferation of early megakaryocyte progenitor cells (*see* Chapter 14). However, only a slight, but clinically insignificant increase in platelets has been noted in patients with chronic renal failure treated in a large phase 3 trial of rHuEPO therapy *(86)*.

Numerous clinical studies have clearly demonstrated that the administration of rHuGM-CSF results in an elevation in circulating neutrophils *(87–93)*. The effect of rHuGM-CSF administration on thrombopoiesis either in bone marrow failure states or chemotherapy-treated patients has, however, been inconclusive *(87–89)*. In fact, thrombocytopenia has been reported as an adverse effect in some clinical studies of patients with cancer or the acquired immune deficiency syndrome (AIDS) receiving rHuGM-CSF *(90,91)*. A phase 3 study designed to determine the effect of rHuGM-CSF on hematological toxicity after chemotherapy and local radiation therapy was closed because of a statistically significant increase in the frequency and duration of thrombocytopenia in patients receiving rHuGM-CSF *(92)*. In addition, a delay in platelet recovery has been observed in patients receiving rHuGM-CSF administration after marrow transplantation from matched unrelated donors *(93)*. Nash et al. *(94)* have investigated the mechanism of thrombocytopenia induced by rGM-CSF. In this canine model, the administration of rGM-CSF reproducibly induced thrombocytopenia associated with shortened platelet survival. The reduced platelet survival was attributed to a nonimmune mechanism, in which GM-CSF administration resulted in an activated monocyte-macrophage system in the liver and spleen that increased the rate of platelet destruction *(94)*. The role of rHuGM-CSF in the treatment of thrombocytopenia, therefore, appears uncertain.

The ability of other lineage-nonspecific human growth factors, such as IL-1, SCF, IL-11, and IL-3, to promote in vivo human thrombopoiesis has also been evaluated. Multiple animal models have been used to investigate the effect of IL-1 on hemopoiesis *(40,95–97)*. Clinical trials with IL-1 have demonstrated its ability to promote platelet production *(95–97)*. The toxicity of IL-1 has been a major impediment to its potentially broad applicability and will likely limit its clinical utility. The in vivo administration of rHuSCF to women with breast cancer was not associated with an elevation in platelet numbers *(98)*.

Studies with rHuIL-3 in a variety of clinical settings of thrombocytopenia appear more promising. In phase 1 trials involving patients with bone marrow failure, as well as those receiving chemotherapy, rHuIL-3 has enhanced platelet production and shortened the duration of thrombocytopenia *(99–101)*. Generally, rHuIL-3 therapy is well tolerated, and toxicity is mild, limited to fever, flu-like symptoms, myalgias, and headache. These effects are dose-dependent, reversible on discontinuation of rHuIL-3, largely controlled by administration of acetaminophen, and not potentiated by the previous administration of chemotherapy. Several reports have shown that in humans, rHuIL-3 effectively stimulated myelopoiesis and thrombopoiesis, resulting in a doubling of peripheral-blood granulocyte and platelet numbers *(99–102)*.

Postmus et al., Biesma et al., and D'Hondt et al. have shown that rHuIL-3 administration after the use of combination chemotherapy regimens resulted in accelerated recovery of platelets and in a decreased requirement for platelet transfusion therapy *(103–105)*. In two of these reports, rHuIL-3 also hastened the recovery of leukocytes and neutrophils. Whether the effect of rHuIL-3 on granulocyte and platelet recovery is entirely caused by the direct action of rHuIL-3 on hemopoietic progenitor cells is at present unknown. Both D'Hondt et al. *(105)* and Postmus et al. *(103)* reported that rHuIL-3 administration was accompanied by an increase in plasma IL-6 levels. IL-6 is a potent megakaryocyte maturation factor and also affects granulopoiesis. It remains possible that some of the in vivo hematologic effects of IL-3 could be in part caused by this secondary elaboration of IL-6.

In a phase 1 study in patients with small-cell lung cancer (SCLC), D'Hondt et al. showed that rHuIL-3 treatment resulted in avoidance of delays in subsequent courses of combination chemotherapy resulting from thrombocytopenia *(105)*. Similarly, Biesma et al. also demonstrated that postponement of chemotherapy for insufficient bone marrow recovery was largely eliminated by rHuIL-3 administration in patients with advanced ovarian cancer who were treated with carboplatin and cyclophosphamide *(104)*. Whether this effect will be observed after the administration of repeated cycles of higher doses of multiple chemotherapeutic agents remains to be determined.

IL-6 is another agent that has been evaluated for clinical thrombopoietic activity. Numerous in vitro and in vivo studies indicate that IL-6 is a potent megakaryocyte maturation factor *(36,41,56)*. Phase 1 clinical studies have now shown that IL-6 is capable of augmenting thrombopoiesis in patients with cancer or myelodysplastic syndromes (MDS) *(106–108)*.

Information about the ability of rHuIL-6 to promote platelet recovery after chemotherapy is also now available *(109,110)*. Veldhuis et al. noted faster platelet recovery associated with the administration of rHuIL-6 after chemotherapy, but neither the depth nor the duration of platelet nadirs was affected by rHuIL-6 administration *(109)*. In addition, rHuIL-6 dose escalation did not prevent the postponement of cycles of chemotherapy. In a phase 1 study, D'Hondt et al. investigated the effectiveness and toxicity of rHuIL-6 in promoting platelet production after carboplatin-containing chemotherapy in patients with advanced ovarian carcinoma *(110)*. In this study, kinetics of platelet recovery was compared in the same patient treated with or without rHuIL-6, as well as with placebo patients during the same cycle. In both comparisons, platelet recovery was accelerated with rHuIL-6. Importantly, the percentage of patients requiring postponement of chemotherapy because of thrombocytopenia was significantly reduced in the group receiving rHuIL-6 *(110)*. This beneficial effect of rHuIL-6 was achieved at doses where systemic symptoms, such as fever, headaches, and myalgias, were controllable with the administration of acetaminophen.

IL-11 has profound effects on in vivo thrombopoiesis. When this growth factor was administered to women with breast cancer, it produced a dose-related increase in mean platelet counts *(111–113)*. In a randomized trial of rHuIL-11 in patients with severe chemotherapy-induced thrombocytopenia, rHuIL-11 was able to significantly reduce the number of patients requiring platelet transfusions compared with a group receiving placebo therapy *(113)*. Furthermore, the duration of clinically significant thrombocytopenia (defined as a platelet count $< 50 \times 10^9/L$) was 7 days in the group receiving rHuIL-11 versus 10 days in the placebo group. Since the safety profile of rHuIL-11 is acceptable, this growth factor appears to be a particularly promising agent for the treatment of chemotherapy-induced thrombocytopenia.

7. Conclusion

The administration of thrombopoietic agents to individuals with thrombocytopenia is likely to result in remarkable effects on platelet production. The use of Mpl ligand in conjunction with rHuIL-3, rHuIL-6, or rHuIL-11, other growth factors now known to promote in vivo thrombopoiesis, has the potential to have an even greater impact on diminishing the incidence of clinically significant, chemotherapy-induced thrombocytopenia.

References

1. Bartley TD, Bogenberger J, Hunt P, et al. Identification and cloning of a megakaryocyte growth and development factor that is a ligand for the cytokine receptor Mpl. *Cell.* 1994; 77: 1117–1124.
2. Lok S, Kaushansky K, Holly RD, et al. Cloning and expression of murine thrombopoietin cDNA and stimulation of platelet production in vivo. *Nature.* 1994; 369: 565–568.
3. Kuter DJ, Beeler DL, Rosenberg RD. The purification of megapoietin: a physiological regulator of megakaryocyte growth and platelet production. *Proc Natl Acad Sci USA* 1994; 91: 11,104–11,108.
4. Wendling F, Maraskovsky E, Debili N, et al. c-Mpl ligand is a humoral regulator of megakaryocytopoiesis. *Nature.* 1994; 369: 571–574.
5. DeSauvage FJ, Hass PE, Spenser SD, et al. Stimulation of megakaryocytopoiesis and thrombopoiesis by the c-Mpl ligand. *Nature.* 1994; 369: 533–538.
6. Hoffman R. Regulation of megakaryocytopoiesis. *Blood.* 1989; 74: 1196–1212.
7. Gordon MS, Hoffman R. Growth factors affecting human thrombocytopoiesis: Potential agents for the treatment of thrombocytopenia. *Blood.* 1992; 80: 302–307.
8. Hollen CW, Henthron J, Koziol JA, Burstein SA. Serum interleukin-6 levels in patients with thrombocytosis. *Leukemia Lymphoma.* 1992; 8: 235–241.
9. Straneva JE, Van Besien KW, Derigs G, Hoffman R. Is interleukin 6 the physiological regulator of thrombopoiesis? *Exp Hematol.* 1992; 20: 47–50.
10. Hollen CW, Henthorn J, Koziol JA, Burstein SA. Elevated serum interleukin-6 levels in patients with reactive thrombocytosis. *Br J Haematol.* 1991; 79: 286–290.
11. Beck JT, Hsu SM, Wijdenes J, et al. Brief report: alleviation of systemic manifestations of Castleman's disease by monoclonal anti-interleukin-6 antibody therapy. *N Engl J Med.* 1994; 330: 602–605.
12. Hoffman R, Briddell RA, van Besien K, et al. Acquired cyclic amegakaryocytic thrombcytopenia associated with an immunoglobulin blocking the action of granulocyte-macrophage colony-stimulating factor. *N Engl J Med.* 1989; 321: 97–102.

13. Gurney AL, Carver-Moore K, de Sauvage FJ, Moore MW. Thrombocytopenia in *c-mpl*-deficient mice. *Science*. 1994; 265: 1445–1447.

14. deSauvage FJ, Luoh SM, Carver-Moore, et al. Deficiencies in early and late stages of megakaryocytopoiesis in TPO-KO mice. *Blood*. 1995; 86: 255a (abstract no 1007).

15. Harker LA, Finch CA. Thrombokinetics in man. *J Clin Invest*. 1969; 48: 963–974.

16. Burstein SA, Adamson JW, Erb SK, Harker LA. Megakaryocytopoiesis in the mouse: response to varying platelet demand. *J Cell Physiol*. 1981; 109: 333–341.

17. Levin J, Levin FC, Metcalf D. The effects of acute thrombopenia on megakaryocyte-CFC and granulocyte-macrophage-CFC in mice: studies of bone marrow and spleen. *Blood*. 1980; 56: 274–283.

18. Ebbe S, Phalen E. Does autoregulation of megakaryocytopoiesis occur? *Blood Cells*. 1979; 5: 123–138.

19. Williams N, Eger RR, Jackson HM, Nelson DJ. Two factor requirement for murine megakaryocyte colony formation. *J Cell Physiol*. 1982; 110: 101–104.

20. Mazur EM. Megakaryocytopoiesis and platelet production: a review. *Exp Hematol*. 1987; 15: 340–350.

21. Debili N, Wendling F, Katz A, et al. The Mpl-ligand or thrombopoietin or megakaryocyte growth and differentiative factor has both proliferative and differentiative activities on human megakaryocyte progenitors. *Blood*. 1995; 86: 2516–2525.

22. Debili N, Wendliing F, Cosman D, et al. The Mpl receptor is expressed on the megakaryocytic lineage from late progenitors to platelets. *Blood*. 1995; 85: 391–401.

23. Hoffman R, Murray LJ, Young JS, Leuns KM, Bruno E. Hierarchical structure of human megakaryocyte progenitor cells. *Stem Cells*. (in press).

24. Kobayashi M, Laver JH, Kato T, Miyazaki H, Ogawa M. Recombinant human thrombopoietin (Mpl ligand) enhances proliferation of erythroid progenitors. Blood. 1995; 86: 2494–2496.

25. Zent CS, Hornkohl A, Arpeally G, et al. Cyclic thrombocytopenia: Thrombopoietin response to spontaneous changes in platelet counts. *Blood*. 1995; 86(suppl 1): 370a (abstract no 1470).

26. Quesenberry PJ, Ihle JN, McGrath E. The effect of interleukin-3 and GM-CSA-2 on megakaryocyte and myeloid clonal colony formation. *Blood*. 1995; 65: 214.

27. Robinson BE, McGrath HE, Quesenberry PJ. Recombinant murine granulocyte macrophage colony stimulating factor has megakaryocyte colony-stimulating activity and augments megakaryocyte colony stimulation by interleukin 3. *J Clin Invest*. 1987; 79: 1648–1652.

28. Kaushansky K, O'Hara PJ, Berkner K, et al. Genomic cloning, characterization and multilineage growth-promoting activity of human granulocyte-macrophage colony-stimulating factor. *Proc Natl Acad Sci USA*. 1986; 83: 3101–3105.

29. Peschel C, Paul WE, Ohara J, Green I. Effects of B cell stimulatory factor-1/interleukin 4 on hematopoietic progenitor cells. *Blood*. 1987; 70: 254–263.

30. Williams N, Jackson H, Iscove NN, Dukes PP. The role of erythropoietin, thrombopoietic stimulating factor, and myeloid colony-stimulating factors on murine megakaryocyte colony formation. *Exp Hematol*. 1984; 12: 734–740.

31. Bruno E, Miller ME, Hoffman R. Interacting cytokines regulate in vitro human megakaryocytopoiesis. *Blood*. 1989; 73: 671–677.

32. Emerson SG, Yang YC, Clark SC, Long MW. Human recombinant granulocyte-macrophage colony stimulating factor and interleukini 3 have overlapping but distinct hematopoietic activities. *J Clin Invest*. 1988; 82: 1282–1287.

33. Bruno E, Cooper RJ, Briddell RA, Hoffman R. Further examination of the effects of recombinant cytokines on the proliferation of human megakaryocyte progenitor cells. *Blood*. 1991; 77: 2339–2346.

34. Quesenberry PJ, McGrath HE, Williams ME, et al. Multifactor stimulation of megakaryocytopoiesis: effects of interleuken 6. *Exp Hematol*. 1991; 19: 35–41.

35. Williams N, Jackson H, Walker F, Oon SH. Multiple levels of regulation of megakaryocytopoiesis. *Blood Cells*. 1989; 15: 123–133.

36. Ishibashi T, Kimura H, Uchida T, et al. Human interleukin 6 is a direct promoter of maturation of megakaryocytes in vitro. *Proc Natl Acad Sci USA.* 1989; 86: 5953–5957.

37. Burstein SA. Interleukin 3 promotes maturation of murine megakaryocytes in vitro. *Blood.* 1986; 67: 1512.

38. Ishibashi T, Burstein SA. Interleukin 3 promotes the differentiation of isolated single megakaryocytes. *Blood.* 1986; 67: 1512–1514.

39. Ishibashi T, Koziol JA, Burstein SA. Human recombinant erythropoietin promotes differentiation of murine megakaryocytes in vitro. *J Clin Invest.* 1987; 79: 286–289.

40. Teramura M, Kobayashi S, Hoshino S, Oshimi K, Mizoguchi H. Interleukin-11 enhances human megakaryocytopoiesis in vitro. *Blood.* 1992; 79: 327–331.

41. Burstein SA, Henthorn J, Mei R, Williams DE. Mast cell growth factor (MGF) promotes human and murine megakaryocytic (MK) differentiation in vitro. *Blood* 1991; 78 (suppl 1): 160a (abstract no 629).

42. Goldman SJ, Lobelenz J, McCarthy K, et al. Recombinant human interleukin-11 (rhIL-11) stimulates megakaryocyte maturation and increases in peripheral platelet numbers in vivo. *Blood.* 1991; 78 (suppl 1): 132a (abstract no 518).

43. Metcalf D, Hilton D, Nicola NA. Leukemia inhibitory factor can potentiate murine megakaryocyte production in vitro. *Blood.* 1991; 77: 2150–2153.

44. Ishibashi T, Miller SL, Burstein SA. Type beta transforming growth factor is a potent inhibitor of murine megakaryocytopoiesis in vitro. *Blood.* 1987; 69: 1737–1741.

45. Dessypris EN, Gleaton JH, Sawyer ST, Armstrong OL. Suppression of maturation of megakaryocyte colony forming unit in vitro by a platelet released glycoprotein. *J Cell Physiol.* 1987; 130: 361–368.

46. Mitjavila MT, Vinci G, Villeval JL, et al. Human platelet alpha granules contain a nonspecific inhibitor of megakaryocyte colony formation: its relationship to type beta transforming growth factor (TGF-beta). *J Cell Physiol.* 1988; 139: 93–100.

47. Gewirtz AM, Calabretta B, Rucinski B, et al. Inhibition of human megakaryocytopoiesis in vitro by platelet factor 4 (PF4) and a synthetic C00H-terminal PF4 peptide. *J Clin Invest.* 1989; 83: 1477–1486.

48. Ganser A, Carlo-Stella C, Greher J, Volkers B, Hoelzer D. Effect of recombinant interferons alpha and gamma on human bone marrow derived megakaryocyte progenitor cells. *Blood.* 1987; 70: 1173–1179.

49. Williams DE, Park LS, Broxmeyer HE, Lu L. Hybrid cytokines as hematopoietic growth factors. *Int J Cell Cloning.* 1991; 9: 542–547.

50. Bruno E, Briddel RA, Cooper RJ, Brandt JE, Hoffman R. Recombinant GM-CSF/IL-3 fusion protein. Its effects on in vitro human megakaryocytopoiesis. *Exp Hematol.* 1992; 20: 494–499.

51. Broudy VC, Lin NL, Kaushansky K. Thrombopoietin (*c-mpl* ligand) acts synergistically with erythropoietin, stem cell factor and interleukin-11 to enhance murine megakaryocytic colony growth and increase megakaryocyte ploidy in vitro. *Blood.* 1995; 85: 1719–1726.

52. Briddell RA, Hoffman R. Cytokine regulation of the human burst-forming unit-megakaryocyte. *Blood.* 1990; 76: 516–522.

53. Briddell RA, Bruno E, Cooper RJ, Brandt JE, Hoffman R. Effect of c-kit ligand on in vitro human megakaryocytopoiesis. *Blood.* 1991; 78: 2854–2859.

54. Warren MK, Conroy LB, Rose JS. The role of interleukin 6 and interleukin 1 in megakaryocytic development. *Exp Hematol.* 1989; 17: 1095–1099.

55. Bruno E, Cooper RJ, Briddell RA, Hoffman R. Effects of recombinant interleukin 11 on human megakaryocyte progenitor cells. *Exp Hematol.* 1991; 19: 378–381.

56. Wallace PM, MacMaster JF, Rillena JR, et al. Thrombocytopoietic properties of oncostatin M. *Blood.* 1995; 86: 1310–1315.

57. Poloni A, Kobari I, Firat H, et al. Ex vivo expansion of megakaryocytic progenitor cells (CFU-MK) in serum-free conditions: the effect of Flt3 ligand, MGDF and G-CSF. *Blood.* 1995; 86: 702a (abstract no 2796).

58. Bruno E, Hoffman R. Effect of interleukin 6 on in vitro human megakaryocytopoiesis: its interaction with other cytokines. *Exp Hematol.* 1989; 17: 1038–1043.

59. Burstein SA, Meir J, Friese P, Turner K. Recombinant human leukemia inhibitory factor (LIF) and interleukin 11 (IL-11) promote murine and human megakaryocytopoiesis in vitro. *Blood.* 1990; 86 (suppl. 1): 450a (abstract no 1789).

60. Liu J, Modrell B, Aruffo A, et al. Interleukin-6 signal transducer gp 130 mediates oncostatin M signaling. *J Biol Chem.* 1992; 267: 16,763–16,766.

61. Bikfalvi A, Han C, Fuhrmann G. Interaction of fibroblast growth factor (FGF) with megakaryocytopoiesis and demonstration of FGF receptor expression in megakaryocytes and megakaryocyte-like cells. *Blood.* 1992; 80: 1905–1913.

62. Brunner G, Nguyen H, Gabrilove JR, Rifkin DB, Wilson EL. Basic fibroblast growth factor expression in human bone marrow and peripheral blood cells. *Blood.* 1993; 81: 631–638.

63. Oliver LJ, Rifkin DB, Gabrilove J, Hannocks MJ, Wilson EL. Long-term culture of human bone marrow stromal cells in the presence of basic fibroblast growth factor. *Growth Factors.* 1990; 3: 231–236.

64. Wilson EL, Rifkin DB, Kelley F, Hannocks MJ, Gabrilove JL. Basic fibroblast growth factor stimulates myelopoiesis in long-term human bone marrow cultures. *Blood.* 1991; 77: 954–960.

65. Bruno E, Cooper RJ, Wilson EL, Gabrilove JL, Hoffman R. Basic fibroblast growth factor promotes the proliferation of human megakaryocyte progenitor cells. *Blood.* 1993; 82: 430–435.

66. Hill RJ, Warren MK, Levin J, Gauldie J. Evidence that interleukin-6 does not play a role in the stimulation of platelet production after induction of acute thrombocytopenia. *Blood.* 1992; 80: 346–351.

67. Navarro S, Debili N, LeCoudic JP, et al. Interleukin-6 and its receptor are expressed by human megakaryocytes: in vitro effect on proliferation and endoreplication. *Blood.* 1991; 77: 461–471.

68. Wickenhauser C, Lorenzen J, Thiele J, et al. Secretion of cytokines (interleukins-1 alpha, -3, and -6 and granulocyte-macrophage colony-stimulating factor) by normal human bone marrow megakaryocytes. *Blood.* 1995; 85: 685–691.

69. Vainchenker W, Chapman J, Deschamps JF, et al. Normal human serum contains a factor(s) capable of inhibiting megakaryocyte colony formation. *Exp Hematol.* 1982; 10: 650–660.

70. Han ZC, Sensebe L, Abgrall JF, Briere J. Platelet factor 4 inhibits human megakaryocytopoiesis in vitro. *Blood.* 1990; 75: 1234–1239.

71. Griffin CG, Grant BW. Effects of recombinant interferons on human megakaryocyte growth. *Exp Hematol.* 1990; 18: 1013–1018.

72. Gewirtz AM, Zhang J, Ratajczak M, et al. Chemokine regulation of human megakaryocytopoiesis. *Blood.* 1995; 86: 2559–2567.

73. Montovani A, Sozzani S. Chemokines. *Lancet.* 1994; 343: 923.

74. Walz A, Baggiolini M. A novel cleavage product of beta-thromboglobulin formed in cultures of stimulated mononuclear cells activates human neutrophils. *Biochem Biophy Res Commun.* 1989; 159: 969–975.

75. Springer TA. Adhesion receptors of the immune system. *Nature.* 1990; 346: 425–434.

76. Long MW. Blood cell cytoadhesion molecules. *Exp Hematol.* 1992; 20: 288–301.

77. Anderson DM, Lyman SD, Baird A, et al. Molecular cloning of mast cell growth factor, a hematopoietin that is active in both membrane bound and soluble forms. *Cell.* 1990; 63: 235–243.

78. Gordon MY, Riley GP, Watt SM, Greaves MF. Compartmentalization of a haematopoietic growth factor (GM-CSF) by glycosaminoglycans in the bone marrow microenvironment. *Nature.* 1987; 326: 403–405.

79. Gospodarowicz D, Ill C. Extracellular matrix and control proliferation of vascular endothelial cells. *J Clin Invest.* 1980; 65: 1351–1364.

80. Gospodarowicz D, Delagado D, Vlodavsky I. Permissive effect of the extracellular matrix on cell proliferation in vitro. *Proc Natl Acad Sci USA.* 1980; 77: 4094–4098.

81. Zucker-Franklin D, Petursson SR. Thrombocytopoiesis—analysis by membrane tracer and freeze-fracture studies on fresh human and cultured mouse megakaryocytes. *J Cell Biol.* 1984; 99: 390–402.

82. Eldor A, Fuks Z, Levine RF, Vlodavsky I. Measurement of platelet and megakaryocyte interaction with the subendothelial extracellular matrix. *Methods Enzymol.* 1989; 169: 76–91.

83. Tablin F, Castro M, Levin RM. Blood platelet formation in vitro. The role of the cytoskeleton in megakaryocyte fragmentation. *J Cell Sci.* 1990; 97: 59–70.

84. Long MW, Briddell R, Walter AW, Bruno E, Hoffman R. Human hematopoietic stem cell adherence to cytokines and matrix molecules. *J Clin Invest.* 1992; 90: 251–255.

85. McDonald TP, Cottrell MB, Clift RE, Cullin MC, Lin FK. High doses of recombinant erythropoietin stimulate platelet production in mice. *Exp Hematol.* 1987; 15: 719–721.

86. Eschbach JW, Abdulhadi MH, Browne JK, et al. Recombinant human erythropoietin in anemic patients with end-stage renal disease. Results of a phase III multi-center clinical trial. *Ann Intern Med.* 1989; 111: 992–1000.

87. Ganser A, Volkers B, Greher J, et al. Recombinant human granulocyte-macrophage colony-stimulating factor in patients with myelodysplastic syndromes: A phaseI/II trial. *Blood.* 1989; 73: 31–37.

88. Vadhan-Raj S, Keating M, Hittelman WN, et al. Effects of recombinant human granulocyte-macrophage colony-stimulating factor in patients with myelodysplastic syndromes. *N Engl J Med.* 1987; 317: 1545–1552.

89. Neumanitis J, Rabinowe S, Singer J, et al. Recombinant granulocyte-macrophage colony-stimulating factor after autologous bone marrow transplantation for lymphoid cancer. *N Engl J Med.* 1991; 324: 1773–.

90. Levine JD, Allan JD, Tessitore JH, et al. Recombinant human granulocyte-macrophage colony-stimulating factor ameliorates zidovudine-induced neutropenia in patients with acquired immune deficiency syndrome (AIDS)/AIDS related complex. *Blood.* 1991; 78: 3148–3154.

91. Lieschke GJ, Maher D, Cebon J, et al. Effects of bacterially synthesized recombinant human granulocyte-macrophage colony-stimulating factor in patients with advanced malignancy. *Ann Intern Med.* 1989; 110: 357–364.

92. Bunn PA Jr, Browly J, Hazaka M, et al. The role of GM-CSF in limited stage SCLC: a randomized phase III study of the Southwest Oncology Group (SWOG). *Proc Am Soc Clin Oncol.* 1992; 11: 292 (abstract no 974).

93. Anasetti C, Anderson G, Applebaum FR, et al. Phase III study of rhGM-CSF in allogeneic marrow transplantation from unrelated donors. *Blood.* 1993; 82 (suppl 1): 454a (abstract no 1799).

94. Nash RA, Burstein SA, Storb R, et al. Thrombocytopenia in dogs induced by granulocyte-macrophage colony-stimulating factor: increased destruction of circulating platelets. *Blood.* 1995; 86: 1765–1775.

95. Smith J II, Longo D, Alvord W, et al. Thrombopoietic effects of IL-1 alpha in combination with high dose carboplatin. *Proc Am Soc Clin Oncol.* 1992; 11: 252 (abstract no 820a).

96. Tewari A, Buhles W Jr, Starnes HF Jr. Preliminary report: effects of interleukin-1 on platelet counts. *Lancet.* 1990; 336: 712–714.

97. Crown J, Jakubowski A, Kemeny N, et al. Phase I trial of recombinant human interleukin-1beta alone and in combination with myeluppressive doses of 5–fluorouracil in patients with gastrointestinal cancer. *Blood.* 1991; 78: 1420–1427.

98. Tong J, Gordon MS, Srour EF, et al. In vivo administration of recombinant methionyl stem cell factor expands the number of human marrow hematopoietic stem cells. *Blood.* 1993; 82: 785–789.

99. Ganser A, Lindemann A, Seipelt G, et al. Effects of recombinant interleukin-3 in normal hemtopoiesis and in patients with bone marrow failure. *Blood.* 1990; 76: 1666–1667.

100. Ganser A, Seipert G, Lindemann A, et al. Effects of recombinant human interleukin-3 in patients with myelodysplastic syndromes. *Blood.* 1990; 76: 455–462.

101. Ottmann OG, Ganser A, Seipert G, et al. Effects of recombinant interleukin-3 on human hematopoietic progenitor and precursor cells in vivo. *Blood.* 1990; 76: 1494–1502.

102. Kurzrock R, Talpaz M, Estrov Z, Rosenblum MG, Gutterman JD. Phase I study of recombinant human interleukin-3 in patients with bone marrow failure. *J Clin Oncol.* 1991; 9: 1241–1250.

103. Postmus PE, Gietma JA, Damsma O, et al. Effects of recombinant interleukin-3 in patients with relapsed small cell lung cancer treatment with chemotherapy: a dose finding study. *J Clin Oncol.* 1992; 10: 1131–1140.

104. Biesma B, Willemse PH, Mulder NH, et al. Effects of interleukin-3 after chemotherapy for advanced ovarian cancer. *Blood.* 1992; 80: 1141–1148.

105. D'Hondt V, Weynants P, Humblet Y, et al. Dose-dependent interleukin-3 stimulation of thrombopoiesis and neutropoiesis in patients with small-cell lung carcinoma before and following chemotherapy: a placebo-controlled randomized phase 1b study. *J Clin Oncol.* 1993; 11: 2063–2071.

106. Weber J, Yang JC, Topalian SL, et al. Phase I trial of subcutaneous interleukin-6 in patients with advanced malignancies. *J Clin Oncol.* 1993; 11: 499–506.

107. Gordon MS, Nemunaitis J, Hoffman R, et al. A phase I trial of recombinant human interleukin-6 on patients with myelodysplastic syndromes and thrombocytopenia. *Blood.* 1995; 85: 3066–3075.

108. Van Gameren MM, Willemse PH, Mulder NH. Effects of recombinant human interleukin-6 in cancer patients: a phase I-II study. *Blood.* 1994; 84: 1434–1444.

109. Veldhuis GJ, Willemse PH, Sleijfer DT, et al. Toxicity and efficacy of escalating dosages of recombinant human interleukin-6 after chemotherapy in patients with breast cancer or non-small cell lung cancer. *J Clin Oncol.* 1995; 13: 2585–2593.

110. D'Hondt V, Humblet Y, Guillaume T, et al. Thrombopoietic effects and toxicity of interleukin-6 in patients with ovarian cancer before and after chemotherapy: a multicentric placebo-controlled, randomized phase Ib study. *Blood.* 1995; 85: 2347–2353.

111. Gorden MS, Hoffman R, Battiato L, et al. Recombinant human interleukin-11 (NEUMEGA™ rhIL-11 growth factor, rhIL-11) prevents severe thrombocytopenia in breast cancer patients receiving multiple cycles of cyclophosphamide and doxorubicin chemotherapy. *Proc Am Soc Clin Oncol.* 1994; 13: 133, 1994 (abstract no 326a).

112. Orazi A, Cooper RJ, Tong J, et al. Effects of recombinant human interleukin eleven (NEUMEGA™rhIL-11 growth factor) on megakaryocytopoiesis in human bone marrow. *Exp Hematol.* 1996 (in press).

113. Elias L, Tepler I, Smith JW, et al. Randomized trial of recombinant human interleukin eleven (NEUMEGA™rhIL-11 growth Factor) in patients with severe chemotherapy-induced thrombocytopenia. *Blood.* 1995; 86: 1979 (abstract no 498a).

III

MOLECULAR BIOLOGY

11

Structure of Thrombopoietin and the Thrombopoietin Gene

Austin L. Gurney and Frederic J. de Sauvage

1. Introduction

The cloning of thrombopoietin (TPO) has been a significant advance for several fields of scientific inquiry. In addition to enabling the production of recombinant TPO, the cloning of TPO cDNA also facilitates examination of TPO itself, a hemopoietic cytokine with unique structural features. Furthermore, the cloning of the TPO gene enables analysis of the mechanisms by which TPO activity is controlled and homeostasis of the platelet is maintained. This chapter summarizes the current understanding of the structure of TPO and the TPO gene.

2. Structure of TPO

Human TPO is a 353 amino-acid protein that includes a predicted 21 amino-acid signal sequence and a 332 amino-acid mature protein *(1,2)*. Comparison of the TPO sequences with the GenBank database reveals homology with erythropoietin (EPO) as well as very weak homologies with interferon (IFN)-α and the neurotrophins. The homology with the neurotrophins is contained within a limited region of 39 amino acids within the amino terminus of TPO (14/39 residues, 36% identity with brain-derived neurotrophic factor [BDNF]) *(3,4)*. This region of homology, which has been termed the "thrombopoietin homologous (TH) domain" of neurotrophins, is also present within EPO and includes residues within the neurotrophins that have been found to be involved in receptor interactions. In contrast, the similarity between TPO and EPO (23% amino acid identity and approx 50% similarity including conservative substitutions) encompasses the full length of EPO and the amino-terminal 153 amino acids of TPO. TPO contains four cysteines, three of which are conserved in EPO. Recombinant forms of TPO that contain this region of EPO homology, termed the "EPO-like domain," but lack the carboxyl domain are able to activate Mpl *(1)*. The homology between TPO and EPO confirms expectations based on the observed similarities between Mpl and the EPO receptor. To what extent this evolutionary relationship will extend into similarities in the signal transduction pathways activated by these receptors or into evolutionary similarities between erythropoiesis and megakaryocytopoiesis are excit-

From: *Thrombopoiesis and Thrombopoietins: Molecular, Cellular, Preclinical, and Clinical Biology*
Edited by: D. J. Kuter, P. Hunt, W. Sheridan, and D. Zucker-Franklin Humana Press Inc., Totowa, NJ

ing questions that are under active study. There is, however, no crossactivation by TPO through the EPO receptor or by EPO through Mpl (unpublished observations).

The carboxy-terminal half of TPO (amino acids 154–332 of the mature protein, 179 amino acids) does not share homology with any other known protein. An analogous large carboxyl domain is not present within EPO or any of the other hemopoietic cytokines that have been identified to date. One possible function for this domain may be to increase the stability or serum half life of TPO. Studies of EPO have found that glycosylation is important for in vivo activity *(1,5–7)*. Unglycosylated forms of EPO are fully able to activate the EPO receptor in vitro, but have greatly decreased half-lives in vivo. The amino terminal EPO-like domain of TPO does not have predicted N-linked glycosylation sites. In contrast, the carboxyl domain of human TPO has six potential N-linked glycosylation sites. Studies of TPO present in serum from aplastic animals confirm that TPO is glycosylated *(1,8,9)*. Immunoprecipitation experiments with human TPO have shown the protein to have an apparent molecular weight of 68–85 kDa, substantially greater than the predicted molecular weight for the unglycosylated molecule of 38 kDa (both on SDS-PAGE gels) *(10)*. The hypothesis that the carboxyl domain has a role in TPO stability is supported by studies of truncated forms of TPO. Experiments measuring the ability of TPO to induce proliferation of the interleukin (IL)-3-dependent cell line Ba/F3 through transfected Mpl demonstrate that the recombinant TPO/EPO domain has a specific activity at least equal to that of full-length TPO (*see* Chapter 13). However, the half life of the TPO/EPO-like domain in vivo is dramatically shorter than full-length TPO. Of interest, perhaps, is the presence of a dibasic motif that may serve as a protease cleavage point located at the carboxyl end of the EPO domain of TPO. The possible role that proteolytic cleavage may play in the regulation of TPO activity has not been fully investigated.

Although a crystal structure of TPO is not yet available, computer modeling suggests that the amino terminal domain forms a four-α-helical bundle configuration similar to that found in other hemopoietic cytokines *(11)*. This structure would be expected to be stabilized by disulfide bonds between cysteines 1 and 4 and cysteines 2 and 3. Mass spectroscopy and peptide protein sequence analysis confirm the presence of these disulfide bonds *(12,13)*. Mutational analysis also substantiates the importance of the cysteines. Replacement of either cysteine 1 or 4 completely abrogates biological activity, whereas disruption of cysteine 2 or 3 greatly diminishes biological activity *(12–14)*.

TPO has been cloned from several mammalian species including human, pig, dog, rat, and mouse *(1,2,4,10)* (Fig. 1). It is expressed in human and mouse as a 1.8-kb mRNA. Expression has been found to be greatest in liver and kidney with detectable mRNA also present in muscle. In each species examined to date, the two-domain structure of TPO has been conserved. The overall identity between TPO of different species is approx 70%. The conservation is substantially greater in the amino terminal EPO-like domain (80–85% identity) than the carboxyl domain (55–67% identity). The lower degree of conservation within the carboxyl domain may reflect the fact that it is not required for activation of Mpl ligand, but may serve solely to provide the molecule with necessary glycosylation. Significantly, the potential N-linked glycosylation sites are well conserved with several sites maintained in all species examined to date. The four cysteines, all located within the EPO-like domain, are absolutely conserved. The high degree of conservation within the EPO-like domain of TPO is reflected in the lack

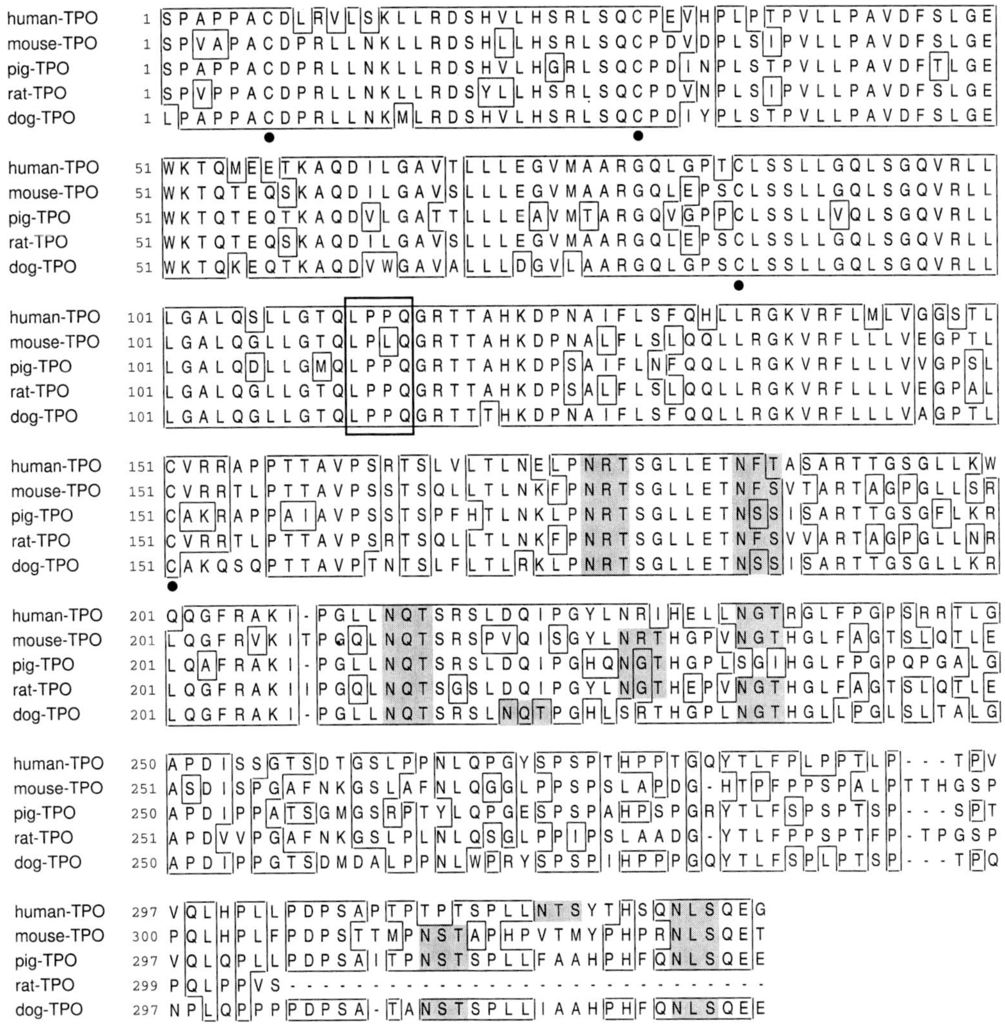

```
human-TPO   1  SPAPPACDLRVLSKLLRDSHVLHSRLSQCPEVHPLPTPVLLPAVDFSLGE
mouse-TPO   1  SPVAPACDPRLLNKLLRDSHLLHSRLSQCPDVDPLSIPVLLPAVDFSLGE
pig-TPO     1  SPAPPACDPRLLNKLLRDSHVLHGRLSQCPDINPLSTPVLLPAVDFTLGE
rat-TPO     1  SPVPPACDPRLLNKLLRDSYLLHSRLSQCPDVNPLSIPVLLPAVDFSLGE
dog-TPO     1  LPAPPACDPRLLNKMLRDSHVLHSRLSQCPDIYPLSTPVLLPAVDFSLGE
                         •                               •

human-TPO  51  WKTQMEETKAQDILGAVTLLLEGVMAARGQLGPTCLSSLLGQLSGQVRLL
mouse-TPO  51  WKTQTEQSKAQDILGAVSLLLEGVMAARGQLEPSCLSSLLGQLSGQVRLL
pig-TPO    51  WKTQTEQTKAQDVLGATTLLLEAVMTARGQVGPPCLSSLLVQLSGQVRLL
rat-TPO    51  WKTQTEQSKAQDILGAVSLLLEGVMAARGQLEPSCLSSLLGQLSGQVRLL
dog-TPO    51  WKTQKEQTKAQDVWGAVALLLDGVLAARGQLGPSCLSSLLGQLSGQVRLL
                                                    •

human-TPO 101  LGALQSLLGTQLPPQGRTTAHKDPNAIFLSFQHLLRGKVRFLMLVGGSTL
mouse-TPO 101  LGALQGLLGTQLPLQGRTTAHKDPNALFLSLQQLLRGKVRFLLLVEGPTL
pig-TPO   101  LGALQDLLGMQLPPQGRTTAHKDPSAIFLNFQQLLRGKVRFLLLVVGPSL
rat-TPO   101  LGALQGLLGTQLPPQGRTTAHKDPSALFLSLQQLLRGKVRFLLLVEGPAL
dog-TPO   101  LGALQGLLGTQLPPQGRTTTHKDPNAIFLSFQQLLRGKVRFLLLVAGPTL

human-TPO 151  CVRRAPPTTAVPSRTSLVLTLNELPNRTSGLLETNFTASARTTGSGLLKW
mouse-TPO 151  CVRRTLPTTAVPSSTSQLLTLNKFPNRTSGLLETNFSVTARTAGPGLLSR
pig-TPO   151  CAKRAPPAIAVPSSTSPFHTLNKFPNRTSGLLETNSSISARTTGSGFLKR
rat-TPO   151  CVRRTLPTTAVPSRTSQLLTLNKFPNRTSGLLETNFSVVARTAGPGLLNR
dog-TPO   151  CAKQSQPTTAVPTNTSLFLTLRKLPNRTSGLLETNSSISARTTGSGLLKR
                •

human-TPO 201  QQGFRAKI-PGLLNQTSRSLDQIPGYLNRIHELLNGTRGLFPGPSRRTLG
mouse-TPO 201  LQGFRVKITPGQLNQTSRSPVQISGYLNRTHGPVNGTHGLFAGTSLQTLE
pig-TPO   201  LQAFRAKI-PGLLNQTSRSLDQIPGHQNGTHGPLSGIHGLFPGPQPGALG
rat-TPO   201  LQGFRAKIIPGQLNQTSGSLDQIPGYLNGTHEPVNGTHGLFAGTSLQTLE
dog-TPO   201  LQGFRAKI-PGLLNQTSRSLNQTPGHLSRTHGPLNGTHGLLPGLSLTALG

human-TPO 250  APDISSGTSDTGSLPPNLQPGYSPSPTHPPTGQYTLFPLPPTLP---TPV
mouse-TPO 251  ASDISPGAFNKGSLAFNLQGGLPPSPSLAPDG-HTPFPPSPALPTTHGSP
pig-TPO   250  APDIPPATSGMGSRPTYLQPGESPSPAHPSPGRYTLFSPSPTSP---SPT
rat-TPO   251  APDVVPGAFNKGSLPLNLQSGLPPIPSLAADG-YTLFPPSPTFP-TPGSP
dog-TPO   250  APDIPPGTSDMDALPPNLWPRYSPSPIHPPPGQYTLFSPLPTSP---TPQ

human-TPO 297  VQLHPLLPDPSAPTPTPTSPLLNTSYTHSQNLSQEG
mouse-TPO 300  PQLHPLFPDPSTTMPNSTAPHPVTMYPHPRNLSQET
pig-TPO   297  VQLQPLLPDPSAITPNSTSPLLFAAHPHFQNLSQEE
rat-TPO   299  PQLPPVS-----------------------------
dog-TPO   297  NPLQPPPPDPSA-TANSTSPLLIAAHPHFQNLSQEE
```

Fig. 1. Comparison of the predicted primary structures of mature TPO from human, mouse, pig, rat, and dog. The predicted amino acid sequences of human, murine, porcine, rat, and canine TPO are aligned. Identical amino acids are boxed, and gaps introduced for optimal alignment are indicated by dashes. Potential N-linked glycosylation sites are indicated by the stippled boxes. The conserved cysteine residues are indicated by dots. The four amino acid deletion found to occur in each species (TPO-2) is outlined by a box.

of strict species specificity observed with TPO. TPO was purified from pig and dog serum using affinity columns containing the extracellular domains of human Mpl and murine Mpl, respectively *(1,2)* (*see* Chapters 8 and 9). Human and murine TPO are each able to activate both human and murine Mpl. There is, however, a limited degree of species specificity apparent in both bioassays and binding studies. Murine TPO has a higher affinity for murine Mpl than does human TPO *(10)*.

Three additional forms of TPO mRNA have been identified *(10,15)*. The alternative splice forms are expressed at similar ratios of abundance in each tissue that expresses

TPO mRNA *(18)*. They differ from TPO by the presence of a 16-base deletion that eliminates four amino acids at positions 112–115 of the mature protein (TPO-2), the presence of a 116-base deletion that results in a frame shift after amino acid 138 and thereby encodes an entirely unrelated and somewhat truncated carboxyl-terminus (TPO-3), or the presence of both of these deletions (TPO-4). The four amino-acid deletion occurs within the amino terminal EPO-like domain at an identical position within murine, porcine, canine, and human TPO. The 116-base deletion originates within the carboxyl end of the EPO-like domain and is similarly conserved between species. Granulocyte colony-stimulating factor (G-CSF) possesses two alternative splice variants that differ by a three amino-acid deletion/insertion *(16)*. Each of these forms of G-CSF is active, although the larger splice form is reported to have a lower specific activity. In contrast, immunoprecipitation experiments suggest that the alternatively spliced forms of TPO are very poorly secreted. To date, no splice variants of EPO have been described. However, a deletion similar to the four amino-acid deletion form of TPO has previously been introduced into EPO to study EPO function *(17)*. A five amino-acid deletion of residues 122–126 altered the loop between α helixes C and D in the proposed structure of EPO and resulted in an inhibition of secretion. The four amino-acid deletion within TPO-2 would be predicted to occur within the analogous loop of the TPO structure. Thus, the significance of this splice form is uncertain. Since mRNA encoding the shorter TPO forms appear to be abundant in each of the species and tissues examined to date, this might represent a novel mechanism for cytokine regulation. The level of TPO activity could potentially be modulated by altering the proportion of TPO mRNA that is spliced to encode each of the isoforms. However, a recent examination of the ratio of TPO mRNA isoforms after induced thrombocytopenia failed to find evidence for regulation *(18)*.

3. TPO Genomic Structure

Southern blot analysis of both human and murine genomic DNA indicates that the gene for TPO is present in a single copy. Genomic clones have been isolated from several species *(10,15,19,20)*. The structure of the human TPO gene has been reported to contain five coding exons and either one or two noncoding upstream exons within 7 kb of the coding exons (Fig. 2). There may be alternative splicing of these upstream exons. The boundaries of all exon/intron junctions are consistent with the consensus (GT/AG) motif established for mammalian genes. The first coding exon contains a 5'-untranslated sequence and the initial four amino acids of the signal peptide. The entire carboxyl-domain and 3'-untranslated domain, as well as approximately 50 amino acids of the EPO-like domain, are encoded within the final exon. Comparison with the EPO gene shows conservation of the boundaries of the coding exons with the exception of the additional carboxyl sequence present within the final exon of the TPO gene. The four amino acids involved in alternatively spliced TPO-2 mRNA are encoded at the 5' junction of the fifth coding exon. The TPO-3 mRNA containing the 116-base deletion arises from the use of an additional splice within the fifth coding exon.

Isolation of genomic clones of TPO greatly facilitates analysis of the possible mechanisms of tissue-specific and regulated control of TPO gene transcription. Two groups have presented distinct localizations of the 5'-end of TPO mRNA. The location of the transcription initiation site within the human TPO gene has been determined by S1

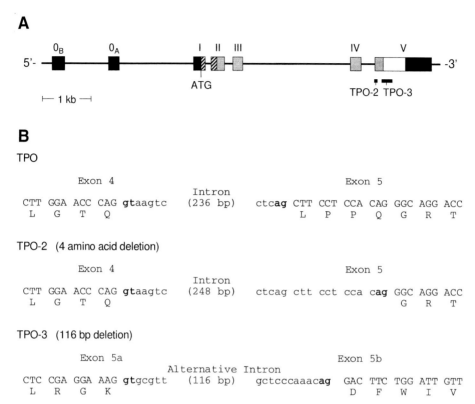

Fig. 2. The human TPO gene. **(A)** Schematic of the TPO gene. Exons within the TPO gene are boxed. The hatched box indicates 5'- and 3'-untranslated sequence as identified in human TPO cDNA. The dark striped boxes indicate the secretion signal. The light stippled boxes indicate sequence encoding the EPO-like domain, and the white box indicates sequence encoding the carboxyl domain. The four amino acids involved in the alternative splice variants of TPO are located at the fifth coding exon and are indicated by black underline as is the 116-base deletion that arises from use of an additional splice within the fifth coding exon. **(B)** Sequence flanking the alternative splice junctions within TPO mRNA. Shown are the splice donor and acceptor sequences at the junction of exon 4 and exon 5 in TPO and TPO-2, as well as the additional splice used in TPO-3. The GT/AG splice donor and acceptor sites are indicated in bold characters.

nuclease mapping of poly(A^+) RNA from normal human liver to a position of –1949 relative to the initiating methionine *(19)*. This corresponds to 279 bases of 5'-untranslated sequence within the TPO mRNA. A second group, using 5'-RACE (5'-rapid amplification of cDNA ends by PCR) with human fetal liver and testis mRNA, has mapped the transcription initiation site to a position of –3431 *(15)*. This corresponds to 544 bases of 5'-untranslated sequence and includes an additional upstream exon. Recent work demonstrates that the putative promoter regions flanking these start sites are each able to drive transcription of heterologous reporter genes in hepatic HepG2 cells, suggesting the presence of two functional TPO promoters *(21)*. This conclusion is supported by the presence of DNase1 hypersensitive sites upstream of each start site. Analysis of the sequence upstream of either of these two reported transcription start sites of the

TPO gene does not reveal obvious TATA- or CAAT-box motifs. Potential regulatory elements including SP-1 and AP-2 motifs are present upstream of the more distal transcription start site, whereas potential ETS and GATA binding sites occur within the proximal promoter region. A consensus ETS binding site located within a 60-bp core promoter region was found to be required for full transcriptional activity of the proximal promoter *(22)*. Footprint analysis reveals the presence of DNA-binding factors present in HepG2 cells that interact with this region. Additional studies will be required to elucidate fully the identity of the sequence elements and transcription factors that are involved in regulating the expression of TPO.

To date, studies of TPO mRNA abundance in several models of thrombocytopenia, including the chronic thrombocytopenia displayed by Mpl-deficient mice, have not found evidence for changes in the abundance of TPO mRNA in the major sites of TPO expression *(18,23)*. One group, using a semiquantitative polymerase chain reaction (PCR) technique, has reported increased expression of TPO mRNA in the bone marrow of thrombocytopenic mice *(24)*. However, recent analysis of mice with targeted disruption of the TPO gene has identified a significant reduction in platelet numbers in mice heterozygous for the wild type TPO gene *(25)* (*see* Chapter 21). This gene dosage effect argues strongly against the existence of functionally significant transcriptional regulation of TPO in response to platelet mass. It remains possible that alterations in TPO expression within local environments of the bone marrow may cause significant alterations in hemopoietic precursor populations or that TPO levels may respond to stimuli other than platelet levels.

The chromosomal location of the human TPO gene has been mapped by fluorescent *in situ* hybridization (FISH) to 3q26–q27 *(10,15,19,20)*. This localization is particularly interesting since structural abnormalities of the long arm of chromosome 3 have been associated with increased numbers of bone marrow megakaryocytes and elevated platelet counts in patients with acute myeloid leukemia (AML) *(26–28)*. Although frequently found in combination with other chromosomal defects, the most common abnormality in these patients is an inversion of 3q between bands 21 and 26.2, suggesting that a gene critical to the regulation of thrombopoiesis may be located at one end of the inverted segment. In fact, elevated TPO activity has been observed in at least some patients with 3q inversion *(29)*. However, recent Southern blot analysis of four cases of the "3q21–q26 syndrome" revealed that the TPO gene was located at least several hundred kilobases from the chromosomal breakpoints *(30)*. These results suggest that TPO may not be involved in the majority of these cases. Although no other cytokine gene has been localized to this locus, it will be of interest to examine this region, since there are several examples in the human and the mouse genome of close tandem linkage of genes encoding growth factors or growth factor receptors. A prominent example is the region of chromosome 5 containing the genes coding for granulocyte-macrophage colony-stimulating factor (GM-CSF), IL-3, IL-4, and IL-5 *(31,32)*.

4. Conclusions

In the past 2 years, there has been remarkable progress toward understanding, at a molecular level, the mechanisms by which production of an important blood lineage, the platelet, is controlled. It is now clear that TPO, a novel member of the hemopoietic cytokine family, acts as the dominant agent to control the expansion and maturation of megakaryocytes. The cloning of TPO has enabled rigorous analysis of the mechanisms by which TPO activity is controlled. The evidence to date from analysis of the TPO gene and promoter suggests that TPO activity is not modulated by alteration in either transcription or alternative splicing. Rather, it appears that TPO activity is regulated by platelet levels through the direct interaction of TPO with Mpl present on the platelet. This represents an elegant feedback mechanism by which the final product of this hemopoietic pathway is able to regulate its own production (*see* Chapter 23). The cloning of TPO has also revealed the existence of a cytokine with unique structural features, notably the large carboxyl domain not present among the other cytokines identified to date. Although the function of this domain is still under investigation, it appears to provide TPO with increased half-life. Future analysis of the structure of TPO and its gene will likely shed further light on TPO action, and may reveal additional mechanisms by which the activity of this important molecule is controlled.

References

1. de Sauvage FJ, Hass PE, Spencer SD, et al. Stimulation of megakaryocytopoiesis and thrombopoiesis by the *c-mpl* ligand. *Nature*. 1994; 369: 533–538.
2. Bartley TD, Bogenberger J, Hunt P, et al. Identification and cloning of a megakaryocyte growth and development factor that is a ligand for the cytokine receptor mpl. *Cell*. 1994; 77: 1117–1124.
3. Li B, Dai W. Thrombopoietin and neurotrophins share a common domain. *Blood*. 1995; 86: 1643–1644.
4. Lok S, Kaushansky K, Holly RD, et al. Cloning and expression of murine thrombopoietin cDNA and stimulation of platelet production in vivo. *Nature*. 1994; 369: 565–568.
5. Narhi LO, Arakawa T, Aoki KH, et al. The effect of carbohydrate on the structure and stability of erythropoietin. *J Biol Chem*. 1991; 266: 23,022–23,026.
6. Spivak JL, Hogans BB. The in vivo metabolism of recombinant human erythropoietin in the rat. *Blood*. 1989; 73: 90–99.
7. Takeuchi M, Takasaki S, Shimada M, Kobata A. Role of sugar chains in the in vitro biological activity of human erythropoietin produced in recombinant Chinese hamster ovary cells. *J Biol Chem*. 1990; 265: 12,127–12,130.
8. Hunt P, Li Y-S, Nichol JL, et al. Purification and biologic characterization of plasma-derived megakaryocyte growth and development factor. *Blood*. 1995; 86: 540–547.
9. Kuter DJ, Beeler DL, Rosenberg RD. The purification of megapoietin: a physiological regulator of megakaryocyte growth and platelet production. *Proc Natl Acad Sci USA*. 1994; 91: 11,104–11,108.
10. Gurney AL, Kuang W-J, Xie M-H, Malloy BE, Eaton DL, de Sauvage FJ. Genomic structure, chromosomal localization, and conserved alternative splice forms of thrombopoietin. *Blood*. 1995; 85: 981–988.
11. Bazan JF. Haemopoietic receptors and helical cytokines. *Immunol Today*. 1990; 11: 350–354.
12. Kato T, Ozawa T, Muto T, et al. Essential structure for biological activity of thrombopoietin. *Blood*. 1995; 86: 365a (abstract no 1448).
13. Linden H, O'Rork C, Hart CE, et al. The structural and functional role of disulfide bonding in thrombopoietin. *Blood*. 1995; 86: 255a (abstract no 1008).
14. Wada T, Nagata Y, Nagahisa H, et al. Characterization of the truncated thrombopoietin variants. *Biochem Biophys Res Commun*. 1995; 213: 1091–1098.
15. Chang MS, McNinch J, Basu R, et al. Cloning and characterization of the human megakaryocyte growth and development factor (MGDF) gene. *J Biol Chem*. 1995; 270: 511–514.
16. Nagata S, Tsuchiya M, Asano S, et al. The chromosomal gene structure and two mRNAs for human granulocyte colony-stimulating factor. *EMBO J*. 1986; 5: 575–581.

17. Boissel JP, Lee WR, Presnell SR, Cohen FE, Bunn HF. Erythropoietin structure-function relationships. Mutant proteins that test a model of tertiary structure. *J Biol Chem.* 1993; 268: 15,983–15,993.

18. Stoffel R, Wiestner A, Skoda RC. Thrombopoietin in thrombocytopenic mice: evidence against regulation at the mrna level and for a direct regulatory role of platelets. *Blood.* 1996; 87: 567–573.

19. Sohma Y, Akahori H, Seki N, et al. Molecular cloning and chromosomal localization of the human thrombopoietin gene. *FEBS Lett.* 1994; 353: 57–61.

20. Foster DC, Sprecher CA, Grant FJ, et al. Human thrombopoietin: Gene structure, cDNA sequence, expression, and chromosomal localization. *Proc Natl Acad Sci USA.* 1994; 91: 13,023–13,027.

21. McCarty JM, K Kaushansky. Functional charaterization of the thrombopoietin promoter. *Blood.* 1995; 86: 364a (abstract no 1446).

22. Ogami K, Kawagishi M, Kudo Y, Kato T, Miyazaki HK, Kawamura K. Promoter analysis and transcriptional regulation of the human thrombopoietin gene. *Blood.* 1995; 86: 365a (abstract no 1447).

23. Fielder PJ, Gurney AL, Stefanich E, et al. Regulation of thrombopoietin levels by c-mpl mediated binding to platelets. *Blood.* 1996; 87: 2154–2161.

24. McCarty JM, Sprugel KH, Fox NE, Sabath DE, Kaushansky K. Murine thrombopoietin mRNA levels are modulated by platelet count. *Blood.* 1995; 86: 3668–3675.

25. de Sauvage FJ, Carver-Moore K, Luoh S-H, et al. Physiological regulation of early and late stages of megakaryocytopoiesis by thrombopoietin. *J Exp Med.* 1996; 183: 651–656.

26. Pintado T, Ferro MT, San Ramon C, Mayayo M, Larana JG. Clinical correlations of the 3q21; q26 cytogenic anomaly. A leukemic or mylodysplastic syndromewith preserved or increased platelet production and lack of response to cytotoxic drug therapy. *Cancer.* 1985; 55: 535–541.

27. Rowley JD, Potter D. Chromosomal banding patterns in acute nonlymphocytic leukemia. *Blood.* 1976; 47: 705–721.

28. Jenkins RB, Tefferi A, Solberg LA, Jr, Dewald GW. Acute leukemia with abnormal thrombopoiesis and inversions of chromosome 3. *Cancer Genet Cytogenet.* 1989; 39: 167–179.

29. Pinto MR, King MA, Goss GD, et al. Acute megakaryoblastic leukaemia with 3q inversion and elevated thrombopoietin (TSF): an autocrine role for TSF? *Br J Haematol.* 1985; 61: 687–694.

30. Suzukawa K, Satoh H, Taniwaki M, Yokota J, Morishita K. The human thrombopoietin gene is located on chromosome 3q26.33-q27, but is not transcriptionally activated in leukemia cells with 3q21 and 3q26 abnormalities (3q21q26 syndrome). *Leukemia.* 1995; 9: 1328–1331.

31. Barlow DP, Bucan M, Lehrach H, Hogan BL, Gough NM. Close genetic and physical linkage between the murine haematopoietic growth factor genes GM-CSF and multi-CSF (IL-3). *EMBO J.* 1987; 6: 617–623.

32. van Leeuwen BH, Martinson ME, Webb GC, Young IG. Molecular organization of the cytokine gene cluster, involving the human IL-3, IL-4, IL-5, and GM-CSF genes, on human chromosome 5. *Blood.* 1989; 73: 1142–1148.

12

Transcription Factors in Megakaryocyte Differentiation and Gene Expression

Ramesh A. Shivdasani

1. Introduction

As discussed in the preceding chapters, thrombopoietin (TPO; Mpl ligand) appears to be unique among hemopoietic growth factors in regulating most, if not all, aspects of megakaryocyte growth and differentiation and platelet production. Other cytokines, also reviewed in an earlier chapter, fulfill various additional roles in these processes (*see* Chapter 10). The pleiotropic effects of endogenous TPO are ultimately mediated through changes in the pattern of gene expression in the effector cell. Indeed, it is believed that during cell differentiation, cytokines provide a permissive (proliferation and/or survival) function, whereas nuclear transcription factors play a deterministic role. General and cell-specific transcription factors combine to establish the lineage-specific pattern of gene expression that defines maturing and terminally differentiated cells, including megakaryocytes. An appreciation of the components that regulate cell-specific gene expression is therefore an integral aspect of a full understanding of mega-karyocyte growth and differentiation. This chapter summarizes the current status of knowledge about transcriptional regulation within megakaryocytes, especially as it pertains to expression of lineage-specific genes and cell differentiation.

2. Lineage-Restricted Transcription Factors Expressed in Megakaryocytes

Obviously, the entire complement of lineage-restricted transcription factors expressed within a given cell type is difficult to characterize. In the future, expressed sequence tags (ESTs) might provide at least a partial profile at a given stage of differentiation. Meanwhile, several studies have revealed a small group of transcription factors whose expression is restricted to selected hemopoietic lineages, including megakaryocytes. For example, the zinc-finger transcription factor GATA-1, the basic helix-loop-helix (bHLH) factor Tal-1 (also called SCL), and the basic-leucine zipper (bZip) factor NF-E2 are coexpressed in megakaryocytes, mast cells, maturing erythroid cells, and multipotential progenitors. This has fueled speculation that these lineages represent the progeny of a common precursor. Indeed, it is likely that expression of each of these transcription

From: *Thrombopoiesis and Thrombopoietins: Molecular, Cellular, Preclinical, and Clinical Biology*
Edited by: D. J. Kuter, P. Hunt, W. Sheridan, and D. Zucker-Franklin Humana Press Inc., Totowa, NJ

Table 1
Lineage-Restricted Transcription Factors Expressed Within
Megakaryocytes

Factor	Protein class	Mode of detection	References
GATA-1	Zinc finger	RNA, protein	*(76,77)*
Tal-1/SCL	bHLH	RNA, *In situ*	*(10,78)*
NF-E2	bZip	RNA (Northern)	*(6,77)*
GATA-2	Zinc finger	RNA (Northern)	*(79)*
PU.1	Ets	*In situ*	*(9)*
Rbtn2	LIM domain	*In situ*	*(80)*
Ets-1	Ets	RNA, protein	*(17)*
Ets-2	Ets	RNA, protein	*(17)*
c-Myc	bHLH-Zip	*In situ*	*(8)*
c-Myb		*In situ*	*(8)*

Abbreviations: bHLH, basic-helix-loop-helix; bZip, basic-leucine zipper.

factors is activated within a common progenitor cell, and then maintained in selected lineages while it is downregulated within monocytes and granulocytes.

A partial list of transcription factors expressed in megakaryocytes is presented in Table 1; a few key aspects deserve comment. First, none of the listed factors is expressed exclusively within this lineage, and expression does not necessarily indicate a function. Second, levels of expression may vary with the stage of differentiation. For example, the GATA-family zinc-finger factor GATA-2 is abundantly expressed in mature megakaryocytes and mast cells, but downregulated with erythroid differentiation (in contrast to GATA-1 and Tal-1, whose levels increase with erythroid differentiation) *(1,2)*. For most transcription factors, the kinetics of their expression as megakaryocytes mature have not been studied extensively. Third, several lines of evidence suggest elements of crossregulation among some of these proteins in other lineages *(3,4)* and point to the complexities underlying the transcriptional control of cell maturation. Finally, although potential target genes for transcriptional regulation by GATA-1 have been identified in both erythroid cells (reviewed in ref. *5*) and megakaryocytes (*see* Section 3.), targets for the other factors listed here remain unknown.

The technical difficulties inherent in the study of megakaryocytes in vivo have limited the analysis of transcription factors expressed in this cell lineage, and much of the preliminary work was done in megakaryocytic cell lines *(6,7)*. *In situ* hybridization analysis has recently provided a better glimpse of hemopoietic expression patterns *(8–10)*. The availability of recombinant TPO (rTPO) to boost megakaryocyte numbers is likely to result in identification of novel transcription factors expressed exclusively within this lineage. However, it is probably naive to consider lineage-restricted transcription factors as the sole determinants of cell differentiation; rather, such factors probably cooperate with widely expressed proteins to coordinate gene expression and cell maturation *(11)*.

3. Regulation of Lineage Markers

The study of elements regulating lineage-specific gene expression has been instructive in many cell types, including hemopoietic cells *(12,13)*. In one fruitful experimen-

Table 2
Megakaryocyte Genes Analyzed for Tissue-Restricted Expression

Gene product	Abbreviation	Function	Assay System
Genes expressed almost exclusively in megakaryocytes			
Platelet glycoprotein IIb	GPIIb	Integrin; fibrinogen recepter	TG; cell lines
Platelet factor-4	PF-4	Granule component	TG; cell lines
Platelet glycoprotein Ibα	GPIbα	vWF receptor subunit	TG; cell lines
Myeloproliferative leukemia	Mpl	TPO receptor	Cell lines
Platelet glycoprotein IX	GPIX	vWF receptor subunit	Cell lines
Platelet glycoprotein Ibβ	GPIbβ	vWF receptor subunit	Cell lines
Selected genes expressed in other cells (as listed) than megakaryocytes			
α2 integrin (epithelial, endothelial, and activated T cells)	α2	Integrin	Cell lines
P-selectin (endothelial cells)	—	Integrin	Cell lines
Platelet-derived growth factor (macrophages, endothelial cells, smooth muscle)	PDGF	Cytokine	Cell lines

Abbreviations: TG, transgenic mice; vWF, von Willebrand factor; TPO, thrombopoietin.

tal approach, analysis of promoters and enhancers of selected genes reveals *cis*-regulatory DNA sequences that direct tissue-specific expression. Study of the *trans*-acting nuclear proteins that interact with these elements can then provide useful insights into the molecular control of cell functions. In particular, those lineage-restricted transcription factors regulating the expression of multiple cell-specific genes may influence cell differentiation through establishment or maintenance of the appropriate genetic program. Within megakaryocytes, investigation has focused on a handful of genes, listed in Table 2, whose expression is largely restricted to this cell lineage. A common theme emerging from these studies implicates transcription factors of the GATA and Ets families in megakaryocyte-specific gene expression. Although considerable information is available about regulation of genes that are expressed in megakaryocytes in addition to other cells, its relevance to megakaryocyte-specific function and differentiation is less clear, and these studies will not be discussed further here.

The most reliable assay for characterization of *cis*-elements that regulate tissue-specific gene expression uses transgenic mice. Although experiments with transfected cell lines have yielded useful leads, these results are frequently misleading or difficult to interpret. This problem is compounded in the case of megakaryocytes, where appropriate immortalized cells are lacking and many investigators have resorted to the use of pluripotential leukemic cell lines. Although results from a few studies have been validated in transgenic mice, genuine tissue specificity has been difficult to establish in vitro. Transfection of DNA into a primary rat marrow culture, as reported recently

(14,15), may circumvent some of the problems associated with transformed cells, but still requires validation in vivo.

3.1. In Vitro Studies

Relatively short 5'-flanking regions of the genes encoding human glycoproteins IIb (GPIIb) and Ibα (GPIbα), platelet factor (PF)-4 , and *c-mpl* can direct expression of reporter constructs in cell lines with megakaryocytic features, including HEL and K562 cells stimulated with differentiating agents *(14–21)*. The minimal upstream region required for this effect, 47–150 bp relative to the transcription start site, typically includes consensus recognition sites for a panoply of DNA-binding proteins. However, canonical sequences known to bind transcription factors of the GATA and Ets families are detected consistently and frequently at multiple sites within these regulatory regions. Further, mutational analyses of selected sites indicate an important role for these transcription factor families in regulating expression of key megakaryocyte marker genes. GATA and Ets recognition sequences, whose function remains to be established, have also been reported in the promoters of the genes encoding human platelet glycoproteins Ibβ and IX *(22,23)* and murine *c-mpl (24)*.

There does not appear to be a stringent spacing requirement between GATA and Ets binding sites in megakaryocyte regulatory regions, the reported distance ranging from 20 to more than 500 bp. Some mutation analyses have suggested a synergistic interaction between GATA and Ets sites; however, a physical association between the relevant DNA-binding proteins or between these factors and the basal transcription apparatus has not been demonstrated. Moreover, although attempts have been made to appreciate the molecular mechanism by which GATA factors contribute to gene activation in erythroid cells *(25,26)* and megakaryocytes *(27)*, this remains an open question. Putative repressor domains within megakaryocyte gene promoters have not been characterized extensively.

In addition to promoters that have been considered to confer tissue-specific expression (as defined in transient transfection of immortalized cell lines), more distant elements with enhancer properties (increased expression in an orientation- and position-independent manner) have been identified approx 500 bp upstream of the human GPIIb gene *(28)* and 120–380 bp upstream of the rat PF-4 gene *(15)*. These elements also contains GATA and Ets binding sites. As discussed above, multiple members of the GATA and Ets transcription factor families may be present in megakaryocytes; although GATA-1 and Ets-1 activity has been demonstrated in cell lines *(17,29)*, there is no clear indication regarding which members regulate megakaryocyte gene expression in vivo.

The above results are shown schematically in Fig. 1A. However, several lines of evidence indicate that GATA and Ets transcription factors alone are unlikely to account for the totality of megakaryocyte-specific gene expression. First, only a few lineage markers have been studied in any detail; a much larger number remain insufficiently characterized. Second, many of the in vitro results more accurately reflect a hemopoietic rather than megakaryocytic preference since the recipient cell lines are actually pluripotential. GATA and Ets elements clearly function within other hemopoietic lineages *(28)*, as well as in nonhemopoietic cell lines *(30)*, and are found in the promoters of genes with a wider pattern of expression, such as P-selectin *(30)* and α_2 integrin *(21)*.

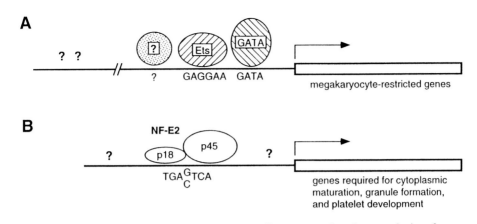

Fig. 1. Schema depicting interactions between lineage-restricted transcription factors and cis-regulatory regions controlling megakaryocyte-specific gene expression. **(A)** Canonical DNA sequences recognized by transcription factors of the GATA and Ets families have been implicated in transcriptional regulation of several lineage-restricted genes, including GPIIb, PF-4, and *c-mpl*. **(B)** The phenotype of mice lacking the transcription factor NF-E2 indicates that this lineage-restricted protein regulates the expression of one or more genes that are critical for aspects of megakaryocyte cytoplasmic maturation related to platelet formation. For details, refer to the text. Regulated gene expression undoubtedly also involves interaction with other proteins whose identity is presently unknown.

Third, quantitative estimates of "enhancer" function in in vitro assays have been modest, and even deletion of GATA and Ets binding sites has resulted in detectable expression above baseline levels *(17)*. Thus, other transcription factors must cooperate with members of the GATA and Ets families in regulating megakaryocyte gene expression. Candidates for this function might include the bZip factor NF-E2, which is discussed below, and the bHLH factor Tal-1/SCL, whose role in megakaryocytes is unknown.

3.2. In Vivo Studies

The regulatory regions of three megakaryocyte genes, rat PF-4, and human GPIIb and GPIbα, have been tested in transgenic mice. Among hemopoietic cells, expression of a reporter gene under the control of 1.1 kb of upstream sequences from the rat PF-4 gene was restricted to megakaryocytes and platelets. Outside the hemopoietic system, much lower levels of expression were detected only in the mineralocorticoid-secreting cells of the adrenal gland *(19)*. Further, placing a conditional oncogene under the control of this promoter in transgenic mice resulted in selective immortalization of immature megakaryocytes and a variable degree of thrombocytopenia *(31,32)*.

Tronik-Le Roux and colleagues have generated transgenic mice in which expression of the herpes simplex virus thymidine kinase gene is governed by 813 bp of sequence derived from the 5'-flank of the human GPIIb gene *(33)*. Treatment of these animals with gancyclovir results in anemia and severe, reversible thrombocytopenia asssociated with decreased numbers of megakaryocytes. The likely explanation for these results is that the toxigene is expressed in a multipotential erythroid-megakaryocyte progenitor, reflecting the postulated expression pattern of human GPIIb; however, higher levels may be present within megakaryocytes relative to erythroid cells. Expression of the reporter was again detected in adrenal tissue; the significance of this observation remains unclear. Finally, 6 kb of DNA, including the entire coding region of the human GPIbα gene and 3 kb of 5'-flanking sequence, directs expression of this gene to megakaryocytes in transgenic mice *(34)*.

In each of the above examples, the level of expression of the transgene has varied among founder strains, no doubt reflecting differences in the site of integration within the recipient genome. Thus, locus control regions (LCRs), which by definition confer position-independent and copy number-dependent expression in transgenic animals *(35)*, have yet to be implicated in the regulation of a megakaryocyte-restricted gene. Moreover, the critical cis-elements within the relatively large stretches of DNA tested have proven difficult to define *(15)*. Finally, detailed examination of the level of reporter gene expression relative to the endogenous PF-4, GPIIb, or GPIbα genes has not been reported; thus, even though the selected segments direct apparently megakaryocyte-restricted gene expression, a role for additional cis-elements in vivo cannot be excluded.

4. Establishment of Megakaryocyte Identity

The GATA sequence motif is invariably present within the regulatory regions of erythroid-expressed genes, and absence of GATA-1 protein leads to arrested erythroid maturation in vivo *(36)*. By analogy, the frequency with which functional GATA and Ets binding sites are detected within the core promoters and enhancers of megakaryocyte genes raises the possibility that transcription factors of the GATA or Ets families play an integral role in megakaryocyte differentiation. Data addressing this question have been intriguing, but inconclusive.

When GATA-1 is overexpressed in the multipotential myeloid cell line 416B, most cells acquire morphologic markers characteristics of megakaryocytes *(37)*. Since low levels of GATA-1 are present within the parental cells, this is presumed to represent a specific effect of overexpression. The related transcription factors GATA-2 and GATA-3, which share homology with GATA-1 largely in their DNA-binding (zinc-finger) domains, have a similar effect, though this appears to be linked to increased expression of endogenous GATA-1 *(38)*. Remarkably, megakaryocyte differentiation is also observed when a truncated product, corresponding to the C-terminal zinc finger of GATA-1 or GATA-2, is tested in this assay *(39)*. The mechanisms for megakaryocyte differentiation in 416B cells are unknown, but these data suggest that binding of critical cis-elements may be more important for the effect than transcriptional activation by a GATA factor. Moreover, it is likely that 416B cells are poised for megakaryocytic differentiation owing to the function of other transcription factors and that GATA-1 provides some additive rather than exclusive deterministic signal.

Similarly, when GATA-1 is overexpressed in transformed multipotential avian hemopoietic cell lines, a spectrum of differentiated cells is observed, including erythrocytes, thrombocytes, and eosinophils. One remarkable aspect of these experiments is that the differentiated cellular phenotype appears to depend in part on the concentration of GATA-1 protein: cells with thrombocytic features express some of the highest levels in comparison with the other lineages *(40)*. However, once again, this differentiation is detected within cells with known thromboblastic potential rather than in truly naive or heterologous cells; this leaves open the possibility that GATA-1 functions at a relatively late stage in thrombocyte maturation rather than in lineage specification *per se.*

Although the above studies implicate GATA-1 as a potential determinant of cell lineage in hemopoiesis, they are subject to all of the caveats that apply to transformed cell lines. Indeed, megakaryocytic differentiation associated with increased GPIIb expression in another hemopoietic cell line is associated with downregulation of endogenous GATA-1 levels *(41)*, concomitant with increased expression of the Ets family factor Egr-1 *(42)*. Further, although in vivo data do not exclude a role for GATA-1 in specification of megakaryocyte cell fate, they argue against an essential function in this regard. Chimeric mice generated using GATA-1-null embryonic stem (ES) cells reveal a significant contribution of these mutant cells to the megakaryocyte lineage, thus indicating that absence of GATA-1 does not interfere with megakaryocyte differentiation or platelet production in vivo *(43)*. Curiously, the number of GATA-1-null megakaryocytes is increased in these chimeric mice; it is unclear whether this reflects a secondary effect of anemia or a primary role for GATA-1 in megakaryocyte homeostasis.

Although all of the above results can be reconciled under the hypothesis that critical GATA-1 functions can be assumed by related proteins in vivo, the role of the GATA family of transcription factors in establishing megakaryocyte identity remains unclear. Targeted disruption of the GATA-2 gene results in a quantitative and broad hemopoietic defect, but specific consequences in megakaryocytes have not been examined in detail *(44)*. Mice lacking GATA-3 protein display ostensibly normal megakaryocytes, yet bleed extensively *(45)*. However, GATA-3 is not known to be expressed within megakaryocytes and other etiologies for the hemorrhage are possible; platelet numbers and function have not been reported.

The role of Ets-1 in megakaryocyte development has proven difficult to test in vivo; although absence of this factor perturbs normal lymphoid cell ratios *(46,47)*, the complete phenotype of Ets-1 knockout mice has not yet been reported. Mice lacking the related hemopoietic-specific transcription factor PU.1 exhibit a maturation arrest of myeloid and lymphoid lineages with normal numbers of megakaryocytes and platelets *(48)*.

Transformation of chicken myeloblasts by the E26 virus, which encodes a fusion oncoprotein consisting of truncated versions of *c-Ets-1* and *c-Myb (49)*, has suggested a role for these factors in hemopoiesis and megakaryocyte differentiation. In this assay, mutation of the viral Myb component results in thrombocytic differentiation of multipotential progenitors, an effect that is blocked by wild-type *v-Myb* and is independent of the Ets component *(50)*. *c-Myb* expression is normally detected in immature hemopoietic precursor cells and declines as the cells mature. Although treatment of human bone marrow cells in vitro with a *c-Myb* antisense oligodeoxynucleotide results in fewer and smaller hemopoietic colonies of all lineages, including megakaryocytes *(51)*, targeted inactivation of the *c-Myb* gene in vivo results in failure of definitive hemopoiesis for all lineages examined except megakaryocytes *(52)*. Thus, the true role of *c-Myb* in megakaryocytopoiesis continues to be controversial.

Ablation of the megakaryocyte lineage has not been a feature of any transcription factor gene knockout reported to date; this indicates that factors essential for determining megakaryocyte identity remain unknown. Another approach toward dissection of differentiation cascades is to identify antecedent transcription factors that might regulate the expression of the lineage-restricted factors discussed above. Although the regulation of expression of the genes encoding GATA-1, Tal-1, and other transcription factors has been explored in other hemopoietic lineages *(1,53–55)*, the elements controlling megakaryocyte expression have not been studied.

5. Control of Megakaryocyte Maturation

Differentiation refers to a linear but complex sequence of events extending from establishment of cell identity to acquisition of a unique, terminal, mature phenotype. In the case of megakaryocytes, important features of the terminal phenotype include a polyploid DNA content and the formation of platelet domains within the cytoplasm. At least one lineage-restricted transcription factor, NF-E2, has been implicated in the latter process.

NF-E2 is a heterodimeric nuclear protein comprised of two polypeptide chains, a hemopoietic-specific 45-kDa subunit (p45) and a widely expressed p18 subunit; both proteins belong to the basic-leucine zipper family of transcription factors *(6,56)*. As discussed above, expression of p45 NF-E2 is restricted to erythroid precursors, megakaryocytes, mast cells, and multipotential progenitors. Mice lacking p45 NF-E2 exhibit profound thrombocytopenia resulting from a maturation arrest of megakaryocytes *(57)*. Megakaryocytes are present in slightly increased numbers, have a polyploid DNA content, and express many lineage-restricted markers, including acetylcholinesterase, GPIIb, Mpl, and PF-4. Moreover, they respond to exogenous TPO with abundant proliferation, but no detectable increase in platelet production in vivo. Ultrastructural analysis reveals an extensive system of demarcation membranes, but a marked reduction in the number of granules and essentially no delimitation of platelet fields in most cells (Fig. 2A).

One interpretation of these results is that, at the very least, NF-E2 function is required for granule formation within megakaryocytes. Megakaryocytic differentiation of the FDC-P2 cell line in response to TPO *(58)*, and of 416B cells after transfection of GATA-1 (Visvader J, personal communication), results in four- to fivefold elevation in the level of p45 NF-E2 transcripts. These data further implicate NF-E2 as an important transcription factor in megakaryocytes. The role of NF-E2 in megakaryocyte gene expression is depicted in Fig. 1B.

Guy and colleagues have recently generated transgenic mice with overexpression of the cell-cycle-associated transcription factor E2F-1 targeted to megakaryocytes *(59)*. In addition to severe thrombocytopenia, mice with high levels of expression harbor increased numbers of megakaryocytes with a significant block in differentiation, presumably related to impaired withdrawal from the mitotic cell cycle (Fig. 2B). These findings implicate E2F-1, a widely expressed transcription factor, as an important participant in the balance between cell replication and postmitotic differentiation. The remarkable resemblance to the megakaryocytes seen in NF-E2-null mice might also suggest a substantial role for NF-E2 in coordinating megakaryocyte cytoplasmic maturation.

Absence of the p18 subunit of the NF-E2 protein complex does not result in thrombocytopenia in vivo *(60)*. p18 NF-E2 belongs to an extended family of small proteins, related to the avian oncogene *Maf*, all of which appear to be widely expressed *(56,61,62)*; alternative heterodimerization of the p45 subunit with other family members probably results in retained function.

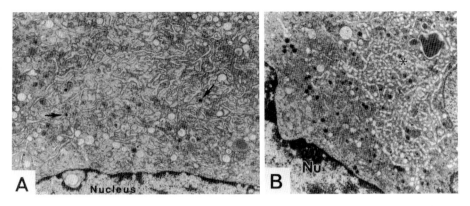

Fig. 2. Electron microscopic phenotype of megakaryocytes from mice lacking the bZip transcription factor NF-E2. **(A)** and transgenic mice selectively overexpressing the cell-cycle-associated transcription factor E2F-1 within megakaryocytes. **(B)** Similarities include paucity of granules, disorganization of excessive demarcation membranes, and absence of platelet fields. In each case, megakaryocyte cytoplasmic maturation is significantly impaired. The ultrastructural similarities, related to a link between megakaryocyte cytoplasmic maturation and withdrawal from the cell cycle, suggest that NF-E2 plays a broad role in cytoplasmic maturation. (Panel A, *57*; Panel B, *59*).

Functional NF-E2 binding sites have not yet been demonstrated within cis-elements known to regulate the expression of megakaryocyte genes. Indeed, it remains unclear whether this lineage-restricted transcription factor controls the expression of multiple genes or of a single product that is essential for granule formation or cytoplasmic maturation. In vitro, NF-E2 has been implicated in the expression of several genes involved in hemoglobin synthesis, including globin chains and heme biosynthetic enzymes; however, absence of p45 NF-E2 has only marginal effects on levels of some of these gene products in vivo *(63)*. Identification of critical NF-E2 target genes within megakaryocytes thus remains an important challenge.

6. The Transcriptional Response to TPO Stimulation

The myriad effects of TPO on megakaryocyte differentiation and platelet production are ultimately mediated by altered gene expression within responding cells. Although critical targets of TPO regulation remain to be elucidated, some steps in the signal transduction pathway toward transcriptional activation have been partially characterized. This topic receives more extended discussion in Chapter 16; a few salient aspects are reviewed here.

As with other hemopoietic cytokine-receptor pairs, TPO-Mpl signaling results in activation of Signal Transducers and Activators of Transcription (STAT) proteins. The latter are a family of latent cytoplasmic transcription factors that are activated in response to exogenous signals in most, if not all, cell types *(64)*. After tyrosine phosphorylation by Janus kinases (JAKs), STAT proteins dimerize and translocate to the cell nucleus where they direct transcriptional responses. At least part of the specificity in the signal derives from selective activation of JAK family kinases in cultured cells and blood platelets: binding of ligand to Mpl initially results in phosphorylation of JAK2 and TYK2, but not of the related kinases JAK1 or JAK3 *(65,66)*. However, the phosphotyrosine pattern induced by TPO binding is both complex (other targets of phosphorylation include Shc, Shc-associated proteins, and Mpl itself) and incompletely revealing (for example, physical association between Mpl and JAK2 is first detected long after phosphorylation of the latter is completed) *(67)*. Additional determinants directing which of the STATs are activated in response to TPO may reside within the cytoplasmic tail of Mpl itself, as exemplified by STAT3-specifying motifs in the ciliary neurotrophic factor (CNTF)/interleukin (IL)-6 cytokine receptor family *(68)*, and different regions of the Mpl cytoplasmic domain may transmit distinct intracellular signals *(69)*.

Various groups have reported phosphorylation of STAT1, STAT3, and STAT5 in response to TPO signaling *(69–72)*. However, some studies have been performed in Mpl-transfected heterologous cells, so that extension of the findings to megakaryocytes in vivo may be speculative. The likely major target, p97/MGF-STAT5, is activated in many cell types in response to diverse extracellular signals, including erythropoietin (EPO), prolactin, growth hormone, IL-2, IL-3, IL-5, granulocyte-macrophage colony-stimulating factor (GM-CSF), and contact with the basement membrane *(73–75)*. This raises the question of how transcription of specific genes is achieved in individual cell types and in response to distinct signals. Physical association with other nuclear proteins likely accounts for much of the specificity, but this question has not been addressed experimentally in megakaryocytes. The lineage-restricted transcription factors discussed above are good candidates for linking ostensibly nonspecific signal transduction pathways with the specific transcriptional response to TPO. This promises to be an exciting area of investigation in the near future.

7. Summary and Conclusions

The program of gene expression that characterizes maturing and terminally differentiated megakaryocytes is established by a combination of general and lineage-restricted transcription factors in response to external signals, including TPO. Little is presently known about how such a genetic program is either established or maintained. Studies in this direction have been facilitated by the identification and characterization of several transcription factors that are expressed only in megakaryocytes and selected other hemopoietic lineages. Further, transcription factors belonging to the GATA and Ets families have been implicated in directing lineage-specific expression of several megakaryocyte marker genes. Another nuclear protein, NF-E2, is important for transcription of genes that coordinate cytoplasmic maturation and platelet formation. However, transcription factors that might specify megakaryocytic fate during hemopoiesis remain elusive. Identification of such proteins as well as of factors that mediate specific gene transcription in response to TPO signaling will continue to be an important focus of investigations in megakaryocyte biology.

Acknowledgments

I am grateful to Jane Visvader for a critical reading of the manuscript and for sharing unpublished data. The author is supported by a Clinical Investigator Development Award from the National Institutes of Health.

References

1. Aplan PD, Nakahara K, Orkin SH, Kirsch IR. The SCL gene product: a positive regulator of erythroid differentiation. *EMBO J.* 1992; 11: 4073–4081.
2. Yamamoto M, Ko LJ, Leonard MW, Beug H, Orkin SH, Engel JD. Activity and tissue-specific expression of the transcription factor NF-E1 multigene family. Genes Dev. 1990; 4: 1650–1662.
3. Briegel K, Lim KC, Plank C, Beug H, Engel JD, Zenke M. Ectopic expression of a conditional GATA-2/estrogen receptor chimera arrests erythroid differentiation in a hormone-dependent manner. *Genes Dev.* 1993; 7: 1097–1109.
4. Osada H, Grutz G, Axelson H, Forster A, Rabbitts TH. Association of erythroid transcription factors: complexes involving the LIM protein RBTN2 and the zinc-finger protein GATA-1. *Proc Natl Acad Sci USA.* 1995; 92: 9585–9589.
5. Orkin SH. GATA-binding transcription factors in hematopoietic cells. *Blood.* 1992; 80: 575–581.
6. Andrews NC, Erdjument-Bromage H, Davidson MB, Tempst P, Orkin SH. Erythroid transcription factor NF-E2 is a haematopoietic-specific basic-leucine zipper protein. *Nature.* 1993; 362: 722–728.
7. Visvader J, Begley CG, Adams JM. Differential expression of the LYl, SCL, and E2A helix-loop-helix genes within the hemopoietic system. *Oncogene.* 1991; 6: 187–194.
8. Emilia G, Donelli S Ferrari S, et al. Cellular levels of mRNA from c-myc, c-myb and c-fes oncogenes in normal myeloid and erythroid precursors of human bone marrow: an in situ hybridization study. *Br J Haematol.* 1986; 62: 287–292.
9. Hromas R, Orazi A, Neiman RS, et al. Hematopoietic lineage- and stage-restricted expression of the ETS oncogene family member PU.1. *Blood.* 1993; 82: 2998–3004.
10. Mouthon MA, Bernard O, Mitjavila MT, Romeo PH, Vainchenker W, Mathieu-Mahul D. Expression of tal-1 and GATA-binding proteins during human hematopoiesis. *Blood.* 1993; 81: 647–655.
11. Shivdasani RA, Orkin SH. The transcriptional control of hematopoiesis. *Blood.* 1996; 87: 4025–4039.
12. Orkin SH. Transcription factors and hematopoietic development. *J Biol Chem.* 1995; 270: 4955–4958.
13 Zhang DE, Hohaus S, Voso MT, et al. Function of PU.1 (Spi-1), C/EBP, and AML1 in early myelopoiesis: regulation of multiple myeloid CSF receptor promoters. *Curr Top Microbiol Immunol.* 1996; 211: 137–147.
14. Block KL, Ravid K, Phung, QH, Poncz M. Characterization of regulatory elements in the 5'-flanking region of the rat GPIIb gene by studies in a primary rat marrow culture system. *Blood.* 1994; 84: 3385–3393.
15. Ravid K, Doi T, Beeler DL, Kuter DJ, Rosenberg RD. Transcriptional regulation of the rat platelet factor 4 gene: interaction between an enhancer/silencer domain and the GATA site. *Mol Cell Biol.* 1991; 11: 6116–6127.
16. Hashimoto Y, Ware J. Identification of essential GATA and Ets binding motifs within the promoter of the platelet glycoprotein Ibα gene. *J Biol Chem.* 1995; 270: 24,532–24,539.
17. Lemarchandel V, Ghysdael J, Mignotte V, Rahuel C, Romeo PH. GATA and Ets cis-acting sequences mediate megakaryocyte-specific expression. *Mol Cell Biol.* 1993; 13: 668–676.
18. Mignotte V, Vigon I, Boucher de Crevecoeur E, Romeo PH, Lemarchandel V, Chretien S. Structure and transcription of the human c-mpl gene (MPL). *Genomics.* 1994; 20: 5–12.
19. Ravid K, Li YC, Rayburn HB, Rosenberg RD. Targeted expression of a conditional oncogene in hematopoietic cells of transgenic mice. *J Cell Biol.* 1993; 123: 1545–1553.
20. Uzan G, Prenant M, Prandini MH, Martin F, Marguerie G. Tissue-specific expression of the platelet GPIIb gene. *J Biol Chem.* 1991; 266: 8932–8939.
21. Zutter MM, Santoro SA, Painter AS, Tsung YL, Gafford A. The human α2 integrin gene pro-

moter. Identification of positive and negative regulatory elements important for cell-type and developmentally restricted gene expression. *J Biol Chem.* 1994; 269: 463–469.

22. Hickey MJ, Roth GJ. Characterization of the gene encoding human platelet glycoprotein IX. *J Biol Chem.* 1993; 268: 3438–3443.

23. Yagi M, Edelhoff S, Disteche CM, Roth GJ. Structural characterization and chromosomal location of the gene encoding human platelet glycoprotein Ib beta. J Biol Chem. 1994; 269: 17,424–17,427.

24. Alexander WS, Dunn AR. Structure and transcription of the genomic locus encoding murine c-Mpl, a receptor for thrombopoietin. *Oncogene.* 1995; 10: 795–803.

25. Martin DI, Orkin SH. Transcriptional activation and DNA binding by the erythroid factor GF-1/NF-E1/Eryf 1. *Genes Dev.* 1990; 4: 1886–1898.

26. Rahuel C, Vinit MA, Lemarchandel V, Cartron JP, Romeo PH. Erythroid-specific activity of the glycophorin B promoter requires GATA-1 mediated displacement of a repressor. *EMBO J.* 1992; 11: 4095–4102.

27. Aird WC, Parvin JD, Sharp PA, Rosenberg RD. The interaction of GATA-binding proteins and basal transcription factors with GATA box-containing core promoters. A model of tissue-specific gene expression. *J Biol Chem.* 1994; 269: 883–889.

28. Prandini MH, Uzan G, Martin F, Thevenon D, Marguerie G. Characterization of a specific erythromegakaryocytic enhancer within the glycoprotein IIb promoter. *J Biol Chem.* 1992; 267: 10,370–10,374.

29. Martin F, Prandini MH, Thevenon D, Marguerie G, Uzan G. The transcription factor GATA-1 regulates the promoter activity of the platelet glycoprotein IIb gene. *J Biol Chem.* 1993; 268: 21,606–21,612.

30. Pan J, McEver RP. Characterization of the promoter for the human P-selectin gene. *J Biol Chem.* 1993; 268: 22,600–22,608.

31. Ravid K, Beeler DL, Rabin MS, Ruley HE, Rosenberg RD. Selective targeting of gene products with the megakaryocyte platelet factor 4 promoter. Proc Natl Acad Sci USA. 1991; 88: 1521–1525.

32. Robinson MO, Zhou W, Hokom M, et al. The tsA58 simian virus 40 large tumor antigen disrupts megakaryocyte differentiation in transgenic mice. *Proc Natl Acad Sci USA.* 1994; 91: 12,798–12,802.

33. Tronik-Le Roux D, Roullot V, Schweitzer A, Berthier R, Marguerie G. Suppression of erythromegakaryocytopoiesis and the induction of reversible thrombocytopenia in mice transgenic for the thymidine kinase gene targeted by the platelet glycoprotein alpha IIb promoter. *J Exp Med.* 1995; 181: 2141–2151.

34. Ware J, Russell SR, Marchese P, Ruggeri ZM. Expression of human platelet glycoprotein Ibα in transgenic mice. J Biol Chem. 1993; 268: 8376–8382.

35. Grosveld F, van Assendelft GB, Greaves DR, Kollias G. Position-independent, high-level expression of the human beta-globin gene in transgenic mice. *Cell.* 1987; 51: 975–985.

36. Pevny L, Simon MC, Robertson E, et al. Erythroid differentiation in chimaeric mice blocked by a targeted mutation in the gene for transcription factor GATA-1. *Nature.* 1991; 349: 257–260.

37. Visvader JE, Elefanty AG, Strasser A, Adams JM. GATA-1 but not SCL induces megakaryocytic differentiation in an early myeloid line. *EMBO J.* 1992; 11: 4557–4564.

38. Visvader J, Adams JM. Megakaryocytic differentiation induced in 416B myeloid cells by GATA-2 and GATA-3 transgenes or 5–azacytidine is tightly coupled to GATA-1 expression. *Blood.* 1993; 82: 1493–1501.

39. Visvader JE, Crossley M, Hill J, Orkin SH, Adams JM. The C-terminal zinc finger of GATA-1 or GATA-2 is sufficient to induce megakaryocytic differentiation of an early myeloid cell line. *Mol Cell Biol.* 1995; 15: 634–641.

40. Kulessa H, Frampton J, Graf T. GATA-1 reprograms avian myelomonocytic cell lines into eosinophils, thromboblasts, and erythroblasts. *Genes Dev.* 1995; 9: 1250–1262.

41. Dai W, Murphy MJ. Downregulation of GATA-1 expression during phorbol myristate acetate-

induced megakaryocytic differentiation of human erythroleukemia cells. *Blood.* 1993; 81: 1214–1221.

42. Cheng T, Wang Y, Dai W. Transcription factor egr-1 is involved in phorbol 12-myristate 13-acetate-induced megakaryocytic differentiation of K562 cells. *J Biol Chem.* 1994; 269: 30,848–30,853.

43. Pevny L, Lin CS, D'Agati V, Simon MC, Orkin SH, Costantini F. Development of hematopoietic cells lacking transcription factor GATA-1. 1995. *Development.* 121: 163–172.

44. Tsai FY, G. Keller G, Kuo FC, et al. An early haematopoietic defect in mice lacking the transcription factor GATA-2. *Nature.* 1994; 371: 221–226.

45. Pandolfi PP, Roth ME, Karis A, et al. Targeted disruption of the GATA3 gene causes severe abnormalities in the nervous system and in fetal liver haematopoiesis. *Nat Genet.* 1995; 11: 40–44.

46. Borles JC, Willerford DM, Grevin D, et al. Increased T-cell apoptosis and terminal B-cell differentiation induced by inactivation of the Ets-1 proto-oncogene. *Nature.* 1995; 377: 635–638.

47. Muthusamy N, Barton K, Leiden JM. Defective activation and survival of T cells lacking the Ets-1 transcription factor. 1995. *Nature.* 377: 639–642.

48. Scott EW, Simon MC, Anastasi J, Singh H. Requirement of transcription factor PU.1 in the development of multiple hematopoietic lineages. *Science.* 1994; 265: 1573–1577.

49. Dexter TM, Spooncer E. Growth and differentiation in the hemopoietic system. *Annu Rev Cell Biol.* 1987; 3: 423–441.

50. Frampton J, McNagny K, Sieweke M, Philip A, Smith G, Graf T. v-Myb DNA binding is required to block thrombocytic differentiation of Myb-Ets-transformed multipotent haematopoietic progenitors. *EMBO J.* 1995; 14: 2866–2875.

51. Gewirtz AM, Calabretta B. A c-myb antisense oligodeoxynucleotide inhibits normal human hematopoiesis in vitro. *Science.* 1988; 242: 1303–1306.

52. Mucenski ML, McLain K, Kier AB, et al. A functional c-myb gene is required for normal murine fetal hepatic hematopoiesis. *Cell.* 1991; 65: 677–689.

53. Bockamp EO, McLaughlin F, Murrell AM, et al. Lineage-restricted regulation of the murine SCL/TAL-1 promoter. *Blood.* 1995; 86: 1502–1514.

54. Skoda RC, Tsai SF, Orkin SH, Leder P. Expression of c-myc under the control of GATA-1 regulatory sequences causes erythroleukemia in transgenic mice. *J Exp Med.* 1995; 181: 1603–1613.

55. Tsai SF, Strauss E, Orkin SH. Functional analysis and in vivo footprinting implicate the erythroid transcription factor GATA-1 as a positive regulator of its own promoter. *Genes Dev.* 1991; 5: 919–931.

56. Andrews NC, Kotkow KJ, Ney PA, Erdjument-Bromage H, Tempst P, Orkin SH. The ubiquitous subunit of erythroid transcription factor NF-E2 is a small basic-leucine zipper protein related to the *v-maf* oncogene. *Proc Natl Acad Sci USA.* 1993; 90: 11,488–11,492.

57. Shivdasani RA, Rosenblatt MF, Zucker-Franklin D, et al. Transcription factor NF-E2 is required for platelet formation independent of the actions of thrombopoietin/MGDF in megakaryocyte development. *Cell.* 1995; 81: 695–704.

58. Nagata Y, Nagahisa H, Aida Y, Okutomi K, Nagasawa T, Todokoro K. Thrombopoietin induces megakaryocyte differentiation in hematopoietic progenitor FDC-P2 cells. *J Biol Chem.* 1995; 270: 19,673–19,675.

59. Guy CT, Zhou W, Kaufman S, Robinson MO. 1996. E2F-1 blocks terminal differentiation and causes proliferation in transgenic megakaryocytes. Mol Cell Biol. 16: 685–693.

60. Kotkow KJ, Orkin SH. Complexity of the erythroid transcription factor NF-E2 as revealed by gene targeting of the mouse p18 NF-E2 locus. *Proc Natl Acad Sci USA.* 1996; 93: 3514–3518.

61. Igarashi K, Kataoka K, Itoh K, Hayashi N, Nishizawa M, Yamamoto M. Regulation of transcription by dimerization of erythroid factor NF-E2 p45 with small Maf proteins. *Nature.* 1994; 367: 568–572.

62. Kataoka K, Igarashi K, Itoh K, et al. Small Maf proteins heterodimerize with Fos and may act as competitive repressors of the NF-E2 transcription factor. *Mol Cell Biol.* 1995; 15: 2180–2190.

63. Shivdasani RA, Orkin SH. Erythropoiesis and globin gene expression in mice lacking the transcription factor NF-E2. *Proc Natl Acad Sci USA*. 1995; 92: 8690–8694.

64. Darnell JE, Kerr IM, Stark GR. Jak-SAT pathways and transcriptional activation in response to IFNs and other extracellular signaling proteins. *Science*. 1994; 264: 1415–1421.

65. Miyakawa Y, Oda A, Druker BJ, et al. Recombinant thrombopoietin induces rapid protein tyrosine phosphorylation of Janus kinase 2 and Shc in human blood platelets. *Blood*. 1995; 86: 23–27.

66. Sattler M, Durstin MA, Frank DA, et al. The thrombopoietin receptor c-MPL activates JAK2 and TYK2 tyrosine kinases. *Exp Hematol*. 1995; 23: 1040–1048.

67. Drachman JG, Griffin JD, Kaushansky K. The c-Mpl ligand (thrombopoietin) stimulates tyrosine phosphorylation of Jak2, Shc, and c-Mpl. *J Biol Chem*. 1995; 270: 4979–4982.

68. Stahl N, Farruggella TJ, Boulton TG, Zhong Z, Darnell JE, Yancopoulos GD. Choice of STATs and other substrates specified by modular tyrosine-based motifs in cytokine receptors. *Science*. 1995; 267: 1349–1353.

69. Gurney AL, Wong SC, Henzel WJ, de Sauvage FJ. Distinct regions of c-Mpl cytoplasmic domain are coupled to the JAK-STAT signal transduction pathway and Shc phosphorylation. *Proc Natl Acad Sci USA*. 1995; 92: 5292–5296.

70. Bacon CM, Tortolani PJ, Shimosaka A, Rees RC, Longo DL, O'Shea JJ. Thrombopoietin (TPO) induces tyrosine phosphorylation and activation of STAT5 and STAT3. *FEBS Lett*. 1995; 370: 63–68.

71. Miyakawa Y, Oda A, Druker BJ, et al. Thrombopoietin induces tyrosine phosphorylation of Stat3 and Stat5 in human blood platelets. *Blood*. 1996; 87: 439–446.

72. Pallard C, Gouilleux F, Benit L, et al. Thrombopoietin activates a STAT5-like factor in hematopoietic cells. *EMBO J*. 1995; 14: 2847–2856.

73. Mui AL, Wakao H, O'Farrell AM, Harada N, Miyajima A. Interleukin-3, granulocyte-macrophage colony stimulating factor and interleukin-5 transduce signals through two STAT5 homologs. 1995. *EMBO J*. 14: 1166–1175.

74. Streuli CH, Edwards GM, Delcommenne M, et al. Stat5 as a target for regulation by extracellular matrix. *J Biol Chem*. 1995; 270: 21,639–21,644.

75. Wakao H, Gouilleux F, Groner B. Mammary gland factor (MGF) is a novel member of the cytokine regulated transcription factor and confers the prolactin response. *EMBO J*. 1994; 13: 2182–2191.

76. Martin DI, Zon LI, Mutter G, Orkin SH. Expression of an erythroid transcription factor in megakaryocytic and mast cell lineages. *Nature*. 1990; 344: 444–447.

77. Romeo PH, Prandini MH, Joulin V, et al. Megakaryocytic and erythrocytic lineages share specific transcription factors. *Nature*. 1990; 344: 447–449.

78. Kallianpur AR, Jordan JE, Brandt SJ. The SCL/TAL-1 gene is expressed in progenitors of both the hematopoietic and vascular systems during embryogenesis. *Blood*. 1994; 83: 1200–1208.

79. Dorfman DM, Wilson DB, Bruns GA, Orkin SH. Human transcription factor GATA-2. Evidence for regulation of preproendothelin-1 gene expression in endothelial cells. *J Biol Chem*. 1992; 267: 1279–1285.

80. Warren, AJ, Colledge WH, Carlton MBL, Evans MJ, Smith AJH, and Rabbitts TH. The oncogenic cysteine-rich LIM domain protein Rbtn2 is essential for erythroid development. *Cell*. 1994; 78: 45–57.

13

The Biological Significance of Truncated and Full-Length Forms of Mpl Ligand

Donald Foster and Pamela Hunt

1. Isolation and Characterization of Plasma-Derived Thrombopoietin

Recently, several groups using three different strategies have isolated cDNA clones coding for molecules with potent effects on megakaryopoiesis and thrombopoiesis (*see* Chapters 8 and 9). These molecules have been isolated from murine, porcine, canine, ovine, rodent, and human origins and have been variously called thrombopoietin (TPO; *1,2*), Mpl ligand *(3)*, megapoietin *(4)*, and megakaryocyte growth and development factor (MGDF) *(5,6)*.

Two of these groups exploited the availability of a recombinant, truncated form of Mpl to purify TPO from biological fluids by Mpl affinity chromatography. The group from Genentech (South San Francisco, CA) *(3)* published a description of purified porcine proteins with a range of molecular weights of 30, 28, and 18 kDa. All the isolated proteins shared identical amino-terminal sequences, indicating that they were all derived from a common precursor molecule. The group from Amgen (Thousand Oaks, CA) *(5)* published the isolation of canine proteins of 31 kDa and 25 kDa, with the 25-kDa form predominating. These molecules had equivalent biological activity on in vitro human megakaryopoiesis *(5,6)* and had identical amino-terminal sequences with the same implication of being derived from a common precursor form.

After the publication of the cloning of TPO by ZymoGenetics (Seattle, WA) *(1)*, Genentech, and Amgen, a group from Kirin Brewery Company (Tokyo, Japan) *(2)*, and Kuter, Beeler, and Rosenberg from the Massachusetts Institute of Technology (MIT) (Boston, MA) *(4)* announced that they too had isolated TPO. Unlike the other investigators, the Kirin and MIT groups purified TPO from aplastic plasma using chromatographic procedures without the use of immobilized Mpl (*see* Chapter 9). The Kirin group was able to purify sufficient protein of 19 kDa from the plasma of irradiated rats to obtain microsequencing results. The MIT laboratory was able to isolate peptides of 37 kDa and 31 kDa from the plasma of thrombocytopenic sheep. On partial sequence analysis, these molecules were also a match to the published sequences of TPO *(1–3,5)*.

From: *Thrombopoiesis and Thrombopoietins: Molecular, Cellular, Preclinical, and Clinical Biology*
Edited by: D. J. Kuter, P. Hunt, W. Sheridan, and D. Zucker-Franklin Humana Press Inc., Totowa, NJ

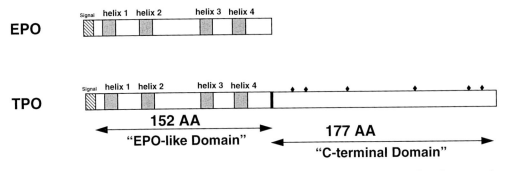

Fig. 1. The domain structures of EPO and TPO. The first 152 amino acids of TPO are partly homologous to EPO structure. The stippled boxes represent the four predicted α-helical regions of both proteins. The second domain of TPO shows no obvious homology with known protein structures. Potential N-linked glycosylation sites are indicated by solid diamonds. The signal peptides are indicated by diagonal stripes.

2. The Predicted Structure of TPO

The cDNAs isolated by the four groups encode the identical protein. The porcine, canine, murine, rodent, and human cDNAs encode proteins with a high degree of sequence and structural conservation (*see* Chapter 11). The encoded polypeptides have a predicted molecular mass of approximately 35,000 kDa (discounting posttranslational glycosylation), nearly twice the predicted encoded size of cytokines, such as the interleukins (IL), granulocyte-macrophage colony-stimulating factor (GM-CSF), and erythropoietin (EPO). A striking feature of TPO from all species examined is a novel two-domain structure with an amino-terminal domain partly homologous with EPO. The C-terminal domain is rich in serine, threonine, and proline residues, and contains as many as seven potential N-linked glycosylation sites. In working nomenclature, the first domain is often called the "EPO-like domain," and the second domain, the "carbohydrate-domain" or "C-terminal domain."

The structure of TPO and its relationship to the structure of EPO are shown in Fig. 1. The amino-terminal domains of murine and human TPO are 21.5 and 23% identical to murine and human EPO, respectively. Allowing for conservative amino acid substitutions, sequence conservation approaches 50%. This domain also has low-level similarities to interferon (IFN) α and β *(1)*. The glycosylation sites present in EPO are not conserved in TPO. Both EPO and TPO have four cysteines, of which three are conserved. These residues include the first and last cysteines of EPO, which have been shown to form a disulfide bond essential for function. Based on sequence alignment, it is predicted that the EPO-like domain of TPO would adopt a four-α-helical structure similar to that determined for IFN β and proposed for EPO *(7)*, and for all the other hemopoietic cytokines for which structural information is available *(8)*.

3. The Potential Role(s) of the C-Terminal Domain

The functional significance of the C-terminal domain is under investigation. No statistically significant structural similarities to entries in the GenBank data base *(9)* have been found for this domain. One notable feature is the abundance of serine, threonine, and proline residues. This domain contains seven potential N-linked glycosylation sites

Table 1
Pharmacokinetic Variables for Recombinant-Derived Forms of Mpl Ligand (Full-Length, Glycosylated TPO and the EPO-Like Domain) in Sprague-Dawley Rats[a]

Variable	EPO-like domain	TPO
Bioavailablilty (sc)	21%	21.9%
$T_{1/2}$ (adsorption)	0.04 h	4.8 h
$T_{1/2}\beta$	1.11 h	14.4h

[a]Data courtesy Ellen Cheung, Amgen Inc., Thousand Oaks, CA.

in murine TPO and six in the human form. The ability of TPO to bind lectin-affinity columns, considered together with data from glucanase digestion experiments, indicates that this region is heavily glycosylated, in both the purified recombinant murine protein produced in baby hamster kidney (BHK) cells *(10)*, or the native protein from canine *(6)*, rodent *(2)*, or ovine *(4)* plasma. One function of this domain might to be to stabilize and enhance the circulating half-life of TPO as has been shown for the glycosylated form of EPO *(11–13)*. Pharmacokinetic analyses of full-length, glycosylated TPO versus the EPO-like domain reveal that the former has roughly 10 times the circulating half-life of the latter after sc administration into rats (Table 1).

The cDNA sequences of murine, rodent, porcine, and human TPO predict that the junction between the EPO-like domain and the C-terminal domain contains a pair of arginine residues resembling a dibasic proteolytic cleavage site *(14)*. It is noteworthy that mature EPO is processed by carboxy terminal proteolytic cleavage at an arginine residue located in a similar position *(15)*. It is possible that TPO is expressed as a precursor protein and undergoes limited proteolysis to generate a mature active ligand. Consistent with this possibility is the observation that sequence conservation between the carboxyl-terminal domains of porcine, murine, rodent, canine, and human TPO is considerably less than that between the corresponding EPO-like domains of these species. It is important to note, however, that actual cleavage at the arginine-arginine (RR) dibasic site has not been documented to date, although several monobasic C-terminal sites appear to be sensitive to proteolysis.

4. The EPO-Like Domain Is Biologically Active

Several reports have indicated that recombinant forms of a truncated human Mpl ligand that include the EPO-like-domain and with much of the C-domain deleted appear to be competent for transduction of a proliferative signal in Mpl-transfected cells *(5)*. These activities were shown to be blocked by addition of soluble Mpl, indicating that the truncated forms are also competent for receptor binding *(5)*. The human EPO-like domain of Mpl ligand is also able to stimulate human megakaryocyte-colony growth and human megakaryocyte maturation in vitro *(5,16–20)*.

To investigate more thoroughly the biological potencies of truncated Mpl ligand forms and to probe the possible biological significance of the C-terminal domain, a series of nested deletion constructs was produced in which varying portions of the C-terminal regions of the murine form were removed by introduction of stop codons (Fig. 2). These truncated forms, together with full-length murine form, were expressed in BHK cells and purified to homogeneity by conventional chromatographic methods (Fig. 3). Pro-

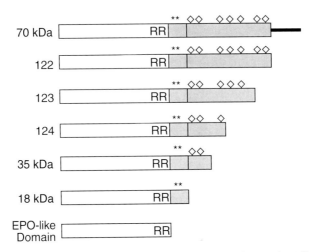

Fig. 2. Nested deletion constructs in the murine TPO second domain. All constructs encode the full EPO-like domain of TPO, shown with RR indicating the end of the domain. The next 18 amino acids encode a region with at least two O-linked carbohydrate sites, indicated by asterisks. Regions of the C-terminal domain were truncated by introduction of stop codons. The approximate location of predicted N-linked glycosylation sites in these constructs is indicated by the branched structures. Constructs labeled 122, 123, and 124 have deleted, respectively, the 3' noncoding region, the final 51 amino acids, or the final 120 amino acids of the second domain.

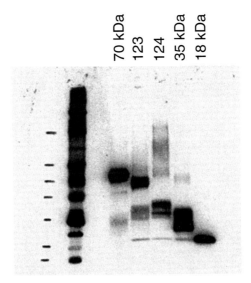

Fig. 3. Sodium dodecyl sulfate-polyacrylamide gel electrophoresis (SDS-PAGE) comparisons of purified truncated Mpl ligand forms. Truncated murine *c-mpl* constructs were transfected into BHK cells, and stable cell lines were established for each construct. The truncated forms were purified to near-homogeneity from the conditioned media by a combination of ion-exchange and hydrophobic interaction chromatographic matrices. One microgram of each preparation was subjected to electrophoresis on SDS-PAGE gels and silver-stained. Data are courtesy of Michele Buddle and Rachel Stevenson, ZymoGenetics, Inc., Seattle, WA.

Table 2
The Effects of Recombinant-Derived Full-Length and
Truncated Forms of TPO on In Vitro Bioactivity[a]

	Specific activities, U/µg	
Form	Murine TPO	Human TPO
70 kDa	129,000	5400
35 kDa	163,000	30,300
18 kDa	517,000	100,000

[a]Full-length (70-kDa) and truncated (35- and 18-kDa) forms of murine and human TPO were purified and concentrations determined by amino-acid composition analyses. Samples of each preparation were assayed in the Ba/F3-mMpl[+] proliferation assay, and their specific activities determined on a U/µg protein basis. Note that the human form has reduced specific activity relative to the murine form when assayed on cells bearing murine *c-mpl*. Data are courtesy of John Forstrom, Michele Buddle, Rachel Stevensen, and Megan Lantry, ZymoGenetics, Inc., Seattle, WA.

tein concentrations were determined by amino-acid composition analysis. Each truncated form was evaluated for relative specific activity in an in vitro proliferation assay using the Ba/F3-Mpl[+] cell line *(1)*. It is clear from examination of the data in Table 2 that on a pure specific-activity basis, truncated forms of Mpl ligand have increased specific activity some 20-fold on a U/µg basis relative to full-length TPO. However, when the effects of molecular weight to convert these comparisons to an equimolar basis are considered, truncation results in an approximately fivefold increase in activity. Intermediate C-terminal truncations produce intermediate increases in specific activity.

These results support an interpretation that not only is the EPO-like domain biologically active in vitro, but it is more active than the full-length protein, suggesting the possibility that proteolytic conversion from the full-length form to shorter truncated forms may reflect a physiologic, regulatory conversion of a less-potent precursor to a form with more active receptor-binding properties.

Full-length Mpl ligand and its EPO-like domain were also tested for in vivo potency, a measure of intrinsic potency, and serum half-life. For this experiment, glycosylated and nonglycosylated versions of both forms of the molecule were produced. Additionally, a truncated form encompassing the EPO-like domain was also conjugated with poly[ethyleneglycol] (PEG), which, like glycosylation, serves to stabilize and increase the circulating half-life of the molecule. The pegylated form of a truncated, *Escherichia coli*-produced, recombinant Mpl ligand, megakaryocyte growth and development factor (PEG-rHuMGDF) is under clinical development by Amgen Inc. *(21)*. These molecules were subcutaneously injected into mice daily for 5 days in graded doses (µg/kg/day). Two days after the last injection, platelet counts were measured in an automatic blood cell analyzer (Fig. 4). Full-length glycosylated Mpl ligand was significantly more effective at increasing platelet counts than the EPO-like domain whether glycosylated or not. Nonetheless, the EPO-like domain appears to contain the active portion of the molecule. When pegylated, the in vivo activity of this portion of the molecule increased substantially. The greater in vivo activity of PEG-rHuMGDF

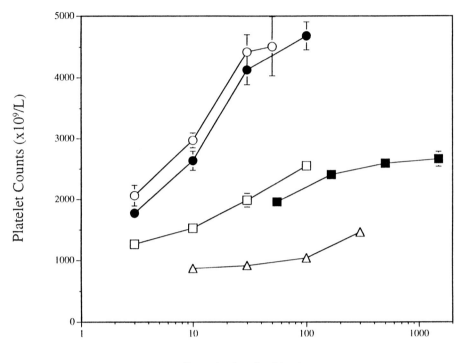

Protein (μg/kg/day)

Fig. 4. The in vivo properties of nonglycosylated, glycosylated, or pegylated recombinant-derived human Mpl ligand or the EPO-like-domain. Mice ($n = 5$–12/group) were injected daily with the indicated version and dose of Mpl ligand for five consecutive days. Two days after the last injection, blood was sampled and platelets measured with a Sysmex automatic cell counter. Open circles, pegylated EPO-like domain; closed circles, full-length glycosylated molecule; open squares, glycosylated EPO-like domain; closed squares, nonglycosylated EPO-like domain; open triangles, nonglycosylated full-length form.

relative to the unconjugated EPO-like domain is in agreement with a longer circulating half-life of the former molecule relative to the latter (Cheung E, unpublished observations). Interestingly, the full-length, nonglycoslated molecule is essentially inert. This is presumably because of the inherent instability of the full-length molecule when produced without glycoslation (Boone T, Narhi L, unpublished observations).

Other studies have also shown that the truncated form of the protein has both megakaryopoietic and thrombopoietic activity in vivo in rodents *(22)* and nonhuman primates *(23,24)*. Administration of the truncated form also ameliorates chemotherapy-induced thrombocytopenia in murine models *(22)* and irradiation-induced thrombocytopenia in rhesus monkeys *(25; see* Chapter 20). Although the EPO-like domain *per se* is clearly effective in vivo, the pegylated version of the molecule, PEG-rHuMGDF, with superior stability in vivo, is significantly more effective in correcting experimentally induced thrombocytopenia in rodents *(21)* and primates *(25–27)*. Early reports on the safety and efficacy of this molecule in human phase 1 clinical trials have also been presented *(28)*.

Human Serum

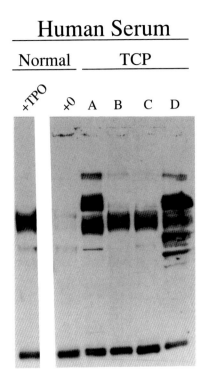

Fig. 5. Precipitation/immunoblotting evaluation of human serum TPO. Sera from normal donors spiked with full-length glycosylated rHuTPO ("TPO") or not spiked ("+0") or thrombocytopenic patients (TCP) were precipitated with beads conjugated with Mpl. Western immunoblotting of the precipitates was developed with rabbit anti-human TPO (generated using the full length, glycosylated form) and the ECL secondary antibody detection system (Amersham, Amersham International, Buckinghamshire, UK). Sera from the normal donor contained 80 pg/mL TPO; sera from the thrombocytopenic patients (A–D) contained 1260, 1690, 1740, and 1800 pg/mL TPO, respectively. Data courtesy Alex Hornkohl, Amgen Inc., Thousand Oaks, CA.

5. Mammalian Cell Produced and Native Forms of Human TPO

5.1. Mammalian Cells Expressing Human TPO Secrete a Full-Length, Highly Glycosylated, 70-kDa Protein

When the full-length cDNA for human TPO is expressed in mammalian cells, such as BHK or Chinese Hamster Ovary (CHO) cells, the primary translation product that accumulates in the conditioned medium under serum-free culture conditions is an approximately 70-kDa protein (Figs. 3 and 5). When this material was purified and biochemically characterized, the N-terminal sequence, mass spectroscopy-derived relative migration (M_r), and the carbohydrate content indicated that it represented the full polypeptide backbone predicted by the cDNA sequence and that it was heavily glycosylated. All six potential N-linked glycosylation sites predicted in the C-terminal domain are modified with complex sugars to varying extents, ranging in occupancy from 6% at asparagine (Asn) to 100–176% at Asn 234 *(29)*. In addition, mass spectroscopy analyses of tryptic peptides indicate the presence of multiple sites of O-linked sugars in both the EPO-like domain and the C-terminal domain. In the EPO-like domain, O-linked modification of the amino-terminal serine (Ser) 1 has been demonstrated. O-linked modification is also seen on at least one of several sites within a peptide including Ser27, threonine (Thr) 37, and Ser47, and on at least one site within a peptide including Thr158, Thr159, and Ser163. This latter peptide is included in a truncated form of Mpl ligand that extends 18 amino acids beyond the EPO-like domain. Within the C-terminal domain, potential glycosylation sites are spaced too tightly together to permit precise and complete determination of glycosylation status. However, the presence of O-linked sugars on peptides, including Thr192, Thr193, and Ser195, as well as on a peptide containing Ser244 have been demonstrated *(29)*.

In contrast to the observation of 70-kDa full-length product in mammalian cell-conditioned medium, lower-molecular-weight forms of the Mpl ligand, most notably 30- and 18-kDa forms, are observed after purification of the product. This purification-associated proteolysis has been particularly associated with purifications that include affinity adsorption to immobilized Mpl, leading to the hypothesis that receptor binding may potentiate proteolysis.

The observation of an intact 70-kDa full-length TPO secreted from mammalian cells in culture also stands in marked contrast to all published accounts of plasma-derived TPO preparations that have had masses ranging from 19 to 31 kDa, depending on species source, and that have all been shown to be C-terminally truncated forms *(2–6)*. This contrast between the apparent predominant form produced by cell culture in vitro and the truncated forms that predominate in various purified preparations also suggests the possibility that degradation occurs during purification and provokes investigation of the form(s) that occur naturally in the circulation.

5.2. Human Serum Contains Multiple Molecular-Weight Species of TPO

The molecular weight of serum-derived human TPO has been evaluated by precipitation/immunoblotting techniques. Sera from normal or thrombocytopenic individuals, containing from 80 pg/mL to as much as 10,000 pg/mL protein, respectively (Chapter 22), were precipitated with beads conjugated with the ligand-binding portion of Mpl. The precipitates were analyzed by immunoblotting using rabbit antihuman TPO as the detection reagent. As a control, purified recombinant human Mpl ligand expressed in CHO cells was added to normal human sera before the precipitation procedure. As shown in Fig. 5, the recombinant form appears as a single, wide band characteristic of a heavily glycosylated molecule of 70 kDa when precipitated with Mpl beads (lane "TPO"). A species of equivalent molecular weight is barely detectable in unspiked normal human sera (lane "+0"). Serum from a thrombocytopenic, chemotherapy-treated breast cancer patient was also tested (lane "A"). In this sample, endogenous TPO appears in two prominent bands, with M_r above and below the recombinant human form. Sera from a thrombocytopenic, chemotherapy-treated patient with acute myeloid leukemia (AML) is in lane "B"; it contains endogenous TPO, which migrates identically to the recombinant form. A sample from a patient with aplastic anemia was evaluated in lane "C"; it also contains endogenous TPO, which migrates identically to the recombinant form. Sera from a thrombocytopenic pediatric patient with amegakaryocytic thrombocytopenia is shown in lane "D." In this case, endogenous TPO was present in many distinct forms ranging from one prominent species with an $M_r > 70$ kDa to several, less abundant, smaller species.

6. Speculation on Proteolytic Cleavage: Potential Role(s) of the C-Terminal Domain

It is clear from the data presented in Table 2 that the EPO-like domain of TPO contains the receptor-binding portion. However, other observations, including those presented in Table 1 and Fig. 4, suggest that the C-terminal domain may play a role in regulating activity in vivo. The fact that mammalian cells secrete primarily full-length 70-kDa TPO and that the 70-kDa form is at least one of the primary forms seen in human serum samples suggests that the protein primarily circulates in this form. Observations that TPO prepared with Mpl-affinity chromatography procedures is primarily a mixture of lower-molecular-weight forms suggest that binding of 70-kDa TPO to the receptor may be associated with proteolytic conversion to smaller species. The fact that these lower-molecular-weight species have higher specific activities than the larger form suggest that proteolysis may play an activity-regulating role, i.e., local proteolysis on receptor binding may be a more potent form that acts locally. One example in support of this hypothesis is the receptor-mediated activation of coagulation Factor VII to Factor VIIa through binding to its physiologic receptor, a member of the Mpl family *(30,31)*. In this example, receptor binding induces a conformational change in the ligand exposing a proteolytic activation site in a conformation favorable for cleavage. The cleaved ligand has its activity vastly enhanced over the uncleaved form, is locally active as a complex with the receptor, and is then susceptible to much more rapid clearance from the circulation than the uncleaved circulating precursor form.

In addition to its possible involvement in the regulation of biological potency and in the regulation of circulating half-life, there is strong evidence that the C-terminal domain may also play a molecular chaperone or folding role in the biosynthesis and cellular secretion of TPO. When truncated constructs consisting of only the EPO-like domain or of the EPO-like domain and small regions of the C-terminal domain are expressed in various mammalian cell hosts (or even in the secretion pathway of heterologous hosts, such as *Saccharomyces cerevisia),* the protein is secreted very poorly and accumulates in intracellular compartments. In contrast, when full-length constructs are expressed in identical host cells, the protein is well secreted, and after a 4-hour labeling period is largely found in the conditioned medium (Lofton-Day C, unpublished observations). The high levels of glycosylation on the C-terminal domain suggest that this effect on secretion of the polypeptide may be comparable to the requirement of many proteins for a glycosylated chaperone component to facilitate transport through various secretory compartments. Perhaps the best documented of these examples is the secretion of the α-factor mating pheromone in yeast. In this case, the α-factor peptide is biosynthesized as a component of a larger precursor that includes a 60 amino-acid glycosylated "leader." This leader peptide remains covalently attached to the mating peptide as the protein traverses through the secretory pathway compartments and is ultimately removed in the trans-Golgi by Kex2 peptidase just before secretion. The α-factor peptide is very poorly secreted if expressed without this glycosylated leader, and extensive structure/function analyses of the leader sequence have documented the requirement for glycosylation to facilitate efficient secretion. It is possible that the glycosylated status of the C-terminal domain of TPO also facilitates movement through the secretory pathway. Alternatively, the full-length protein may fold more efficiently and be more efficiently secreted than poorly folded, truncated forms that might be retained by endoplasmic reticulum retention mechanisms and turned over from the endoplasmic reticulum lumen.

7. Conclusions

The discovery and elucidation of the structure of TPO have provided a unique series of opportunities to correlate the novel two-domain structure of this cytokine with its rapidly evolving biological profile. It has been clearly demonstrated that the EPO-like domain is sufficient to provoke a full spectrum of biological responses both in vitro and in vivo. Although the EPO-like domain has some fivefold higher specific activity than full-length TPO in in vitro proliferation assays, it is cleared from the circulation much more rapidly in vivo and fails to elicit a full thrombopoietic response in animals. This lack of potency in vivo is likely attributable to altered pharmacokinetics and can be functionally corrected by derivitization with poly[ethyleneglycol] (PEG-rHuMGDF), which significantly improves the half-life and potency profile of the EPO-like domain. The C-terminal domain certainly plays a role in the biosynthesis and efficient secretion of Mpl ligand from mammalian cells in culture, probably reflecting a significant role of this domain physiologically.

The fact that TPO is initially synthesized and primarily circulates in a relatively stable two-domain precursor form with relatively lower specific activity than the cleaved form(s), taken together with the observation that cleavage of the precursor form to the smaller forms is associated with binding to Mpl, suggests that proteolysis of full-length TPO is likely to be a physiologically important process by which its effects are regulated. One simple model is that the C-terminal domain initially plays a role in biosynthesis of TPO. After secretion from the liver, the second domain primarily regulates the activity of Mpl ligand and promotes stability in the circulation. An obvious prediction of this model is that a mutated "uncleavable" full-length molecule should display severely impaired in vivo activity while retaining a normal pharmacokinetic profile. Several experimental approaches to exploring these models are currently under way.

References

1. Lok S, Kaushansky K, Holly RD, et al. Cloning and expression of murine thrombopoietin cDNA and stimulation of platelets in vivo. *Nature*. 1994; 369: 565–568.
2. Kato T, Ogami K, Shimada Y, et al. Purification and characterization of thrombopoietin. *J Biochem*. 1995; 118: 229–236.
3. de Sauvage FJ, Hass PE, Spencer DD, et al. Stimulation of megakaryopoiesis and thrombopoiesis by the c-mpl ligand. *Nature*. 1994; 369: 533–538.
4. Kuter DJ, Beeler DL, Rosenberg RD. The purification of megapoietin: A physiological regulator of megakaryocyte growth and platelet production. *Proc Natl Acad Sci USA* 1994; 91: 1104–1108.
5. Bartley TD, Bogenberger J, Hunt P, et al. Identification and cloning of a megakaryocyte growth and development factor (MGDF) that is a ligand for the cytokine receptor Mpl. *Cell*. 1994; 77: 1117–1124.
6. Hunt P, Li YS, Nichol JL. Purification and biologic charaterization of plasma-derived megakaryocyte growth and development factor. *Blood*. 1995; 86: 540–547.

7. Boissel J-P, Lee W-R, Presnell SR, Cohen FE, Bunn HF. Erythropoietin structure-function relationships: Mutant proteins that test a model of tertiary structure. *J Biol Chem*. 1993; 268: 15,983–15,993.

8. Manavalan P, Swope DL, Withy RM. Sequence and structural relationships in the cytokine family. *J Protein Struct*. 1992; 11: 321–331.

9. Benson D, Lipman DJ, Ostell J. GenBank. *Nucleic Acids Res*. 1993; 21: 2963–2965.

10. Osborn SG, Walker K, Evans SJ, Buddle MM, Forstrom JW. Characterization of murine thrombopoietin. The Protein Society Symposium. June 9–13, 1994. San Diego, CA.

11. Spivak JL, Hogans BB. The in vivo metabolism of recombinant human erythropoietin in the rat. *Blood*. 1989; 73: 90–99.

12. Takeuchi M, Takasaki S, Shimada M, Kobata A. Role of sugar chains in the in vitro biological activity of human erythropoietin produced in recombinant Chinese hamster ovary cells. *J Biol Chem*. 1990; 265: 12,127–12,130.

13. Narhi LO, Arakawa T, Aoki KH, et al. The effect of carbohydrate on the structure and stability of erythropoietin. *J Biol Chem*. 1991; 266: 23,022–23,026.

14. Barr PJ. Mammalian subtilisins: the long-sought dibasic processing endoproteases. *Cell*. 1994; 66: 1–3.

15. Recny MA, Scoble HA, Kim Y. Structural characterization of natural human urinary and recombinant DNA-derived erythropoietin. Identification of desarginien/66 erythropoietin. *J Biol Chem*. 1987; 262: 17,156–17,163.

16. Eaton DL, Gurney A, Malloy WJ, et al. Biological activity of human thrombopoietin (TPO), the c-mpl ligand, and TPO variants and the chromosomal location of TPO. *Blood*. 1994; 84: 241a (abstract no 948).

17. Choi ES, Hokom MM, Bartley T, et al. Recombinant human megakaryocyte growth and development factor (rhuMGDF), a ligand for c-mpl, produces functional human platelets in vitro. *Stem Cells*. 1995; 13: 317–322.

18. Hunt P. The physiological role and therapeutic potential of Mpl-ligand. *Stem Cells*. 1995; 13: 579–587.

19. Debili N, Wendling F, Katz A, et al. The Mpl-ligand (Mpl-L) or thrombopoietin (TPO) or megakaryocyte growth and differentiative factor (MGDF) has both direct proliferative and differentiative activities on human megakaryocyte progenitors. *Blood*. 1995; 86: 2516–2525.

20. Nichol JL, Hokom MM, Hornkohl A, et al. Megakaryocyte growth and development factor: Analyses of in vitro effects on human megakaryopoiesis and endogenous serum levels during chemotherapy-induced thrombocytopenia. *J Clin Invest*. 1995; 95: 2973–2978.

21. Hokom M, Lacey D, Kinstler O, et al. Pegylated megakaryocyte growth and development factor abrogates the lethal thrombocytopenia associated with carboplatin and irradiation in mice. *Blood*. 1995; 86: 4486–4492.

22. Ulrich TR, delCastillo J, Yin S, et al. Megakaryocyte growth and development factor ameliorates carboplatin-induced thrombocytopenia in mice. *Blood*. 1995; 86: 971–976.

23. Farese AM, Hunt P, Boone TC, McVittie T. Recombinant human megakaryocyte growth and development factor stimulates thrombocytopoiesis in normal nonhuman primates. *Blood*. 1995; 86: 54–59.

24. Harker L, Hunt P, Marzac U, et al. Regulation of platelet production and function by megakaryocyte growth and development factor (MGDF). *Blood*. 1996; 87: 1833–1845.

25. Farese A, Hunt P, Grab L, MacVittie T. Combined administration of recombinant human megakaryocyte growth and development factor and granulocyte colony-stimulating factor enhances multilineage hematopoietic reconstitution in nonhuman primates after radiation-induced marrow aplasia. *J Clin Invest*. 1996; 97: 2145–2151.

26. Harker LA, Marzec UM, Kelly AB, Hanson SR. Enhanced hematopoietic regeneration in primate model of myelosuppressive chemotherapy by pegylated recombinant human megakaryocyte growth and development factor (PEG-rHuMGDF) in combination with granulocyte colony-stimulating factor (G-CSF). *Blood*. 1995; 86: 497a (abstract no 1976a).

27. Harker LA, Hunt P, Marzec UM, Kelly AB, Tomer A, Sanson SR. Dose-response effects of pegylated human megakaryocyte growth and development factor (PEG-rHuMGDF) on platelet production and function in nonhuman primates. Blood. 1995; 86: 256a (abstract no 1012).

28. Basser R, Clarke K, Fox R, et al. Randomized, double-blind, placebo-controlled phase I trial of pegylated megakaryocyte growth and development factor (PEG-rHuMGDF) administered to patients with advanced cancer before and after chemotherapy-Early results. *Blood*. 1995; 86: 257a (abstract no 1014).

29. Linden H, O'Rork C, Hart CE, Stamm M, Hoffman R, Kaushansky K. The structural and functional role of disulfide bonding in thrombopoietin. *Blood*. 1995; 86: 255a (abstract no 1008).

30. Sakai T, Lund-Hansen T, Paborsky L, Pedersen AH, Kisiel W. Binding of human factors VII and VIIa to a human bladder carcinoma cell line (J82). Implications for the initiation of the extrinsic pathway of blood coagulation. *J Biol Chem*. 1989; 264: 9980–9988.

31. Bazan JF. Structural design and molecular evolution of a cytokine receptor superfamily. *Proc Natl Acad Sci USA* 1990; 87: 6934–6938.

IV

Cellular Biology

14

In Vitro Effects of Mpl Ligand on Human Hemopoietic Progenitor Cells

Najet Debili, Elisabeth Cramer, Françoise Wendling, and William Vainchenker

1. Introduction

Regulation of megakaryocytopoiesis and platelet production is a complex phenomenon. Numerous pleiotropic cytokines act in vivo and in vitro on megakaryocytopoiesis *(1)*. Historically, studies on the regulation of megakaryocytopoiesis were largely dominated by the concept of humoral regulation, a concept based on the model of erythropoiesis and formulated before the development of clonal assays for hemopoietic progenitors. In animals, induction of acute thrombocytopenia with antiplatelet antibodies rapidly results in an increase of platelet production as measured by isotopic techniques. Simultaneously, an increase in platelet size, along with megakaryocyte number, size, ploidy, and cytoplasmic maturation is observed. Despite 35 years of work, the factor responsible for these effects (thrombopoietin, TPO) *(2)* could not be purified to homogeneity *(3)*, and TPO was thought to be a late differentiation factor that acted synergistically with other growth factor, such as interleukin (IL)-3 *(1,3)* (*see* Chapter 5).

Another humoral growth factor, designated megakaryocyte colony-stimulating activity (MK-CSA), also had been defined in humans and large animals. The name MK-CSA is derived from its capacity to induce megakaryocyte colony formation. It is present in plasma, serum, and urine of thrombocytopenic patients with disorders of platelet production owing to reduced megakaryocyte number. Megakaryocyte colony-stimulating factor (MK-CSF) is found in the plasma of aplasic patients or in patients after radiation therapy or chemotherapy *(4–7)*. MK-CSA is also found in large amounts in the plasma of irradiated dogs or pigs *(8)*. In contrast to TPO, MK-CSA is not found in the plasma of patients with immune thrombocytopenic purpura (ITP) *(4,6,9)*. Therefore, it was hypothesized that the regulation of MK-CSA synthesis was dependent on the megakaryocyte compartment and not on the platelet mass *(1)*.

Despite their apparently differing properties, MK-CSA and TPO could be the same molecule for several reasons: first, assays for TPO/MK-CSA have been constructed without taking into account possible species specificity. Second, the hypothesis that MK-CSA was regulated by the size of the megakaryocyte compartment and not by the platelet mass was based only on results obtained by screening plasma of aplastic or ITP

From: *Thrombopoiesis and Thrombopoietins: Molecular, Cellular, Preclinical, and Clinical Biology*
Edited by: D. J. Kuter, P. Hunt, W. Sheridan, and D. Zucker-Franklin Humana Press Inc., Totowa, NJ

patients. Third, MK-CSA had been tested initially only on unpurified progenitors, a situation in which a large number of cytokines are concurrently synthesized by marrow cells. When plasmas from irradiated dogs or pigs or from patients with aplastic anemia were tested on purified CD34[+] cells, MK-CSA was low. Finally, there was a major activity not related to IL-6 that acted on megakaryocyte maturation and ploidy *(10–12)*. Therefore, it could not be excluded that the activity detected in aplastic plasma was related to a cytokine acting on more mature megakaryocyte progenitors and the late stages of megakaryocytopoiesis. The properties of such a cytokine are very similar to those that define TPO.

The proto-oncogene *c-mpl* has been demonstrated to encode a protein (Mpl) that is the receptor for a humoral cytokine regulating megakaryocytopoiesis. Its ligand was isolated by three independent groups *(13–15)*. The purified protein was termed Mpl ligand *(14)*, TPO *(15)*, or megakaryocyte growth and development factor (MGDF) *(13)*. Of note, two other groups succeeded in the isolation of the same molecule by purifying a thrombopoietic factor from the plasma of markedly thrombocytopenic animals (rats or sheep) using a conventional purification strategy *(16,17)* (*see* Chapter 9).

Strong evidence that Mpl ligand is the homeostatic regulator of platelet production is provided by *c-mpl* and TPO knockout mice, each of which has a severe, but not lethal, thrombocytopenia *(18,19)*. In addition, preliminary results suggest that Mpl ligand level in the plasma was regulated by the platelet mass *(20)*.

In this chapter, the in vitro effects of Mpl ligand on megakaryocyte differentiation will be described by summarizing the literature and describing experimental results. It will be shown that, although its major action is on megakaryocytopoiesis and thrombopoiesis, Mpl ligand also acts on other hemopoietic lineages.

2. Mpl Ligand Is a Potent Mitotic Factor for Megakaryocyte Progenitors

2.1. Background

When the recombinant molecule was produced, early results clearly showed that Mpl ligand acted on murine and human megakaryocyte progenitors. Kaushansky et al. showed that Mpl ligand was able to induce murine megakaryocyte colony formation from marrow or spleen *(21)*. de Sauvage et al. *(14)* and Bartley et al. *(13)* also demonstrated that Mpl ligand was able to induce megakaryocyte differentiation from peripheral blood mononuclear or CD34[+] cells. These preliminary in vitro biological effects were subsequently extended. In serum-depleted culture, Mpl ligand was able to support megakaryocyte colony formation from unfractionated marrow cells with about a twofold higher plating efficiency (15–20 colonies/10^5 marrow cells) than observed with IL-3-stimulated cultures *(22)*. However, colonies were composed of few (<20) megakaryocytes. An additive effect on the plating efficiency was observed when IL-3 and Mpl ligand were combined, resulting in a marked increase in the number of megakaryocytes per colony. Moreover, stem cell factor (SCF), IL-11, and erythropoietin (EPO) markedly synergized with Mpl ligand to enhance the number of megakaryocyte colonies. Using human CD34[+] peripheral blood cells, Hunt et al. have shown that Mpl ligand is able to induce megakaryocyte colony formation in a dose-dependent manner with a high plating efficiency (4%) *(23)*. As in the mouse experiments, IL-3 had a much less potent effect (approximately 0.7% cloning efficiency). The number of megakaryocytes per colony was much higher when IL-3 was used: 25% of the colonies contained more than 20 megakaryocytes, whereas with Mpl ligand alone, megakaryocyte colony size did not exceed 20 megakaryocytes. In combination with a suboptimal concentration of Mpl ligand, IL-3 had an additive effect in the induction of megakaryocyte colonies, whereas the combination of Mpl ligand and SCF was highly synergistic and gave the highest plating efficiency.

However, none of these results totally excluded the possibility that a portion of Mpl ligand activity is either indirect and related to a synergistic effect with other cytokines present in serum-containing cultures, or synthesized by either accessory cells or CD34$^+$ cells.

2.2. Classes of Megakaryocyte Progenitor Cells

To assist in delineation of the precise biological activity of Mpl ligand on human megakaryocyte progenitors, these cells can be purified according to their differentiation stage and used in serum-free experiments at limiting cell dilution.

Megakaryocyte progenitors can be divided according to their in vitro properties into three main types that may correspond to different stages of differentiation. Megakaryocyte burst-forming cells (MK-BFC) produce colonies composed of more than 50 cells organized in subcolonies, based on the same model as erythroid burst-forming cell (E-BFC)-derived colonies. In humans, such colonies are quantitated after 21 days in culture *(24)*. A more proliferative progenitor, called high-proliferative potential megakaryocyte colony-forming cell (HPP-MK-CFC), has been described recently *(25)*. Megakaryocyte colony-forming cells (MK-CFC) are progenitor cells that differ from the previous ones by a lower capacity for proliferation. In humans, they give rise to megakaryocytes in short-term liquid culture, to colonies composed of 3–50 cells after 12 days in culture *(26,27)*, or to small megakaryocyte colonies in 7 days *(28)*. The third type is the mature MK-CFC. These cells have been identified in the mouse as a megakaryocyte progenitor with a density lower than 1.050 g/mL. They give rise, in 2–3 days, to colonies composed of a few megakaryocytes with a high ploidy *(29)*. In humans, these cells give rise to megakaryocytes in short-term liquid culture *(30)*. The developmental stage of this last progenitor is extremely close to the transitional cell that has switched toward an endomitotic process.

The use of differentiation markers has recently permitted a more precise delineation of the compartments of megakaryocyte progenitors. Megakaryocyte progenitors express the CD34 antigen, as do the majority of myeloid progenitors. Among CD34$^+$ cells, myeloid progenitors can be fractionated by stage of differentiation according to their phenotype. Immature progenitors are found in the CD38$^{-/low}$, Thy1$^+$, or HLA-DR^{low} cell fraction, whereas their more mature counterparts are found in the complementary fraction (CD38$^+$, Thy1$^-$, HLA-DR$^+$). For the megakaryocyte lineage, GPIIb/IIIa (CD41/CD61) is expressed on a subset of CD34$^+$ cells that corresponds to mature MK-CFC *(30)*, the megakaryocyte analog of erythroid colony-forming cells (E-CFC) in the erythroid lineage. This CD34$^+$ CD41$^+$ cell population is a nearly pure population of megakaryocyte progenitors, and therefore facilitates limiting dilution experiments *(31)*.

2.2.1. Mpl Ligand Acts Directly on Late Megakaryocyte Progenitors

In an initial set of experiments, the recombinant Mpl ligand molecule was tested in liquid cultures of unfractionated CD34$^+$ marrow cell populations. A plateau for the absolute number of megakaryocytes obtained at day 12 of culture was reached at 5 ng/

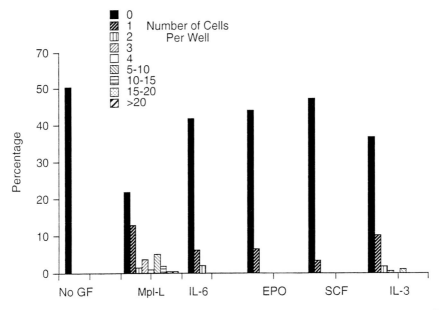

Fig. 1. Distribution of the number of cells per well observed on day 5 of culture at limiting cell dilution in the absence of growth factor (No GF), or the presence of Mpl ligand (10 ng/mL), IL-6 (100 U/mL), EPO (1 U/mL), SCF (50 ng/mL), or IL-3 (100 U/mL). Individual marrow CD34+CD41+ cells were deposited in a 96-well tissue-culture plate using a cell sorter. The number of cells per well was evaluated at day 5 of culture under an inverted microscope. Results are expressed as a percentage (one experiment, 288 wells for each growth factor). IL = interleukin; EPO = erythropoietin; SCF = stem cell factor.

mL. In subsequent experiments, Mpl ligand was used at 10 ng/mL, and CD34+CD41+ cells were plated at 1–50 cells/100 μL vol in the presence of Mpl ligand alone. A linear relationship was found between the average number of cells per well and the number of cells plated. Experiments were performed at 1 cell/well. The percentage of positive wells and the number of cells per well were determined on day 5 of culture. In the absence of growth factor, all wells were devoid of cells. In contrast, in the presence of Mpl ligand alone, the percentage of positive wells was 60%. Among these positive wells, approximately 50% contained an individual large megakaryocyte, whereas in the others, proliferation had occurred since clones containing up to 25 identifiable megakaryocytes were present (Fig. 1). These experiments clearly demonstrate that Mpl ligand can act directly on mature megakaryocyte progenitors.

2.2.2. Mpl Ligand Stimulates Mitosis of Late Megakaryocyte Progenitors as Efficiently as Cytokine Combinations

In subsequent experiments, the effects of Mpl ligand were compared with those of either SCF, IL-6, EPO, or IL-3 (Fig. 1). In the presence of SCF, IL-6, or EPO, the percentage of positive wells was <10%, and no clone contained more than three cells. In contrast, in the presence of IL-3, the cloning efficiency was 20% with clones containing as many as eight cells. Therefore, the effects of Mpl ligand alone were compared with those of a combination of cytokines. The combination of IL-3, IL-6, and SCF could be totally replaced by Mpl ligand, resulting in the same plating efficiency

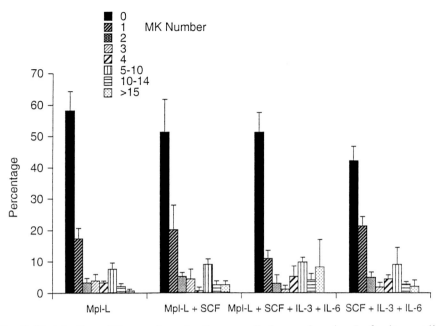

Fig. 2. Distribution of the number of cells per well observed on day 5 of culture at limiting-cell dilution in the presence of Mpl ligand (10 ng/mL), the combination of Mpl ligand (10 ng/mL), and SCF (50 ng/mL); of Mpl ligand (10 ng/mL), SCF (50 ng/mL), IL-3 (100 U/mL), and IL-6 (100 U/mL); and of SCF (50 ng/mL), IL-3 (100 U/mL), and IL-6 (100 U/mL). Culture methods same as for Fig. 1. IL = interleukin; SCF = stem cell factor.

and the same number of cells per clone (Fig. 2). Recent experiments suggest that the combination of SCF plus IL-6 is highly synergistic. In contrast, this combination of cytokines together with Mpl ligand did not change the cloning efficiency, but increased the number of cells per clone twofold.

2.2.3. Mpl Ligand Acts Directly on Earlier Megakaryocyte Progenitors

Similar experiments were performed on the CD34+CD41− cell populations obtained from marrow or cord blood. This population contains all types of megakaryocyte progenitors (including some late megakaryocyte progenitors) as well as all types of myeloid progenitors. When such cells were grown in liquid culture at a concentration of 1×10^4 cells/mL in the presence of Mpl ligand alone, megakaryocytes were obtained in suspension. Megakaryocyte differentiation was detected slightly later (beginning at day 5) than in cultures of CD34+CD41+ cells, and megakaryocytes were present until day 20 of culture. Megakaryocytes can also be obtained from CD34+CD38low cells in serum-free culture stimulated by Mpl ligand alone, as recently reported *(32)*.

Suspension cultures were also performed at limiting dilution using cord blood or marrow CD34+CD41− cell subsets stimulated by Mpl ligand alone. Surprisingly, the percentage of positive wells (containing 1 cell or more/well) was very high for such a heterogenous population both for the marrow (24%) and cord blood cells (15%). Some wells contained morphologically identifiable megakaryocytes, and half of these contained only one large megakaryocyte. The distribution of cell numbers in the clones differed from the CD34+CD41+ cell population, since rare wells contained as many as

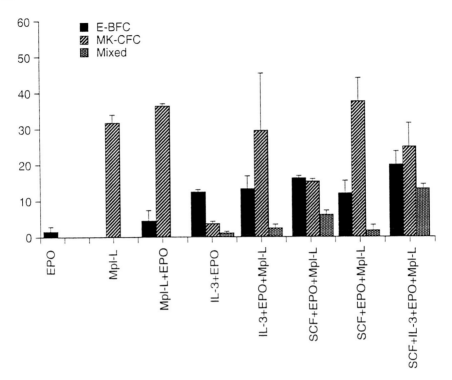

Fig. 3. Cloning efficiency of CD34$^+$CD38^{low} cells in the presence of Mpl ligand and different combination of cytokines. CD34$^+$CD38^{low} cells (500/mL) were cultured in a serum-free fibrin-clot assay and colonies were enumerated at day 12 of culture after immunoenzymatic staining for CD61. The mixed colonies are erythroid-megakaryocyte colonies. Results are expressed as the number of colonies per 1000 cells. IL = interleukin; EPO = erythropoietin; SCF = stem cell factor.

50 cells for marrow cells and several hundred cells for cord blood cells. These large clones were composed of small cells only, and their cellular nature is currently under investigation. These results demonstrate that Mpl ligand can act directly on CD34$^+$CD41$^-$ megakaryocyte progenitors, but also suggest that Mpl ligand may also act on other myeloid progenitors.

It was not possible to identify precisely the nature of the other cell types present using this limiting dilution technique, since most clones did not contain enough cells for immunophenotyping. Therefore, the CD34$^+$CD38$^{-/low}$ marrow cell population was plated. This population is greatly enriched in primitive hemopoietic progenitors (including megakaryocyte progenitors) with a high proliferative capacity using the serum-free fibrin clot technique. Mpl ligand alone was able to induce megakaryocyte colony formation from this cell population with the same efficiency as the combination of SCF, IL-3, and IL-6 plus EPO (Fig. 3). However, colonies grown with Mpl ligand alone were smaller.

2.2.4. SCF and IL-3 Have Synergistic or Additive Effects on Early Megakaryocyte Progenitors

In serum-free semisolid assays, addition of SCF or IL-3 (or a larger combination of cytokines) to Mpl ligand did not significantly increase the plating efficiency of mega-

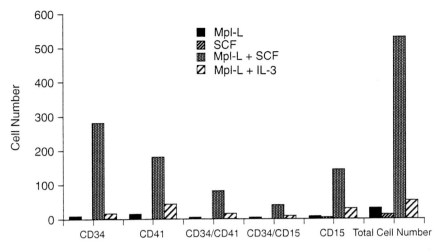

Fig. 4. Synergistic effect of Mpl ligand plus SCF or IL-3 on the growth of hemopoietic cells from cord blood CD34⁺CD38^low cells. Cord blood CD34⁺CD38^low cells (1×10^4 cells per mL) were grown in serum-free suspension cultures. After 11 days of culture, cells were characterized by flow cytometry using double staining of CD34 and either CD41 or CD15. The absolute number of cells with each phenotype was calculated from the flow cytometric analysis and the total number of cells. Using IL-3 alone, the number of cells grown was too low to permit flow cytometry analysis. SCF = stem cell factor; IL = interleukin.

karyocyte progenitors (Fig. 3). However, these cytokine combinations allowed the detection of mixed colonies that were undetectable in the presence of Mpl ligand alone. The size of megakaryocyte colonies stimulated by a combination of cytokines was larger than when stimulated by Mpl ligand alone, and some colonies contained several hundred cells. Therefore, if a combination of cytokines did not have a great effect on plating efficiency, it had a marked synergistic effect on the proliferation of megakaryocyte progenitors. This phenomenon is better investigated in liquid culture where the absolute number of megakaryocytes can be evaluated by flow cytometry. A marked synergistic effect of Mpl ligand with SCF or IL-3 was observed using marrow CD34⁺CD41⁻ cells. The synergistic effect between Mpl ligand and SCF was even greater on cord blood CD34⁺CD41⁻ or CD34⁺CD38^low cells (Fig. 4).

2.3. Discussion

These data demonstrate that Mpl ligand has a direct proliferative effect on megakaryocyte progenitors and is indeed an MK-CSA for individual cells. The only limitation concerns the serum-free medium that includes bovine serum albumin (BSA), which may contain undefined growth factors *(33)*. In addition, insulin was initially included in the components of the medium, but its removal did not abolish the proliferative effects of Mpl ligand. In agreement with Kaushansky et al. *(34)*, of the other cytokines tested, only IL-3 had a direct, but modest, MK-CSA. However a combination of three cytokines (SCF, IL-3, and IL-6) had about the same stimulating capacity as Mpl ligand. Therefore, Mpl ligand is not obligatory for induction of proliferation and differentiation of megakaryocyte progenitors.

These results also show that Mpl ligand directly acts on earlier megakaryocyte progenitors, including those with a CD34+CD38low phenotype, even though late progenitors are the predominant target cells of this growth factor. This result agrees with those obtained by Zeigler et al. in the mouse, which demonstrate that Mpl ligand could induce megakaryocyte differentiation from murine primitive hemopoietic cells (Sca+ CD41+) *(35)*. In addition, in our experiments, Mpl ligand induced a near-maximal plating efficiency of the megakaryocyte progenitors. Synergy with other cytokines, such as IL-3 and SCF, was mainly observed on proliferation of megakaryocyte progenitors leading to a marked increase in the number of megakaryocytes per colony. For example, the addition of SCF is able to increase approximately 10-fold the number of megakaryocytes obtained from CD34+CD41− or CD38low cells compared with Mpl ligand alone. Other investigators have shown that IL-3 also has an additive effect with Mpl ligand on the cloning efficiency of megakaryocyte progenitors *(36,37)*.

A close relationship exists between erythropoiesis and megakaryocytopoiesis *vis à vis* the effects of EPO and Mpl ligand. For example, Mpl ligand has a direct effect on CD34+CD41+ cells, a cell type that has many similarities with E-CFC. Addition of other cytokines at optimal concentration moderately increases or does not enhance Mpl ligand activity. In contrast, despite a direct action on more immature megakaryocyte progenitor cells (CD34+CD41−), the effects of the ligand are markedly synergized by addition of IL-3 and SCF, as has been observed for the action of EPO on E-BFC. However, as suggested by the experiments performed at limiting dilution on CD34+CD41− cells, Mpl ligand may have a broader activity than EPO, since it is able to induce the proliferation of some primitive megakaryocyte progenitors as well as nonmegakaryocyte progenitors. In contrast, EPO does not appear to act on primitive progenitors, including those of the erythroid series. In conclusion, in the hierarchy of megakaryocyte progenitors, the predominant target for Mpl ligand is a relatively mature cell population, although this cytokine also acts on more primitive cells.

3. Mpl Ligand Induces Megakaryocyte Differentiation and Maturation

3.1. Background

The presumptive humoral regulator of platelet production, called TPO *(2)*, has always been considered a cytokine primarily acting on megakaryocyte terminal differentiation. Terminal differentiation can be schematically divided into three simultaneous processes: induction of platelet protein synthesis; polyploidization; and cytoplasmic maturation. Therefore, it is important to know if Mpl ligand possesses these differentiative properties.

3.2. Mpl Ligand Induces Expression of Platelet Glycoproteins

Permanent cell lines are appropiate systems for the investigation of whether a growth factor is capable of inducing a differentiation program and the synthesis of lineage-specific proteins. Indeed, the role of growth factors on the differentiation of normal hemopoietic cells remains controversial. Certain investigators have observed that lineage specific growth factors are able to induce differentation, whereas others have noted that the differentiation properties of growth factors are the consequence of their antiapoptotic and proliferative effects on committed cells permitting the differentiation program to take place. The detection of GPIIb/IIIa, GPIb, and other platelet proteins after growth of CD34+ cells in the presence of Mpl ligand *(13,14,23,31)* can be ascribed by either mechanism.

In some permanent cell lines, it has been demonstrated that EPO enables the synthesis of erythroid-specific proteins. Indeed, it is able to induce the synthesis of globin in murine Ba/F3 cells transfected with the Epo-R gene *(38)* and to enhance erythroid differentiation of a pluripotent human cell line (UT7) in competition with granulocyte-macrophage colony-stimulating factor (GM-CSF) *(39)*, suggesting that its effects are not simply a result of apoptosis inhibition.

Using the CMK cell line, Kaushansky et al. have shown that Mpl ligand was able to induce the expression of GPIb *(21)*. Other investigators have shown that Mpl ligand slightly increases the expression of GPIIb/IIIa in three new megakaryocyte cell lines (HU-3, CMK-6, and CMS) *(36,40)*. A large panel of cell lines that express Mpl have been tested. Some of them, such as Mo-7E, UT7, TF1, and Ba/F3-mpl, are factor-dependent for their proliferation, whereas others, such as HEL, CMK, LAMA, DAMI, KU812, AP217, and ELF 153, are factor-independent *(41,42)*. The two factor-dependent cell lines, UT7 and TF1, could not be grown in the presence of Mpl ligand, whereas Mo-7E could be cultured for several weeks in the presence of Mpl ligand. Recently, it was reported that in serum-free conditions, Mpl ligand is more of a survival factor for Mo-7E than a proliferative factor *(43)*. However, we were unable to induce or enhance the expression of GPIIb or GPIbα in any of these cell lines. A similar result has been observed by other investigators *(44)*. These data can be interpreted either as the absence of a differentiating signal in the Mpl ligand transduction pathway or the consequence of leukemic origin of these cell lines. To complete these studies further, normal *c-mpl* cDNA was tranduced into the UT7 and TF1 cell lines, which are both inducible by EPO, using a retroviral vector. After retroviral transfer, both cell lines expressed high levels of *c-mpl* and proliferated in the presence of Mpl ligand. In addition, Mpl ligand was able to induce rapidly the expression of GPIIb/IIIa *(45)*. A similar approach was reported by Porteu et al. with the UT7 cells *(45)*. An enhancement of megakaryocyte proteins associated with the induction of GATA-1 and NF-E2 was observed in the murine FDC-P2 cell line stimulated by Mpl ligand *(46)*.

Thus, these data clearly demonstrate that the Mpl ligand is able to induce the expression of platelet proteins. Such Mpl ligand-dependent cell lines will provide models to study the differentiation signaling pathway associated with activation of Mpl.

3.3. Mpl Ligand Stimulates Endomitosis of Normal Human Megakaryocytic Cells

It had been assumed that the humoral factor TPO was primarily involved in the control of the endomitotic rate *(3)*. "Megapoietin" has been purified from the plasma of thrombocytopenic sheep based on its ability to induce polyploidization *(47)*; the amino acid sequence of the megapoietin polypeptide is identical to that of Mpl ligand *(16)*. In the mouse, when marrow cells were grown in the presence of Mpl ligand, megakaryocytes achieved high ploidy with a 64*N* modal distribution *(21,22)*. Consequently, the properties of Mpl ligand in human systems were tested. Using marrow or cord blood CD34+ cells, approximately 50% of the megakaryocytes had a ploidy >4*N* whatever the cytokines used *(10,48,49)*. Initially, CD34+CD41+ cells were grown under serum-free conditions, and megakaryocyte ploidy was determined by flow cytometry at day 5

Marrow CD34+ CD41+ cells

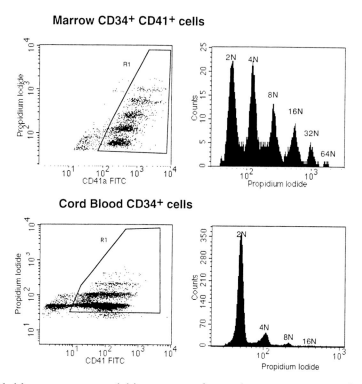

Cord Blood CD34+ cells

Fig. 5. Ploidy cytograms and histograms of megakaryocytes grown from marrow CD34+CD41+ cells and from neonatal cord blood CD34+ cells in serum-free conditions in the presence of Mpl ligand alone. At day 5 (adult marrow) or day 12 (cord blood) of culture, double staining with an anti-FITC CD41 monoclonal antibody (MAb) and propidium iodide was performed. Ploidy histograms were performed in the R1 gate, which corresponded to CD41+ cells.

of culture. Typical cytograms and ploidy histograms are illustrated in Fig. 5. The majority of megakaryocytes belonged to the 2N, 4N, and 8N ploidy classes, but megakaryocytes with DNA content of 32N or 64N were always observed. In addition, in some experiments, 128N megakaryocytes (1%) were also detected. When ploidy levels were determined later in culture, no increase was observed. Addition of EPO, IL-6, IL-3, or SCF to Mpl ligand did not significantly change the ploidy histogram or the mean ploidy (Fig. 6). When cultures were stimulated by three cytokines (IL-6, IL-3, and SCF) instead of the Mpl ligand, ploidy histograms were similar to those observed with Mpl ligand alone, except that 64N and 128N classes were never detected. In experiments using CD34+CD41− cells as the target, the mean ploidy was even lower than when using the CD34+CD41+ cell population. When CD34+ cord blood cells were cultured in these conditions, megakaryocytes had an even lower ploidy (approximately 90% were 2N or 4N megakaryocytes) (Fig. 5), confirming previous results with aplastic plasma *(49)*.

Recently, another group has reported very similar results *(50)*. Nichol et al. have also shown that megakaryocytes grown from blood CD34+ cells had a different ploidy than marrow megakaryocytes *(37)*. However, the modal ploidy (8N) observed in this latter study was closer to that of marrow megakaryocytes than previous results *(49)*. In

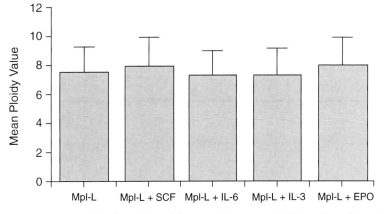

Fig. 6. Effects of a combination of cytokines on the mean ploidy value of megakaryocytes grown from marrow CD34⁺CD41⁺ cells. Experiments were performed as in Fig. 5, and the mean ploidy value was calculated from the ploidy histograms (mean of three experiments). IL = interleukin; EPO = erythropoietin; SCF = stem cell factor.

the Nichol et al. study, an increase in ploidy was observed from day 8 to day 12 of culture in Mpl ligand *(37)*.

3.4. Mpl Ligand Induces Cytoplasmic Maturation of Normal Human Megakaryocytes

Using electron microscopy, cytoplasmic maturation was studied on progeny of CD34⁺CD38⁺ cells grown in serum-free conditions. A large proportion of cultured cells (contaminated only with a few macrophages and basophils) were mature megakaryocytes. One population contained resting megakaryocytes and was composed of large cells, with a smooth surface (Fig. 7A). The nuclei displayed several round lobes, with abundant euchromatin and prominent nucleoli. These cells had a large cytoplasm with numerous randomly scattered α-granules, only a few microvesicular bodies, and a well-developed demarcation membrane system with randomly distributed open cisternae. Some dense granules, suggesting the terminal maturation stage, were also present (Fig. 7B). These cells closely mimic normal megakaryocytes maturing in the bone marrow, and accounted for more than half of this cell population. However, a minority of megakaryocytes had alterations of organelle development, including the demarcation membrane system (insufficient number or composed of flat cisternae) and granules (insufficient number of storage granules).

A second population of megakaryocytes had not been observed by us before the use of Mpl ligand. These cells have a tighter nucleus with closely apposed nuclear lobes and an indented shape. Their surfaces are uneven, often bristled with thin pseudopods. An average of 15% of the megakaryocytes in this population were undergoing active platelet shedding. Since it is rare to encounter evidence of platelet shedding either in fresh bone marrow cells or in culture in the absence of Mpl ligand, these experiments provided an excellent opportunity to observe the process of producing platelets. The following events occurred leading to eventual platelet shedding:

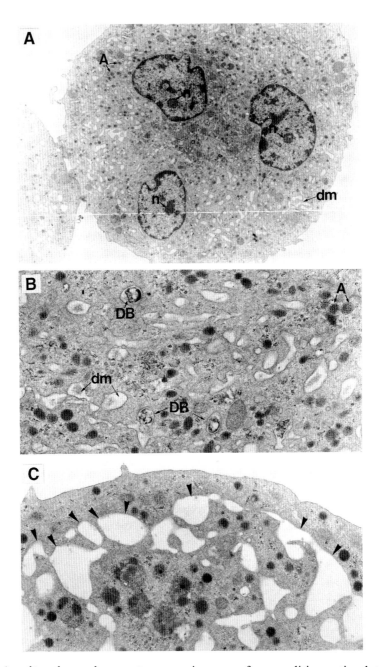

Fig. 7. A cultured megakaryocyte grown in serum-free conditions stimulated by Mpl ligand for 7 days from CD34⁺CD38⁺ cells. **(A)** The cell nucleus displays several lobes with abundant euchromatin and nucleoli. The cytoplasm has matured with numerous α-granules and a well-developed demarcation membrane system (dm) (magnification: ×5310). **(B)** The extent of maturation is attested to by the presence of dense bodies (DB) (A = α-granules, dm = demarcation membrane) (magnification: ×20,580). **(C)** Demarcation membrane alignment along the cell periphery is the first event, which seems to precede platelet shedding (arrowheads).

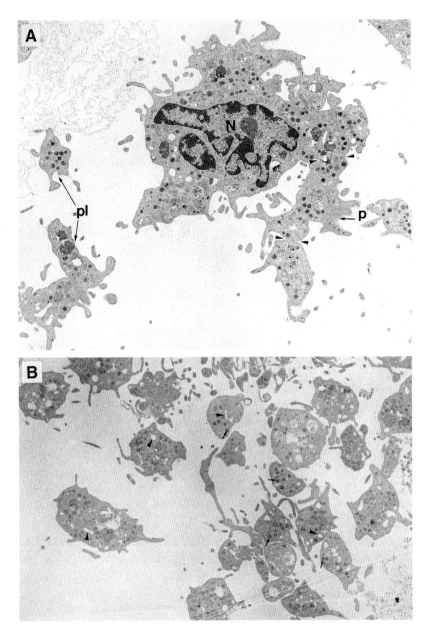

Fig. 8. Platelet shedding in culture. **(A)** This mature megakaryocyte is in the process of shedding platelets. The nucleus displays condensed heterochromatin and its shape is indented. The cell surface is bristled and the cytoplasm extends long expansions or proplatelets (P) from the tip of which newborn platelets (pl) detach (magnification: ×7260). **(B)** Numerous platelet-sized cytoplasmic fragments are liberated into the culture medium. They contain the usual platelet organelles, such as α-granules (arrow) and OCS (arrowheads) (magnification: ×6270).

- Alignment and dilatation of the peripheral demarcation membranes (Fig. 7C);
- Uncoiling of cytoplasmic sheets and cytoplasmic expansions leading to proplatelet formation (Fig. 8A);

- Elongation of these cytoplasmic extentions with the appearance of a central longitudinal bundle of microtubules;
- Appearance of microtubules transverse to the long axis of the cytoplasmic extensions on both sides of a constriction zone and widening of a central vacuole;
- Finally, detachment from the constriction zones (Fig. 8A) of platelets of various sizes into the culture medium.

The ultrastructure and functionality of these platelets have been described by others *(51,52)*. We have studied this cell population by electron microscopy and immuno-electron microscopy (Fig. 8B). The shape and size of the shed platelets produced in vitro were generally uneven. They contain the usual organelles encountered in peripheral blood platelets (e.g., α-granules, open canalicular system [OCS], and microtubules). The microtubules were often elongated along one side of the platelets rather than circumferential, but transverse sections of microtubules were frequently observed at the extremities of the platelets corresponding to those observed on each side of the constriction areas of proplatelets. Platelet glycoproteins GPIb, GPIIb–IIIa, and P-selectin were distributed according to the classical blood platelet distribution (e.g., plasma membrane, OCS, and α-granule membrane, respectively).

3.5. Discussion

These results with human cells support further the contention that Mpl ligand is also a differentiative factor, having most of the properties assigned to TPO. Mpl ligand clearly induced cytoplasmic maturation mimicking that of normal marrow. Some megakaryocytes have maturation abnormalities, but it cannot be excluded that this is owing to culture conditions that are subject to improvement. No other single cytokine was able to induce this degree of maturation, although a combination of several cytokines was able to induce cytoplasmic maturation to a similar degree *(48)*. These results, in contrast to those of Kaushansky et al. *(34)*, indicate that most effects of Mpl ligand can be replaced by a combination of nonlineage-specific cytokines. This result may explain why the TPO knockout mice have a profound, but not lethal, thrombocytopenia. Platelet shedding was also observed in culture with a much higher frequency than with any other cytokine. However, it is presently unclear whether this is owing to a direct effect of Mpl ligand on this process or only the consequence of improved megakaryocyte cytoplasmic maturation. The results of Choi et al. showing efficient proplatelet formation after Mpl ligand deprivation suggest that the effects of Mpl ligand are mainly indirect *(51,52)* (*see* Chapter 17). Recent studies have also shown that Mpl ligand acts on platelet function and induces tyrosine phosphorylation of several cellular proteins *(53,54)*. Both in vivo and in vitro, Mpl ligand primes platelets that become more sensitive to activating agents, such as thrombin. A large part of this in vivo activity is because of the presence of circulating young platelets, which are intrinsically more active than older platelets. Thus, it may be concluded that Mpl ligand acts directly on different stages of megakaryocytopoiesis: on megakaryocyte progenitors, especially on the late subsets, by inducing their proliferation; on megakaryoblasts, by inducing their polyploidization; on mature megakaryocytes, by inducing cytoplasmic maturation and consequently promoting proplatelet formation; and on platelets, by sensitizing them to agonists.

It is important to note that with human cells it was not possible to obtain in vitro a ploidy distribution close to that of marrow megakaryocytes. This can be interpreted in several ways. First, our culture conditions may not be optimal or not permissive for endomitosis. Second, Mpl ligand alone may not be sufficient to induce a high endomitosis rate and another still unidentified factor may be missing. Third, there may be a balance between proliferation and endomitosis. Mpl ligand in vitro may induce a proliferation at the level of progenitor cells to the detriment of endomitosis. In other models of differentiation, there is increasing evidence that induction of proliferation inhibits differentiation. For example, induction of differentiation by EPO requires a delay in the G1 phase *(55)*. Recent experiments showing that increasing concentrations of Mpl ligand may inhibit polyploidization and proplatelet formation, and our limiting-dilution experiments in which very large megakaryocytes were observed in clones that contained one to two cells only, further support this last hypothesis. To test this hypothesis, it will be valuable to investigate the effects of a low concentration of Mpl ligand on polyploidization of the $CD34^+CD41^+$ cell population, which has only residual proliferative capacities.

4. Mpl Ligand Is a Synergistic Growth Factor for Primitive Hemopoietic Progenitors

In vivo and in vitro studies have shown that Mpl ligand also has an effect on the erythroid lineage. Injection of Mpl ligand in vivo accelerates erythroid recovery after myelosuppression or marrow transplantation in the mouse *(56,57)*. In humans, Mpl ligand added in vitro has E-CFC-promoting activity in synergy with EPO, especially on progenitors derived from neonatal cord blood *(58)*. Mpl ligand has more than a bilineage spectrum, since Mpl has been detected in murine fetal cells with a primitive phenotype ($AA41^+Sca^+$) *(35)* and *c-mpl* transcripts were detected in human primitive hemopoietic cells with a $CD34^+CD38^{low}$ phenotype *(42)* or in $CD34^+$ cells that remained quiescent after IL-3 stimulation *(59)*.

In liquid cultures of $CD34^+CD38^{low}$ cord blood cells, Mpl ligand has a synergistic effect with SCF, granulocyte colony-stimulating factor (G-CSF), EPO, IL-3, or various combinations of cytokines, and increases the number of total cells, $CD34^+$ cells, and mature cells (Fig. 4). Similar results were observed at limiting dilution. Therefore, Mpl ligand has a marked direct proliferative effect on primitive neonatal progenitors, increasing megakaryocyte differentiation, but also potentiating the differentiation pattern induced by other growth factors (EPO for the erythroid lineage, SCF for basophils, and G-CSF for granulocytes; our unpublished data).

The ligand was also tested in clonogenic assays of $CD34^+CD38^{low}$ adult marrow cells. A combination of SCF, IL-3, EPO, and Mpl ligand significantly increased the cloning efficiency of E-BFC, granulocyte-macrophage colony-forming cells (GM-CFC), and mixed colony-forming cells (GEMM-CFC). This effect was enhanced in coculture with the murine stromal cell line MS-5. These results further demonstrate that Mpl ligand acts on primitive hemopoietic cells in synergy with a large number of cytokines.

An effect on primitive hemopoietic cells has also been observed in the mouse *(60)*. The factor does not seem to have a direct effect on long-term repopulating cells *(61)*. In humans, Murray et al. have reported that Mpl ligand acts directly on $CD34^+ Thy-1^+$ cells inducing their entry into the cell cycle *(62)*. These in vitro results are in agreement with recent studies of hemopoiesis in knockout mice *(18,19)*. Indeed, although platelets are the only mature cells that are decreased in the blood, hemopoietic progenitors from all myeloid lineages are greatly diminished in the marrow and spleen in both types of mice. This indicates that Mpl ligand might modulate the proliferation of several hemopoietic progenitors, and that later in differentiation, Mpl ligand may be required for the expansion and differentiation of cells of the megakaryocyte lineage.

Acknowledgments

This work has been supported by the Institut National de la Santé et de la Recherche Médicale (INSERM) and the Institut Gustave Roussy and by grants from the Association pour la Recherche Contre le Cancer (ARC) and la Ligue Nationale Contre le Cancer. The authors thank Monique Titeux (U362) and Josette Guichard (U91) for excellent technical assistance, and are grateful to Sam Burstein for improving the manuscript.

References

1. Hoffman R. Regulation of megakaryocytopoiesis. *Blood.* 1989; 74: 1196–1212.
2. Kelemen E, Cserhati I, Tanos B. Demonstration and some properties of human thrombopoietin in thrombocythemic sera. *Acta Haematol.* 1958; 20: 350–355.
3. McDonald TP. Thrombopoietin: its biology, purification, and characterization. *Exp Hematol.* 1988; 16: 201–205.
4. Hoffman R, Mazur E, Bruno E, Floyd V. Assay of an activity in the serum of patients with disorders of thrombopoiesis that stimulates formation of megakaryocytic colonies. *N Engl J Med.* 1981; 305: 533–538.
5. Kawakita M, Enomoto K, Katayama N, Kishimoto S, Miyake T. Thrombopoiesis- and megakaryocyte colony-stimulating factors in the urine of patients with idiopathic thrombocytopenic purpura. *Br J Haematol.* 1981; 48: 609–615.
6. Mazur EM, South K. Human megakaryocyte colony-stimulating factor in sera from aplastic dogs: partial purification, characterization, and determination of hematopoietic cell lineage specificity. *Exp Hematol.* 1985; 13: 1164–1172.
7. De Alarcon PA, Schmieder JA, Gringrich R, Klugman MP. Pattern of response of megakaryocyte colony-stimulating activity in the serum of patients undergoing bone marrow transplantation. *Exp Hematol.* 1988; 16: 316–319.
8. Mazur EM. Megakaryocytopoiesis and platelet production: a review. *Exp Hematol.* 1987; 15: 340–350.
9. Kawakita M, Ogawa M, Goldwasser E, Miyake T. Characterization of human megakaryocyte colony-stimulating factor in the urinary extracts from patients with aplastic anemia and idiopathic thrombocytopenic purpura. *Blood.* 1983; 61: 556–560.
10. Debili N, Hegyi E, Navarro S, et al. In vitro effects of hematopoietic growth factors on the proliferation, endoreplication, and maturation of human megakaryocytes. *Blood.* 1991; 77: 2326–2338.
11. Straneva JE, Goheen MP, Hiu SL, Bruno E, Hoffman R. Terminal cytoplasmic maturation of human megakaryocytes in vitro. *Exp Hematol.* 1986; 14: 919–929.
12. Straneva JE, Yang HH, Hui SL, Bruno E, Hoffman R. Effects of megakaryocyte colony-stimulating factor on terminal cytoplasmic maturation of human megakaryocytes. *Exp Hematol.* 1987; 15: 657–663.
13. Bartley TD, Bogenberger J, Hunt P, et al. Identification and cloning of a megakaryocyte growth and development factor that is a ligand for the cytokine receptor Mpl. *Cell.* 1994; 77: 1117–1124.
14. de Sauvage FJ, Hass PE, Spencer SD, et al. Stimulation of megakaryocytopoiesis and thrombopoiesis by the *c-Mpl* ligand. *Nature.* 1994; 369: 533–538.
15. Lok S, Kaushansky K, Holly RD, et al. Cloning and expression of murine thrombopoietin cDNA and stimulation of platelet production in vivo. *Nature.* 1994; 369: 565–568.

16. Kuter DJ, Beeler DL, Rosenberg RD. The purification of megapoietin: a physiological regulator of megakaryocyte growth and platelet production. *Proc Natl Acad Sci USA.* 1994; 91: 11,104–11,108.

17. Sohma Y, Akahori H, Sebi N,et al. Molecular cloning and chromosomal localization of the human thrombopoietin gene. *FEBS Lett.* 1994; 353: 57–61.

18. Gurney AL, Carver-Moore K, de Sauvage FJ, Moore MW. Thrombocytopenia in *c-mpl*-deficient mice. *Science.* 1994; 265: 1445–1447.

19. de Sauvage FJ, Luoh SM, Carver-Moore K, et al. Deficiencies in early and late stages of megakaryocytopoiesis in TPO-KO mice. *Blood.* 1995; 86: 255a (abstract no 1007).

20. Wendling F, Maraskovsky E, Debili N, et al. c-mpl ligand is a humoral regulator of megakaryocytopoiesis. *Nature.* 1994; 369: 571–574.

21. Kaushansky K, Lok S, Holly RD, et al. Promotion of megakaryocyte progenitor expansion and differentiation by the *c-Mpl* ligand thrombopoietin. *Nature.* 1994; 369: 568–571.

22. Broudy VC, Lin NL, Kaushansky K. Thrombopoietin (*c-mpl* Ligand) acts synergistically with erythropoietin, stem cell factor, and interleukin-11 to enhance murine megakaryocyte colony growth and increases megakaryocyte ploidy in vitro. *Blood.* 1995; 85: 1719–1726.

23. Hunt P, Li YS, Nichol JL, et al. Purification and biologic characterization of plasma-derived megakaryocyte growth and development factor. *Blood.* 1995; 86: 540–547.

24. Briddell RA, Brandt JE, Straneva JE, Srour EF, Hoffman R. Characterization of the human burst-forming unit-megakaryocyte. *Blood.* 1989; 74: 145–151.

25. Long MW. Population heterogeneity among cells of the megakaryocyte lineage. *Stem Cells.* 1993; 11: 33–40.

26. Mazur EM, Hoffman R, Chasis J, Marchesi S, Bruno E. Immunofluorescent identification of human megakaryocyte colonies using an antiplatelet glycoprotein antiserum. *Blood.* 1981; 57: 277–286.

27. Vainchenker W, Bouguet J, Guichard J, Breton-Gorius J. Megakaryocyte colony formation from human bone marrow precursors. *Blood.* 1979; 54: 940–945.

28. Levene RB, Williams NT, Lamaziere JM, Rabellino EM. Human megakaryocytes. IV. Growth and characterization of clonable megakaryocyte progenitors in agar. *Exp Hematol.* 1987; 15: 181–189.

29. Chatelain C, De Bast M, Symann M. Identification of a light density murine megakaryocyte progenitor (LD-CFU-M). *Blood.* 1988; 72: 1187–1192.

30. Debili N, Issaad C, Masse JM, et al. Expression of CD34 and platelet glycoproteins during human megakaryocytic differentiation. *Blood.* 1992; 80: 3022–3035.

31. Debili N, Wendling F, Katz A, et al. The Mpl-Ligand or thrombopoietin or megakaryocyte growth and differentiative factor has both direct proliferative and differentiative activities on human megakaryocyte progenitors. *Blood.* 1995; 86: 2516–2525.

32. Guerriero R, Testa U, Gabbianelli M, et al. Unilineage megakaryocytic proliferation and differentiation of purified hematopoietic progenitors in serum-free liquid culture. *Blood.* 1995; 86: 3725–3736.

33. Correa PN, Eskinazi D, Axelrad AA. Circulating erythroid progenitors in polycythemia vera are hypersensitive to insuline-like growth factor-1 in vitro: studies in an improved serum-free medium. *Blood.* 1994; 83: 99–112.

34. Kaushansky K, Broudy VC, Lin N, et al. Thrombopoietin, the Mpl ligand, is essential for full megakaryocyte development. *Proc Natl Acad Sci USA.* 1995; 92: 3234–3238.

35. Zeigler FC, de Sauvage F, Widmer HR, et al. In vitro megakaryocytopoietic and thrombopoietic activity of *c-mpl* ligand (TPO) on purified murine hematopoietic stem cells. *Blood.* 1994; 84: 4045–4052.

36. Banu N, Wang J, Deng B, Groopman JE, Avraham H. Modulation of megakaryocytopoiesis by thrombopoietin: the c-Mpl ligand. *Blood.* 1995; 86: 1331–1338.

37. Nichol JL, Hokom MM, Hornkohl A, et al. Megakaryocyte growth and development factor. Analyses of in vitro effects on human megakaryopoiesis and endogenous serum levels during chemotherapy-induced thrombocytopenia. *J Clin Invest.* 1995; 95: 2973–2978.

38. Chiba T, Nagata Y, Kishi A, et al. Induction of erythroid-specific gene expression in lymphoid cells. *Proc Natl Acad Sci USA*. 1993; 90: 11,593–11,597.

39. Hermine O, Mayeux P, Titeux M, et al. Granulocyte-macrophage colony-stimulating factor and erythropoietin act competitively to induce two different programs of differentiation in the human pluripotent cell line UT-7. *Blood*. 1992; 80: 3060–3069.

40. Morgan D, Soslau G, Brodsky I. Differential effects of thrombopoietin (mpl) on cell lines MB-02 and HU-3 derived from patients with megakaryoblastic leukemia. *Blood*. 1994; 84: 330a (abstract no 1306).

41. Debili N, Wendling F, Cosman D, et al. The Mpl receptor is expressed in the megakaryocytic lineage from late progenitors to platelets. *Blood*. 1995; 85: 391–401.

42. Methia N, Louache F, Vainchenker W, Wendling F. Oligodeoxynucleotides antisense to the protooncogene *c-mpl* specifically inhibit in vitro megakaryocytopoiesis. *Blood*. 1993; 82: 1395–1401.

43. Ritchie A, Vadhan-Raj S, Broxmeyer HE. Thrombopoietin suppresses growth factor withdrawal-induced apoptosis and behaves as a survival factor for the human growth factor-dependent cell line M07e. *Blood*. 1995; 86: 21a (abstract no 71).

44. Quentmeier H, Zaborski M, Graf G, Ludwig WD, Drexler HG. Expression of receptor MPL and proliferative effects of ligand thrombopoietin (TPO) on human leukemia cell lines. *Blood*. 1995; 86: 371a (abstract no 1472).

45. Porteu F, Rouyez M-C, Cocoult L, et al. Functional regions of the mouse thrombopoietin receptor cytoplasmic domain: evidence for a critical region which is involved in differentiation and can be complemented by erythropoietin. *Mol Cell Biol*. 1996; 16: 2473–2482.

46. Nagata Y, Nagahisa H, Aida Y, Okutomi K, Nagasawa T, Todokoro K. Thrombopoietin induces megakaryocyte differentiation in hematopoietic progenitor FDC-P2 cells. *J Biol Chem*. 1995; 270: 19,673–19,675.

47. Kuter DJ, Rosenberg RD. Appearance of a megakaryocyte growth-promoting activity, megapoietin, during acute thrombocytopenia in the rabbit. *Blood*. 1994; 84: 1464–1472.

48. Debili N, Massé JM, Katz A, Guichard J, Breton-Gorius J, Vainchenker W. Effects of the recombinant hematopoietic growth factors interleukin-3, interleukin-6, stem cell factor, and leukemia inhibitory factor on the megakaryocytic differentiation of CD34+ cells. *Blood*. 1993; 82: 84–95.

49. Hegyi E, Nakazawa M, Debili N, et al. Developmental changes in human megakaryocyte ploidy. *Exp Hematol*. 1991; 19: 87–94.

50. Angchaisuksiri P, Carlson PL, Dessypris EN. Effects of thrombopoietin on megakaryocyte colony formation and ploidy by human CD34 positive cells in a serum-free system. *Blood*. 1995; 86: 369a (abstract no 1464).

51. Choi E, Nichol J, Hokom M, Hornkohl A, Hunt P. Recombinant human MGDF (rhuMGDF), a ligand for *c-mpl*, produces functional platelets from megakaryocytes in vitro. *Blood*. 1994; 84: 242a (abstract no 954).

52. Choi ES, Nichol JL, Hokom MM Hornkohl AC, Hunt P. Platelets generated in vitro from proplatelet-displaying human megakaryocytes are functional. *Blood*. 1995; 85: 402–413.

53. Chen J, Herceg-Harjacek L, Groopman JE, Grabarek J. Regulation of platelet activation in vitro by the *c-Mpl* ligand, thrombopoietin. *Blood*. 1995; 86: 4054–4062.

54. Miyakawa Y, Oda A, Druker BJ, et al. Recombinant thrombopoietin induces rapid protein tyrosine phosphorylation of Janus kinase 2 and Shc in human blood platelets. *Blood*. 1995; 86: 23–27.

55. Carroll M, Zhu Y, D'Andrea AD. Erythropoietin-induced cellular differentiation requires prolongation of the G1 phase of the cell cycle. *Proc Natl Acad Sci USA*. 1995; 92: 2869–2873.

56. Fibbe WE, Heemskerk DP, Laterveer L, et al. Accelerated reconstitution of platelets and erythrocytes after syngenic transplantation of bone marrow cells derived from thrombopoietin pretreated donor mice. *Blood*. 1995; 86: 3308–3313.

57. Kaushansky K, Broudy VC, Grossmann A, et al. Thrombopoietin expands erythroid progenitors, increases red cell production, and enhances erythroid recovery after myelosuppressive therapy. *J Clin Invest*. 1995; 96: 1683–1687.

58. Kobayashi M, Laver JH, Kato T, Miyazaki H, Ogawa M. Recombinant human thrombopoietin (Mpl Ligand) enhances proliferation of erythroid progenitors. *Blood.* 1995; 86: 2494–2499.
59. Berardi AC, Wang A, Levine JD, Lopez P, Scadden DT. Functional isolation and characterization of human hematopoietic stem cells. *Science.* 1995; 267: 104–108.
60. Ku H, Kaushansky K, Ogawa M. Thrombopoietin, the ligand for the MPL receptor, synergises with steel factor and other early-acting cytokines in supporting proliferation of primitive hematopoietic progenitors of mice. *Blood.* 1995; 86: 256a (abstract no 1010).
61. Sitnicka E, Lin N, Fox N, et al. The effect of thrombopoietin on the proliferation and differentiation of murine hematopoietic stem cells. *Blood.* 1995; 86: 419a (abstract no 1664).
62. Murray LJ, Luens KM, Bruno E, et al. The effects of thrombopoietin on human hematopoietic stem cells. *Blood.* 1995; 86: 256a (abstract no 1009).

15

Effect of Cytokines on the Development of Megakaryocytes and Platelets

An Ultrastructural Analysis

Dorothea Zucker-Franklin

1. Introduction

Megakaryocytes have intrigued investigators for more than a century, an interest spurred enormously by the availability of newer methods of analysis during the past 25 years. More recently, the discovery of hemopoietic cytokines culminating in the identification, purification, and manufacture of recombinant human thrombopoietins (rHuTPOs) has driven research on megakaryocytes and thrombocytopoiesis to an unprecedented level.

At the outset of considering the effect of various cytokines on megakaryocyte development and thrombocytopoiesis, it should be mentioned that ultrastructural analyses have remained crucial in determining whether the development of this cell lineage under the influence of one or another cytokine used either in vivo or in vitro is normal. Second, it should be kept in mind that marked differences exist between species. For instance, although mouse and human megakaryocytes are morphologically almost indistinguishable, guinea pig megakaryocytes do not seem to demarcate platelet territories in the same manner (unpublished observations). The peculiarities of bovine platelets have already been pointed out elsewhere in this volume (*see* Chapter 2). For instance, they do not have an organelle that is comparable to the open canalicular system (OCS). Therefore, the present discussion will be limited to observations made with megakaryocytes of humans, nonhuman primates, and mice. Furthermore, it should be realized that the methods used to culture megakaryocytes have included various semisolid media and liquid suspension. The impact of these media on megakaryocyte development, especially the terminal phase of such development, is significant. When considering studies done in vivo, it is important to consider whether the cytokines were administered to healthy individuals with normal platelet counts or whether they were given to animals or patients who were thrombocytopenic. It will not be possible to do justice to all the studies that have contributed to our understanding of megakaryocytopoiesis by these various techniques. Rather, this chapter is meant to delineate the advantages and shortcomings of methods in current use, primarily to preclude misinter-

From: *Thrombopoiesis and Thrombopoietins: Molecular, Cellular, Preclinical, and Clinical Biology*
Edited by: D. J. Kuter, P. Hunt, W. Sheridan, and D. Zucker-Franklin Humana Press Inc., Totowa, NJ

pretation or overinterpretation of observations made following the introduction of newly defined cytokines. Finally, the reader should familiarize him/herself thoroughly with the normal ultrastructure of megakaryocytes and platelets in the context of which the observations presented here will be better appreciated. Though much has been published on this subject *(1–4)*, this material will not be reiterated in this chapter. The reader should turn to the chapter by Jackson et al. (Chapter 1) for this purpose.

2. Methods Used to Assess Megakaryocyte Proliferation and Thrombocytopoiesis

2.1. Megakaryocytopoiesis in Semisolid Media

Since the growth of hemopoietic colonies was first observed in semisolid media, such as agar and methyl cellulose *(5)*, it is not surprising that the generation of megakaryocytes from undefined precursors was also first observed in such media *(6,7)*. These cells were recognized by their size, ploidy, and reaction to antiplatelet antibodies *(8)*. In most of these early reports, megakaryocyte colonies consisted of only a few cells, they were often found mixed with other hemopoietic cells *(9)*, and with the exception of a single study *(10)*, platelet formation was not observed or mentioned.

The cytokines needed for megakaryocyte development were known to be released by mitogen-stimulated mononuclear cells present in the feeder layers of such cultures *(11)* or derived from media conditioned by tissue-culture cell lines, such as WEHI-3. It was appreciated in the early 1980s that multilineage hemopoietic growth factors, particularly interleukin (IL)-3 and granulocyte-macrophage colony-stimulating factor (GM-CSF), were essential for megakaryocyte colony formation *(12,13)*, and that other factors, such as erythropoietin (EPO) *(14,15)* and TPO *(16)*, prepared in a variety of ways had potentiating effects *(17)*.

Considerable impetus to in vitro studies of megakaryocytopoiesis came from the development of the plasma and fibrin clot systems *(8,10,18,19)*, which, among other advantages, permitted more precise titration and definition of added nutrients and cytokines. However, in these systems, platelet formation has not been observed, and few ultrastructural studies have been done. Megakaryocyte colonies are detected by light microscopy and on the basis of their positivity for GPIIb/IIIa in humans and the presence of acetylcholinesterase (AChE) in the mouse. It is possible that platelet formation is precluded in such colonies by virtue of the fact that plasma clots must contain thrombin and that fibrin itself is a strong megakaryocyte/platelet agonist that would undoubtedly lead to "viscous metamorphosis" of any megakaryocyte fragments that might be elaborated (Fig. 1). Therefore, although plasma and fibrin clot systems are very useful in the assessment of the effect of various cytokines on the proliferation of the megakaryocyte lineage in general, they are probably inadequate for the evaluation of terminal differentiation, i.e., the production of normal platelets.

2.2. Suspension Cultures

Commencing with buffy coat specimens of whole bone marrow, it became possible to observe megakaryocyte differentiation to the point of platelet production in a modi-

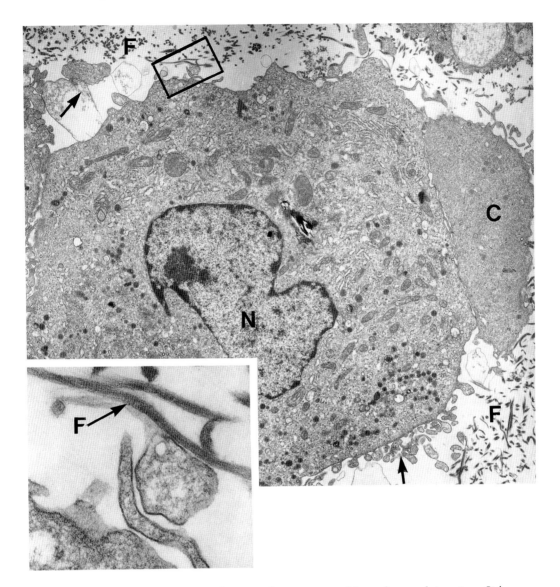

Fig. 1. Human megakaryocyte from a colony generated in a plasma clot system. It is surrounded by strands of fibrin (F). The piece of cytoplasm (C) adjacent to the cell is almost devoid of organelles. Arrows indicate fragments that could be mistaken for platelets on light microscopy. The demarcated area is seen at higher resolution in the inset, which shows fibrin (F) and cytoplasmic fragments to better advantage. N, nucleus. (Magnification, ×6000. Magnification of inset, ×44,000.)

fied Dexter system *(20,21)*. Although such cultures contained hydrocortisone and serum, no other growth factors were added, and it must therefore be assumed that the necessary cytokines were elaborated by other bone marrow cells present in this system. At weekly intervals, half of the medium and suspended cells were removed, and the medium was replaced with an equal volume of fresh medium. In this fashion, it became possible to assess platelet production and to perform electron microscopic analyses on the harvested samples for up to 20 weeks. Therefore, it seems clear that a suspension culture system is preferable over semisolid media if terminal differentiation and the assessment of normal platelet structure/function relationships are the goal of the study. At a time when it has become possible to initiate cultures with well-defined precursor or progenitor cells and a host of purified multilineage and lineage-specific hemopoietic growth factors has become available, suspension cultures may be the only valid in vitro system in which thrombocytopoiesis can be analyzed.

2.3. In Vivo Studies

Although *a priori* no in vivo study will ever be able to elucidate the precise mechanism(s) whereby any particular cytokine functions, in the final analysis, it is the only method that can assess the beneficial and/or adverse effects of administered cytokines in a clinical setting. However, here again, multiple variables need to be measured. For instance, the administration of a cytokine to a healthy animal with a normal platelet count, which has been done in many studies, may accelerate megakaryocyte proliferation and cause thrombocytosis. However, this may lead to the production of aberrant megakaryocytes within a short time (*see* Section 3.2.). Few investigators have done ultrastructural studies on the megakaryocytes of such animals, and bone marrow samples have not been analyzed when cytokines are administered for more than a few days. The effects of megakaryopoietic cytokines in thrombocytopenic individuals may be dependent on the etiology of the thrombocytopenia. Experimental bone marrow ablation by immunologic or pharmaceutical means may affect megakaryocyte progenitors to the point where administered cytokines may not be effective, even when their stimulatory role has been fully established in vitro. In addition, the adhesive, cohesive, and secretory functions of platelets elicited with the help of cytokines, including various preparations of TPO, need to be tested to safeguard against undue activation that could lead to a thromboembolic state. Many of these studies are in progress at this time (reviewed in Chapter19). Data derived from clinical and preclinical trials are accumulating rapidly. Only a few examples will be used to illustrate where ultrastructural analyses may help to interpret observations made by other means.

3. Cytokines with Multilineage Effects: IL-3, IL-6, and IL-11, and GM-CSF

Although a host of other hemopoietic growth factors, such as *c-kit* ligand (stem cell factor [SCF]) and leukemia-inhibitory factor (LIF) deserve consideration, these have pleiotropic and often indirect effects. For instance, LIF will increase the platelet count, but this is probably attributable to its stimulation of acute-phase reactants *(22,23)*. The four factors listed above are chosen for discussion because they have been studied by the largest number of investigators. Even when limiting the discussion to these four factors, it is difficult to reach definitive conclusions regarding their precise roles, since the published studies are not comparable in methods and reagents used. Relevant studies have been done either in different culture systems, in different species, or the results have been measured by different parameters. Only in the last few years have cultures been initiated with purified progenitors, such as CD34[+] cells, and in serum-free media. Many of the earlier studies have not been repeated in purified systems. However, such a system is necessary to establish whether one or another cytokine is truly essential for megakaryocyte production and platelet release. In this context the major cellular events that must be evaluated during megakaryocyte development are: progenitor proliferation; nuclear endoreduplication; cytoplasmic maturation, which encompasses the synthesis of platelet-specific proteins packaged into granules; orderly development of demarcation membranes delimiting platelet territories; and the elaboration of platelets of normal size and function. Therefore, what will be presented in this section is a consensus impression gained from a review of the literature modified to some extent by a limited number of ultrastructural analyses that have been conducted in the author's laboratory.

3.1. Interleukin-3 and Granulocyte-Macrophage Colony-Stimulating Factor

IL-3 and GM-CSF will be considered together because the majority of investigators have done comparative studies of their single as well as combined use. There is little question that IL-3 increases megakaryocyte colony formation in vitro *(23–27)* and that it will increase the platelet count when it is administered in vivo *(28,29)*, even when administered to selected patients with amegakaryocytic thrombocytopenia *(30)*. However, most data, including our own observations, suggest that IL-3 primarily supports the early stages of megakaryocyte proliferation and differentiation. When endogenous TPO is neutralized in suspension cultures initiated with mouse bone marrow specimens, from which morphologically recognizable megakaryocytes have been eliminated, IL-3 will not only increase the number of AChE-positive cells, but it will also generate ultrastructurally identifiable megakaryocytes *(31)*. However, the cytoplasm of the cells remains immature and platelet territories do not form (Figs. 2 and 3). These observations were even more definitive in a culture system initiated with human CD34[+] cells *(32)*. In liquid cultures, IL-3 markedly expanded all cell lineages, but particularly the CD34[+]Thy-1[+] Lin[-] cell population, and in the presence of TPO increased the number of mature platelet-producing megakaryocytes to a level way beyond what had been attained previously *(33)*.

The synergistic role of IL-3 and GM-CSF in stimulating megakaryocytopoiesis is also well recognized *(34,35)*. However, these studies were not done in cultures initiated with purified progenitors, and therefore, the contributing effect of cytokines secreted by other cells cannot be assessed. IL-3 has, however, proven to be essential in some in vitro culture systems. The important role of IL-3 in megakaryocyte development in vitro, especially when the cells are grown in serum-free media, has been observed by numerous investigators. It is believed to substitute for other essential components missing in such media. However, supplementation with IL-3 or GM-CSF is unnecessary when aplastic serum is added to the culture medium. The addition of IL-3 or GM-CSF to in vitro cultures or its administration in vivo does not seem to affect the ultrastructure of the cells. The administration of GM-CSF to nonhuman primates resulted in increased megakaryocyte ploidy and size, but did not increase the number of megakaryocytes or platelets *(36)*. GM-CSF did not appear to cause any ultrastructural abnormalities (unpublished observations). IL-3 is known to be very pleiotropic and has been shown to be clinically

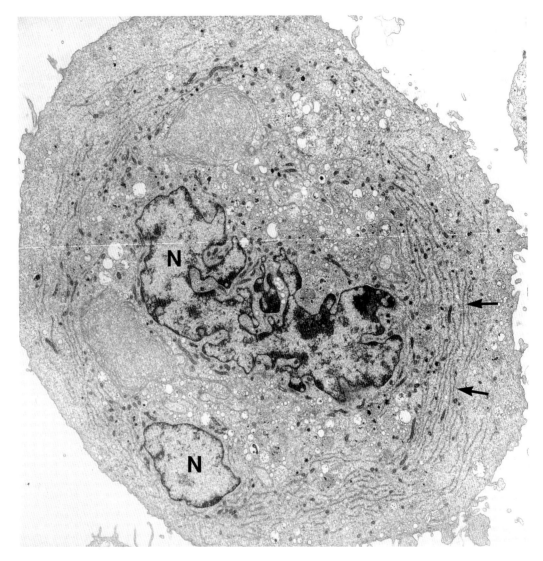

Fig. 2. Mouse megakaryocyte from a serum-free suspension culture to which only IL-3 had been added (for details *see* ref. *31*). The cell is very large and has a polyploid nucleus (N). The abundance of rough endoplasmic reticulum (ER) (arrows) indicates the immaturity of the cytoplasm. Demarcation membranes and platelet territories are rudimentary. No fragmentation is evident. (Magnification, ×5500.)

effective in some patients with thrombocytopenia. The administration of IL-3 to healthy rhesus monkeys did not increase the ploidy of their megakaryocyte or their circulating platelet count *(37)*. Moreover, mice made IL-3-deficient by gene-targeting techniques have normal megakaryocyte and platelet levels (Dranoff G, Mulligan R, personal communication).

3.2. Interleukin-6

Among the hemopoietic growth factors and interleukins known to influence megakaryocyte development, IL-6 has the most potent thrombopoietic activity of

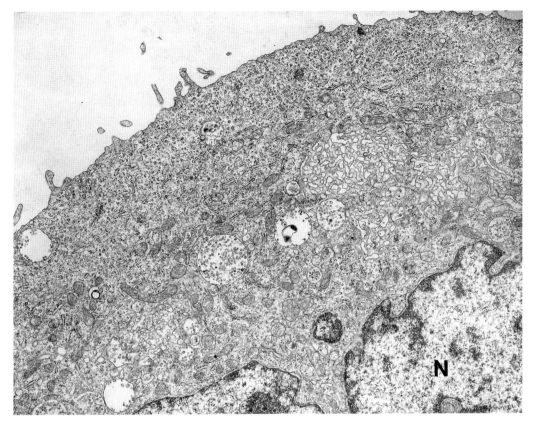

Fig. 3. Detail of a mouse megakaryocyte grown in suspension culture to which IL-3 and s-Mpl have been added. Although there are many ribosomes and profiles of RER, apparently s-Mpl inhibited the development of granules and demarcation membranes. N, nucleus. (for details of experiments, *see* ref. *31.*) (Magnification, ×9500.)

any agent except TPO *(38–45)*. At the same time, it has also been recognized that the response to IL-6 is not specific and that it could even be detrimental. IL-6 is an acute phase reactant, primarily synthesized by hepatocytes, but also by many other cells (for review, *see* ref. *46*). It has a wide range of biological activities. Megakaryocytes not only have receptors for this cytokine, but they also synthesize it *(47)*. Administration of IL-6 to animals with normal platelet counts results in a remarkable increase in the number, size, and ploidy of megakaryocytes. Although this is usually accompanied by an increase in platelet count, the mean platelet volume does not increase, and large platelets, commonly seen in compensatory thrombocytosis, are not observed. Ultrastructural studies prompted by these observations revealed markedly aberrant megakaryocytes *(48)* described in the legends to Figs. 4–7. These abnormalities persisted for several days after administration of IL-6 had been discontinued. There was excessive membrane formation without clearcut demarcation of platelet territories, a marked decrease in granules, and aberrant distribution of heterochromatin in the nucleus.

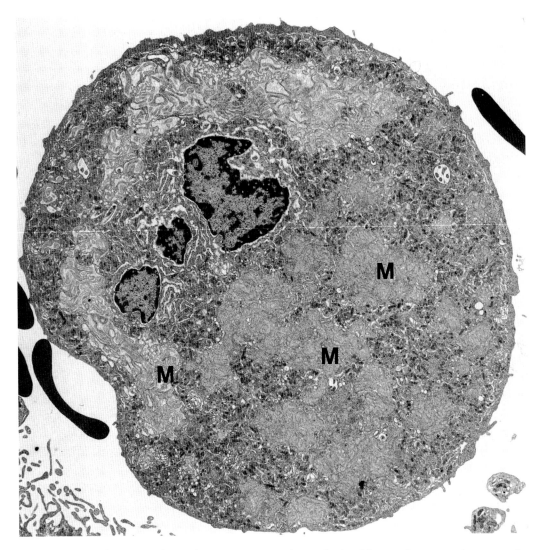

Fig. 4. Megakaryocyte from a bone marrow specimen obtained from a rhesus monkey treated with IL-6 for 5 days. Vast areas of cytoplasm are occupied by membranes (M). Platelet fields are not demarcated (magnification, ×4800) *(48)*.

There is other evidence to suggest that megakaryocytopoiesis stimulated by IL-6 may be abnormal. Following cessation of IL-6 administration to mice, a "rebound thrombocytopenia" was noted *(29)*. On the basis of the ultrastructural studies illustrated in Figs. 4–7, it is possible that these large, abnormal megakaryocytes fail to develop platelet territories and are therefore not able to produce platelets. Bone marrow specimens obtained from rhesus monkeys 5 days after a 7-day course of IL-6 seemed to have a decreased number of megakaryocytes, and many of the cells appeared degenerate (Fig. 7). Moreover, a study carried out on the function of platelets obtained from dogs treated with IL-6 showed that the platelets were activated as mea-

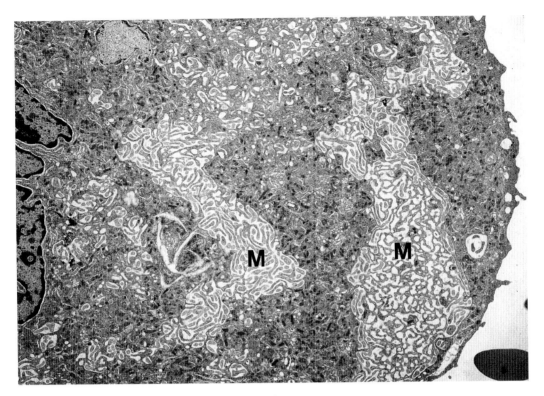

Fig. 5. Detail of a megakaryocyte from the bone marrow of a rhesus monkey treated with 30 µg/kg/day of IL-6 for only 3 days. The development of excessive membranes (M) which are not organized into demarcation membrane systems (DMS) is readily apparent. (Magnification, ×4500.)

sured by surface P-selectin, and reactivity to thrombin and platelet activating factor (PAF) *(49)*. For all these reasons, the clinical use of IL-6 to stimulate thrombopoiesis must be viewed with reservation.

3.3. Interleukin-11

IL-11 was recognized only a few years ago (reviewed in ref. *50*) and was cloned in 1990 *(51)*. Although its receptor has not yet been characterized, it is known that, like IL-6, this pleiotropic cytokine also uses GP130 as a signaling pathway, and therefore, it is likely to stimulate a variety of hemopoietic cells. However, by itself, IL-11 does not seem able to stimulate megakaryocyte progenitors to form colonies in serum-free cultures *(52)*. Nevertheless, it has excellent synergy with IL-3 that can be observed with precursors derived from mouse as well as from human bone marrow *(53–55)* (also *see* Table 1). IL-11 also increases the ploidy of megakaryocytes, a response that does not occur in response to IL-3 alone *(56)*.

In a serum-free liquid culture system initiated with CD34⁺Lin⁻ cells to which defined cytokines and TPO were added, the proliferation and maturation of megakaryocytes, including the formation of platelets, were observed in the absence of

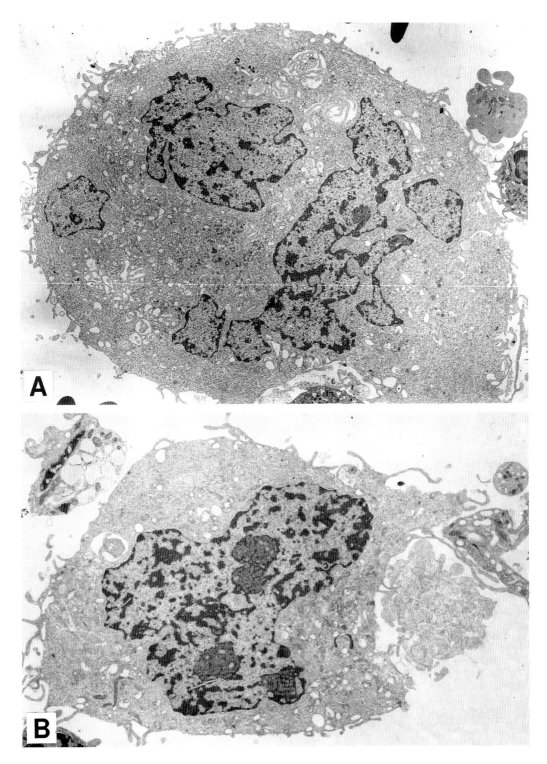

Fig. 6. Megakaryocytes from rhesus monkey treated with IL-6 for 7 days. **(A)** A very large megakaryocyte with a hypersegmented nucleus that is undoubtedly of high ploidy shows hardly any granules or demarcated platelet territories. (Magnification, ×2500.) The cell depicted in **B** taken from the same specimen as A illustrates unusual nucleolar masses, heterochromatin scattered throughout the nucleus, lack of granules, and aberrant fragmentation *(48)*. (Magnification, ×5000.)

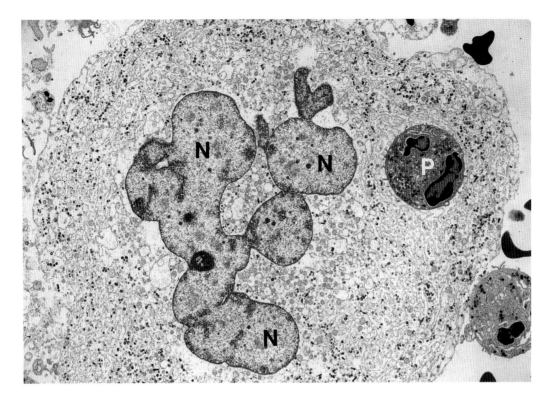

Fig. 7. Necrotic megakaryocyte from the bone marrow of a rhesus monkey 5 days after IL-6 had been administered for 7 days. Such cells were commonly encountered. The nucleus (N) appears apoptotic, but whether apoptosis or necrosis leads to cell death has not yet been established. There is abundant emperipolesis. A neutrophil (P) is seen in the cytoplasm of this megakaryocyte. It is common to see several well-preserved leukocytes within the cytoplasm of the megakaryocytes. (Magnification, ×4000.)

added IL-11, leading to the conclusion that IL-11 is not essential for this process *(32)*. An interesting observation was made in a serum-free system initiated with mouse precursors devoid of recognizable megakaryocytes, in which endogenous TPO was inhibited with soluble Mpl and in which IL-3 and IL-11 were the only cytokines added to the medium. Here, cells that had attained the size and ploidy of megakaryocytes showed sparse granulation, but gave rise to platelet-sized fragments not readily observed in the absence of IL-11, i.e., when IL-3 was the only added cytokine (Fig. 8). On the basis of this observation, it seems possible that IL-11 may play a role during terminal differentiation. This may hold true particularly when there is a demand for accelerated platelet production. In fact, the administration of IL-11 as a single agent, even to animals with a normal platelet count, will stimulate megakaryocytopoiesis and platelet production *(52,57,58)*. Since IL-11 is an acute-phase reactant, this effect may come about indirectly, e.g., by increasing the circulating level of IL-6. Perhaps of greater clinical importance are the observations suggesting that recombinant human (rHu) IL-11 is able to increase the plate-

Table 1
Effects of IL-3, IL-6, IL-11, and GM-CSF on
Proliferation of Megakaryocyte Colony-Forming
Cells (MK-CFC) in Serum-Depleted Fibrin Clot
Cultures Initiated with CD34+DR+ Cells[a]

Cytokines	Colonies
No addition	0.0 ± 0.0
IL-3 (125.0 pg/mL)	3.4 ± 0.9
IL-3 (1.0 ng/mL)	11.3 ± 0.8
GM-CSF (25.0 pg/mL)	1.0 ± 0.6
GM-CSF (200.0 pg/mL)	2.6 ± 0.4
IL-6 (30.0 ng/mL)	0.5 ± 0.4
IL-6 (60.0 ng/mL)	0.3 ± 0.2
IL-11 (1/1600)	1.2 ± 0.7
IL-11 (1/800)	1.0 ± 0.4
IL-11 (1/400) + IL-3 (125 pg/mL)	4.5 ± 0.9
IL-11 (1/400) + IL-3 (1.0 ng/mL)	125.0 ± 2.6

[a]Modified from ref. 56.
    IL = interleukin; GM-CSF = granulocyte-macrophage colony-stimulating factor.

let count in myelosuppressed nonhuman primates as well as in dogs after sublethal irradiation (59). As of this writing, IL-11 is undergoing clinical trials (see Chapter 10). Since this cytokine appears to be equally effective in stimulating proliferation and differentiation of the myeloid and erythroid cell series, it is conceivable that it will turn out to be a useful agent when multilineage hemopoietic recovery is required after iatrogenic marrow ablation (58).

4. Thrombopoietins

It would be redundant to reiterate here the 40-year history of theoretical considerations, stepwise scientific exploration, and other observations that have led to the identification of TPO and its receptor. They have been addressed in detail elsewhere in this volume (see Chapters 5–9). Even though some megakaryocyte and platelet production still appears to be possible in mice whose c-mpl and tpo genes have been disrupted by genetic manipulation (ref. 60 and Chapter 21), there is little doubt that both Mpl and its ligand are essential for normal platelet production. The synergistic or additive roles of EPO and the cytokines discussed above have also become well recognized. Therefore, only a few observations will be presented here to widen the scope of data provided by others. The first of these are derived from a study carried out in a serum-free mouse marrow culture system, in which any TPO that might have been elaborated by cells present in the culture was neutralized by addition of soluble receptor (s-Mpl). In such cultures, megakaryocytes had very few granules and lacked demarcated platelet fields (Fig. 8A). On the other hand, when the same serum-free cultures were supplemented with TPO, the majority of

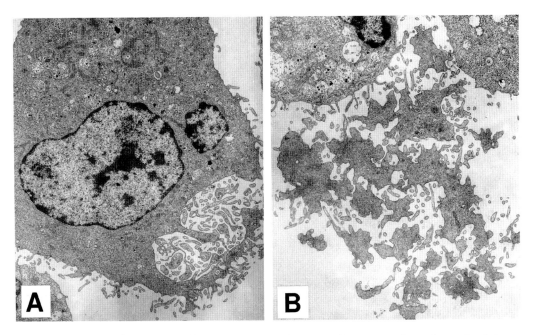

Fig. 8. Details of mouse megakaryocytes grown in serum-free suspension cultures for 5 days in the presence of IL-3, IL-11, and s-Mpl (for experimental detail, *see* ref. *55*). **(A)** A cell that is sparsely granulated and has no demarcated platelet fields is, nevertheless, undergoing fragmentation (magnification, × 3000). **(B)** Fragments of a megakaryocyte almost completely devoid of the organelles normally seen in platelets (magnification, ×4700) *(55)*.

cells showed normal granulation, platelet territories, and fragmentation (Fig. 9). Moreover, a serum-free culture system initiated with purified CD34+Lin− bone marrow cells to which only TPO was added yielded normal megakaryocytes that released platelets by day 11 of culture (Fig. 10). By day 13, many megakaryocyte nuclei were denuded of cytoplasm, probably having shed their platelets (Fig. 10B). The role of TPO in these phenomena is incontrovertible. Of interest in relation to the other studies reported in this volume was the finding that addition of IL-3 to serum-free mouse marrow cultures, even when TPO was blocked with s-Mpl, led to the proliferation of megakaryocytes. However, although such cells had an abundance of ribosomes and rough endoplasmic reticulum (RER), they lacked granules and demarcation membranes (Fig. 11). Since these cells are very large, polyploid, and AChE-positive *(55)*, it could lead to the erroneous conclusion that TPO is not essential for megakaryocyte differentiation, once again underscoring the necessity for ultrastructural analyses. Since IL-3 stimulates the proliferation of myeloid lineages that are able to elaborate cytokines, such as IL-6 and GM-CSF, including monocytes and granulocytes, it is likely that partial development of megakaryocytes was stimulated. Indeed when such cytokines where inhibited with anti-Gp-130, an antibody directed against the shared signal transducing unit, and in addition TPO was neutralized with s-Mpl, the cultures were completely devoid of recognizable megakaryocytes (Fig. 12). Similar results were obtained by others who were

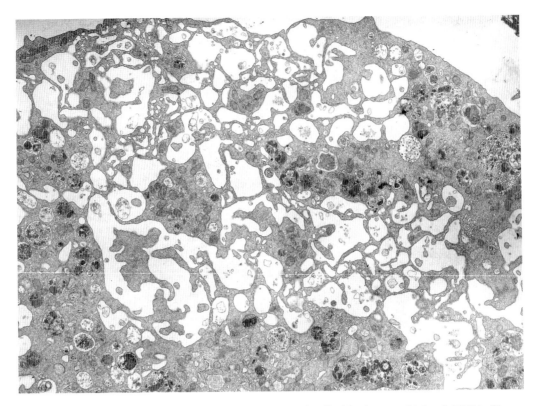

Fig. 9. Detail of a mouse megakaryocyte grown in serum-free liquid culture to which only TPO had been added. There is abundant formation of platelet territories and fragmentation *(31)* (magnification, ×8000).

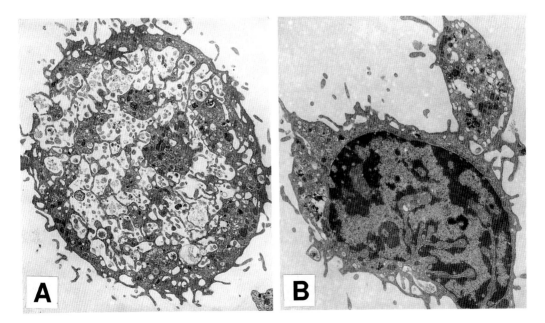

Fig. 10. Cells derived from an 11-day, serum-free liquid culture system initiated with purified CD34⁺ cells. **(A)** Section through the pole of a megakaryocyte above or below the plane of the nucleus shows platelet territories. Only TPO and IL-6 (10 ng/mL) had been added to the medium (magnification, ×4200). **(B)** "Denuded" megakaryocyte nucleus to which a large fragment of cytoplasm is still attached in a 13-day culture initiated with CD34⁺ cells to which only TPO had been added *(32)* (magnification, ×6000).

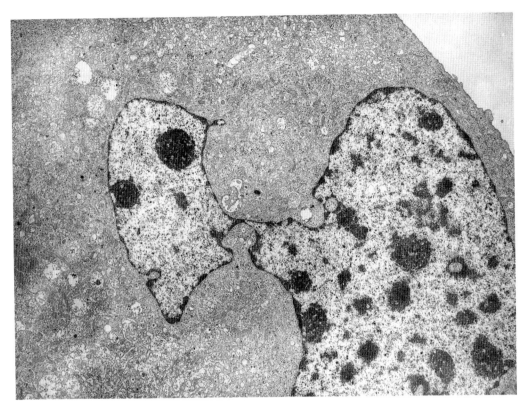

Fig. 11. Detail of a mouse megakaryocyte from a 5-day serum-free culture to which IL-3 and IL-11 had been added and TPO had been blocked with s-Mpl. This is a very large, but immature nucleus. Although ribosomes and RER are present in the cytoplasm, there are very few granules and no platelet territories *(32)* (magnification , ×4700).

able to inhibit megakaryocyte development in vitro using antisense oligonucleotides specific for *c-mpl (61)* (*see* Chapters 6 and 7).

5. Conclusion

It has become clear that most cytokines considered in this chapter have a pleiotropic effect on many different hemopoietic cell types, including the megakaryocyte/platelet lineage. This probably holds also true for TPO, since the Mpl receptor is found on many hemopoietic stem cells, in particular those belonging to the erythroid series. It is likely that, in vitro and in vivo, normal platelets cannot be elaborated in the absence of TPO. What has also become clear is that multiple variables must be measured to conclude whether or not any particular cytokine stimulates normal megakaryocyte development that eventuates in the release of platelets that are normal in number, size, and function. Examples have been provided to show that ultrastructural analysis is a necessary component of this assessment. On the other hand, it must be recognized that no matter how pure and biochemically well-defined the reagents or culture conditions have become, in the final analysis, the beneficial or adverse effects of any agent will

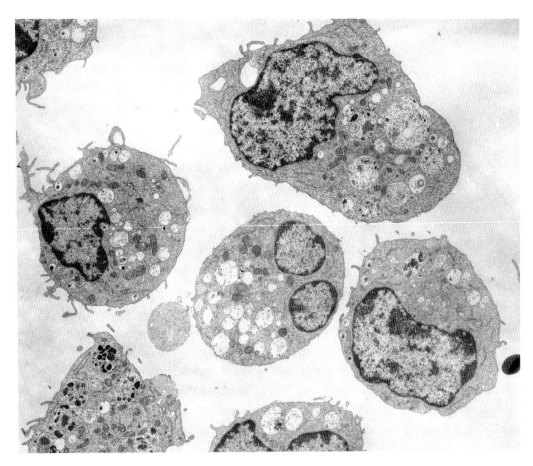

Fig. 12. Representative sample of mouse progenitor cells grown in serum-free medium in the presence of IL-3, anti-GP130, and in which the possible presence of TPO was neutralized with s-Mpl. There are no recognizable megakaryocytes. These cells have the appearance of cells belonging to the monocyte/macrophage series *(55)* (magnification , ×5000).

have to be determined in the clinical arena. In many instances, it may be possible to enhance benefits or ameliorate adverse effects by the use of combination therapy requiring lower doses of each individual agent—a strategy that has proven to be effective in other areas of medicine.

Acknowledgment

The author acknowledges George Grusky's expert help in preparing the illustrations for this chapter.

References

1. Zucker-Franklin D. The ultrastructure of megakaryocytes and platelets. In: Gordon AS (ed). *Regulation of Hematopoiesis*. New York: Appleton-Century-Crofts; 1970: 1553–1586.
2. Zucker-Franklin D. Megakaryocytes and platelets. In: Zucker-Franklin D, Greaves MF, Grossi CE, Marmont AM (eds). *Atlas of Blood Cells, Function and Pathology*. Philadelphia: Edi Ermes Milan & Febiger; 1989: 623–693.

3. Zucker-Franklin D. Platelet morphology and function. In: Williams WJ, Beutler E, Erslev AJ, Lichtman MA (eds). *Hematology*. New York: McGraw Hill; 1990: 1172–1181.
4. Zucker-Franklin D, Stahl C, Hyde P. Antigenic dissimilarity between platelet and megakaryocyte surface membranes. In: Levine RF, Williams N, Levin J, Evatt BL (eds). *Megakaryocyte Development and Function. Progress in Clinical and Biological Research*. New York: Alan R. Liss; 1986: 215: 259–264.
5. Metcalf D. *Hematopoietic Colonies: In Vitro Cloning of Normal and Leukemia Cells*. New York: Springer Verlag; 1977.
6   Metcalf D, MacDonald HR, Odartchenko N, Sordat B. Growth of mouse megakaryocyte colonies in vitro. *Proc Natl Acad Sci USA*. 1975; 72: 1744–1748.
7. Nakeff A, van Noord MJ, Blansjaar N. Electron microscopy of megakaryocytes in thin-layer agar cultures of mouse bone marrow. *J Ultrastruct Res*. 1974; 49: 1–10.
8. Mazur EM, Hoffman R, Chasis J, Marchesi S, Bruno E. Immunofluorescent identification of human megakaryocyte colonies using an antiplatelet glycoprotein antiserum. *Blood*. 1981; 57: 277–286.
9. Fauser M, Messner HA. Identification of megakaryocytes, macrophages, and eosinophils in colonies of human bone marrow containing neutrophilic granulocytes and erythroblasts. *Blood*. 1979; 53: 1023–1027.
10. McLeod DL, Shreve MM, Axelrad AA. Induction of megakaryocyte colonies with platelet formation in vitro. *Nature*. 1976; 261: 492–494.
11. Geissler D, Konwalinka G, Peschel C, Grunewald K, Odavic R, Braunsteiner H. A regulatory role of activated T lymphocytes on human megakaryocytopoiesis in vitro. *Br J Haematol*. 1985; 60: 233–238.
12. Quesenberry PJ, Ihle JN, McGrath E. The effect of Interleukin-3 and GM-CSA-2 on megakaryocyte and myeloid clonal colony formation. *Blood*. 1985; 65: 214–217.
13. Williams N, Eger RR, Jackson HM, Nelson DJ. Two factor requirement for murine megakaryocyte colony formation. *J Cell Physiol*. 1982; 110: 101–104.
14. Sakaguchi M, Kawakita M, Matsushita J, Shibuya K, Koishihara Y, Takatsuki K. Human erythropoietin stimulates murine megakaryopoiesis in serum-free culture. *Exp Hematol*. 1987; 15: 1028–1034.
15. Williams N, Jackson H, Iscove NN, Dukes PP. The role of erythropoietin, thrombopoietic stimulating factor, and myeloid colony-stimulating factors on murine megakaryocyte colony formation. *Exp Hematol*. 1984; 12: 734–740.
16. McDonald TP, Shadduck RK. Comparative effects of thrombopoietin and colony-stimulating factors. *Exp Hematol*. 1982; 10: 544–550.
17. Evatt BL, Kellar KL, Ramsey RB. Thrombopoietin: past, present and future. In: Levine RF, Williams N, Levin J, Evatt BL (eds). *Megakaryocyte Development and Function. Progress in Clinical and Biological Research*. New York: Alan R. Liss; 1986; 215: 143–155.
18. Mazur EM, Hoffman R, Bruno E. Regulation of human megakaryocytopoiesis. An in vitro analysis. *J Clin Invest*. 1981; 68: 733–741.
19. McLeod DL, Shreeve MM, Axelrad AA. Improved plasma culture system for production of erythrocytic colonies in vitro: quantitative assay method for CFU-E. *Blood*. 1974; 44: 517–534.
20. Dexter TM, Allen TD, Lajtha LG. Conditions controlling the proliferation of haematopoietic stem cells in vitro. *J Cell Physiol*. 1977; 91: 335–344.
21. Zucker-Franklin D, Petursson S. Thrombocytopoiesis—analysis by membrane tracer and freeze-fracture studies on fresh human and cultured mouse megakaryocytes. *J Cell Biol*. 1984; 99: 390–402.
22. Mayer P, Geissler K, Ward M, Metcalf D. Recombinant human leukemia inhibitory factor induces acute phase proteins and raises the blood platelet counts in nonhuman primates. *Blood*. 1993; 81: 3226–3233.
23. Briddell RA, Bruno E, Cooper RJ, Brandt JE, Hoffman R. Effect of c-kit ligand on in vitro human megakaryocytopoiesis. *Blood*. 1991; 78: 2854–2859.

24. McNiece IK, McGrath HE, Quesenberry PJ. Granulocyte colony-stimulating factor augments in vitro megakaryocyte colony formation by Interleukin-3. *Exp Hematol.* 1988; 16: 807–810.

25. Debili N, Masse JM, Katz A, Guichard J, Breton-Gorius J, Vainchenker W. Effects of the recombinant hematopoietic growth factors interleukin-3, interleukin-6, stem cell factor, leukemia inhibitory factor, on the megakaryocytic differentiation of CD34$^+$ cells. *Blood.* 1993; 82: 84–95.

26. Ishibashi T, Burstein SA. Interleukin-3 promotes the differentiation of isolated single megakaryocytes. *Blood.* 1986; 67: 1512–1514.

27. Williams N, Sparrow R, Gill K, Yasmeen D, McNiece I. Murine megakaryocyte colony stimulating factor: its relationship to Interleukin-3. *Leukemia Res.* 1985; 9: 1487–1496.

28. Ganser A, Lindemann A, Seipelt G, et al. Effects of recombinant human interleukin-3 in patients with normal hematopoiesis and in patients with bone marrow failure. *Blood.* 1990; 76: 666–676.

29. Carrington PA, Hill RJ, Stenberg PE, et al. Multiple *in vivo* effects of interleukin-3 and interleukin-6 on murine megakaryocytopoiesis. *Blood.* 1991; 77: 34–41.

30. Guinan EC, Lee YS, Lopez KD, et al. Effects of Interleukin-3 and granulocyte-macrophage colony-stimulating factor on thrombopoiesis in congenital amegakaryocytic thrombocytopenia. *Blood.* 1993; 81: 1691–1698.

31. Kaushansky K, Broudy VC, Lin N, et al. Thrombopoietin, the Mpl ligand, is essential for full megakaryocyte development. *Proc Natl Acad Sci USA* 1995; 92: 3234–3238.

32. Murray LJ, Bruno E, Zucker-Franklin D, et al. Thrombopoietin induction of megakaryocytopoiesis from purified subpopulations of human CD34$^+$ cells including primitive CD34$^+$ Thy-1$^+$ Lin$^-$ cells. *Exper Hematol.* (in press).

33. Bruno E, Murray LJ, Zucker-Franklin D, et al. Further definition of the cellular target of human thrombopoietin. *Blood.* 1995; 86: 365a (abstract no 1449).

34. Robinson BE, McGrath HE, Quesenberry PJ. Recombinant murine granulocyte macrophage colony-stimulating factor has megakaryocyte colony-stimulating activity and augments megakaryocyte colony stimulation by Interleukin 3. *J Clin Invest.* 1987; 79: 1648–1652.

35. Bruno E, Briddell R, Hoffman R. Effect of recombinant and purified hematopoietic growth factors on human megakaryocyte colony formation. *Exp Hematol.* 1988; 16: 371–377.

36. Stahl CP, Winton EF, Monroe MC, et al. Recombinant human granulocyte-macrophage colony-stimulating factor promotes megakaryocyte maturation in nonhuman primates. *Exp Hematol.* 1991; 19: 810–816.

37. Stahl CP, Winton EF, Monroe MC, et al. Differential effects of sequential, simultaneous, and single agent interleukin-3 and granulocyte-macrophage colony-stimulating factor on megakaryocyte maturation and platelet response in primates. *Blood.* 1992; 80: 2479–2485.

38. Ishibashi T, Kimura H, Shikama Y, et al. Interleukin 6 is a potent thrombopoietic factor in vivo in mice. *Blood.* 1989; 74: 1241–1244.

39. Bruno E, Hoffman R. Effect of interleukin 6 on *in vitro* human megakaryocytopoiesis: its interaction with other cytokines. *Exp Hematol.* 1989; 17: 1038–1043.

40. Koike K, Nakahata T, Kubo T, et al. Interleukin-6 enhances murine megakaryocytopoiesis in serum-free culture. *Blood.* 1990; 75: 2286–2291.

41. Kimura H, Ishibashi T, Ushida T, Maruyama Y, Friese P, Burstein SA. Interleukin-6 is a differentiation factor for human megakaryocytes in vitro. *Eur J Immunol.* 1990; 20: 1927–1931.

42. Hill RJ, Warren MK, Stenberg P, et al. Stimulation of megakaryocytopoiesis in mice by human recombinant interleukin-6. *Blood.* 1991; 77: 42–48.

43. Imai T, Koike K, Kubo T, et al. Interleukin-6 supports human megakaryocyte proliferation and differentiation in vitro. *Blood.* 1991; 78: 1969–1974.

44. Leven RM, Rodriguez A. Immunomagnetic bead isolation of megakaryocytes from guinea-pig bone marrow: effect of recombinant interleukin-6 on size, ploidy and cytoplasmic fragmentation. *Br J Haematol.* 1991; 77: 267–273.

45. Burstein SA, Downs T, Friese P, et al. Thrombocytopoiesis in normal and sublethally irradiated dogs: response to human interleukin-6. *Blood.* 1992; 80: 420–428.

46. Kishimoto T. The biology of Interleukin-6. *Blood.* 1989; 74: 1–10.
47. Navarro S, Debili N, Le Couedic JP, et al. Interleukin-6 and its receptor are expressed by human megakaryocytes: in vitro effects on proliferation and endoreduplication. *Blood.* 1991; 77: 461–471.
48. Stahl CP, Zucker-Franklin D, Evatt BL, Winton EF. Effects of human interleukin-6 on megakaryocyte development and thrombocytopoiesis in primates. *Blood.* 1991; 78: 1467–1475.
49. Peng J, Friese P, George JN, Dale GL, Burstein SA. Alteration of platelet function in dogs mediated by interleukin-6. *Blood.* 1994; 83: 398–403.
50. Du XX, Williams DA. Interleukin-11: a multifunctional growth factor derived from the hematopoietic environment. *Blood.* 1994; 83: 2023–2030.
51. Paul SR, Bennet F, Calvetti JA, et al. Molecular cloning of a cDNA encoding interleukin-11, a stromal cell-derived lymphopoietic and hematopoietic cytokine. *Proc Natl Acad Sci USA.* 1990; 87: 7512–7516.
52. Yonemura Y, Kawakita M, Masuda T, Fujimoto K, Takastuki K. Effect of recombinant human interleukin-11 on rat megakaryopoiesis and thrombopoiesis in vivo: comparative study with interleukin-6. *Br J Haematol.* 1993; 84: 16–23.
53. Teramura M, Kobayashi S, Hoshino S, et al. Interleukin-11 enhances human megakaryocytopoiesis in vitro. *Blood.* 1992; 79: 327–331.
54. Burstein SA, Mei-RL, Henthorn J, et al. Leukemia inhibitory factor and interleukin-11 promote maturation of murine and human megakaryocytes in vitro. *J Cell Physiol.* 1992; 153: 305–312.
55. Zucker-Franklin D, Kaushansky K. The effect of thrombopoietin on the development of megakaryocytes and platelets: an ultrastructural analysis. *Blood.* 1996; 88: 1632–1638.
56. Bruno E, Briddell RA, Cooper RJ, Hoffman R. Effects of recombinant interleukin-11 on human megakaryocyte progenitor cells. *Exp Hematol.* 1991; 19: 378–381.
57. Neben TY, Loebelenz J, Hayes L, et al. Recombinant human interleukin-11 stimulates megakaryocytopoiesis and increases peripheral platelets in normal and splenectomized mice. *Blood.* 1993; 81: 901–908.
58. Leonard JP, Quinto CM, Kozitza MK, Neben TY, Goldman SJ. Recombinant human IL-11 stimulates multilineage hematopoietic recovery in mice after a myelosuppressive regimen of sublethal irradiation and carboplatin. *Blood.* 1994; 83: 1499–1506.
59. Nash RA, Seidel K, Storb R, et al. Effects of rhIL-11 on normal dogs and after sublethal radiation. *Exp Hematol.* 1995; 23: 389–396.
60. Gurney Al, Carver Moore K, de Sauvage FJ, Moore MW. Thrombocytopenia in c-mpl-deficient mice. *Science.* 1994; 265: 1445–1447.
61. Methia N, Louache F, Vainchenker W, Wendling F. Oligodeoxynucleotides antisense to the proto-oncogene c-mpl specifically inhibit in vitro megakaryocytopoiesis. *Blood.* 1993; 82: 1395–1401.

16

The Thrombopoietin Receptor, Mpl, and Signal Transduction

Kenneth Kaushansky, Virginia C. Broudy, and Jonathan G. Drachman

1. Introduction

The recent identification and characterization of the *c-mpl* proto-oncogene set off an explosive wave of research that has advanced our understanding of megakaryocyte and platelet biology immensely. The mechanisms by which this member of the hemopoietic cytokine receptor family promotes the survival, proliferation, and differentiation of megakaryocytic progenitors resulting in platelet production will require many years, if not decades, to unravel completely. However, important insights into this process have already been achieved, and will be reviewed in this chapter.

As discussed in detail in Chapters 6 and 7, studies with a retroviral complex termed myeloproliferative leukemia virus (MPLV) *(1)* led to the identification of a novel transforming gene, *v-mpl (2)*. Sequence analysis suggested that the oncogene encoded a truncated hemopoietic growth factor receptor, an impression reinforced on cloning of the human cellular homolog *c-mpl*, two years later *(3)*. The following year, using a chimeric interleukin (IL)-4 receptor/Mpl protein, Skoda and coworkers reported that the cytoplasmic domain of the molecule could support proliferation of an IL-3-dependent cell line *(4)*. Similar results and conclusions were obtained with a granulocyte colony-stimulating factor (G-CSF) receptor/Mpl fusion protein *(5)*. However, since a myriad of hemopoietic cytokine receptors support proliferation in the host cells used in these experiments, the results did not give any indication of the cellular system(s) in which Mpl normally plays a role.

2. Mpl Is Displayed on Multiple Hemopoietic Cell Types

2.1. Receptors on Megakaryocytes and Leukemic Cell Lines

Human *c-mpl* was first isolated from HEL cells, a biphenotypic cell line initially derived from a patient with erythroleukemia *(6)*. Since HEL cells display features of both the erythroid and megakaryocytic lineages *(7,8)*, it was postulated that Mpl might be involved in the proliferation or differentiation of either of these two cell types. Verification of this hypothesis came from studies designed to identify its pattern of expression.

The first report suggesting the importance of *c-mpl* in megakaryocyte biology came from the work of Methia and colleagues. By surveying an extensive number of leuke-

From: *Thrombopoiesis and Thrombopoietins: Molecular, Cellular, Preclinical, and Clinical Biology*
Edited by: D. J. Kuter, P. Hunt, W. Sheridan, and D. Zucker-Franklin Humana Press Inc., Totowa, NJ

mic lines, these investigators found that only those cells that displayed, or could be induced to display, markers of megakaryocytic differentiation (UT-7, Mo-7E, TF-1, PMA-treated HEL, DAMI, and KU812) contained detectable levels of c-mpl transcripts (9). Moreover, of all the normal tissues and cell types tested, only megakaryocytes, their precursors (CD34+ marrow cells or fetal liver cells), or their progeny (platelets) displayed significant levels of c-mpl-specific mRNA. However, perhaps the most compelling data that Mpl plays a critical role in megakaryopoiesis came from experiments using antisense oligodeoxynucleotides designed to eliminate c-mpl expression in CD34+ human marrow cells. In the presence of antisense, but not sense c-mpl oligodeoxynucleotides, megakaryocyte colony formation was significantly reduced (9). In contrast, erythroid and myeloid colony formation were unaffected. These data clearly indicated that Mpl plays a critical role in megakaryocyte development.

More recently, these findings have been confirmed and extended. The first extensive study of c-mpl expression was reported by Vigon and colleagues, who could detect c-mpl transcripts from marrow cells of nearly 50% of patients with myeloid malignancies of all histologic subtypes (10). In contrast, patients with lymphoid malignancies displayed levels of c-mpl transcripts similar to normal individuals. A more extensive study has recently been completed that supports these conclusions and suggests that in many cases, the leukemic cells can proliferate in response to thrombopoietin (TPO) (11).

2.2. Receptors on Other Hemopoietic Cells

A number of lines of evidence now point to the presence of Mpl on more primitive, normal hemopoietic cells in animals and humans. As noted, c-mpl mRNA is found in CD34+ human hemopoietic cells, a population of cells that contains both stem and progenitor cells (9). An enriched population of primitive human hemopoietic cells that includes long-term culture-initiating cells also contains c-mpl mRNA (12). Flow cytometric analysis using the M1 monoclonal antibody (MAb) that recognizes human Mpl documented its presence on the surface of megakaryocytic and pluripotent human hemopoietic cell lines, and confirmed that normal human megakaryocytes and platelets exhibit Mpl on the cell surface (13). To identify the types of progenitor cells that express Mpl, CD34+ cells were sorted into M1+ and M1- fractions, and cultured in vitro. Megakaryocyte colony-forming cells (MK-CFC) were enriched in the CD34+M1+ fraction of cells in comparison to the CD34+M1- fraction. In addition, the ratio of erythroid burst-forming cells (E-BFC) to granulocyte-macrophage colony-forming cells (GM-CFC) was modestly increased in the CD34+M1+ population of cells, suggesting that some erythroid progenitor cells may also express Mpl.

The concept that primitive hemopoietic cells and erythroid progenitor cells can display functional Mpl is supported by experiments demonstrating that these cells can proliferate in the presence of TPO. Highly purified murine hemopoietic stem cells (lineage negative, low Hoechst 33344/low Rhodamine 123 fluorescence cells that are capable of lymphohemopoietic reconstitution in lethally irradiated hosts [14]) were cultured at single-cell density in the presence of hemopoietic growth factors with or without TPO. The addition of TPO to stem cell factor (SCF) or to IL-3 shortened the time to first cell division and increased the cloning efficiency of these primitive cells (15). These data suggest that TPO, in concert with SCF or IL-3, can act directly on purified hemopoietic stem cells. Other investigators have provided data indicating that the effects of TPO on erythropoiesis may be direct. Early erythroid E-BFC progeny, generated after 5–6 days culture of human marrow CD34$^+$ cells, were plucked and replated in the presence of TPO, erythropoietin (EPO), or both cytokines. The combination of TPO plus EPO increased the number of erythroid colonies above that seen with EPO alone *(16)*. These experiments argue that Mpl display is not restricted to the megakaryocytic lineage; functional protein is also expressed in primitive hemopoietic stem cells and in progeny of E-BFC. Thus, taken together, these reports indicate that Mpl is displayed by multiple types of hemopoietic cells and that by binding to its receptor, TPO can influence many aspects of blood cell development.

3. Functional Organization of Mpl

Members of the cytokine receptor superfamily share a number of structural features *(17,18)*. All the cytokine receptor family members contain a single 18–22 amino acid transmembrane domain that divides the polypeptide into a ligand-binding, amino-terminal extracytoplasmic domain, and an intracellular domain, responsible for initiating signal transduction. The extracellular domains of these receptors contain one or two 200 amino acid modules that contain four spatially conserved cysteine residues near the amino-terminal region, and a tryptophan-serine-X-tryptophan-serine (WSXWS) motif near the carboxyl-terminus. The sites on the extracellular domain of Mpl that bind TPO are presently unknown, but are likely to conform to a model of cytokine–receptor interactions based extensively on the homodimeric interaction of the growth hormone receptor with its ligand *(19)*.

Cytokine receptors that do not display an intrinsic tyrosine kinase motif are arranged in two general patterns, homodimeric and heteromeric. Receptors for growth hormone, EPO, and G-CSF are believed to represent examples of the former, and receptors for most of the interleukins, granulocyte-macrophage colony-stimulating factor (GM-CSF), oncostatin M (OSM), leukemia-inhibitory factor (LIF), ciliary neurotrophic factor (CNTF), and cardiotrophin 1 are examples of the latter. In either case, ligand contact with the first receptor subunit is stabilized by interaction with the second receptor subunit, thereby increasing the affinity of ligand–receptor binding and bringing the two receptor subunits into close proximity. This is believed to represent the critical first step in receptor signaling.

Two cDNA clones encoding human Mpl, Mpl-P and Mpl-K, were originally described *(3)*. These two isoforms are identical in the extracellular and transmembrane domains, but diverge after a common nine amino-acid sequence in the cytoplasmic domain. This is because of alternate splicing of the 10th intron. In the P form, intron 10 is removed, and exons 11 and 12 complete the coding sequence of the isoform. In contrast, failure to splice out intron 10 results in the K form, in which the gene sequence after the ninth amino acid of the intracytoplasmic domain continues until a termination codon is reached 57 residues further downstream. Since most of the K form carboxyl-terminal coding sequence is derived from intron 10 *(20)*, and the P form is derived from exons 11 and 12, the two intracytoplasmic domains are almost entirely distinct. However, Mpl-P mRNA is the predominant form found in hemopoietic tissues *(3)*. Two additional isoforms of human *c-mpl* that predict a 24-amino-acid deletion near the WSXWS box of the second of the 200-amino-acid modules have now been described *(21)*, as have several isoforms of murine *c-mpl*, including one predicted to encode a soluble receptor *(4,5,22)*. A genetically engineered soluble isoform of Mpl inhibits the biological activity of TPO *(23)*.

4. Evidence that *c-mpl* Encodes the Entire Receptor for TPO
4.1. GP130 Is Not a TPO Receptor β-Chain

Because some of the members of the cytokine receptor superfamily function as heteromeric or as trimeric receptor complexes, the question arose whether signal transduction by Mpl requires a second subunit. Although transfection of Ba/F3 cells with *c-mpl* confers the ability to proliferate in response to TPO *(24)*, these cells are known to express receptors for a number of hemopoietic cytokines. Certain of the IL-6 family of cytokines, which include IL-6, IL-11, OSM, CNTF, and cardiotropin 1, can promote megakaryocyte maturation in vitro *(25–27)*, and can increase platelet counts in vivo. Signal transduction by the IL-6 family of cytokines requires the GP130 receptor subunit. Also, binding of IL-6 to its receptor triggers association and homodimerization of GP130, and initiation of signal transduction *(28)*. The GP130 receptor subunit is widely expressed *(29)*. For these reasons, we investigated whether the effects of TPO on megakaryopoiesis require the participation of the GP130 receptor subunit. The addition of the neutralizing anti-GP130 MAb RX187 did not impair the ability of TPO to promote MK-CFC growth, megakaryocyte cytoplasmic maturation, or megakaryocyte nuclear endoreduplication *(30)*. In contrast, IL-11-induced MK-CFC growth and nuclear endoreduplication were inhibited by the antibody, demonstrating its neutralizing activity. These results indicate that signal transduction mediated by Mpl can occur when the activity of GP130 has been neutralized and imply that GP130 does not function as a β-chain for the TPO receptor.

4.2. β_c Is Not a TPO Receptor β-Chain

IL-3 is a potent stimulant of MK-CFC proliferation *(31)*, unlike the IL-6 family of cytokines that predominantly affects megakaryocyte maturation. The receptor for IL-3 is an αβ heterodimer, and shares a common β chain (β_c) with the receptors for GM-CSF and IL-5 *(32)*. To determine whether the TPO receptor might employ β_c, baby hamster kidney (BHK) cells were transfected with human *c-mpl* or with human *c-mpl* plus β_c. No increment in the proliferation response to TPO was found in cells cotransfected with *c-mpl* plus β_c, in comparison to cells transfected with *c-mpl* alone. These results suggest that signal transduction via Mpl does not require coexpression of the common β chain.

The spectrum of biological activity of TPO and EPO overlap; TPO and EPO can synergistically promote MK-CFC and E-CFC growth in vitro *(33,34)*, and treatment of mice with TPO can accelerate both platelet and red blood cell recovery after chemoradiotherapy *(34,35)*. A point mutation in the EPO receptor extracellular domain causes receptor dimerization and constitutive activation, suggesting that the EPO receptor normally functions as a homodimer *(36)*. However, because of the extensive homology between the N-terminal domain of TPO and EPO, the conserved location of three of the four cysteine residues in these two molecules, and the predicted similar tertiary structure of TPO and EPO *(24,37)*, the ability of TPO and EPO to crosscompete for receptor binding to Ba/F3 cells engineered to express functional receptors for both of these cytokines was tested *(38)*. No crosscompetition was detected, indicating that the Mpl and the EPO receptor do not share a ligand-binding subunit that is present in limiting quantity in these cells. Of note, crosscompetition studies provided the initial evidence that the IL-3 and GM-CSF receptors share a common β-subunit *(32)*.

A soluble form of Mpl (s-Mpl) has been generated by truncation at the junction of the extracellular and transmembrane regions. sMpl affixed to a support matrix has been used to purify endogenous TPO from the plasma of thrombocytopenic animals *(37,39)*, demonstrating that this form of Mpl is sufficient to bind TPO. sMpl can also compete with cell-surface Mpl for TPO binding and can neutralize the biological effects of TPO in vitro *(23,24)*. These observations demonstrate that sMpl can bind TPO in the absence of a receptor β-chain, at least with low affinity. Binding of G-CSF to a chimeric receptor consisting of the G-CSF receptor extracellular domain and the Mpl transmembrane and cytoplasmic domain can stimulate cell proliferation *(5)*. Taken together, these studies suggest that Mpl, similar to the receptors for EPO, G-CSF, and growth hormone *(14)*, is likely to function as a homodimer.

5. Ligand Binding Kinetics of Mpl

Radiolabeled TPO binds specifically to normal human megakaryocytes and platelets *(38)*, as well as to hemopoietic cell lines with megakaryocytic features. Equilibrium binding experiments demonstrate that platelets display a single class of high-affinity receptors for TPO with a binding affinity of approximately 100–200 pM with an estimated 25 receptors/platelet. Although full Scatchard analysis of TPO binding to soluble receptor in solution has not been published to date, studies using the BIAcore device have suggested that TPO binds to immobilized soluble Mpl with low affinity *(22)*. The mechanism responsible for the discrepancy in binding affinity of cell-surface protein and immobilized soluble Mpl receptor is not clear at present, but likely relates to the differing techniques employed for its quantitation. Additional investigation of TPO-Mpl-binding kinetics and affinity crosslinking, in an attempt to demonstrate receptor dimerization directly, may provide further insights into these issues.

6. The Structure of the Mpl Cytoplasmic Domain

6.1. Cytoplasmic Domain of Murine and Human Mpl

The members of the cytokine receptor superfamily have tremendous variability in the length of their cytoplasmic domains (from 54–569 residues) and share little primary sequence homology. Within this range, Mpl has a relatively short cytoplasmic region, smaller than those of most hemopoietic growth factor receptors (e.g., GP130, 277 residues; EPO-R, 235 residues; β$_c$, 434 residues), but longer than the IL-2-Rγ signaling subunit (86 residues). The cytoplasmic domain encoded by the transforming gene *v-mpl* is virtually identical to that of the murine *c-mpl* proto-oncogene, indicating that this portion of the receptor was captured by the MPLV. The only amino acid change is a deletion of the final two residues (Gln-Pro), resulting in a 119 amino-acid cytoplasmic domain instead of the wild-type 121 residues. As noted above, cloning of the human *c-mpl* revealed two distinct cytoplasmic variants, termed P and K forms. When the amino acid sequences are compared, the P form is 91% identical to murine Mpl and includes 122 amino acids. As judged by Northern blotting experiments, this is the predominant form of the receptor in most human tissues and has been shown to be active in TPO signaling *(40)*. In contrast, the K form contains only 66 intracytoplasmic amino acids, is not homologous to the murine molecule, and does not contain the box 1/box 2 motifs that are conserved in all cytokine receptors capable of signaling *(41)*. Because most functional and biological studies have been performed with the P form, subsequent information about human Mpl will pertain to this molecule.

6.2. Subdomain Structure–Function Relationships

Although some of the cytokine receptors lack enzymatic function, they activate signal transduction pathways that cause rapid changes in proliferation and differentiation. It is useful to think of the cytoplasmic domain as a scaffolding on which elaborate signaling complexes are assembled. Through spontaneous and engineered receptor mutations, the subdomains responsible for association with and activation of various signaling molecules have been mapped. This information has been extensively studied for the EPO and G-CSF receptors, and for the GP130 and β_c receptor subunits (Fig. 1).

Deletion analysis of various receptors has demonstrated that only the first 50–70 cytoplasmic amino acids are necessary for proliferation *(40,42,43)*. This region contains two short motifs (each 6–12 amino acids in length), which are relatively conserved across the cytokine receptor superfamily *(42)*. These motifs, termed box 1 and box 2, are essential for activation of the Janus tyrosine kinases (*see* Section 7.2.2.), and disruption of either of these motifs eliminates proliferation (reviewed in *44,45*). Homologous regions have been identified within the Mpl cytoplasmic domain. Box 1 is characterized by a pair of appropriately spaced prolines (PXXP) after 17 cytoplasmic residues; box 2 is located from amino acid 52–61 of the cytoplasmic domain and has LEIL as its core amino acid sequence *(40)*. Substitution and deletion mutations of *c-mpl* and *v-mpl* have demonstrated that these two elements are necessary and sufficient to support proliferation *(40,46)*.

Downstream of box 1/box 2, subdomains can be identified that activate other signaling pathways. For the β_c signaling subunit, a region between amino acids 626 and 763 has been identified as critical for Ras and MAP kinase activation, as well as SHC and β_c tyrosine phosphorylation *(43,47)*. This region is not absolutely required for growth in complete media, but is critical for proliferation under serum-free conditions *(48)*. Therefore, this portion of the β_c receptor may initiate a signal that can also be generated by a component of fetal bovine serum (FBS). Similarly, terminal truncations of the EPO receptor support proliferation in complete media, but not under serum-free conditions (Krystal G, ASH 1995). The terminal 20 residues of Mpl may be necessary for SHC phosphorylation, transcription of the early response gene, *fos*, and may initiate a similar proliferative signal in the absence of supplemental serum *(46)*.

Studies of the G-CSF receptor and GP130 receptor subunit have identified regions near the carboxyl-terminus of each molecule that direct expression of various lineage-specific genes. The terminal 120 amino acids of the G-CSF receptor are required for myeloperoxidase and leukocyte esterase expression *(49)*. Similarly, the carboxyl-terminus of the GP130 receptor subunit is necessary for expression of various acute-phase proteins and β-fibrinogen in hepatocytes and fibroblasts *(50,51)*. It is expected that Mpl has similar downstream loci that direct aspects of megakaryocytic differentiation, but they have not yet been defined.

A truncated form of the EPO receptor has been described that confers increased proliferative potential in response to EPO *(52)*. Additionally, a familial form of

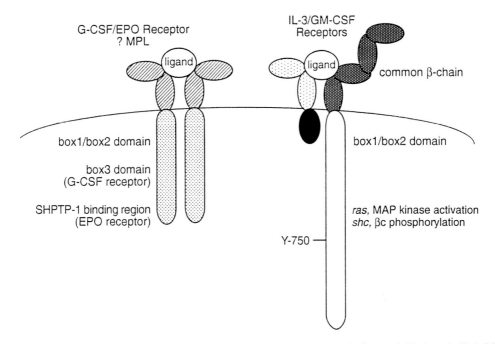

Fig. 1. Stylized drawing of the EPO and G-CSF receptors (left), and IL-3 and GM-CSF receptors (right). Functional domains of the cytoplasmic portions are indicated.

polycythemia has been defined, in which the terminal portion of the EPO receptor is missing *(53)*. It now appears that this carboxyl-terminus encodes the binding site for a specific phosphotyrosine phosphatase, SHPTP-1 *(54,55)*. In the full-length receptor, SHPTP-1 presumably dephosphorylates the activated signaling molecules and turns off the proliferative response. Similar negative regulatory regions may also exist for Mpl.

7. TPO Binding Induces Tyrosine Phosphorylation

An important element of cytokine signaling is the activation of cellular tyrosine kinases. Ligand binding initiates a cascade of tyrosine phosphorylation, which affects numerous proteins. Experiments with inhibitors have confirmed the physiological relevance of tyrosine phosphorylation. Addition of sodium vanadate, a phosphatase inhibitor, prolongs cytokine-induced proliferation, whereas inhibitors of tyrosine kinases, such as genistein, block the proliferative effect (reviewed in *56*). Thus, it was not surprising to find that TPO and Mpl also signal through this mechanism. It has been shown that TPO causes phosphotyrosine incorporation at the same concentrations necessary for development of megakaryocytes in bone-marrow culture experiments *(57)*. Cytokine-dependent cell lines, engineered to express Mpl, undergo dramatic increases in phosphotyrosine content in response to TPO *(57–59)*. Cell lines with megakaryocytic features *(60,61)* and normal human platelets *(62,63)* express native Mpl and also demonstrate a TPO-induced increase in tyrosine phosphorylation. Thus, it is likely that activation of tyrosine kinases is a critical early step in TPO signaling.

7.1. The Kinetics of Tyrosine Phosphorylation

After TPO stimulation, tyrosine phosphorylation is evident within 1 minute, peaks between 5 and 10 minutes, and then decreases by 60 minutes *(57,59)*. The rapidity of this process suggests that the signaling complex is partially preassembled at or near the cell membrane, since there is insufficient time for protein synthesis or molecular translocation between cellular compartments. Turning off the signaling cascade may occur through receptor downregulation, activation of tyrosine phosphatases, or both. At this time, the role of specific phosphatases or of protein internalization in Mpl receptor regulation is unknown.

7.2. Phosphorylation Substrates and Their Significance

7.2.1. Mpl

One of the common targets of tyrosine phosphorylation is the cytokine receptor itself. Tyrosine phosphorylation of the receptor creates docking sites for signaling molecules that contain the *src* homology 2 (SH2) domain *(64)*. In this manner, intricate signaling complexes can be assembled, using the receptor as a molecular scaffold. The amino acids immediately surrounding a phosphotyrosine residue provide specificity for distinct SH2-containing proteins. Consistent with this pattern among cytokine receptors, Mpl is itself tyrosine phosphorylated after TPO stimulation *(57,59)*. There are five tyrosine residues in the cytoplasmic tail of the protein that are potential targets for phosphorylation, but the precise residue(s) involved has not yet been determined.

Of considerable interest, Western blot analysis of cellular lysates derived from both naturally occurring Mpl-bearing cells and engineered cell lines identifies Mpl proteins of several sizes, ranging from 70–95 kDa. This variation likely represents at least some degree of posttranslational modification (e.g., glycosylation), since the engineered cells were transfected with a single cDNA. Despite this variation in size, only the largest form (95 kDa) is tyrosine phosphorylated *(57,59)*. Since the relative molecular weight (M_r) of the phosphorylated form is significantly larger than that predicted for the unmodified polypeptide, it suggests that posttranslational modification is necessary for receptor function.

7.2.2. JAK Family Proteins

There are four members of the Janus family of tyrosine kinases—JAK1, JAK2, JAK3, and TYK2—each with an approximate molecular weight of 130 kDa. These proteins associate with conserved membrane-proximal motifs, box 1 and box 2, in many cytokine receptors *(44,45)*. Upon ligand binding, two or more receptor subunits associate, bringing two JAK molecules into close proximity. It is believed that one JAK molecule can activate its neighbor through tyrosine phosphorylation *(44,45)*. Mutation or deletion of the box 1/box 2 elements eliminates both JAK phosphorylation and proliferation.

The precise combination of JAKs used by each cytokine receptor imposes one level of signaling specificity. Several recent reports, as well as our unpublished data, indicate that TPO can activate JAK2 and TYK2 in certain cell lines and human platelets *(62,65,66)*. Others have found that only JAK2 is tyrosine phosphorylated *(59,67)*. This discrepancy may be important because, in general, receptors that function as homodimers activate only a single JAK family member. If two distinct JAKs are necessary for TPO signaling, this poses several possibilities: first, both JAK2 and TYK2 can physically associate with Mpl; second, the TPO receptor includes two distinct subunits, each of which associates with a specific kinase; third, only one JAK is physiologically important for TPO activation, whereas the other is incidentally or indirectly phosphorylated. These issues should be clarified in the near future.

7.2.3. Signal Transducer and Activator of Transcription (STAT) Molecules

Despite some similarities in signal transduction, molecular mechanisms must exist to explain the distinct differentiation programs that result from stimulation with various cytokines. The STAT family of transcription factors is a leading candidate to fulfill this role (reviewed in *45,68*). Under basal conditions, these signaling molecules reside in the cytoplasm and display no DNA binding affinity. After cytokine stimulation, specific STAT proteins are tyrosine phosphorylated (probably by JAKs), allowing dimerization via the intrinsic STAT-SH2 domain *(69,70)*. These events have two important consequences: translocation to the nucleus, and DNA binding to specific promoter regions *(68,69)*. Phosphorylation of specific STAT family members appears to be regulated, in part, by the STAT–receptor interaction *(70)*.

A number of studies have examined the STATs activated by TPO. There is substantial evidence that STAT5 is tyrosine phosphorylated and activated *(60,65,66,71)*. This same protein, originally recognized as a prolactin-stimulated transcription factor, has now been implicated in signaling by the EPO and IL-3 receptors as well *(72,73)*. Several investigators have now reported tyrosine phosphorylation of STAT3 in both cell lines and platelets in response to TPO *(59,62,65,71)*. Additionally, two groups have reported involvement of STAT1 in TPO signaling *(46,66)*. The roles played by each of these transcription factors in TPO-induced megakaryocytopoiesis will be determined in the next few years.

7.2.4. Adapter Proteins

A growing number of adapter or linker signaling molecules have been described that have no enzymatic function, but contain two or more binding sites, such as SH2 and SH3 domains, or phosphotyrosine residues. Such proteins can function as molecular bridges linking the receptor to various signaling pathways. This paradigm has been best studied for activation of RAS and mitogen activated protein kinase (MAPK). It is believed that a complex of SHC, GRB2, VAV, SOS, and the newly discovered p145[SHIP] act in concert to stimulate RAS through exchange of GTP for GDP *(74)*.

Several of these adapter proteins have been implicated in TPO signaling owing to tyrosine phosphorylation. SHC and an associated 145-kDa protein (SHIP) are both phosphorylated in response to TPO *(57,58,61,63)*. In addition, TPO has been shown to induce tyrosine phosphorylation of SOS *(61)*, and the proto-oncogenes VAV *(58,59,61)* and c-CBL *(61)* as well as association of SHC and GRB2 *(61,63)*. Together, these observations provide a plausible connection between Mpl and activation of the RAS/MAP kinase pathway. In fact, several reports have shown tyrosine phosphorylation of MAP kinase in response to TPO *(59,66)*. Finally, truncation of the terminal 20 residues of the Mpl cytoplasmic domain abrogates SHC phosphorylation and transcription of *c-fos*, an early response gene *(46)*. However, the truncated receptor still supports proliferation of the transfected cells. Studies such as these will undoubtedly lead to a better understanding of TPO signaling pathways.

Mpl Signaling: current model

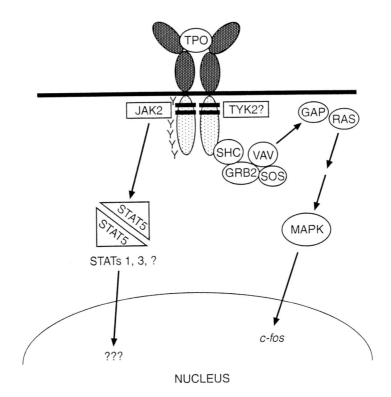

Fig. 2. A model of our current understanding of how Mpl transduces its proliferative signal. Clear evidence for phosphorylation of JAK2 and TYK2 has been provided, although the physiologic relevance of the latter is not yet certain. Also, phosphorylation of STAT5 and possibly STAT1 and STAT3, SHC, and other members of the MAPK pathway likely involved in the response to TPO is shown.

8. Ultimate Targets of TPO

Interaction of TPO and its receptor sets in motion the lineage-specific maturation and proliferation of megakaryocytes. Much has been learned about the earliest steps in signaling, including Mpl homodimerization and tyrosine phosphorylation of numerous signaling molecules (Fig. 2). These studies have shown remarkable overlap between the signaling pathways of TPO and those of other cytokines. It is, however, becoming clear that specificity also exists at the level of the JAK and STAT family members that are activated. Several other transcription factors have been found in megakaryocytes, including SCL/TAL, GATA-1, and RBTN-2, but their role in TPO signaling is as yet unknown (*see* Chapter 12). Some combination of these signaling proteins or activation of novel transcription factors not yet discovered must be sufficient to direct megakaryocyte growth and development. At a molecular level, future research will clarify the changes in cell cycle control that permit nuclear replication without cell division (nuclear endoreduplication) and the complex cytoplasmic processes that result in platelet production. Understanding these cellular events will be a major challenge.

References

1. Wendling F, Varlet P, Charon M, Tambourin P. MPLV: a retrovirus complex inducing an acute myeloproliferative leukemic disorder in mice. *Virology.* 1986; 149: 242–246.
2. Souyri M, Vigon I, Penciolelli JF, Heard JM, Tambourin P, Wendling F. A putative truncated cytokine receptor gene transduced by the myeloproliferative leukemia virus immortalizes hematopoietic progenitors. *Cell.* 1990; 63: 1137–1147.
3. Vigon I, Mornon JP, Cocault L, et al. Molecular cloning and characterization of *MPL*, the human homolog of the *v-mpl* oncogene: identification of a member of the hematopoietic growth factor receptor superfamily. *Proc Natl Acad Sci USA.* 1992; 89: 5640–5644.
4. Skoda RC, Seldin DC, Chiang MK, Peichel CL, Vogt TF, Leder P. Murine *c-mpl*: a member of the hematopoietic growth factor receptor superfamily that transduces a proliferative signal. *EMBO J.* 1993; 12: 2645–2653.
5. Vigon I, Florindo C, Fichelson S, et al. Characterization of the murine Mpl proto-oncogene, a member of the hematopoietic cytokine receptor family: molecular cloning, chromosomal location and evidence for a function in cell growth. *Oncogene.* 1993; 8: 2607–2615.
6. Martin P, Papayannopoulou T. HEL cells: a new human erythroleukemia cell line with spontaneous and induced globin expression. *Science.* 1982; 216: 1233–1235.
7. Long MW, Heffner CH, Williams JL, Peters C, Prochownik EV. Regulation of megakaryocyte phenotype in human erythroleukemia cells. *J Clin Invest.* 1990; 85: 1072–1084.
8. Hong Y, Martin JF, Vainchenker W, Erusalimsky JD. Inhibition of Protein Kinase C suppresses megakaryocytic differentiation and stimulates erythroid differentiation in HEL cells. *Blood.* 1994; 87: 123–131.
9. Methia N, Louache F, Vainchenker W, Wendling F. Oligodeoxynucleotides antisense to the proto-oncogene c-mpl specifically inhibit in vitro megakaryocytopoiesis. *Blood.* 1993; 82: 1395–1401.
10. Vigon I, Dreyfus F, Melle J, et al. Expression of the *c-mpl* proto-oncogene in human hematologic malignancies. *Blood.* 1993; 82: 877–883.
11. Matsumura I, Kanakura Y, Kato T, et al. Growth response of acute myeloblastic leukemia cells to recombinant human thrombopoietin. *Blood.* 1995; 86: 703–709.
12. Berardi AC, Wang A, Levine JD, Lopez P, Scadden DT. Functional isolation and characterization of human hematopoietic stem cells. *Science.* 1995; 267: 104–108.
13. Debili N, Wendling F, Cosman D, et al. The mpl receptor is expressed in the megakaryocytic lineage from late progenitors to platelets. *Blood.* 1995; 85: 391–401.
14. Wolf NS, Kone A, Priestley GV, Bartelmez SH. In vivo and in vitro characterization of long-term repopulating primitive hematopoietic cells isolated by sequential Hoechst 33342-rhodamine 123 FACS selection. *Exp Hematol.* 1993; 21: 614–622.
15. Sitnicka E, Lin N, Priestley GV, et al. The effect of thrombopoietin on the proliferation and differentiation of murine hematopoietic stem cells. *Blood.* 1995; 86: 419a (abstract no 1664).
16. Kobayashi M, Laver JH, Kato T, Miyazaki H, Ogawa M. Recombinant human thrombopoietin (mpl ligand) enhances proliferation of erythroid progenitors. *Blood.* 1995; 86: 2494–2499.
17. Bazan JF. Structural design and molecular evolution of a cytokine receptor superfamily. *Proc Natl Acad Sci USA.* 1990; 87: 6934–6938.
18. Cosman D. The hematopoietin receptor superfamily. *Cytokine.* 1993; 5: 95–106.
19. Wells JA. Binding in the growth hormone receptor complex. *Proc Natl Acad Sci USA.* 1996; 93: 1–6.
20. Alexander WS, Dunn AR. Structure and transcription of the genomic locus encoding murine *c-Mpl*, a receptor for thrombopoietin. *Oncogene.* 1995; 10: 795–803.
21. Kiladjian JJ, Hetet G, Briere J, Grandchamp B, Gardin C. New mRNA isoforms of the *c-mpl* receptor in human platelets. *Blood.* 1995; 86: 366a (abstract no 1451).
22. Lofton-Day C, Buddle M, Berry J, Lok S. Differential binding of murine thrombopoietin to two isoforms of murine c-mpl. *Blood.* 1995; 86: 594a (abstract no 2362).
23. Kaushansky K, Broudy VC, Lin N, et al. Thrombopoietin, the Mpl-ligand, is essential for full megakaryocyte development. *Proc Natl Acad Sci USA.* 1995; 92: 3234–3238.

24. Lok S, Kaushansky K, Holly RD, et al. Cloning and expression of murine thrombopoietin cDNA and stimulation of platelet production in vivo. *Nature.* 1994; 369: 565–568.

25. Ishibashi T, Kimura H, Uchida T, Kariyone S, Friese P, Burstein SA. Human interleukin 6 is a direct promoter of maturation of megakaryocytes in vitro. *Proc Natl Acad Sci USA.* 1989; 86: 5953–5957.

26. Burstein SA, Mei RL, Henthorn J, Friese P, Turner K. Leukemia inhibitory factor and interleukin-11 promote maturation of murine and human megakaryocytes in vitro. *J Cell Physiol.* 1992; 153: 305–312.

27. Wallace PM, MacMaster JF, Rillema JR, Peng J, Burstein SA, Shoyab M. Thrombocytopoietic properties of oncostatin M. *Blood.* 1995; 86: 1310–1315.

28. Kishimoto T, Akira S, Narazaki M, Taga T. Interleukin-6 family of cytokines and gp130. *Blood.* 1995; 86: 1243–1254.

29. Saito M, Yoshida K, Hibi M, Taga T, Kishimoto T. Molecular cloning of a murine IL-6 receptor-associated signal transducer, gp130, and its regulated expression in vivo. *J Immunol.* 1992; 148: 4066–4071.

30. Broudy VC, Lin N, Fox N, Taga T, Saito M, Kaushansky K. Thrombopoietin stimulates CFU-Meg proliferation and megakaryocyte maturation independently of cytokines that signal through the gp130 receptor subunit. *Blood.* 1996 (in press).

31. Segal GM, Stueve T, Adamson JW. Analysis of murine megakaryocyte colony size and ploidy: effects of interleukin-3. *J Cell Physiol.* 1988; 137: 537–544.

32. Miyajima A, Mui AL, Ogorochi T, Sakamaki K. Receptors for granulocyte-macrophage colony-stimulating factor, interleukin-3 and interleukin-5. *Blood.* 1993; 82: 1960–1974.

33. Broudy VC, Lin NL, Kaushansky K. Thrombopoietin (c-mpl ligand) acts synergistically with erythropoietin, stem cell factor, and interleukin-11 to enhance murine megakaryocyte colony growth and increases megakaryocyte ploidy in vitro. *Blood.* 1995; 85: 1719–1726.

34. Kaushansky K, Broudy VC, Grossmann A, et al. Thrombopoietin expands erythroid progenitors, increases red cell production, and enhances erythroid recovery after myelosuppressive therapy. *J Clin Invest.* 1995; 96: 1683–1687.

35. Ulich TR, del Castillo J, Yin S, et al. Megakaryocyte growth and development factor ameliorates carboplatin-induced thrombocytopenia in mice. *Blood.* 1995; 86: 971–976.

36. Watowich SS, Yoshimura A, Longmore GD, Hilton DJ, Yoshimura Y, Lodish HF. Homodimerization and constitutive activation of the erythropoietin receptor. *Proc Natl Acad Sci USA.* 1992; 89: 2140–2144.

37. de Sauvage FJ, Hass PE, Spencer SD, et al. Stimulation of megakaryocytopoiesis and thrombopoiesis by the c-Mpl ligand. *Nature.* 1994; 369: 533–538.

38. Broudy VC, Lin N, Fox N, Atkins H, Iscove N, Kaushansky K. Hematopoietic cells display high affinity receptors for thrombopoietin. *Blood.* 1995; 86: 593a (abstract no 2361).

39. Bartley TD, Bogenberger J, Hunt P, et al. Identification and cloning of a megakaryocyte growth and development factor that is a ligand for the cytokine receptor Mpl. *Cell.* 1994; 77: 1117–1124.

40. B'enit L, Courtois G, Charon M, Varlet P, Dusanter-Fourt I, Gisselbrecht S. Characterization of mpl cytoplasmic domain sequences required for myeloproliferative leukemia virus pathogenicity. *J Virol.* 1994; 68: 5270–5274.

41. Mignotte V, Vigon I, Boucher de Crevecoeur E, Roméo PH, Lemarchandel V, Chrétien S. Structure and transcription of the human *c-mpl* gene (MPL). *Genomics.* 1994; 20: 5–12.

42. Murakami M, Narazaki M, Hibi M, et al. Critical cytoplasmic region of the interleukin 6 signal transducer gp130 is conserved in the cytokine receptor family. *Proc Natl Acad Sci USA.* 1991; 88: 11,349–11,353.

43. Sakamaki K, Miyajima I, Kitamura T, Miyajima A. Critical cytoplasmic domains of the common beta subunit of the human GM-CSF, IL-3 and IL-5 receptors for growth signal transduction and tyrosine phosphorylation. *EMBO J.* 1992; 11: 3541–3549.

44. Ihle JN, Witthuhn B, Tang B, Yi T, Quelle FW. Cytokine receptors and signal transduction. *Baillieres Clin Haematol.* 1994; 7: 17–48.

45. Ihle JN. Cytokine receptor signalling. *Nature*. 1995; 377: 591–594.

46. Gurney AL, Wong SC, Henzel WJ, de Sauvage FJ. Distinct regions of c-Mpl cytoplasmic domain are coupled to the JAK-STAT signal transduction pathway and Shc phosphorylation. *Proc Natl Acad Sci USA*. 1995; 92: 5292–5296.

47. Sato N, Sakamaki K, Terada N, Arai K, Miyajima A. Signal transduction by the high-affinity GM-CSF receptor: two distinct cytoplasmic regions of the common beta subunit responsible for different signaling. *EMBO J*. 1993; 12: 4181–4189.

48. Inhorn RC, Carlesso N, Durstin M, Frank DA, Griffin JD. Identification of a viability domain in the granulocyte/macrophage colony-stimulating factor receptor beta-chain involving tyrosine-750. *Proc Natl Acad Sci USA*. 1995; 92: 8665–8669.

49. Fukunaga R, Ishizaka-Ikeda IE, Nagata S. Growth and differentiation signals mediated by different regions in the cytoplasmic domain of granulocyte colony-stimulating factor receptor. *Cell*. 1993; 74: 1079–1087.

50. Baumann H, Gearing D, Ziegler SF. Signaling by the cytoplasmic domain of hematopoietin receptors involves two distinguishable mechanisms in hepatic cells. *J Biol Chem*. 1994; 269: 16297–16304.

51. Baumann H, Symes AJ, Comeau MR, et al. Multiple regions within the cytoplasmic domains of the leukemia inhibitory factor receptor and gp130 cooperate in signal transduction in hepatic and neuronal cells. *Mol Cell Biol*. 1994; 14: 138–146.

52. D'Andrea AD, Yoshimura A, Youssoufian H, Zon LI, Koo JW, Lodish HF. The cytoplasmic region of the erythropoietin receptor contains nonoverlapping positive and negative growth-regulatory domains. *Mol Cell Biol*. 1991; 11: 1980–1987.

53. de la Chapelle A, Traskelin AL, Juvonen E. Truncated erythropoietin receptor causes dominantly inherited benign human erythrocytosis. *Proc Natl Acad Sci USA*. 1993; 90: 4495–4499.

54. Tauchi T, Feng GS, Shen R, et al. Involvement of SH2-containing phosphotyrosine phosphatase Syp in erythropoietin receptor signal transduction pathways. *J Biol Chem*. 1995; 270: 5631–5635.

55. Klingmuller U, Lorenz U, Cantley LC, Neel BG, Lodish HF. Specific recruitment of SH-PTP1 to the erythropoietin receptor causes inactivation of JAK2 and termination of proliferative signals. *Cell*. 1995; 80: 729–738.

56. Miyajima A, Mui AL, Ogorochi T, Sakamaki K. Receptors for granulocyte-macrophage colony-stimulating factor, interleukin-3, and interleukin-5. *Blood*. 1993; 82: 1960–1974.

57. Drachman JG, Griffin JD, Kaushansky K. The c-Mpl ligand (thrombopoietin) stimulates tyrosine phosphorylation of Jak2, Shc, and c-Mpl. *J Biol Chem*. 1995; 270: 4979–4982.

58. Dorsch M, Fan PD, Bogenberger J, Goff SP. TPO and IL-3 induce overlapping but distinct protein tyrosine phosphorylation in a myeloid precursor cell line. *Biochem Biophys Res Commun*. 1995; 214: 424–431.

59. Mu SX, Xia M, Elliot G, et al. Megakaryocyte growth and development factor and interleukin-3 induce patterns of protein-tyrosine phosphorylation that correlate with dominant differentiation over proliferation of mpl-transfected 32D cells. *Blood*. 1995; 86: 4532–4543.

60. Pallard C, Gouilleux F, Bénit L, et al. Thrombopoietin activates a STAT5-like factor in hematopoietic cells. *EMBO J*. 1995; 14: 2847–2856.

61. Sasaki K, Odai H, Hanazono Y, et al. TPO/c-mpl ligand induces tyrosine phosphorylation of multiple cellular proteins including proto-oncogene products, Vav and c-Cbl, and Ras signaling molecules. *Biochem Biophys Res Commun*. 1995; 216: 338–347.

62. Ezumi Y, Takayama H, Okuma M. Thrombopoietin, c-Mpl ligand, induces tyrosine phosphorylation of Tyk2, JAK2, and STAT3, and enhances agonists-induced aggregation in platelets in vitro. *FEBS Lett*. 1995; 374: 48–52.

63. Miyakawa Y, Oda A, Druker BJ, et al. Recombinant thrombopoietin induces rapid protein tyrosine phosphorylation of Janus kinase 2 and Shc in human blood platelets. *Blood*. 1995; 86: 23–27.

64. Izuhara K, Harada N. Interleukin-4 (IL-4) induces protein tyrosine phosphorylation of the IL-4 receptor and association of phosphatidylinositol 3-kinase to the IL-4 receptor in a mouse T cell line, HT2. *J Biol Chem*. 1993; 268: 13097–13102.

65. Miyakawa Y, Oda A, Druker BJ, et al. Thrombopoietin induces tyrosine phosphorylation of Stat3 and Stat5 in human blood platelets. *Blood*. 1996; 87: 439–446.

66. Sattler M, Durstin MA, Frank DA, et al. The thrombopoietin receptor c-MPL activates JAK2 and TYK2 tyrosine kinases. *Exp Hematol*. 1995; 23: 1040–1048.

67. Tortolani PJ, Johnston JA, Bacon CM, et al. Thrombopoietin induces tyrosine phosphorylation and activation of the Janus kinase, JAK2. *Blood*. 1995; 85: 3444–3451.

68. Darnell JE Jr, Kerr IM, Stark GR. Jak-STAT pathways and transcriptional activation in response to IFNs and other extracellular signaling proteins. *Science*. 1994; 264: 1415–1421.

69. Shuai K, Horvath CM, Huang LH, Qureshi SA, Cowburn D, Darnell JE Jr. Interferon activation of the transcription factor Stat91 involves dimerization through SH2-phosphotyrosyl peptide interactions. *Cell*. 1994; 76: 821–828.

70. Heim MH, Kerr IM, Stark GR, Darnell JE Jr. Contribution of STAT SH2 groups to specific interferon signaling by the Jak-STAT pathway. *Science*. 1995; 267: 1347–1349.

71. Bacon CM, Tortolani PJ, Shimosaka A, Rees RC, Longo DL, O'Shea JJ. Thrombopoietin (TPO) induces tyrosine phosphorylation and activation of STAT5 and STAT3. *FEBS Lett*. 1995; 370: 63–68.

72. Mui AL, Wakao H, O'Farrell AM, Harada N, Miyajima A. Interleukin-3, granulocyte-macrophage colony stimulating factor and interleukin-5 transduce signals through two STAT5 homologs. *EMBO J*. 1995; 14: 1166–1175.

73. Pallard C, Gouilleux F, Charon M, Groner B, Gisselbrecht S, Dusanter-Fourt I. Interleukin-3, erythropoietin, and prolactin activate a STAT5-like factor in lymphoid cells. *J Biol Chem*. 1995; 270: 15,942–15,945.

74. Downward J. The GRB2/Sem-5 adaptor protein. *FEBS Lett*. 1994; 338: 113–117.

17

Regulation of Proplatelet and Platelet Formation In Vitro

Esther Choi

1. Historical Perspectives on Mechanisms of Platelet Production

Nearly a century ago, James Homer Wright published a landmark paper describing the origin of platelets from megakaryocytes *(1)*. He observed cytoplasmic pseudopodia from megakaryocytes that had the same staining characteristics as platelets and concluded that the pseudopodia produced future platelets. Today, it is universally accepted that platelets are derived from megakaryocytes. However, a subject of controversy has been the mechanism of platelet release from megakaryocytes. The various extant theories include those of platelet budding, megakaryocyte cytoplasmic fragmentation, and pseudopodia or "proplatelet" formation from megakaryocytes. Three theories will be discussed. Following the theories will be a review of the four most commonly used in vitro proplatelet assay systems and the effect of various substances on these systems.

1.1. Platelet Budding

Several investigators have observed platelet-sized projections on the surface of megakaryocytes examined by scanning electron microscopy and have postulated that platelets may "bud" from megakaryocytes *(2,3)*. However, when these platelet buds were examined by transmission electron microscopy, no platelet organelles were observed *(2,3)*. Based on these results, platelet budding seems to be an unlikely mechanism by which platelets are released from megakaryocytes.

1.2. Cytoplasmic Fragmentation

The theory of cytoplasmic fragmentation was based largely on early observations of transmission electron micrographs in which "platelet territories," a concentration of platelet-specific organelles surrounded by membranes inside megakaryocyte cytoplasm, were seen *(4)*. The collection of internal membranes was named the demarcation membrane system (DMS) to indicate its role in delineating preformed platelets within the megakaryocyte cytoplasm *(4)*. Various other names describing this phenomenon include cytoplasmic fragmentation, cytoplasmic dissolution, and megakaryocyte fragmentation. It has been proposed that fragmentation of the megakaryocyte plasma membrane results in platelet release *(5,6)*; however, the specific signals necessary to

From: *Thrombopoiesis and Thrombopoietins: Molecular, Cellular, Preclinical, and Clinical Biology*
Edited by: D. J. Kuter, P. Hunt, W. Sheridan, and D. Zucker-Franklin Humana Press Inc., Totowa, NJ

induce plasma-membrane dissolution as well as the mechanisms of plasma-membrane fragmentation remain unknown. Radley and Haller pointed out that these platelet territories lacked microtubule bundles, a hallmark of platelet structure, and that the term DMS may be a misnomer. They suggested that "invagination membrane system" may be a more accurate term to describe the extensive membrane system that exists in megakaryocytes, based on important experiments in which more membrane material was observed inside the cytoplasm of megakaryocytes when "proplatelets" were induced to retract compared with proplatelet-displaying megakaryocytes *(7)*. In addition, it was postulated that if platelets are actually formed within the megakaryocyte cytoplasm, then the megakaryocyte cytoplasm should be filled with platelet plasma membrane as well as platelet-specific organelles *(8)*. This is not the case. Careful examinations of freeze-fractured megakaryocytes were done intentionally looking for platelet plasma membranes similar to those found in platelet aggregates, and no membranes resembling those of platelet aggregates were observed. It was concluded that mature, intact platelets do not form inside the megakaryocyte cytoplasm, but may form from megakaryocyte processes as postulated by Radley et al. or some variation of this mechanism *(8) (see* Section 1.3.). To date, there is no direct evidence that these platelet field-containing megakaryocytes fragment directly to produce functional platelets.

1.3. Proplatelet Formation

The term "proplatelet" is commonly used to describe long cytoplasmic extensions of megakaryocytes that contain platelet-specific organelles within an area defined by constriction points *(9)*. Proplatelets are the presumed intermediate structure between megakaryocytes and platelets, and have been observed both in vitro *(10–21)* and in vivo *(22–25)*. Several names have been used by different investigators to describe this phenomenon, including pseudopodia formation, proplatelet formation, proplatelet-like formation, and cytoplasmic process formation. For the sake of simplicity, the word "proplatelet" will be used to describe the megakaryocyte processes. There are two variations of the proplatelet theory, and both variations support the in vivo observations of megakaryocyte process formation.

The first classical "proplatelet" theory was proposed by Becker and De Bruyn, who suggested that megakaryocytes extend long, sinuous processes that fragment into various pieces; these pieces subsequently undergo further fragmentation to yield platelets *(9)*. This theory does not exclude the possibility of platelet-field formation within the megakaryocyte cytoplasm before process formation. A second theory, commonly known as the "flow model" proposed by Radley and Haller, states that megakaryocytes extend processes with attenuation points along the length with platelet "beads" between the constrictions *(7)*. These constrictions were proposed to mark the point of platelet break-off. The platelet-sized "beads" have been demonstrated to express platelet-specific glycoproteins on the plasma membrane *(21,26)* and to contain platelet-specific organelles *(11,13,27–29)*, as well as microtubule coils *(11,21)*. The radical aspect of this theory, incongruent with the platelet-field theory, was that the DMS appeared to serve as a membrane reservoir for the nascent proplatelet plasma membrane instead of delineating future platelet fields. The timing of microtubule coil formation has also been a subject of controversy. Since rodent megakaryocyte processes contain microtubule coils *(11)*, it was proposed that microtubule coils may be responsible for the beaded appearance of proplatelets *(30)*. Recently, it has been shown that proplatelets of human megakaryocytes contain microtubule coils *(21)*. However, other researchers have not observed microtubule coil formation within platelet beads of proplatelets derived from mouse bone marrow explants *(31)*. The differences in these results may be owing to variations in cell sources, species of cells, or culture systems, or the stability of these proplatelet structures in culture.

Increasing evidence in recent years supports the flow model of Radley and Haller *(7)*. Proplatelets have been observed by many investigators, both in vivo as well as in vitro. In vivo, proplatelets have been documented crossing the venous sinusoid endothelial wall, reaching into the vessel lumen *(22–25)*. It has been proposed that the shear force of the blood flow in the lumen aids in fragmentation of proplatelets into individual platelets *(22–24)*. It is important to note here that the site of platelet fragmentation has not yet been conclusively determined and remains a subject of controversy. Both the bone marrow *(22–24)* and the lung *(25,32,33)* have been proposed as sites for platelet shedding. In vitro, with the aid of various assays using megakaryocytes isolated from humans *(16,21,26)*, the guinea pig *(10,11,14,15,17)*, rat *(12,34,35)*, mouse *(18–20,27)*, and cow *(29)*, many investigators have documented proplatelet formation and regulation. Four of the most commonly used in vitro assays, using recently developed reagents and techniques allowing the study of proplatelet formation in highly enriched cell populations, will be discussed.

Regarding other tissue-culture techniques, caution must be exercised when interpreting data from these in vitro systems, since the physiological relevance of these cultures is not clear. These systems are tools that may be used to help dissect the complex physiology of thrombopoiesis.

2. Current Proplatelet Assays

Plasma and serum have been used widely as culture supplements in proplatelet systems. In this chapter, the term "plasma" will be used to indicate the platelet-poor liquid fraction of whole blood drawn with an anticoagulant. In our laboratory, the platelet-poor plasma is routinely ultracentrifuged to remove any platelet particles that may be present. The term "serum" will be used to indicate the liquid fraction of clotted whole blood drawn without an anticoagulant that contains platelet granular products.

2.1. Rat Proplatelet Assay

The procedure for a proplatelet assay using megakaryocyte progenitors megakaryocyte colony-forming cells (MK-CFC) isolated from rat bone marrow uses flushed rat bone marrow cells subjected to a Percoll density gradient *(35)*. Nonadherent, light-density cells containing megakaryocytes and MK-CFC can be collected, and megakaryocytes further selected by panning with a monoclonal antibody (MAb) directed against rat GPIIb/IIIa (clone designation p55). The final enriched population contains approximately 7% MK-CFC.

Cells collected by this technique are incubated in 10% fetal bovine serum (FBS) and various cytokines, and scored for proplatelets 3–4 days later *(35)*. In the absence of cytokines, proplatelets are not observed, but in the presence of erythropoietin (EPO) (5 U/mL) and interleukin (IL)-6 (50 ng/mL), as many as 30% of the cells are proplatelet-displaying megakaryocytes (PF-MK).

2.2. Mouse Proplatelet Assay

Megakaryocytes can also be isolated from mouse bone marrow *(20)*. In this procedure, total bone marrow cells are also subjected to a Percoll density gradient, followed by collection of megakaryocyte-rich interphase fractions and selection of megakaryocytes using a mouse platelet antiserum. Antibody-labeled megakaryocytes can be isolated using immunomagnetic beads (Dynabeads, Dynal, Oslo, Norway) coated with antirabbit IgG secondary antibodies.

The yield (70%) and purity (95%) of megakaryocytes achieved with this method are impressive, and it is a relatively fast and efficient method for isolation of large quantities of megakaryocytes suitable for molecular and biochemical studies. When plated in the presence of IL-6 (20 ng/mL), as many as 33% of megakaryocytes formed proplatelets under serum-free conditions. The disadvantages of this method include the presence of large immunomagnetic beads in the purified population of megakaryocytes. These beads are difficult to separate from megakaryocytes after the purification, making morphological studies a problem.

2.3. Guinea Pig Proplatelet Assay

Intact megakaryocytes can also be isolated from guinea pig long bones using metabolic inhibitors to prevent cell–cell aggregation by means of calcium- and magnesium-free Hank's buffer, with adenosine, theophylline, sodium citrate, and sodium bicarbonate (CATCH) *(36)*. This CATCH buffer has hence been used widely to isolate intact megakaryocytes *(14)*. Bone marrow cells are subjected to sequential bovine serum albumin (BSA)-step gradients, and the final fraction contains 80–95% pure megakaryocytes with 90% viability. When these megakaryocytes are cultured in a serum-free medium, 21–29% of the cells spontaneously develop proplatelets *(14)*. The advantages of this system include the serum-free conditions, the purity of megakaryocytes, and the relatively large number of megakaryocytes obtained per animal (i.e., between 1×10^5 and 2×10^5) *(14)*. Although the physiological relevance of these serum-free culture systems may be obscure, they may yield insight into the preprogrammed cellular events that megakaryocytes are destined to undergo without influence from plasma and serum components.

2.4. Human Proplatelet System

Although human bone marrow megakaryocytes have been used to demonstrate in vitro proplatelet formation *(16)*, a consistent and large-scale system wherein human proplatelets are observed has been lacking until recently. An in vitro culture system was recently described in which relatively pure (90–95%) CD34+ cells are plated in the presence of irradiated dog plasma (a source of endogenous thrombopoietin [TPO]), supplemented with 10% normal platelet-poor human heparinized plasma, and cultured for for 8 days *(21,37)*. The mixed population of megakaryocytes and other cells produced are selected on a BSA-step gradient to enrich for mature megakaryocytes. The enriched mature megakaryocytes (an average of 87.7 ± 2.4% GPIIb/GPIIIa+) are then replated in the absence of irradiated dog plasma to induce proplatelet formation in the presence or absence of 10% normal human heparinized platelet-poor plasma. Under these conditions, up to 50% of megakaryocytes displayed proplatelets after 2–3 days of replating *(21)* (Fig. 1). The proplatelet-displaying cultures yielded platelet-sized

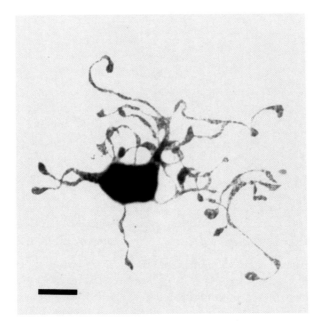

Fig. 1. A human megakaryocyte displaying proplatelets. Proplatelets are megakaryocyte cytoplasmic extensions with constrictions along the length. The areas between the constriction points are platelet-sized and contain platelet-specific organelles. Human megakaryocytes were generated from CD34+ cells *(26)*, selected on a bovine serum albumin (BSA) step gradient and replated to observe proplatelet formation *(21)*. Supernatant of a proplatelet-displaying mega-karyocyte culture was cytocentrifuged onto glass slides and stained with modified Wright-Giemsa (bar = 10 μm).

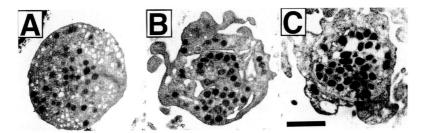

Fig. 2. In vitro-derived platelets are morphologically normal. Electron micrographs of in vitro-derived platelets **(A,B)**, and in vivo-derived (plasma) platelets **(C)**. **(A)** An in vitro-derived platelet with a smooth contour, resembling a resting platelet. **(B)** An in vitro-derived platelet with ruffled contour, resembling an actived platelet. **(C)** An in vivo (plasma)-derived platelet with activated morphology (bar = 1 μm) *(37)*. (Reprinted with permission from ref. *21*.)

fragments morphologically and functionally comparable to those of plasma-derived platelets *(21)* (Figs. 2–4).

The advantages of this culture system include the availability of CD34+ cells and MK-CFC that are easily manipulated in liquid culture, as well as the ability to observe the complete spectrum of thrombopoiesis in vitro. The disadvantages of this culture

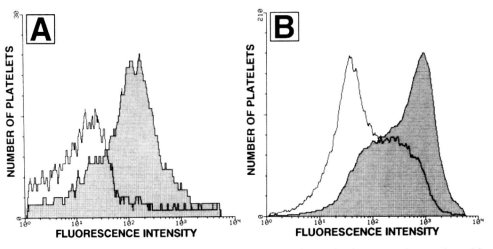

Fig. 3. In vitro-derived platelets are functionally normal: P-selectin expression on thrombin-activated platelets. In vitro-derived **(A)** or in vivo-derived **(B)** platelets were fixed before (unshaded) or after (shaded) 60 seconds of thrombin (1 U/mL) activation. Platelets were incubated with anti-P-selectin (anti-CD62) antibody followed by goat antimouse FITC for flow cytometric analyses. (Reprinted with permission from ref. *21*.)

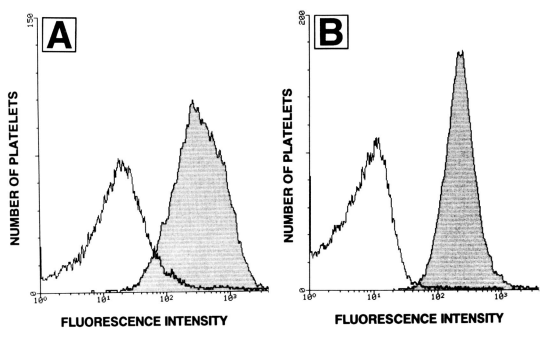

Fig. 4. In vitro-derived platelets are functionally normal: Activation-dependent GPIIb/IIIa expression on adenosine diphosphate (ADP)-stimulated platelets. In vitro-derived **(A)** and in vivo-derived **(B)** platelets were incubated with PAC-1 antibody (gift of Sanford Shattil) before (unshaded) or after (shaded) 15 minutes of ADP (40 μ*M*) incubation. (Reprinted with permission from ref. *21*.)

system include the labor-intensive CD34[+] cell-isolation procedure and the relatively high cost of supplies and reagents.

3. Regulators of Proplatelet Formation In Vitro

3.1. Extracellular Matrix Glycoproteins and Proteoglycans

Basement membrane secreted by cultured bovine endothelial cells has enhancing activities on human bone marrow-derived megakaryocyte proplatelet formation and allows adherence and flattening of megakaryocytes *(16)*. Matrigel, a commercially available murine basement membrane containing a variety of extracellular matrix proteins and proteoglycans, has been shown to have proplatelet-enhancing activity on bovine bone marrow-derived megakaryocytes in the presence of 10% heat-inactivated FBS, 2 mM L-glutamine and antibiotics in Dulbecco's Minimal Essential Medium (DMEM) *(29)*. Another study used guinea pig bone marrow megakaryocytes cultured under serum-free conditions on either untreated or Matrigel-coated tissue-culture plastic *(14)*. Under serum-free conditions, no difference in the number of proplatelet-bearing megakaryocytes was observed between the two substrates *(14)*. Under serum-containing conditions, however, Matrigel does enhance the proplatelet-forming activity of megakaryocytes because of its ability to bind serum proteins (i.e., prothrombin) that can inhibit proplatelet formation. An antibody (LM609) specifically directed to a shared epitope of $\alpha_v\beta_3$ and the cell adhesion tetrapeptide RGDS (arginine-glycine-aspartate-serine) was used to show that guinea pig megakaryocyte-proplatelet formation on collagen type 1 (CN I) is mediated through the $\alpha_v\beta_3$ vitronectin (VN) receptor *(17)*. By adding purified VN (50 µg/mL) in a serum-free proplatelet culture system that enhanced proplatelet development, it was demonstrated that VN played an important role in mediating guinea pig megakaryocyte-proplatelet formation *(17)*. On the other hand, others have reported that purified extracellular matrix molecules, including fibrinogen, fibronectin, VN, and CN I, had no statistically significant effects on guinea pig megakaryocyte-proplatelet formation under serum-free conditions *(14)*.

Matrigel was also evaluated in a human proplatelet assay system, and results of these experiments help to clarify these apparently conflicting data *(37)*. Initially, more proplatelets were observed on the bottom of Matrigel-coated wells. However, because of the buoyancy of cells displaying proplatelets in the tissue-culture medium, the proplatelet-displaying cultures were centrifuged in the presence 0.38% formaldehyde and 100 mM ethylenediamine tetraacetic acid (EDTA), to fix all the proplatelet-displaying cells to the bottom of the 96-well plate before scoring. In this case, approx 45% of megakaryocytes plated on Matrigel, as well as those plated on untreated tissue-culture plastic, displayed proplatelets. It is likely that the cell-adhesion molecules present in Matrigel (e.g., fibronectin) aid in the anchoring of proplatelet-displaying megakaryocytes to the bottom of the well, making them seem more numerous. These cells were also studied over time to measure the kinetics and extent of proplatelet formation on the two substrates, and no significant differences between the two surfaces were detected (unpublished observations).

Extracellular matrix, because of its relative localization in the bone marrow, most likely plays an important role during thrombopoiesis. However, in vitro studies must be interpreted carefully. VN is abundant in human plasma and serum, and is found at an average of 0.35 mg/mL *(38)*. Therefore, any proplatelet assay system containing 10%

serum or plasma would have approximately 35 μg/mL VN, possibly saturating the $\alpha_v\beta_3$ receptors on the megakaryocyte cell surface. Thus, even though $\alpha_v\beta_3$ may play an important role during proplatelet formation, it is not surprising to observe no significant difference in proplatelet formation when megakaryocytes are plated on Matrigel compared with megakaryocytes plated on untreated tissue-culture plastic.

Serglycin is a chondroitin-sulfated glycosaminoglycan in platelets and bone marrow-derived mast cells that allows proplatelet formation in the presence of the plasma protein prothrombin and serum thrombin *(15)*. Although serglycin by itself does not enhance proplatelet formation, it has been shown to be able to complex with thrombin to prevent thrombin from exerting its inhibitory effects on megakaryocyte-proplatelet formation. It has been demonstrated that at least a 100-fold excess of serglycin is necessary to complex all thrombin molecules fully *(15)*. When administered in vivo in combination with a suboptimal dose of IL-6, serglycin mildly, but specifically increased the circulating platelet number to approximately 1400×10^9/L, but not when IL-6 was administered alone *(15)*. In the plasma of rabbits rendered thrombocytopenic by antiplatelet serum administration, the level of serglycin increased 28 times above normal level *(15)*. This may in part explain the description of proplatelet activity in thrombocytopenic plasma *(10)*.

Heparin and heparan-sulfate proteoglycans also permit proplatelet formation in the presence of thrombin, probably because of their direct inhibitory effects on thrombin activity and the inhibitory roles in the conversion of prothrombin to thrombin *(15)*. The mechanisms of thrombin inhibition by serglycin and heparin that allow proplatelet formation are different *(15)*. Heparin has also been shown to act as a cofactor for antithrombin III, preventing thrombin from inhibiting proplatelet formation by both guinea pig and human megakaryocytes *(37,39)*.

3.2. Plasma Proteins

Thrombin is a well-studied platelet agonist found in serum and has been shown to induce megakaryocyte proplatelet retraction *(18)*. Since the first demonstration of this phenomenon, it has been shown that the inhibitory action of serum on proplatelet formation is primarily owing to the presence of thrombin or its precursor molecule, prothrombin *(14)*. The addition of 0.05% human serum was shown to induce statistically significant inhibition of proplatelet formation using guinea pig bone marrow-derived megakaryocytes incubated under serum-free conditions. Megakaryocytes in culture are capable of converting prothrombin to thrombin, the active form responsible for inhibiting proplatelet formation *(14)*. The presence of thrombin receptors on megakaryocytes was demonstrated by specific binding of ^{125}I-thrombin *(37)* (Fig. 5). Thrombin has been shown to cause retraction of proplatelets as well as prevention of proplatelet formation by mouse *(18)*, guinea pig *(14,15,39)*, and human *(37)* megakaryocytes. This observation may explain the reduced platelet counts in postsurgical patients or in patients with disseminated intravascular coagulation (DIC) whose thrombin levels are elevated (Kuter D, personal communication). As yet, there is no direct evidence to support this hypothesis. Thrombin is one of the most potent inhibitors of proplatelet formation identified to date, and its mechanism of action on proplatelets may be similar to its inhibitory action on neurites, possibly by altering the intracellular calcium level or by altering other second messenger systems that result in microtubule dissolution *(14,18)*.

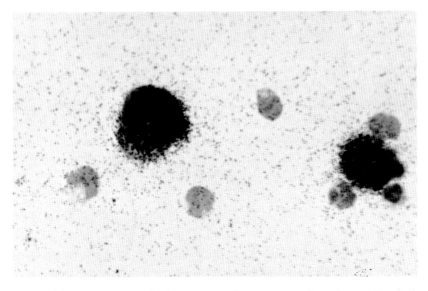

Fig. 5. Thrombin receptors on the human megakaryocyte surface. Autoradiography using ^{125}I-thrombin on megakaryocytes generated in culture. Other cell types (smaller in size) in the same field do not bind ^{125}I-thrombin (bar = 20 μm). (Reprinted with permission from ref. *37.*)

3.3. Cytoskeleton-Disrupting Agents

Proplatelets contain both microtubules and actin filaments *(23,27)*, and many drugs that disrupt the dynamic equilibrium of the cytoskeleton have been shown to modulate proplatelet formation significantly. Colchicine, an alkaloid of *Colchicum autumnale* (meadow saffron), binds to tubulin and prevents its polymerization into microtubules, and can completely inhibit proplatelet formation at a concentration of 0.1 mg/mL *(12)*. Vincristine, an alkaloid of *Vinca rosea* (periwinkle plant), causes rapid disassembly of the mitotic spindle. When added to megakaryocyte cultures, vincristine at a concentration of 0.2 μg/mL causes a significant retraction of proplatelets and at 20 μg/mL completely prevents formation of proplatelets *(7)*. Paclitaxel, extracted from the yew (*Taxus*), binds to microtubules, prevents normal depolymerization, and results in abnormally long microtubules. Paclitaxel at a concentration of 10 μ*M* added to megakaryocyte cultures results in long and thick processes with significantly reduced beading *(11)*. Nocodazole is another microtubule-disrupting agent that prevents vesicular transport. Nocodazole, at 1 μg/mL, causes retraction of proplatelets *(11)*.

Cytochalasins are fungal metabolites that prevent actin from polymerizing by binding to the plus end of actin filaments. Cytochalasin B, at 2 μg/mL, has a mild inhibitory effect on proplatelet formation from rat megakaryocytes *(12)*. However, it has been reported that cytochalasin B (10 μg/mL) and cytochalasin D (2 μg/mL) enhanced proplatelet formation from guinea pig bone marrow-derived megakaryocytes *(10)*. Another study reported that the addition of cytochalasin D (at 1 μg/mL) to guinea pig megakaryocytes plated on collagen gels accelerated the formation of abnormal proplatelets that largely lacked a beaded appearance *(11)*. Taken together, these studies suggest a close interrelationship between microtubules and actin filaments in the formation of nascent platelets.

Retraction of proplatelets is observed when the culture dishes are cooled from 37°C. Although not directly proven, this retraction is most likely due to the instability of microtubules at lower temperatures *(14)*.

3.4. Cytokines and Growth Factors

In the last few years, a number of laboratories have reported on the proplatelet-enhancing activities of cytokines, including IL-3, IL-6, EPO, and various combinations of these factors. In 1991 it was reported that IL-6 at a concentration of 10 ng/mL added to megakaryocyte cultures plated on CN I under serum-free conditions appeared to enhance proplatelet formation *(40)*. Later it was reported that IL-6 acted synergistically with either EPO or IL-3 to enhance proplatelet formation by rat megakaryocytes *(34)*, and similar results were published for IL-3 and EPO using mouse bone marrow-derived megakaryocytes *(20)*. However, it has also been reported that in a serum-free proplatelet culture system using guinea pig megakaryocytes, IL-3, IL-6, EPO, granulocyte-macrophage colony-stimulating factor (GM-CSF), platelet factor (PF)-4, and transforming growth factor (TGF)-β did not have any significant effects, positive or negative, on proplatelet formation *(14)*. It was first noted by this laboratory that factors expressed in mammalian cell lines, especially Chinese Hamster Ovary (CHO) cells, had a nonspecific proplatelet-promoting activity, possibly owing to the high proteoglycan content in the conditioned medium *(14)*. Because of this observation, extreme measures were taken to use only the highest purity of cytokines in the assays. Recently, IL-6 at concentrations as great as 50 ng/mL, IL-11 at concentrations as great as 50 ng/mL, and EPO at concentrations as great as 20 U/mL were used in a human proplatelet assay set up to avoid proteoglycan contamination. In this case, IL-6 and IL-11 were purified from *Escherichia coli* cultures, and EPO was expressed in CHO cells, but was highly purified (clinical grade). In these studies, there were no significant effects of these cytokines, positive or negative, on proplatelet formation *(41)*.

3.5. Thrombopoietins

With the discovery of the Mpl ligand, TPO, it seemed likely that this cytokine would have an effect on proplatelet formation. Megakaryocyte growth and development factor (MGDF) is a recombinant-derived, truncated, and pegylated version of Mpl ligand. When added to mature human megakaryocytes and observed for proplatelet formation, MGDF has a concentration-dependent inhibitory effect on proplatelet formation *(41)*. The maximum inhibition of proplatelet formation is achieved with an MGDF concentration of 50–100 ng/mL. This unexpected result is specific, since the addition of either s-Mpl (the soluble form of the TPO receptor) *(41)* or anti-TPO antibody reversed the inhibition (unpublished observation). Recombinant HuMGDF-mediated inhibition of proplatelet formation is also observed using guinea pig bone marrow-derived megakaryocytes, demonstrating that this is not a species-specific phenomenon (unpublished observations). Endogenous TPO purified from thrombocytopenic canine plasma also induces inhibition of proplatelet formation, indicating that it is not a recombinant molecule-related phenomenon (unpublished observations).

TPO is structurally similar to EPO, the primary cytokine for erythrocyte growth and development. EPO and TPO have also been compared with respect to their roles in growth and differentiation of erythrocytes and megakaryocytes, respectively. With re-

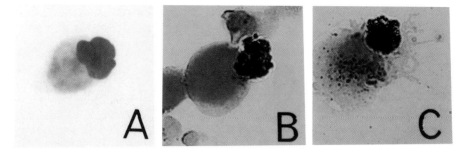

Fig. 6. Mature human megakaryocytes undergo apoptosis. Human megakaryocytes were generated from CD34⁺ cells in liquid culture, in the presence of 10 ng/mL MGDF and 10% normal human platelet-poor heparinized plasma *(26)*. At day 8 of culture, all cells were collected, and mature megakaryocytes were enriched on a BSA step gradient. The selected mature megakaryocytes were replated in fresh culture medium in 10% human plasma only, without MGDF, to induce maximal proplatelet formation. Proplatelet-displaying cultures were harvested 2 days after the replate and stained for 3'-OH DNA ends using ApopTag-peroxidase (Oncor, Gaithersburg, MD), according to manufacturer's instructions. The dark brown nuclear staining demonstrates megakaryocytes undergoing apoptosis **(B,C)**. These cells were counterstained with methyl green. For morphological analysis, cells from the same culture were stained with modified Wright-Giemsa *(21)* **(A)**.

gard to terminal differentiation, it was reported that erythrocytes lose their dependence on EPO between days 8 and 10 in culture, demonstrating that EPO is not necessary for end-cell production *(42)*. It is clear that TPO is also not necessary for proplatelet and platelet production in vitro and may, in fact, be inhibitory for proplatelet formation *(41)*. Another report indicating that TPO is not essential for the final production of platelets is based on observations from TPO "knock-out" mice: mice heterozygous for the TPO gene have decreased platelet numbers proportional to decreased megakaryocyte numbers, indicating that platelet shedding is TPO independent and is an inherent property of mature megakaryocytes *(43)*.

Although the mechanism of TPO-mediated inhibition of proplatelet formation remains to be elucidated, it has been speculated that the growth factor may have antiapoptotic activity during proplatelet formation (Choi E and Hunt P, unpublished observation). When enriched cultured human megakaryocytes were incubated in high concentrations of MGDF, these megakaryocytes formed significantly fewer proplatelets compared with megakaryocytes incubated without MGDF. These megakaryocytes continued to grow larger in culture, maintaining their cell morphology over several days (Choi E and Hunt P, unpublished observation). In this case, MGDF may be acting as an antiapoptotic factor, preventing megakaryocytes from proplatelets and consequently, platelets. Human megakaryocytes matured in culture from progenitor cells and replated without MGDF displayed apoptotic nuclei when stained with ApopTag (Oncor Inc., Gaithersburg, MD), and a maximum average of 4.6 ± 0.8% apoptotic megakaryocytes of total cells (that were 86 ± 5% GPIb- and GPIIb-positive) were detected between 11 and 14 days of total culture time (Chen J-L and Choi E, unpublished observation). Most of the apoptotic nuclei were found in mature megakaryocytes, as characterized by Wright-Giemsa staining (Fig. 6). More than 10 years ago, it was reported that in normal mouse bone marrow, megakaryocytes undergoing apoptosis composed approximately 0.3% of recognizable megakaryocytes *(44)*. The effect of MGDF on megakaryocyte apoptosis and the relationship of this phenomenon to proplatelet formation is currently under study.

In summary, the mechanism of platelet release from megakaryocytes remains a poorly understood area of megakaryocyte biology. It probably involves some aspects of the platelet field theory as well as the proplatelet theory. Currently, it is reasonable to speculate that platelet-specific granules cluster within the megkaryocyte cytoplasm, that the massive cytoskeletal rearrangement organizes proplatelet formation, that the DMS serves as a membrane reservoir for proplatelet extensions, and that these extensions migrate through the endothelial vessel wall reaching the blood flow where actual breaking of the platelet-sized beads of proplatelets occurs. However, the precise mechanism of platelet release from megakarycytes still remains to be determined.

Much recent work in vitro has demonstrated that morphologically and functionally normal platelets can be made in vitro. A practical outcome of this work is that larger-scale bioreactors might be able to produce clinically significant amounts of platelets for transfusion. Furthermore, in vitro and in vivo analyses of platelet formation from megakaryocytes have documented that this process is not promoted by TPO and, indeed, high levels of TPO are inhibitory to the production of platelets from megakaryocytes. How platelets actually form from megakaryocytes remains unknown, but the biological systems described herein may lead to a greater understanding of this process.

Acknowledgment

I thank Jen-Li Chen for outstanding technical assistance and her absolute devotion to this project.

References

1. Wright JH. The origin and nature of blood plates. *Boston Med Surg J*. 1906; 154: 643–645.
2. Ihzumi T, Hattori A, Sanada M, Muto M. Megakaryocyte and platelet formation: a scanning electron microscope study in mouse spleen. *Arch Histol Jpn*. 1977; 40; 305–320.
3. Djaldetti M, Fishman P, Bessler H, Notti I. SEM observations on the mechanism of platelet release from megakaryocytes. In: Gastpar H, Kuhn K, Marx R (eds). *Collagen-Platelet Interaction*. Stuttgart, Germany; FK Schattauer Verlag GmbH; 1978: 611–620.
4. Yamada F. The fine structure of the megakaryocyte in the mouse spleen. *Acta Anat*. 1957; 29: 267–290.
5. Albrecht M. Studien zur thrombocytenbildung und megakaryocyten in menschlichen knochenmarkskulturen. *Acta Haematol*. 1957; 17: 160–168.
6. Kosaki G, Inosita K, Okuma H. Human thrombocytogenesis. In: Didisheim P (ed). *Platelets, Thrombosis and Inhibitors*. Stuttgart-New York: FK Shattauer, 1974: 97.
7. Radley JM, Haller CJ. The demarcation membrane system of the megakaryocyte: a misnomer? *Blood*. 1982; 60: 213–219.
8. White JG: Delivery of platelets from megakaryocytes. In: *Molecular Biology and Differentiation of Megakaryocytes*. New York: Wiley-Liss; 1990: 25–39.
9. Becker RP, De Bruyn PP. The transmural passage of blood cells into myeloid sinusoids and the entry of platelets into the sinusoidal circulation; a scanning electron microscopic investigation. *Am J Anat*. 1976: 145; 183–205.
10. Leven RM, Yee MK. Megakaryocyte morphogenesis stimulated in vitro by whole and partially fractionated thrombocytopenic plasma: a model system for the study of platelet formation. *Blood*. 1987; 69: 1046–1052.

11. Tablin F, Castro M, Leven RM. Blood platelet formation in vitro. The role of the cytoskeleton in megakaryocyte fragmentation. *J Cell Sci.* 1990; 97: 59–70.

12. Handagama PJ, Feldman BF, Jain NC, Farver TB, Kono CS. In vitro platelet release by rat megakaryocytes: effect of metabolic inhibitors and cytoskeletal disrupting agents. *Am J Vet Res.* 1987; 48: 1142–1146.

13. Smith CM II, Burris SM, White JG. High frequency of elongated platelet forms in guinea pig blood: ultrastructure and resistance to micropipette aspiration. *J Lab Clin Med.* 1990; 115: 729–737.

14. Hunt P, Hokom MM, Wiemann B, Leven RM, Arakawa T. Megakaryocyte proplatelet-like process formation in vitro is inhibited by serum prothrombin, a process which is blocked by matrix-bound glycosaminoglycans. *Exp Hematol.* 1993; 21: 372–381.

15. Hunt P, Hokom MM, Hornkohl A, Wiemann B, Rohde M, Arakawa T. The effect of the platelet-derived glycosaminoglycan serglycin on in vitro proplatelet-like process formation. *Exp Hematol.* 1993; 21: 1295–1304.

16. Caine YG, Vlodavsky I, Hersh M, et al. Adhesion, spreading and fragmentation of human megakaryocytes exposed to subendothelial extracellular matrix: a scanning electron microscopy study. *Scan Electron Microsc.* 1986; 111: 1087–1094.

17. Leven RM, Tablin F. Extracellular matrix stimulation of guinea pig megakaryocyte proplatelet formation in vitro is mediated through the vitronectin receptor. *Exp Hematol.* 1992; 20: 1316–1322.

18. Radley JM, Hartshorn MA, Green SL. The response of megakaryocytes with processes to thrombin. *Thromb Haemost.* 1987; 58: 732–736.

19. Radley JM, Rogerson J, Ellis SL, Hasthorpe S. Megakaryocyte maturation in long-term marrow culture. *Exp Hematol.* 1991; 19: 1075–1078.

20. An E, Ogata K, Kuriya S, Nomura T. Interleukin-6 and erythropoietin act as direct potentiators and inducers of in vitro cytoplasmic process formation on purified mouse megakaryocytes. *Exp Hematol.* 1994; 22: 149–156.

21. Choi ES, Nichol JL, Hokom MM, Hornkohl AC, Hunt P. Platelets generated in vitro from proplatelet-displaying human megakaryocytes are functional. *Blood.* 1995; 85: 402–413.

22. Behnke O. An electron microscope study of the rat megakaryocyte. II. Some aspects of platelet release and microtubules. *J Ultrastruct Res.* 1969; 26: 111–129.

23. Scurfield G, Radley JM. Aspects of platelet formation and release. *Am J Hematol.* 1981; 10: 285–296.

24. Lichtman MA, Chamberlain JK, Simon W, Santillo PA. Parasinusoidal location of megakaryocytes in marrow: a determinant of platelet release. *Am J Hematol.* 1978; 4: 303–312.

25. Tavassoli M, Aoki M. Migration of entire megakaryocytes through the marrow-blood barrier. *Br J Haematol.* 1981; 48: 25–29.

26. Choi ES, Hokom M, Bartley T, et al. Recombinant human megakaryocyte growth and development factor (rHuMGDF), a ligand for c-Mpl, produces functional human platelets in vitro. *Stem Cells.* 1995; 13: 317–322.

27. Radley JM, Scurfield G. The mechanism of platelet release. *Blood.* 1980; 56: 996–999.

28. Radley JM. Ultrastructural aspects of platelet production. *Prog Clin Biol Res.* 1986; 215: 387–398.

29. Topp KS, Tablin F, Levin J. Culture of isolated bovine megakaryocytes on reconstituted basement membrane matrix leads to proplatelet process formation. *Blood.* 1990; 76: 912–924.

30. Leven RM. Megakaryocyte motility and platelet formation. *Scanning Microsc.* 1987; 1: 1701–1709.

31. Radley JM, Hartshorn MA. Megakaryocyte fragments and the microtubule coil. *Blood Cells.* 1987; 12: 603–614.

32. Bessis M. The thrombocytic series. In: Masson and Cie (eds). *Living Blood Cells and Their Ultrastructure.* New York: Springer-Verlag; 1973: 367–411.

33. Trowbridge EA, Martin JF, Slater DN. Evidence for a theory of physical fragmentation of megakaryocytes implying that all platelets are produced in the pulmonary circulation. *Thromb Res.* 1982; 28: 461–475.

34. Inoue H, Ishii H, Tsutsumi M, et al. Growth factor-induced process formation of megakaryocytes derived from CFU-MK. *Br J Haematol.* 1993; 85: 260–269.

35. Miyazaki H, Inoue H, Yanagida M, et al. Purification of rat megakaryocyte colony-forming cells using a monoclonal antibody against rat platelet glycoprotein IIb/IIIa. *Exp Hematol.* 1992; 20: 855–861.

36. Levine RF, Fedorko ME. Isolation of intact megakaryocytes from guinea pig femoral marrow. Successful harvest made possible with inhibitors of platelet aggregation: enrichment achieved with a two-step separation technique. *J Cell Biol.* 1976; 69: 159–172.

37. Choi ES, Hokom MM, Nichol JL, Hornkohl A, Hunt P. Functional human platelet generation in vitro and regulation of cytoplasmic process formation. *CRC Sci.* 1995; 318: 387–393.

38. Preissner KT, Wassmuth R, Muller-Berghaus G. Physicochemical characterization of human S-protein and its function in the blood coagulation system. *Biochem. J.* 1995; 231: 349–355.

39. Hokom M, Choi E, Nichol J, Hornkohl A, Arakawa T, Hunt P. Regulation of proplatelet formation in guinea pig and human megakaryocytes. In: Abraham NG, Shadduck RK, Levine AS, Takaku F (eds). *Molecular Biology of Haematopoiesis*, vol. 3. Andover: Intercept; 1993: 15–30.

40. Leven RM, Rodriguez A. Immunomagnetic bead isolation of megakaryocytes from guinea-pig bone marrow: effect of recombinant interleukin-6 on size, ploidy and cytoplasmic fragmentation. *Br J Haematol.* 1991; 77: 267–273.

41. Choi ES, Hokom MM, Chen J-L, et al. The role of megakaryocyte growth and development factor (MGDF) in terminal stages of thrombopoiesis. *Blood.* 1995; 86: 285a (abstract no 1125).

42. Wickrema A, Krantz SB, Winkelmann JC, Bondurant MC. Differentiation and erythropoietin receptor gene expression in human erythroid progenitor cells. *Blood.* 1992; 80: 1940–1949.

43. de Sauvage FJ, Luoh S-M, Carver-Moore K, et al. Deficiencies in early and late stages of megakaryocytopoiesis in TPO-KO mice. *Blood.* 1995; 86: 255a (abstract no 1007).

44. Radley JM, Haller CJ. Fate of senescent megakaryocytes in the bone marrow. *Br J Haematol.* 1983; 53: 277–287.

18

In Vitro Effects of Mpl Ligands on Platelet Function

Laurence A. Harker, Ulla M. Marzec, and Christopher F. Toombs

1. Introduction

Mpl ligands amplify platelet production by inducing dose-dependent megakaryocyte development from early marrow Mpl^+ hemopoietic progenitors and stimulating subsequent megakaryocyte proliferation and endoreduplication *(1–9)*. Mpl ligands are megakaryocyte-lineage-dominant, as illustrated by the complete dependence of hemopoietic stem cells on Mpl ligand for megakaryocyte development in vitro *(9)*. Pegylated megakaryocyte growth and development factor (PEG-rHuMGDF) produces log-linear dose–response effects in baboons with respect to peak peripheral platelet counts; platelet mass turnover; and marrow megakaryocyte volume, ploidy, number, and mass *(10–13)*. Blood levels of the endogenous Mpl ligand, thrombopoietin (TPO), sustain constant peripheral platelet concentrations by regulating megakaryocytopoiesis to compensate for changing peripheral requirements *(10,11,14–20)*. For example, during thrombocytopenia in patients receiving marrow-ablative chemotherapy, endogenous TPO levels increase by several orders of magnitude, thereby stimulating megakaryocytopoiesis, and TPO levels normalize as the peripheral platelet counts return to baseline (owing to platelet transfusional therapy or hemopoietic recovery) *(21,22)*. Plasma TPO levels do not appear to be physiologically modulated by gene transcription *(23,24)*, as evidenced by the fact that TPO mRNA levels in heterozygote knockout mice remain half that of the normal mice, despite thrombocytopenia *(23)*. One important regulatory feedback mechanism determining TPO levels in blood involves competitive binding of unbound TPO with Mpl on platelets and megakaryocytes followed by internalization and degradation *(8,25–27)*. Thus, the plasma levels of unbound TPO are reciprocally related to the peripheral platelet mass and perhaps megakaryocyte mass *(5,6,8,22,23,28,29)* (*see* Chapter 23).

Mpl ligand triggers intracellular signaling pathways in hemopoietic progenitors that mediate megakaryocyte development, proliferation, and differentiation, including tyrosine phosphorylation of membrane-associated kinases and adapter molecules, i.e., the Shc-Ras-raf1-MEK-ERK pathway, the JAK2-STAT3 and 5 pathway, PLC-γ, and Mpl itself *(9,30–36)* (*see* Chapter 16). Mpl ligand initiates similar signaling in resting platelets (Fig. 1) *(31,32,37–41)*. In addition to these effects on signaling pathway proteins, Mpl ligand appears to enhance platelet aggregation induced by sensitizing platelets to

From: *Thrombopoiesis and Thrombopoietins: Molecular, Cellular, Preclinical, and Clinical Biology*
Edited by: D. J. Kuter, P. Hunt, W. Sheridan, and D. Zucker-Franklin Humana Press Inc., Totowa, NJ

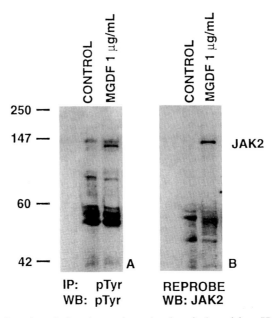

Fig. 1. Tyrosine phosphorylation in resting platelets induced by rHuMGDF. **(A)** Western blot of antiphosphotyrosine immunoprecipitates from human platelets before and after a 5-minute incubation with rHuMGDF at 1 μg/mL. Detection of tyrosine-phosphorylated proteins shows the appearance of a band with apparent molecular weight of 130 kDa that was not present in untreated control platelets. **(B)** A reproduction of the same blot shown in **(A)** that has been stripped and reprobed with an antibody to the JAK2 member of the Janus kinase family of tyrosine kinases, indicating that rHuMGDF produces activation of JAK2 in platelets.

low concentrations of physiologic agonists that induce platelet aggregation in vitro (adenosine diphosphate [ADP], collagen, and thrombin receptor agonist peptide, $TRAP_{1-6}$) *(31,32,38,42–47)*. Mpl ligand-induced enhancement of platelet reactivity could theoretically predispose patients receiving recombinant Mpl ligand therapy to thrombo-occlusive risk, so it is important to evaluate the evidence linking Mpl ligand with thrombus formation. This concern is particularly relevant for older cancer patients receiving chemotherapy since these patients are inherently prone to thrombosis because of the coincident association of malignancy and atherosclerotic disease *(48)*. It is therefore important to determine if Mpl ligand modulates platelet function.

2. In Vitro Platelet Aggregation Studies

In vitro platelet aggregation is studied by recording increases in light transmission through stirred opalescent suspensions of platelet-rich plasma (PRP) maintained at 37°C and prepared within 1 hour of blood drawing. The platelet count in the PRP is standardized, e.g., 300×10^9/L, by adding an appropriate proportion of autologous platelet-poor plasma *(49–51)*. The physiologic agonists inducing platelet aggregation include ADP, collagen, $TRAP_{1-6}$, and epinephrine. These agonists are tested at doses spanning the range of responsiveness. For studies assessing the enhancing effects of Mpl ligand, low doses of agonist have been used in combination with specific Mpl ligands:

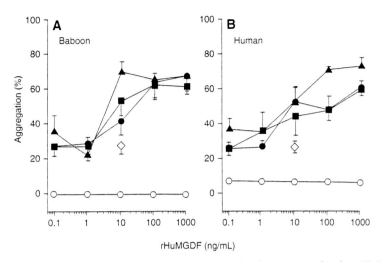

Fig. 2. In vitro enhancement of baboon and human platelet aggregation by rHuMGDF. The addition of rHuMGDF in final concentrations of 0.1–1000 ng/mL (○, 3-minute preincubation without agonist) to PRP fails to induce aggregation of either resting baboon **(A)** or human **(B)** platelets. In contrast, rHuMGDF (final concentration 10–1000 ng/mL) significantly enhances the aggregatory responses of minimally effective concentrations of ADP (●, 3 minute preincubation, 3.9 μM ADP; $p < 0.05$; ■, 30 minute preincubation, 3.9 μM ADP), collagen (1.9 μg/mL for baboon platelets and 0.3 μg/mL for human platelets; ▲, 3 minute preincubation, 1.9 μg/mL/ 0.3 mg/mL; $p < 0.05$). Excess s-Mpl (100 μg/mL) blocks the enhancing effects of rHuMGDF on agonist-induced platelet aggregation (◇, 10-fold excess MPL receptor; 3.9 μM ADP). The error bars depict ±1 SD.

1. rHuTPO, a recombinant full-length glycosylated human Mpl ligand expressed in mammalian cells *(5,6)*;
2. Recombinant human megakaryocyte growth and development factor (rHuMGDF) *(21,22,51,52)*, a nonglycosylated truncated amino-terminal polypeptide of human Mpl ligand encompassing the erythropoietin (EPO)-like domain; and
3. Poly[ethylene glycol]-derivitized rHuMGDF (PEG-rHuMGDF), similar to rHuMGDF in biologic activity, except for its prolonged persistence in plasma resulting from pegylation *(53)*.

Other hemopoietic growth factors evaluated in this assay system include erythropoietin (rHuEPO), granulocyte colony-stimulating factor (rHuG-CSF), stem cell factor (rHuSCF), and interleukins (rHuIL)-3, 6, and 11 *(45,54,55)*.

Although Mpl ligands alone do not induce platelet aggregation in vitro (Figs. 2–4), Mpl ligands significantly enhance platelet aggregatory responses to low-dose ADP, collagen, and TRAP$_{1-6}$ when platelets are prepared from mice, rabbits, dogs, nonhuman primates, and humans *(32,43,44,46–48)*. The effects of adding rHuMGDF to platelets prepared from blood of baboons and humans are illustrated in Fig. 2. No platelet aggregation is induced by rHuMGDF alone for concentrations ranging from 1 ng/mL to 1000 ng/mL ($p < 0.05$). However, in vitro aggregatory responses of baboon platelets to low-dose ADP (3.9 μM) or collagen (1.9 μg/mL) are enhanced twofold by 10 ng/mL rHu-MGDF (Fig. 2A; $p < 0.05$ in all cases), effects that are blocked by 10-fold excess

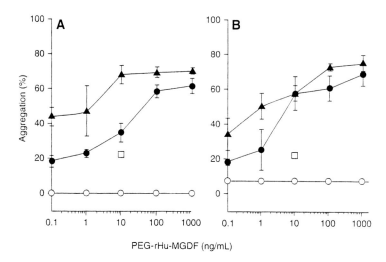

Fig. 3. Effects of PEG-rHuMGDF on normal baboon and human platelet aggregation in vitro. The addition of PEG-rHuMGDF in final concentrations of 0.1–1000 ng/mL (○) to PRP fails to induce aggregation of either resting baboon **(A)** or human **(B)** platelets. In contrast, PEG-rHuMGDF (final concentration 10–1000 ng/mL) significantly enhances the aggregatory responses of low ADP (4.9 μM for baboon and 2.2 μM for human platelets; ●; $p < 0.05$), and collagen (1.5 $\mu g/mL$ for baboon platelets and 0.3 $\mu g/mL$ for human platelets; ▲; $p < 0.05$). Tenfold excess s-Mpl receptor blocks the enhancing effects of PEG-rHuMGDF (□). The error bars depict ± 1 SD.

soluble Mpl (s-Mpl) receptor (Fig. 2A). Similar enhancement of aggregatory responses and blockade by excess s-Mpl receptor are observed for human platelets using ADP and collagen, although concentrations of 0.3 $\mu g/mL$ collagen and 100 ng/mL rHuMGDF are required to achieve maximal enhancing effects (Fig. 2B).

The effects of PEG-rHuMGDF on platelet function in vitro have also been assessed in aggregation studies on resting platelets obtained from normal untreated baboons and humans (Fig. 3). No platelet aggregation is induced in vitro by concentrations of PEG-rHuMGDF ranging from 0.1–1000 ng/mL ($p > 0.05$). However, in vitro aggregatory responses by baboon platelets to low-dose ADP or collagen are enhanced twofold by 10 ng/mL PEG-rHuMGDF (Fig. 3A; $p < 0.05$ in all cases), effects that are blocked by 10-fold excess s-Mpl receptor (Fig. 3A). PEG-rHuMGDF enhances human platelet aggregatory responses in vitro with blocking of enhanced aggregation by excess s-Mpl receptor using ADP or collagen at concentrations of 0.3 $\mu g/mL$ collagen. One hundred nanograms per milliliter PEG-rHuMGDF are required to achieve maximal enhancing effects (Fig. 3B).

Similarly, enhanced aggregatory responses by human platelets stimulated by low-dose ADP or other physiologic platelet agonists are also produced by rHuTPO (Fig. 4). Other hemopoietic cytokines are reported to enhance aggregatory responses initiated by low-dose ADP, including rHuG-CSF *(55)*, rHuEPO *(45,54,55)*, SCF *(54)*, and IL-3, 6, and 11 *(44)*. Augmentation of the platelet aggregation response by rHuG-CSF has been reported to be dependent on the G-CSF receptor, since the enhancement was abolished by monoclonal antibodies (MAb) to rHuG-CSF receptor *(55)*. As shown in

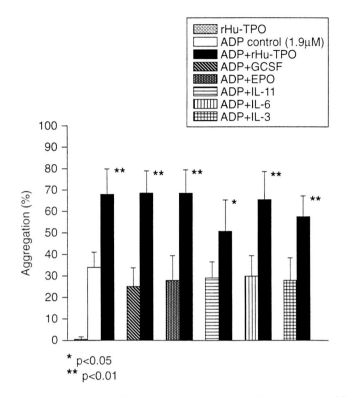

Fig. 4. Effects of rHuTPO, G-CSF, EPO, IL-11, IL-6, and IL-3 on normal human platelet aggregation in vitro. The addition of rHuTPO in final concentrations of 100 ng/mL to PRP does not induce aggregation of resting human platelets (smallest bar at extreme left). However, rHuTPO (100 ng/mL) significantly enhances the aggregate responses of minimally effective concentrations of ADP (1.9 μM; open bar; $p < 0.05$). By contrast, other hemopoietic cytokines do not enhance platelet aggregation induced by low-dose ADP. Recombinant HuTPO also augments aggregation to low-dose ADP in the presence of other hemopoietic cytokines, but without added or synergistic responses.

Fig. 4, the results have been inconsistent when the effects of other hemopoietic cytokines on platelet aggregation are examined.

Epinephrine is a weak platelet agonist that produces a characteristic biphasic aggregation response. The biphasic response is produced by rapid primary wave aggregation (platelet adhesion), and slower secondary wave aggregation that occurs as a result of secretion of platelet granule contents and thromboxane A_2 production. Platelet aggregation to epinephrine is enhanced by rHuMGDF (Fig. 5). However, the nature of the change is not an increase in sensitivity to lower concentrations of epinephrine. Rather, the data indicate that secondary aggregation occurs more rapidly to the extent that the response appears monophasic, similar to the response obtained with collagen or thrombin. These data may suggest that Mpl ligand enhances granule secretion or thromboxane A_2 biosynthesis by unknown means.

The explanation for variable outcomes regarding platelet responsiveness to aggregating agents in the presence of other hemopoietic cytokines is not known.

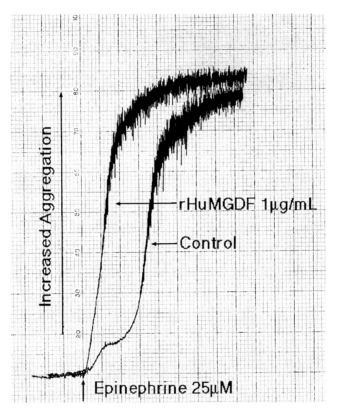

Fig. 5. Recombinant HuMGDF enhances epinephrine-induced aggregation of human PRP. The addition of the weak platelet agonist epinephrine (25 μM) to human PRP (250 × 10^9 platelets/L) results in a biphasic aggregation pattern produced by rapid primary wave aggregation and slow secondary wave aggregation. The addition of rHuMGDF (1 μg/mL) converts the biphasic epinephrine response to a monophasic response.

3. In Vitro Platelet Activation Marker Studies

Platelet activation by physiologic agonists involves membrane expression of functional fibrinogen receptors, secretory granule membrane domains, and phosphatidyl-serine-rich binding sites. The appearance of these activation markers on platelet membrane surfaces may be evaluated by flow cytometry using epitope-specific MAb for conformationally altered glycoprotein (GP)IIb/IIIa ligand-induced binding sites (LIBS) *(56,57)* or the secretory granule membrane, P-selectin (CD62) *(58,59)*. Binding of Annexin V detects membrane phospholipid changes *(60–62)*. Annexin V, a member of the multigene family of calcium-dependent phospholipid binding proteins, exhibits high affinity for phosphatidylserine-rich, negatively charged phospholipid platelet-membrane surfaces that promote assembly of the macromolecular coagulation enzyme complexes *(63–65)*.

No enhanced expression of activation markers on resting platelets has been observed for either rHuMGDF or PEG-rHuMGDF in vitro or ex vivo during administration of rHuMGDF (Tables 1 and 2; $p > 0.1$ in all cases compared with baseline controls) *(13,51)*.

Table 1
Absent Expression of Platelet Activation Markers During Administration of rHuMGDF Dosing of Baboons[a]

Duration of rHuMGDF therapy	Binding Sites/Platelet, $\times 10^3$					
	LIBS		P-selectin		Annexin V	
	Resting	Activated	Resting	Activated	Resting	Activated
Baseline	2.19 ± 0.06	20.0 ± 0.6	0.28 ± 0.06	2.60 ± 0.05	0.96 ± 0.42	130 ± 67
7 days	1.04 ± 0.32	—	0.26 ± 0.09	—	0.88 ± 0.17	—
14 days	0.8 ± 0.37	15.7 ± 0.37	0.24 ± 0.02	4.40 ± 0.01	$0.95 - 0.27$	398 ± 81
28 days	0.9 ± 0.17	—	0.38 ± 0.12	—	0.94 ± 0.22	—

[a]rHuMGDF = recombinant human megakaryocyte growth and development factor; LIBS = ligand-induced binding site.

Recombinant HuMGDF was administered to baboons by sc injection. PRP was prepared for flow cytometry analysis on the indicated day.

Table 2
The Effects of PEG-rHuMGDF on In Vivo and Ex Vivo Platelet Function in Baboons[a]

	Baseline	5 µg/kg/day		
		Day 3	Day 7	Day 28
Bleeding time (min)	3.5 ± 1.0	2.5 ± 1.4^b	2.8 ± 1.5	3.5 ± 0.7
Platelet aggregation AC$_{50}$				
ADP (µM)	4.3 ± 1.25	4.2 ± 1.2	3.1 ± 0.9	4.0 ± 0.8
Collagen (µg/mL)	2.83 ± 1.88	1.1 ± 0.9^b	4.1 ± 3.2	4.6 ± 2.7
TRAP$_{1-5}$ (µM)	120 ± 83	17 ± 15^c	100 ± 47	88 ± 61
Expression of activation markers				
LIBS (binding sites)	2204 ± 655	2640 ± 730	1120 ± 160^c	1000 ± 390
LIBS-activated (binding sites)	$22,500 \pm 4750$			
P-selectin (binding sites)	250 ± 48	235 ± 50	260 ± 60	280 ± 40
P-selectin activated (binding sites)	2735 ± 1080			
Annexin V (binding sites)	1500 ± 1260	2034 ± 1805	5060 ± 4500	1035 ± 320
Annexin V activated (binding sites)	$332,000 \pm 125,000$			

ADP = adenosine diphosphate; LIBS = ligand-induced binding site.

[a]Pegylated rHuMGDF was administered to baboons by daily subcutaneous injection. Platelet-rich plasma (PRP) was prepared. Bleeding time was measured using standard techniques.
[b]$p < 0.01$.
[c]$p < 0.001$.

In contrast, full-length glycosylated TPO induces resting platelets to express ligand-induced binding site (LIBS) and P-selectin, but not to bind with Annexin V (Fig. 5, Table 3). The capacity of rHuTPO to induce resting platelets to express functional fibrinogen receptors may represent the mechanism whereby it augments platelet re-

Table 3
Platelet Activation with Cytokines (%)

	LIBS, 7%			P-selectin, 3%			Annexin V, 4%		
	Mean	SD	p	Mean	SD	p	Mean	SD	p
Baseline	10.0	7.6	—	4.0	1.0	—	4.4	1.4	—
Maximal activation	90.1	4.9	0.0001	83.3	3.2	0.0001	93.1	3.2	0.0001
TPO (100 ng/mL)	33.6	10.2	0.0004	9.5	2.5	0.0200	7.7	5.7	NS
EPO (U/mL)	9.5	6.4	NS	4.3	1.5	NS	4.5	2.0	NS
TPO/EPO	39.3	10.5	0.0001	10.6	2.6	0.0150	5.6	1.5	NS
G-CSF (100 ng/mL)	11.5	8.7	NS	4.6	—	—	5.5	—	—
G-CSF/TPO	35.7	13.6	0.0018	11.9	—	—	7.2	—	—

EPO = erythropoietin; G-CSF = granulocyte colony-stimulating factor; TPO = full-length, glycosylated human thrombopoietin; NS = not significant

sponsiveness to physiologic agonists. These findings are concordant with the recent report by Wun et al. *(44)* that rHuTPO (500 U/mL) induces the expression of P-selectin (CD62) by human platelets when concurrently incubated with rHuEPO (2 U/mL), rHuSCF (100 ng/mL), rHuIL-3 (100 ng/mL), rHuG-CSF (100 ng/mL), or rHuIL-11 (100 ng/mL) ($p < 0.05$ in all cases). This enhanced expression of activation epitopes is dependent on Mpl, as shown by the reduction in P-selectin expression when the platelets are preincubated with s-Mpl (2 µg/mL).

4. Mechanism(s) for Enhancement of Platelet Function In Vitro

Hemopoietic progenitors express receptors transmitting intracellular signaling induced by TPO, EPO, rHu-CSF, rHuGM-CSF, or IL-1, 3, 4, 5, 6, 7, 9, and 11) *(66–69)*. The typical cytokine receptor comprises two polypeptide chains exhibiting ligand-specific and signal-transducer domains that undergo dimerization after ligand binding (Mpl, EPO receptors, and G-CSF receptors undergo homodimerization), thereby giving rise to conformational alterations in the intracellular domains, and consequent activation of signaling pathways *(9,69,70–72)*. Although Mpl exhibits considerable homology with EPO receptor, megakaryocyte development in vitro is absolutely dependent on Mpl ligand *(9)*. The intracellular signaling induced in Mpl+ progenitors by Mpl ligand mediates proliferation and differentiation with suppression of apoptosis *(9,30,31)*. These signaling responses involve tyrosine phosphorylation of membrane-associated kinases and adapter molecules *(9,30–36)* (*see* Chapter 16).

In vitro, Mpl ligand initiates intracellular signaling in resting platelets similar to that produced in megakaryocyte progenitors and hemopoietic cell lines *(31,32,37–40)*. Consequently, Mpl-induced signaling responses in resting platelets may represent, in part, residual signaling events related to remnant proliferation and differentiation pathways in platelets, the cytoplasmic products of marrow megakaryocytes. Thus, the ability of Mpl ligand to initiate residual proliferative intracellular signaling, i.e., JAK2-STAT3 and 5 pathway, and the Sho-Ras-rafl-MAPK pathway, do not necessarily link Mpl ligand to activation signaling of platelet hemostatic functions. The mechanism(s) that enhance aggregatory responses may include nonspecific interreceptor facilitation pathways shared by Mpl-dependent proliferative signaling and agonist receptor-dependent functional signaling capable of inducing the expression of fibrinogen receptors by resting platelets (Table 3).

Other hemopoietic cytokines, including rHuG-CSF *(55)* and rHuSCF *(54)*, are reported to enhance platelet aggregation in a fashion similar to Mpl ligand, in that they initiate intracellular signaling related to proliferation and differentiation, but do not induce platelet aggregation in the absence of physiologic agonists *(73)*. These reports of enhanced agonist-induced platelet aggregation by other hemopoietic cytokines are consistent with the more general possibility that function-related signaling in platelets may be nonspecifically enhanced by some signaling sequences shared with cytokine-dependent intracellular proliferative signaling. The reported inconsistencies regarding the effects of other cytokines on the enhancement of platelet aggregation will require additional investigation involving well-characterized reagents, methodology, and analysis. For example, it is now evident that the specific Mpl ligand used in such experiments may significantly alter the results.

In vivo, the effects of Mpl ligand on platelet recruitment in the formation of thrombi has been evaluated using two clinically appropriate experimental animal models and species: cyclic flow reductions in stenotic arteries of rabbits *(74,75)*; and the accumulation of ^{111}In-platelets and ^{125}I-fibrin in thrombi forming on thrombogenic segments interposed in arteriovenous shunts in baboons, for fixed periods under standard flow conditions without systemic anticoagulation *(76,77)*. There is no direct evidence in either of these models for enhanced platelet deposition onto thrombogenic surfaces independent of the elevated peripheral platelet count *per se* (*see* Chapter 19).

In summary, in vitro rHuTPO induces signaling pathways in resting platelets that are clearly related to cellular proliferation and differentiation events in megakaryocyte precursors and megakaryocytes. This initiates the functional expression of fibrinogen receptors, giving rise to enhanced platelet aggregation responses induced by low concentrations of physiologic agonists. It remains to be established whether this enhancement of aggregation in vitro represents nonspecific sharing of signaling pathways between proliferation/differention and functional processes, or represents the induction of function-specific signaling. It also remains unresolved whether the in vitro effects of Mpl ligand on fibrinogen receptor expression and platelet aggregation are specific for rHuTPO or represent a property shared by other Mpl ligands and other hemopoietic cytokines. In any event, the present evidence indicates that these laboratory phenomena do not contribute significantly to thrombogenesis in quantitative experimental animal models of thrombosis *(13,66,76)*.

References

1. Gewirtz AM. Human megakaryocytopoiesis. *Semin Hematol.* 1986; 23: 27–42.
2. Greenberg SM, Kuter DJ, Rosenberg RD. In vitro stimulation of megakaryocyte maturation by megakaryocyte stimulatory factor. *J Biol Chem.* 1987; 262: 3269–3277.
3. Vainchenker W, Kieffer N. Human megakaryocytopoiesis: In vitro regulation and characterization of megakaryocytic precursor cells by differentiation markers. *Blood Rev.* 1988; 2: 102–107.
4. Lok S, Kaushansky K, Holly RD, et al. Cloning and expression of murine thrombopoietin cDNA and stimulation of platelet production in vivo. *Nature.* 1994; 369: 565–568.
5. Kaushansky K, Lok S, Holly RD, et al. Promotion of megakaryocyte progenitor expansion and differentiation by the *c-Mpl* ligand thrombopoietin. *Nature.* 1994; 369: 568–571.

6. deSauvage FJ, Hass PE, Spencer SD, et al. Stimulation of megakaryocytopoiesis and thrombopoiesis by the *c-Mpl* ligand. *Nature.* 1994; 369: 533–538.

7. Wendling F, Maraskovsky E, Debili N, et al. *c-Mpl* ligand is a humoral regulator of megakaryocytopoiesis. *Nature.* 1994; 369: 571–574.

8. Kuter DJ, Beeler DL, Rosenberg RD. The purification of megapoietin: A physiological regulator of megakaryocyte growth and platelet production. *Proc Natl Acad Sci USA.* 1994; 91: 11,104–11,108.

9. Kaushansky K. Thrombopoietin: the primary regulator of platelet production. *Blood.* 1995; 86: 419–431.

10. Paulus JM. *Platelet Kinetics: Radioisotopic, Cytological, Mathematical and Clinical Aspects.* Amsterdam: North-Holland; 1971.

11. Harker LA. Kinetics of thrombopoiesis. *J Clin Invest.* 1968; 47: 458–465.

12. Tomer A, Harker LA, Burstein SA. Flow cytometric analysis of normal human megakaryocytes. *Blood.* 1988; 71: 1244–1252.

13. Harker LA, Marzec UM, Hunt P, et al. Dose–response effects of pegylated human megakaryocyte growth and development factor (PEG-rHuMGDF) on platelet production and function in nonhuman primates. *Blood.* 1996; 88: 511–521.

14. Odell TT, McDonald TP, Detwiler TC. Stimulation of platelet production by serum of platelet-depleted rats. *Proc Soc Exp Biol Med.* 1961; 108: 428–431.

15. Evatt BL, Shreiner DP, Levin J. Thrombopoietic activity of fractions of rabbit plasma: studies in rabbits and mice. *J Lab Clin Med.* 1974; 83: 364–371.

16. Mazur EM, South K. Human megakaryocyte colony-stimulating factor in sera from aplastic dogs: partial purification, characterization and determination of hematopoietic cell lineage specificity. *Exp Hematol.* 1985; 13: 1164–1172.

17. Tayrien G, Rosenberg RD. Purification and properties of a megakaryocyte stimulatory factor present both in the serum-free conditioned medium of human embryonic kidney cells and in thrombocytopenic plasma. *J Biol Chem.* 1987; 262: 3262–3268.

18. Harker L. Regulation of thrombopoiesis. *Am J Physiol.* 1970; 218: 1376–1380.

19. Breton-Gorius J, Vainchenker W. Expression of platelet proteins during the in vitro and in vivo differretiation of megakaryocytes and morphological aspects of their maturation. *Semin Hematol.* 1986; 23: 43–67.

20. Choi ES, Hokom M, Bartley T, et al. Recombinant human megakaryocyte growth and development factor (rHuMGDF), a ligand for *c-Mpl*, produces functional human platelets in vitro. *Stem Cell.* 1995; 13: 317–322.

21. Bartley TD, Bogenberger J, Hunt P, et al. Identification and cloning of a megakaryocyte growth and development factor that is a ligand for the cytokine receptor Mpl. Cell. 1994; 77: 1117–1124.

22. Nichol JL, Hokom MM, Hornkohl A, et al. Megakaryocyte growth and development factor. Analyses of in vitro effects on human megakaryopoiesis and endogenous serum levels during chemotherapy-induced thrombocytopenia. *J Clin Invest.* 1995; 95: 2973–2978.

23. de Sauvage FJ, Luoh S-M, Carver-Moore K, et al. Deficiencies in early and late stages of megakaryocytopoiesis in TPO-KO mice. *Blood.* 1995; 86: 255a (abstract no 1007).

24. Stoffel R, Wiestner A, Skoda RC. Thrombopoietin in thrombocytopenic mice: evidence against regulation at the mRNA level and for a direct regulatory role of platelets. *Blood.* 1996; 87: 567–573.

25. Kuter DJ, Rosenberg RD. Appearance of a megakaryocyte growth-promoting activity, megapoietin, during acute thrombocytopenia in the rabbit. *Blood.* 1994; 84: 1464–1472.

26. Fielder PJ, Hass P, Nagle M, et al. Human platelets as a model for the binding, internalization, and degradation of thrombopoietin (TPO). *Blood.* 1995; 86: 365a (abstract no 1450).

27. Shieh J-H, Chen Y-F, Molineux G, McNiece IK. MGDF binding to platelets: upmodulation by MGDF in vitro. *Blood.* 1995; 86: 369a (abstract no 1465).

28. Gurney AL, Carver-Moore K, de Sauvage FJ, Moore MW. Thrombocytopenia in c-mpl-deficient mice. *Science.* 1994; 265: 1445–1447.

29. Chang M, Suen Y, Meng G, et al. Regulation of TPO mRNA expression and protein production: TPO gene regulation appears post transcriptional, and endogenous levels are inversely correlated to megakaryocyte mass and circulating platelet count. *Blood.* 1995; 368a (abstract no 1460).

30. Morella KK, Bruno E, Kumaki S, et al. Signal transduction by the receptors for thrombopoietin (*c-mpl*) and interleukin-3 in hematopoietic and nonhematopoietic cells. *Blood.* 1995; 86: 557–571.

31. Drachman JG, Griffin JD, Kaushansky K. The c-mpl ligand (thrombopoietin) stimulates tyrosine phosphorylation of Jak2, Shc and c-mpl. *J Biol Chem.* 1995; 270: 4979–4982.

32. Toombs CF, Young CH, Glaspy JA, Varnum BC. Recombinant human megakaryocyte growth and development factor (MGDF) moderately enhances ex vivo platelet aggregation by a receptor dependent mechanism. *Thromb Res.* 1995; 80: 23–33.

33. Ritchie A, Vadhan-Raj S, Broxmeyer HE. Thrombopoietin suppresses growth factor withdrawal induced apoptosis and behaves as a survival factor for the human growth factor-dependent cell line MO7e. *Blood.* 1995; 86: 21a (abstract no 71).

34. Katagiri T, Miyazawa K, Tauchi T, et al. Comparative analysis of signal transduction pathways between thrombopoietin and erthropoietin involves RAS- and JAK-stat-mediated signalings. *Blood.* 1995; 86: 21a (abstract no 591).

35. Hill RJ, Zozulya S, Lu J, Hollenbach P, Bogenberger J, Gishizky M. Stimulation of the *c-MPL* receptor with megakaryocyte growth and development factor (MGDF) induces SHC-dependent differentiation phenotype in hematopoietic cells. *Blood.* 1995; 86: 696a (abstract no 3006).

36. Ohashi H, Morita H, Misaizu T, et al. Receptor activation mechanisms and signal transduction pathways of truncated or chimeric c-mpl. *Blood.* 1995; 86: 905a (abstract).

37. Kato T, Ogami K, Shimada Y, et al. Purification and characterization of thrombopoietin. *J Biochem.* 1995; 118: 229–236.

38. Miyakawa Y, Oda A, Druker BJ, et al. Thrombopoietin induces tyrosine phosphorylation of Stat3 and StatS in human blood platelets. *Blood.* 1996; 87: 439–446.

39. Miyakawa Y, Oda A, Drucker BJ, et al. Thrombopoietin (TPO) induces tyrosine phosphorylation of Stat3 and Stat5 in human blood platelets. *Blood.* 1995; 86: 700a (abstract no 2788).

40. Miyakawa Y, Oda A, Druker BJ, et al. Recombinant thrombopoietin induces rapid protein tyrosine phosphorylation of Janus kinase 2 and Shc in human blood platelets. *Blood.* 1995; 86: 23–27.

41. Hammond WP, Kaplan A, Kaplan S, Kaushansky K. Thrombopoietin (TPO) activates platelets in vitro. *Blood.* 1994; 84: 534a (abstract no 2121).

42. Peng J, Friese P, Woff RF, et al. Relative reactivity of platelets from thrombopoietin- and interleukin-6-treated dogs. *Blood.* 1995; 86: 370a (abstract no 1467).

43. Torii Y, Nishiyama U, Yuki C, Akahori H, Kato T, Miyazaki H. Effects of thrombopoietin on platelet aggregation in normal thrombocythemic and thrombocytopenic mouse model in vitro and ex vivo. *Blood.* 1995; 86: 370a (abstract no 1469).

44. Wun T, Paglieroni T, Kaushansky K, Hammond W, Foster D. ThrombopoietIn (Tpo) is synergistic with other hematopoietic growth factors (HGF) for platelet activation. *Blood.* 1995; 86: 912a (abstract no 3635).

45. Ault KA, Mitchell J, Knowles C. Recombinant human thrombopoietin augments spontaneous and ADP induced platelet activation both in vitro and in vivo. *Blood.* 1995; 86: 367a (abstract no 1456).

46. Oda A, Miyakawa Y, Druker BJ, et al. Thrombopoietin primed human platelet aggregation induced by shear stress, ADP, epinephrine, serotonin, vasopressin or collagen in a manner independent of thromboxane production. *Blood.* 1995; 86: 21a (abstract no 72).

47. Chen J, Herceg-Harjacek L, Groopman JE, Grabarek J. Regulation of platelet activation in vitro by the *c-mpl* ligand, thrombopoietin. *Blood.* 1995; 86: 285a (abstract no 1124).

48. Rickles FR, Levine M, Edwards RL. Hemostatic alterations in cancer patients. *Cancer Metastasis Rev.* 1992; 11: 237–248.

49. Hanson SR, Pareti Fl, Ruggeri ZM, et al. Effects of monoclonal antibodies against the platelet glycoprotein IIb/IIIa complex on thrombosis and hemostasis in the baboon. *J Clin Invest.* 1988; 81: 149–158.

50. Cadroy Y, Hanson SR, Kelly AB, et al. Relative antithrombotic effects of monoclonal antibodies targeting different platelet glycoprotein-adhesive molecule interactions in non-human primates. *Blood.* 1994; 83: 3218–3224.

51. Harker LA, Hunt P, Marzec UM, et al. Regulation of platelet production and function by megakaryocyte growth and development factor (MGDF) in nonhuman primates. *Blood.* 1995; 87: 1833–1844.

52. Chang MS, McNinch J, Basu R, et al. Cloning and characterization of the human megakaryocyte growth and development factor (MGDF) gene. *J Biol Chem.* 1995; 270: 511–514

53. Hunt P, Li YS, Nichol JL, et al. Purification and biologic characterization of plasma-derived megakaryocyte growth and development factor. *Blood.* 1995; 86: 540–547.

54. Grabarek J, Groopman JE, Lyles YR, et al. Human kit ligand (stem cell factor) modulates platelet activation in vitro. *J Biol Chem.* 1994; 269: 21,718–21,724.

55. Shimoda K, Okamura S, Harada N, Kondo S, Okamura T, Niho Y. Identification of a functional receptor for granulocyte colony-stimulating factor on platelets. *J Clin Invest.* 1993; 91: 1310–1313.

56. Frelinger AL III, Cohen I, Plow EF, et al. Selective inhibition of integrin function by antibodies specific for ligand-occupied receptor conformers. *J Biol Chem.* 1990; 265: 6346–6352.

57. Scharf RE, Tomer A, Marzec UM, Teirstein PS, Ruggeri ZM, Harker LA. Activation of platelets in blood perfusing angioplasty-damaged coronary arteries. Flow cytometric detection. *Arteriosclerosis Thromb.* 1992; 12: 1475–1487.

58. Stenberg PE, McEver RP, Shuman MA, Jacques YV, Bainton DF. A platelet alpha-granule membrane protein (GMP-140) is expressed on the plasma membrane after activation. *J Cell Biol.* 1985; 101: 880–886.

59. Thiagaraian P, Tait JF. Binding of annexin V/placental anticoagulant protein I to platelets. Evidence for phosphatidylserine exposure in the procoagulant response of activated platelets. *J Biol Chem.* 1990; 265: 17,420–17,423.

60. Dachary-Prigent J, Freyssinet J-M, Pasquet J-M, Carron J-C, Nurden AT. Annexin V as a probe of aminophospholipid exposure and platelet membrane vesiculation: A flow cytometry study showing a role for free sulfhydryl groups. *Blood.* 1993; 81: 2554–2565.

61. Harker LA, Hanson SR. Platelet factors predisposing to arterial thrombosis. *Bailliere's Clin Haematol.* 1994; 7: 499–522.

62. Funakoshi T, Heimark R, Hendrickson L, McMullen BA, Fujikawa K. Human placental anticoagulant protein: Isolation and characterization. *Biochemistry.* 1987; 26: 5572–5578.

63. Iwasaki A, Suda M, Nakao H, et al. Structure and expression of cDNA for an inhibitor of blood coagulation isolated from human placenta: A new lipocortin-like protein. *J Biochem.* 1987; 102: 1261–1273.

64. Creutz CE. The annexins and exocytosis. *Science.* 1992; 258: 924–931.

65. Zwaal RFA, Comfurius P, Bevers EM. Platelet procoagulant activity and microvesicle formation Its putative role in hemostasis and thrombosis. *Biochim Biophys Acta.* 1992; 1180: 1–8.

66. Nicola NA. Cytokine pleiotrophy and redundancy: a view from the receptor. *Stem Cells.* 1994; 12: 3–12.

67. Kaushansky K, Karplus PA. Hematopoietic growth factors: understanding functional diversity in structural terms. *Blood.* 1993; 82: 3229–3240.

68. Bazan JF. Haemopoietic receptors and helical cytokines. *Immunol Today.* 1990: 11: 350–354.

69. Cosman D, Lyman SD, Idzerada RL, et al. A new cytokine receptor superfamily. *Trends Biochem Sci.* 1990; 15: 265–270.

70. Cadena DL, Gill GN. Receptor tyrosine kinases. *FASEB J.* 1992; 6: 2332–2337.

71. Yarden Y, Ullrich A. Growth factor receptor tyrosine kinases. *Annu Rev Biochem.* 1988; 57: 443–478.

72. Taniguchi T. Cytokine signaling through nonreceptor protein tyrosine kinases. *Science.* 1995; 268: 251–255.

73. Broudy VC, Lin N, Fox N, Taga T, Kaushansky K. Thrombopoietin stimulates CFU-MEG proliferation and megakaryocyte maturation independently of cytokines that signal through the gp130 receptor subunit. *Blood.* 1995; 86: 10a (abstract no 28).

74. Folts J. An in vivo model of experimental arterial stenosis, intimal damage and periodic thrombosis. *Circulation.* 1991; 83: 3–14.

75. Golino P, Ambrosio G, Pascucci I, Ragni M, Russolillo E, Chiariello M: Experimental carotid stenosis and endothelial injury in the rabbit: an in vivo model to study intravascular platelet aggregation. *Thromb Haemost.* 1992; 67: 302–305.

76. Harker LA, Kelly AB, Hanson SR. Experimental arterial thrombosis in non-human primates. *Circulation.* 1991; 83: 41–55.

77. Kelly AB, Marzec UM, Krupski W, Bass A, Cadroy Y, Hanson SR, Harker LA. Hirudin interruption of heparin-resistant arterial thrombus formation in baboons. *Blood.* 1991; 77: 1006–1012.

V

PRECLINICAL BIOLOGY

19

In Vivo Dose–Response Effects of Mpl Ligands on Platelet Production and Function

Laurence A. Harker, Christopher F. Toombs, and Richard B. Stead

1. Introduction

Thrombopoietin (TPO), the endogenous Mpl ligand, maintains peripheral platelet concentrations at a constant level by modulating megakaryocytopoiesis to compensate for changing peripheral requirements *(1–6)*. TPO promotes platelet production by inducing hemopoietic stem cells to differentiate into megakaryocytes *(3,7–9)*, and stimulating megakaryocyte proliferation and endoreduplication *(3,10–16)*, thereby amplifying formation of the megakaryocyte cytoplasmic substrate destined for fragmentation and release as functional circulating platelets *(17–19)* (*see* Chapters 14–17).

Megakaryocytopoiesis is regulated by plasma levels of unbound TPO *(12,13,15,20–22)*. For example, endogenous TPO levels increase several orders of magnitude in patients with thrombocytopenia after marrow-ablative chemotherapy as the platelet counts decrease, and subsequently return to baseline after the peripheral platelet counts are normalized by platelet transfusion therapy or hemopoietic recovery *(20,23)*. Feedback modulation of plasma TPO levels does not appear to involve regulation of gene transcription *(24)*, since TPO mRNA levels in heterzygote TPO "knockout" mice remain half that of the normal mice *(22)*, despite thrombocytopenia (*see* Chapter 21). One important feedback mechanism modulating plasma levels of unbound TPO involves its competitive binding with Mpl of peripheral platelets, marrow megakaryocytes, and possibly soluble Mpl (s-Mpl) in the plasma *(15,24,25)*, leading to reciprocal changes in plasma unbound TPO levels and peripheral platelet counts and/or number of marrow megakaryocytes (*see* Chapter 23).

Several recombinant Mpl ligands have been developed that induce the development of megakaryocytes from early hemopoietic progenitors, and stimulate megakaryocyte proliferation in vitro and in vivo *(20,23,26)*. Recombinant human megakaryocyte growth and development factor (rHuMGDF) *(20,23,27,28)* is a recombinant, nonglycosylated, truncated form of Mpl ligand encompassing the erythropoietin (EPO)-like domain expressed in *Escherichia coli*. Poly[ethylene glycol]-derivitized rHuMGDF (PEG-rHuMGDF) exhibits 10-fold greater biologic activity in vivo than rHuMGDF, at least in part because of its persistence in plasma resulting from pegylation *(29)*.

From: *Thrombopoiesis and Thrombopoietins: Molecular, Cellular, Preclinical, and Clinical Biology*
Edited by: D. J. Kuter, P. Hunt, W. Sheridan, and D. Zucker-Franklin Humana Press Inc., Totowa, NJ

It is important to review the evidence reporting enhanced platelet responsiveness induced by these Mpl ligands, since increased platelet deposition at the site of vascular injury could predispose patients to thrombo-occlusive events. Older cancer patients undergoing chemotherapy are particularly prone to thrombosis because of the prothrombotic effects of malignancy and coincident association with atherosclerotic disease *(30)*. The evidence implicating Mpl ligands in this predisposition to thrombosis is derived from in vitro and ex vivo *(31–33)* laboratory testing of platelet function. Thus, the clinical significance of these findings is uncertain.

To establish the relevance of these in vitro findings, it is important to evaluate the effects of Mpl ligands on platelet recruitment during the formation of thrombi in vivo using clinically appropriate experimental animal models. The findings from two model systems are discussed here. These models consist of cyclic flow reductions in stenotic arteries of rabbits *(34,35)* and the accumulation of ^{111}In-platelets and ^{125}I-fibrin in thrombi forming on thrombogenic segments, interposed in arteriovenous shunts in baboons for fixed periods under standard flow conditions without systemic anticoagulation *(36,37)*.

2. Pharmacokinetics of Mpl Ligands

Blood levels of endogenous TPO in normal nonhuman primates and humans average approximately 100 pg/mL *(29,38,39)* (*see* Chapter 22).

Animal pharmacokinetic studies have been reported for both rHuMGDF and PEG-rHuMGDF. In baboons, the sc daily injection of 5 µg/kg nonpegylated rHuMGDF produces peak plasma levels 2 hours postinjection of 1300 ± 300 pg/mL and trough levels of 300 ± 65 pg/mL. The terminal half-life of rHuMGDF in baboons is 8 ± 3 hours *(40)*.

Subcutaneous daily injections of PEG-rHuMGDF (0.05, 0.10, 0.50, and 2.50 µg/kg) in baboons produce dose-dependent increases in steady-state plasma trough levels (Fig. 1). The ratio of area-under-the-curve (AUC) plasma concentrations on day 4 versus day 1 in animals receiving 0.5 µg/kg/day is 2.65 ± 0.88, indicating approximately two- to threefold accumulation after daily dosing of PEG-rHuMGDF. Average steady-state trough levels, calculated from predose concentrations obtained after 4 days of dosing 0.10, 0.50, and 2.50 µg/kg/day are 0.03 ± 0.01, 0.26 ± 0.13, and 1.52 ± 0.51 ng/mL, respectively, indicating that plasma levels are directly proportional to dose. The terminal half-life for PEG-rHuMGDF averages 22 ± 10 hours. Peak blood levels are achieved after sc injections of PEG-rHuMGDF in baboons within 6 hours *(29)*.

In baboons, sc administration of PEG-rHuMGDF at 0.50 µg/kg/day achieves steady-state trough plasma levels of 0.26 ± 0.13 ng/mL; however, 10-fold more rHuMGDF (5 µg/kg/day) produces similar trough levels averaging 0.30 ± 0.06 ng/mL ($p = 0.3$). As predicted by these results, injection of 0.5 µg/kg/day PEG-rHuMGDF increases platelet production comparable to that resulting from 5 µg/kg/day nonpegylated rHuMGDF (Fig. 2; $p = 0.87$). Thus, this approximately 10-fold increase in efficacy of PEG-rHuMGDF over rHuMGDF is largely attributable to delayed clearance and prolonged plasma half-life of PEG-rHuMGDF compared with nonpegylated rHu-MGDF *(29,40)* (*see* Chapter 13).

In rabbits, iv injections of rHuMGDF (0.1, 1.0, or 10.0 µg/kg/day) produce linearly dose-dependent, incremental increases in plasma levels measured 10 minutes after iv injections, i.e., 0.05 ± 0.02, 0.98 ± 0.07, and 21.3 ± 2.35 ng/mL, respectively *(41)*.

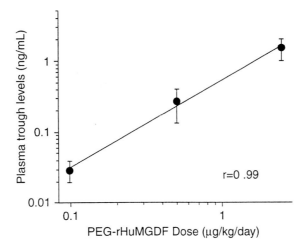

Fig. 1. Dose–response of sc administration of PEG-rHuMGDF on plasma trough levels in baboons. After attaining steady-state conditions with 4 days of dosing, there is a log-log (linear) relationship between steady-state trough levels and dose at 0.1, 0.5, and 2.5 µg/kg/day; (correlation coefficient 0.99, $p < 0.001$) *(29)*.

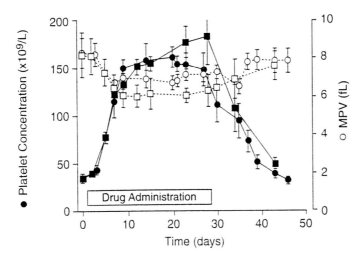

Fig. 2. Relative effectiveness of PEG-rHuMGDF and rHuMGDF in baboons. The time course of peripheral platelet counts for PEG-rHuMGDF (0.5 µg/kg/day, ●) is compared with the findings obtained using nonpegylated rHuMGDF (5 µg/kg/day, ■). The bar indicates duration of dosing. The circulating levels (not shown) and increase in platelet counts are statistically equivalent ($p = 0.3$ by ANOVA). The modest reduction in mean platelet volume (MPV; ○, PEG-rHuMGDF; □, rHuMGDF) for each agent is also similar. The variance about the mean represents ±1 SD *(29)*.

3. Effects of Mpl Ligands on Platelet Production

PEG-rHuMGDF increases the concentration of circulating platelets in a log-linear dose-dependent manner over a 50-fold dose range (0.05–2.5 µg/kg/day), reaching peak values 2 weeks after initiating daily sc injections in baboons (Fig. 3) *(29)*.

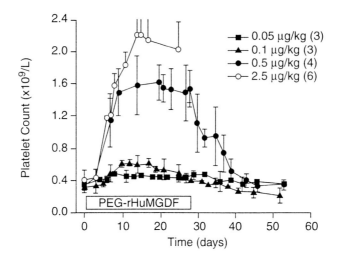

Fig. 3. Dose–response of PEG-rHuMGDF on the concentration of peripheral platelets in baboons. Daily sc injections of PEG-rHuMGDF (0.05, 0.10, 0.50, and 2.50 µg/kg/day) increase the peripheral platelet counts incrementally in a log-linear dose–response manner ($r = 0.94$; $p < 0.01$). The bar indicates duration of dosing. After discontinuing PEG-rHuMGDF treatment, the platelet counts return to baseline within 2 weeks. The variance about the mean represents ±1 SD *(29)*.

Although the mean platelet volume (MPV) decreases as the platelet count increases (8.0 ± 1.2 versus 6.6 ± 0.8 fL; $p < 0.01$), there is no relationship between the extent of change in MPV and the dose of PEG-rHuMGDF administered. After discontinuing therapy, platelet counts decrease to baseline values and MPV normalizes within 2 weeks (Fig. 2). Platelet lifespan in baboons is normal during therapy with PEG-rHuMGDF or rHuMGDF (Fig. 4). Platelet mass turnover, a steady-state measure of the rate at which platelet mass enters the peripheral circulation, increases in a log-linear dose-dependent manner (Fig. 5). The ultrastructure of platelets produced during treatment with Mpl ligands is indistinguishable from native platelets, except for a slightly smaller size (Fig. 6) *(40)*.

In baboons, PEG-rHuMGDF increases megakaryocyte number, size, and ploidy in a log-linear, dose-dependent fashion (Fig. 7A), comparing basal measurements with findings obtained after 3 days and 14 days of treatment (Tables 1 and 2). Since mean megakaryocyte volumes and ploidy attain predictable maximum values within 3 days of beginning therapy, megakaryocyte ploidy is an accurate early measure reflecting Mpl ligand stimulation of megakaryocytopoiesis *(29,40)*.

In baboons, PEG-rHuMGDF produces log-linear, dose-dependent expansion of the number of marrow megakaryocytes (Fig. 7B). Total megakaryocytopoiesis, estimated as marrow megakaryocytes mass, represents the product of the total number of megakaryocytes and their mean volume *(42,43)*. PEG-rHuMGDF increases megakaryocyte mass in a log-linear, dose-dependent manner (Fig. 7B), and this increase correlates with increased platelet mass turnover (Tables 1 and 2). In determining marrow megakaryocyte numbers, the ratio of marrow megakaryocytes to marrow nucleated eryth-

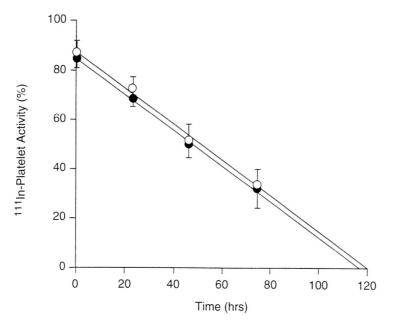

Fig. 4. Platelet life-span during rHuMGDF administration in baboons. The disappearance times of autologous ^{111}In-platelets are compared before (○) and after (●) the administration of rHuMGDF. The mean time in circulation for the ^{111}In-platelets is determined by γ function analysis. Variation about the mean is shown for ±1 SD *(40)*.

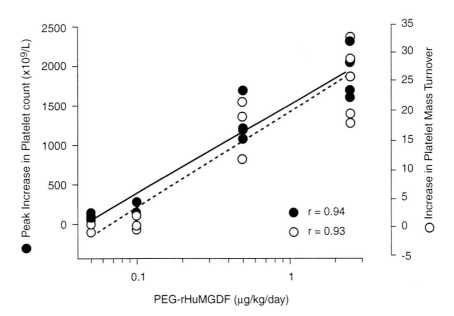

Fig. 5. Log-linear dose–response of PEG-rHuMGDF on platelet production in baboons. The increases in platelet count (● and solid line) and platelet mass turnover (○ and dashed line) corresponding to each dosing regimen are directly related to the logarithm of the dose ($n = 4$) ($r = 0.93$; $p = 0.01$ by ANOVA) *(29)*.

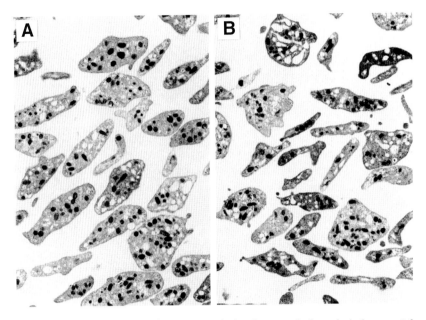

Fig. 6. Effects of rHuMGDF on ultrastructural platelet morphology in baboons. After attaining a fivefold increase in the concentration of circulating platelets, their morphology was elevated using transmission electron microscopy. **(A)** Low-power views of platelets before initiating rHuMGDF therapy. **(B)** The appearance at the same magnification after 28 days of rHuMGDF therapy. The platelets in B appear to be somewhat smaller *(40)*.

roid forms is measured by relating nucleated GPIIb/IIIa-positive megakaryocytes to nucleated glycophorin A⁺ (GPA) erythroid cells (using monoclonal antibodies [MAb] to GPIIb/IIIa and erythroid-specific GPA, respectively) together with supravital nuclear staining *(44)*.

Despite the reported effects of TPO on hemopoietic stem cells and committed erythroid progenitors (*see* Chapter 14), peripheral leukocyte, neutrophil, or erythrocyte counts do not change significantly during PEG-rHuMGDF or rHuMGDF administration in baboons (Fig. 8) or rabbits (29,40,41).

4. Effects of Mpl Ligands on Platelet Function Ex Vivo

Mpl ligands trigger platelet intracellular signaling reactions via tyrosine kinases and enhance aggregatory responses in vitro *(31–33,45–51)*. However, similar proaggregatory effects are reportedly produced by other hemopoietic growth factors, including recombinant human granulocyte colony-stimulating factor (rHuG-CSF) and rHu stem cell factor (rHuSCF) *(52,53)*, without the emergence of thrombotic complications during extensive investigational and therapeutic use *(54,55)*. Thus, because of uncertainty regarding the significance of Mpl ligand-induced enhancement of platelet aggregation in vitro, studies have been done to assess *ex vivo* aggregatory responsiveness of platelets obtained from animals receiving Mpl ligands *(29,40,41)*. Since Mpl ligands elevate peripheral platelet counts, evaluating their effects on platelet function *per se* in animals requires that the testing be independent of the circulating platelet concentration.

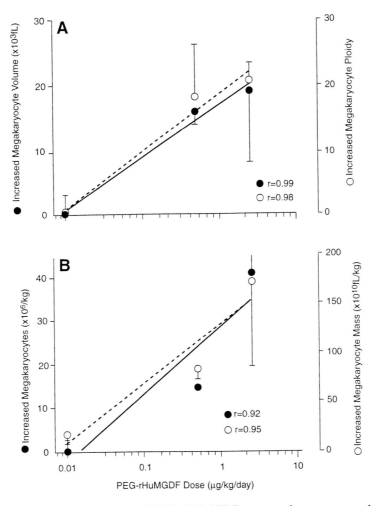

Fig. 7. Log-linear-dose–response of PEG-rHuMGDF on megakaryocyte number, volume, and ploidy and marrow mass in baboons. The increase in megakaryocyte volume and ploidy induced by PEG-rHuMGDF by day 3 is directly related to the logarithm of the dose ($r = 0.99$, solid line; $p < 0.0001$) **(A)**. The calculated marrow megakaryocyte mass on day 28 is also directly related to the logarithm of the dose **(B)** *(29)*. Megakaryocyte mass is the product of megakaryocyte number and volume (dashed line).

In rabbits, iv bolus injection of rHuMGDF (10 µg/kg) enhances platelet responsiveness to adenosine diphosphate (ADP)-induced aggregation ex vivo. Figure 9A illustrates the ex vivo aggregation of rabbit platelets from a representative animal immediately before and 10 minutes after the iv injection of 10 µg/kg rHuMGDF. In contrast, no enhancement of platelet responsiveness is observed in rabbits receiving lower doses (0.1 or 1.0 µg/kg) of rHuMGDF. To characterize more fully the change in platelet sensitivity to agonists, the concentration of agonist producing half-maximal platelet aggregation (AC_{50}) can be calculated by fitting a sigmoid curve to a plot of percentage platelet aggregation as a function of agonist concentration *(56)*. In rabbits treated with

Table 1
Effects of PEG-rHuMGDF on Megakaryocyte Production[a]

	Megakaryocytes				
Time-point	Ratio of megakaryocytes: nucleated erythrocytes	Number, $\times 10^6$ megas/kg	Volume, $\times 10^3$ fL	Mass, $\times 10^{10}$ fL/kg	Fold increase
Baseline	1.62 ± 0.26	8.59 ± 1.4	28.7 ± 2.1	24.6 ± 4.3	1.0
3 days	1.85 ± 0.94	9.8 ± 5.01	52.7 ± 10.0	51.0 ± 23.3	2.1
7 days	3.24 ± 1.96	17.2 ± 10.4	46.9 ± 2.84	80.4 ± 48.6	3.3
14 days	3.53 ± 1.73	18.7 ± 9.19	49.3 ± 4.02	94.3 ± 50.6	3.8
28 days	6.06 ± 2.15	31.8 ± 2.15	48.9 ± 4.0	161 ± 122	6.5

rHuMGDF = recombinant human megakaryocyte growth and development factor; fL = femtoliter.
[a]Baboons were administered 0.5 μg/kg/day of PEG-rHuMGDF for 28 days, and megakaryocyte number, volume, and mass were determined *(40)*.

Table 2
Effects of 28 days of Administration of 0.5 μg/kg/day of PEG-rHuMGDF on Platelet Production in Baboons *(40)*

	Platelets			
Time-point	Concentration, 10^9/L	Volume, 10^3/fL	Mass turnover, 10^5 fL platelets/μL/day	Fold increase
Baseline	0.35 ± 0.05	8.2 ± 1.2	6.30 ± 1.91	1.0
3 days	0.49 ± 0.08	6.2 ± 0.08	7.29 ± 2.01	1.2
7 days	1.21 ± 0.13	6.0 ± 0.7	17.8 ± 3.22	2.8
14 days	1.21 ± 0.13	6.0 ± 0.7	17.8 ± 3.22	2.8
14 days	1.54 ± 0.20	6.0 ± 0.7	22.6 ± 4.34	3.6
28 days	1.83 ± 0.40	6.0 ± 0.7	31.8 ± 5.42	5.0

rHuMGDF = recombinant human megakaryocyte growth and development factor; fL = femtoliter

10 μg/kg rHuMGDF intravenously, the AC_{50} for ADP decreases significantly from a baseline value of 4.48 ± 1.89–1.95 ± 1.02 μM 10 min after acute dosing (Fig. 9B; $p < 0.05$), whereas no change in AC_{50} was found for rabbits receiving the 0.1- or 1.0-μg/kg doses of rHuMGDF *(41)*.

In baboons, ex vivo platelet-aggregatory responsiveness to physiologic agonists is transiently enhanced in platelet-rich plasma (PRP) (platelet count 300×10^9/L) obtained from baboons receiving rHuMGDF *(40)* or PEG-rHuMGDF *(29)*. Platelet-aggregatory responsiveness to ADP, collagen, and thrombin receptor agonist peptide ($TRAP_{1-6}$) is transiently increased after injecting rHuMGDF (5 μg/kg) subcutaneously on day 1 (Fig. 10), and after 3 days of PEG-rHuMGDF sc injections of 0.05–2.5 μg/kg (Fig. 11). The concentration of agonists inducing half-maximal aggregation decreases significantly, following PEG-rHuMGDF for ADP from 5.4 ± 1.4–3.0 ± 0.9 μM ($p = 0.04$), for collagen from 2.8 ± 1.9–1.1 ± 0.9 μg/mL ($p = 0.03$); and for $TRAP_{1-6}$ from 120 ± 83–17 ± 15 μM ($p = 0.001$) for PEG-rHuMGDF. These effects are blocked by 10-fold excess s-Mpl *(40)*. Interestingly, no significant augmentation of platelet aggre-

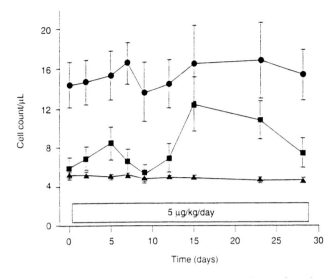

Fig. 8. Effects of PEG-rHuMGDF on peripheral concentrations of erythrocytes, leukocytes, and neutrophils in baboons. No significant changes are produced in other peripheral blood cell counts by PEG-rHuMGDF, including red cells ($\times$ 10^{12}/L; $\blacktriangle$), white cells ($\times$ 10^9/L; $\bullet$) or neutrophils ($\times$ 10^9/L; $\blacksquare$), comparing basal and final values ($p > 0.1$ in all cases). The transient increase in absolute neutrophil count on day 14 is attributable to the effects of performing platelet survival time measurements.

gation ex vivo is observed after 7 days of therapy with either rHuMGDF or PEG-rHuMGDF, despite continued administration (Figs. 10 and 11). In PRP prepared from baboons receiving Mpl ligand, the in vitro addition of Mpl ligand does not induce platelet aggregation in the absence of physiologic agonists (data not shown).

The period of increased platelet responsiveness ex vivo coincides with the entry of newly formed platelets that supplant older platelets and expand the intravascular platelet pool severalfold. The observed enhancement of functional responses may be attributable, at least in part, to the as-yet undefined hyperreactivity exhibited by newly formed platelets *(57)*. "Young" platelets are reported to have "enhanced" hemostatic *(58,59)* and aggregatory *(60)* function ex vivo, without spontaneously expressing activation epitopes *(61,62)*, features reproduced by the cohort of platelets entering the circulation between days 3 and 7 after PEG-rHuMGDF therapy is begun.

5. Effects of Mpl Ligands on Platelet Recruitment into Forming Thrombus

To evaluate further the potential risk of Mpl ligand enhancing thrombosis, animal experiments were done to determine the effects of Mpl ligand on platelet-dependent thrombus formation in vivo. Results are presented from experiments done in a rabbit model of cyclic flow reductions in stenotic arteries *(34,35)* and in a baboon model of ^{111}In-platelet and ^{125}I-fibrin deposition in thrombus forming on thrombogenic segments interposed in arteriovenous shunts *(36,37)*.

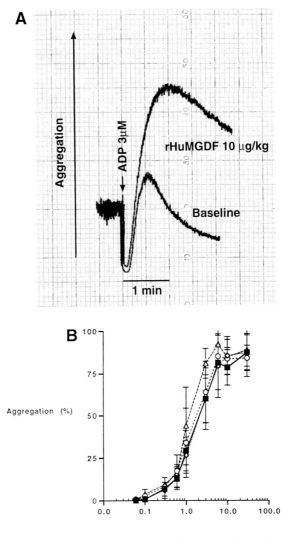

Fig. 9. (A) Effects of rHuMGDF on ex vivo platelet aggregation in rabbit PRP. PRP is prepared by differential centrifugation of whole blood collected in 3.8% citrate anticoagulant (1 part anticoagulant to 9 parts blood), adjusted to a platelet count of 250×10^9/L by dilution with platelet-poor plasma (PPP), and platelet aggregometry induced by ADP is measured optically. The response of PRP to 3.0 μM adenosine diphosphate (ADP) is shown before iv administration of rHuMGDF (baseline). The data show that in the same animal, aggregation to 3 μM ADP is enhanced after the iv administration of rHuMGDF (10 μg/kg). **(B)** The approach to estimating AC_{50} is illustrated. Increases in light transmission are recorded on an analog chart recorder and converted to percent platelet aggregation using the optical absorbance of PRP as a reference for 0% platelet aggregation and the optical absorbance of PPP as a reference for 100% platelet aggregation. The AC_{50} value for the agonist concentration producing half-maximal aggregation is calculated from regression parameters by solving the logistic function for an *x* value (agonist concentration) using the computer-determined parameters and the *y* value (percent aggregation) set to one-half maximal aggregation for the particular experiment *(56)*. ■, control; ◇, 1 ng/mL rHuMGDF; ○, 3 ng/mL rHuMGDF; △, 10 ng/mL rHuMGDF.

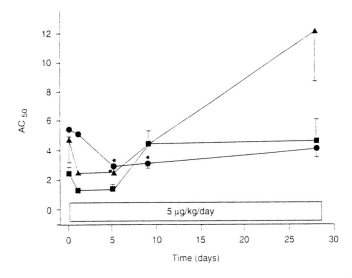

Fig. 10. Effects of rHuMGDF on baboon platelet aggregation ex vivo. During rHuMGDF dosing, aggregatory responses of platelets obtained from baboons receiving rHuMGDF (5 µg/kg/day) were evaluated by comparing the AC_{50} for various agonists. The AC_{50}s for ADP (●), collagen (■), and $TRAP_{1-6}$ (▲) were decreased significantly during the first week of rHuMGDF administration ($p < 0.05$ in all cases shown by asterisk). Thereafter, there was no detectable enhancement of platelet aggregatory reactivity *(40)*.

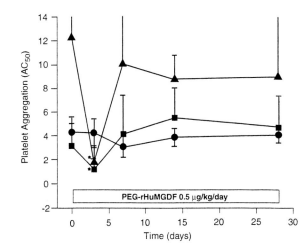

Fig. 11. Effects of PEG-rHuMGDF on platelet aggregation ex vivo. During PEG-rHuMGDF dosing, aggregatory responses of platelets obtained from baboons receiving PEG-rHuMGDF (0.5 µg/kg/day) are shown. The concentrations of collagen (■, µg/mL) and $TRAP_{1-6}$ (▲, ×10µ*M*) inducing half-maximal platelet aggregation are decreased during the first week of PEG-rHuMGDF administration ($p < 0.05$, asterisks). Thereafter, there is no detectable enhancement of platelet aggregatory reactivity *(29)*. ●, ADP, µ*M*.

5.1. Rabbit Carotid Artery Thrombosis

The cyclic flow reduction (CFR) model developed by Folts et al. *(63)* successfully identifies both inhibitors *(64,65)* and promoters *(66–69)* of platelet thrombus formation in vivo and ex vivo. Increase of either CFR frequency or slope indicates increased in vivo platelet deposition. Platelet-activating agents, such as epinephrine, increase both the frequency of CFR and the rate at which blood flow is reduced from maximal to near zero (increase in slope). Thus, the CFR model is an appropriate system for assessing the potential prothrombotic effects of recombinant Mpl ligands in vivo.

In this model, a common carotid artery is isolated, fitted with a Doppler flow probe, constricted with an external plastic constrictor to create critical stenosis (diameter ranging from 1.2–1.8 mm), and damaged by repeated clamping with forceps, thereby promoting deposition of platelet thrombus and platelet thromboembolism *(34,63)*. Carotid blood flow is progressively constricted with sequentially more narrow constrictors until the vessel lumen is reduced sufficiently to produce platelet-thromboembolic cyclic reductions in blood flow *(35,41)*. After establishing a pattern of steady oscillations in blood flow over time from peak to near zero, a series of CFRs is recorded for approximately 30–40 minutes (control period). The rabbits then receive iv bolus injections of rHuMGDF at 0.1, 1.0, or 10 µg/kg. After rHuMGDF administration, CFRs are monitored for an additional 30–40 minutes. Separate positive and negative control animals are prepared by administering iv infusions of epinephrine (0.2 µg/kg/minute) to enhance CFR, or the combination of aspirin (5 mg/kg) and ketanserin (0.25 mg/kg) to abolish CFR (Fig. 12A). Data obtained from the CFR model can be quantitatively assessed by calculating the mean slope for each experiment. Slope is calculated by drawing a tangent to the steady decline in blood flow during each cycle. Comparison to the mean CFR slope, measured at baseline, can be used to assess enhanced or attenuated CFR formation after the administration of test substance. In this series of experiments, control CFR slope averages -3.00 ± 0.88 mL/minute/minute. Epinephrine (0.2 µg/kg/minute) increases the mean slope to -4.38 ± 1.75 mL/minute/minute ($p < 0.05$), whereas the combination of aspirin and ketaserin abolishes CFRs (mean slope of -0.38 ± 0.75 mL/minute/minute; $p < 0.01$).

Figure 13A illustrates a baseline series of CFRs followed by a series of CFRs produced after iv bolus rHuMGDF at 10 µg/kg. Recombinant HuMGDF (0.1, 1.0, or 10 µg/kg) does not affect CFR slope (Fig. 13B) when compared with CFR slope obtained during the baseline study, indicating that rHuMGDF neither enhances nor diminishes platelet recruitment in the formation of platelet thromboembolus as assessed by CFR slope analysis. Thus, because the CFR model sensitively detects both enhancement and diminution of platelet-dependent thrombosis, the negative results with rHuMGDF constitute persuasive relevant evidence regarding the absence of detectable prothrombotic effects produced by rHuMGDF.

5.2. ^{111}In-Platelet Deposition and ^{125}I-Fibrin Accumulation in Thrombus in Baboons

The possibility that PEG-rHuMGDF might amplify thrombus formation has also been examined by comparing basal versus treatment measurements of ^{111}In-platelet deposition and ^{125}I-fibrin accumulation on thrombogenic segments interposed in chronic ar-

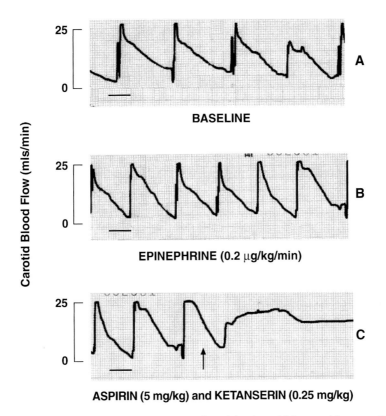

Fig. 12. Representative tracings of CFRs produced in the rabbit carotid artery. Time reference of 1 minute (horizontal bar) appears in the lower left section of each panel, and the time scale is equivalent between panels. The data indicate that the relative rate of thrombus formation is enhanced by the infusion of epinephrine **(B)** relative to the preinfusion baseline **(A)**. **(C)** CFRs were readily inhibited in the same animal by the combined administration of aspirin and ketanserin (arrow).

teriovenous femoral shunts in baboons *(29,40)*. In measuring the effects of PEG-rHuMGDF on thrombus formation, the relative accumulations of ^{111}In-platelets and ^{125}I-fibrin are determined in thrombus forming on segments of two well-characterized thrombogenic surfaces, endarterectomized homologous aorta (EA) and knitted prosthetic vascular grafts (VG). These thrombogenic segments are interposed in arteriovenous shunts for fixed periods under standard-flow conditions without systemic anticoagulation *(36,37)*. Since these models are highly reproducible, they are well suited for repeated comparisons. The number of deposited platelets in normal baboons is directly related to the peripheral platelet count over the range of $100–800 \times 10^9$ cells/L *(57)*.

Platelet accumulation into thrombus forming on these thrombogenic segments remains proportional to the elevation in peripheral platelet count after PEG-rHuMGDF administration, both early, between days 5 and 7 when platelet counts are increasing and ex vivo platelet aggregation is enhanced, and 3 and 4 weeks later when platelet counts are at steady state without enhanced platelet responsiveness ex vivo (Figs. 14

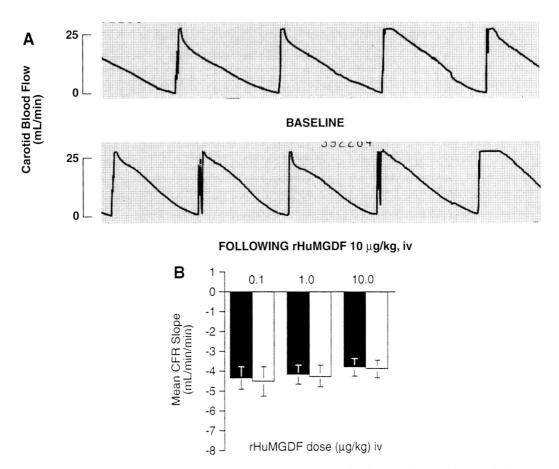

Fig. 13. (A) Representative tracings of CFRs produced in the rabbit carotid artery before (upper panel) and after (lower panel) administration of rHuMGDF. Time reference of 1 minute as for Fig. 12. The data indicate that the CFR pattern is not altered after the iv administration of 10 μg/kg rHuMGDF (lower panel) relative to the baseline period (upper panel). **(B)** Individual experiments indicate that rHuMGDF did not affect the CFR slope at any dose tested. ■, baseline; □, following rHuMGDF.

and 15). Platelet deposition was 4.4 ± 1.3 vs $12 \pm 2 \times 10^9$ platelets on VG ($p < 0.0001$), and 1.8 ± 0.1 versus $10 \pm 2 \times 10^9$ platelets on EA ($p < 0.0001$) at baseline vs during steady-state thrombocytosis. Moreover, no increase in fibrin accumulation is observed (1.5 ± 0.8 mg versus 2.5 ± 0.63 mg for VG, and 1.8 ± 1.2 mg versus 1.9 ± 1.3 for EA; $p > 0.5$ in both cases; data not shown). Thus, platelet deposition is directly proportional to the peripheral platelet count, and no excessive thrombotic responses are attributable to Mpl ligand enhancement of platelet function *per se*.

In summary, these models demonstrate that thrombus formation does not exceed that predicted by the peripheral platelet concentration, indicating that the in vitro and ex vivo evidence suggesting enhanced platelet responsiveness is not evident when performing direct measurements of thrombus formation in vivo. The risk of developing thrombo-occlusive events in patients receiving PEG-rHuMGDF should

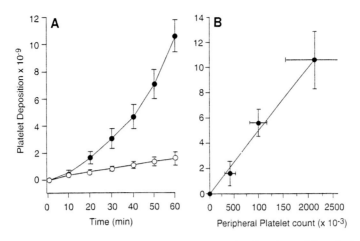

Fig. 14. Effects of rHuMGDF on endarterectomy platelet deposition in baboons. The deposition of ^{111}In-platelets onto endarterectomized segments of homologous aorta (EA) interposed in arteriovenous femoral shunts is measured in real time by γ camera imaging and expressed as total platelets deposited. **(A)** Basal measurements are obtained (○) when the peripheral platelet count is $350 \pm 50 \times 10^9$ platelets/L; this is compared with deposition observed after elevating the concentration of peripheral platelets up to fivefold by rHuMGDF therapy (●). **(B)** The dependence of platelet deposition on the peripheral platelet count, including findings from other control studies involving animals with normal platelet counts ($350 \pm 51 \times 10^9$ platelets/L) and modestly elevated counts ($660 \pm 230 \times 10^9$ platelets/L) *(40)*.

be small, because PEG-rHuMGDF therapy is intended for thrombocytopenic patients, and any thrombotic risk is directly related to the concentration of circulating platelets, a response that should be readily controlled by the dose and duration of PEG-rHuMGDF therapy.

6. Summary

Mpl ligands stimulate platelet production by inducing dose-dependent megakaryocyte development from early marrow hemopoietic progenitors and subsequent proliferation and endoreduplication in vivo. Trough blood levels are directly proportional to dose, and log-linear biologic responses are produced when PEG-rHuMGDF is administered to baboons over the dose range of 0.05–2.5 μg/kg/day with respect to peak peripheral platelet counts, platelet mass turnover, and marrow megakaryocyte volume, number, and mass. Other forms of Mpl ligand are reported to produce similar increases in platelet count in various species.

Whereas Mpl ligands do not directly induce platelet aggregation in vitro, they transiently enhance aggregatory responsiveness of platelets from several different species to physiologic agonists both in vitro and ex vivo. However, platelet recruitment into forming thrombi is not augmented by Mpl ligands when evaluated in quantitative rabbit or baboon models of platelet-dependent thrombus formation, except for the effect of platelet concentration *per se*. These findings indicate that

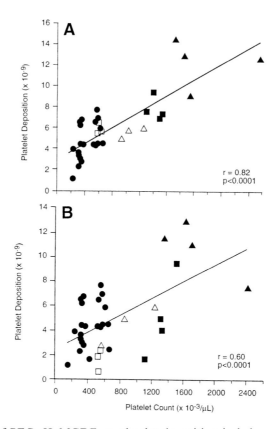

Fig. 15. Effects of PEG-rHuMGDF on platelet deposition in baboons. The deposition of ^{111}In-platelets onto segments of VG **(A)** and EA **(B)** interposed in arteriovenous femoral shunts is measured. **(A)** The relationship between peripheral platelet count and platelet deposition on segments of VG is shown at the baseline (peripheral platelet count of approximately $350 \pm 50 \times 10^9$ platelets/L) and subsequent measurements performed on days 7 and 28 after elevation of platelet count up to fivefold. **(B)** The relationship of platelet deposition onto segments of EA aorta for the baseline peripheral platelet count and the platelet count on days 7 and 28 is depicted *(29)*. ●, controls; □, 0.1 µg/kg PEG-rHuMGDF, day 28; ■, 0.5 µg/kg PEG-rHuMGDF, day 28; △, 0.5 µg/kg PEG-rHuMGDF, day 7; ▲, 5 µg/kg rHuMGDF.

appropriate dosing of recombinant Mpl ligand after marrow suppression will prevent thrombocytopenia without increasing the risk of platelet-dependent thrombo-occlusive complications.

References

1. Odell TT, McDonald TP, Detwiler TC. Stimulation of platelet production by serum of platelet-depleted rats. *Proc Soc Exp Biol Med.* 1961; 108: 428.
2. Harker LA. Kinetics of thrombopoiesis. *J Clin Invest.* 1968; 47: 458–465.
3. Paulus JM. *Platelet Kinetics: Radioisotopic, Cytological, Mathematical and Clinical Aspects.* Amsterdam: North-Holland, 1971.
4. Evatt BL, Shreiner DP, Levin J. Thrombopoietic activity of fractions of rabbit plasma: Studies in rabbits and mice. *J Lab Clin Med.* 1974; 83: 364–371.

5. Mazur E, South K. Human megakaryocyte colony-stimulating factor in sera from aplastic anemia dogs: partial purification, characterization and determination of hematopoietic cell lineage specificity. *Exp Hematol*. 1985; 13: 1164–1172.

6. Tayrien G, Rosenberg RD. Purification and properties of a megakaryocyte stimulatory factor present both in the serum-free conditioned medium of human embryonic kidney cells and in thrombocytopenic plasma. *J Biol Chem*. 1987; 262: 3262–3268.

7. Gewirtz AM. Human megakaryocytopoiesis. *Semin Hematol*. 1986; 23: 27–42.

8. Greenberg SM, Kuter DJ, Rosenberg RD. In vitro stimulation of megakaryocyte maturation by megakaryocyte stimulatory factor. *J Biol Chem*. 1987; 262: 3269–3277.

9. Vainchenker W, Kieffer N. Human megakaryocytopoiesis: in vitro regulation and characterization of megakaryocytic precursor cells by differentiation markers. *Blood Rev*. 1988; 2: 102–107.

10. Tomer A, Harker LA, Burstein SA. Flow cytometric analysis of normal human megakaryocytes. *Blood*. 1988; 71: 1244–1252.

11. Lok S, Kaushansky K, Holly RD, et al. Cloning and expression of murine thrombopoietin cDNA and stimulation of platelet production in vivo. *Nature*. 1994; 369: 565–568.

12. Kaushansky K, Lok S, Holly RD, et al. Promotion of megakaryocyte progenitor expansion and differentiation by the *c-Mpl* ligand thrombopoietin. *Nature*. 1994; 369: 568–571.

13. deSauvage FJ, Hass PE, Spencer SD, et al. Stimulation of megakaryocytopoiesis and thrombopoiesis by the *c-Mpl* ligand. *Nature*. 1994; 369: 533–538.

14. Wendling F, Maraskovsky E, Debili N, et al. *c-Mpl* ligand is a humoral regulator of megakaryocytopoiesis. *Nature*. 1994; 369: 571–574.

15. Kuter DJ, Beeler DL, Rosenberg RD. The purification of megapoietin: a physiological regulator of megakaryocyte growth and platelet production. *Proc Natl Acad Sci USA*. 1994; 91: 11,104–11,108.

16. Kaushansky K. Thrombopoietin: the primary regulator of platelet production. *Blood*. 1995; 86: 419–431.

17. Harker LA. Regulation of thrombopoiesis. *Am J Physiol*. 1970; 218: 1376–1380.

18. Breton-Gorius J, Vainchenker W. Expression of platelet proteins during the in vitro and in vivo differentiation of megakaryocytes and morphological aspects of their maturation. *Semin Hematol*. 1986; 23: 43–67.

19. Choi ES, Hokom M, Bartley T, et al. Recombinant human megakaryocyte growth and development factor (rHuMGDF), a ligand for *c-Mpl*, produces functional human platelets in vitro. *Stem Cell*. 1995; 13: 317–322.

20. Nichol JL, Hokom MM, Hornkohl A, et al. Megakaryocyte growth and development factor. Analyses of in vitro effects on human megakaryopoiesis and endogenous serum levels during chemotherapy-induced thrombocytopenia. *J Clin Invest*. 1995; 95: 2973–2978.

21. Gurney AL, Carver-Moore K, de Sauvage FJ, Moore MW. Thrombocytopenia in c-mpl-deficient mice. *Science*. 1994; 265: 1445–1447.

22. de Sauvage FJ, Luoh S-M, Carver-Moore K, et al. Deficiencies in early and late stages of megakaryocytopoiesis in TPO-KO mice. *Blood*. 1995; 86: 255a (abstract no 1007).

23. Bartley TD, Bogenberger J, Hunt P, et al. Identification and cloning of a megakaryocyte growth and development factor that is a ligand for the cytokine receptor Mpl. *Cell*. 1994; 77: 1117–1724.

24. Stoffel R, Wiestner A, Skoda RC. Thrombopoietin in thrombocytopenic mice: evidence against regulation at the mRNA level and for a direct regulatory role of platelets. *Blood*. 1996; 87: 567.

25. Kuter DJ, Rosenberg RD. Appearance of a megakaryocyte growth-promoting activity, megapoietin, during acute thrombocytopenia in the rabbit. *Blood*. 1994; 84: 1464–1472.

26. Nichol JL, Hornkohl ES, Choi ES, et al. Enrichment and characterization of peripheral blood-derived megakaryocyte progenitors that mature in short-term liquid culture. *Stem Cells*. 1994; 12: 494–505.

27. Chang MS, McNinch J, Basu R, et al. Cloning and characterization of the human megakaryocyte growth and development factor (MGDF) gene. *J Biol Chem*. 1995; 270: 511–514.

28. Hunt P, Li Y-S, Nichol JL, et al. Purification and biologic characterization of plasma-derived megakaryocyte growth and development factor. *Blood.* 1995; 86: 540–547.

29. Harker LA, Marzec UM, Hunt P, et al. Dose-response effects of pegylated human megakaryocyte growth and development factor (PEG-rHuMGDF) on platelet production and function in nonhuman primates. *Blood.* 1996; 88: 511–521.

30. Rickles FR, Levine M, Edwards RL. Hemostatic alterations in cancer patients. *Cancer Metastasis Rev.* 1992; 11: 237–248.

31. Toombs CF, Young CH, Glaspy JA, Varnum BC. Recombinant human megakaryocyte growth and development factor (MGDF) moderately enhances ex vivo platelet aggregation by a receptor dependent mechanism. *Thromb Res.* 1995; 80: 23–33.

32. Drachman JG, Griffin JD, Kaushansky K. The c-mpl ligand (thrombopoietin) stimulates tyrosine phosphorylation of Jak2, Shc and c-mpl. *J Biol Chem.* 1995; 270: 4979–4982.

33. Hammond WP, Kaplan A, Kaplan S, Kaushansky K. Thrombopoietin (TPO) activates platelets in vitro. *Blood.* 1994; 84: 534a (abstract no 2121).

34. Folts JD. An in vivo model of experimental arterial stenosis, intimal damage and periodic thrombosis. *Circulation.* 1991; 83: 3–14.

35. Golino P, Ambrosio G, Pascucci I, Ragni M, Russolillo E, Chiariello M. Experimental carotid stenosis and endothelial injury in the rabbit: an in vivo model to study intravascular platelet aggregation. *Thromb Haemost.* 1992; 67: 302–305.

36. Harker LA, Kelly AB, Hanson SR. Experimental arterial thrombosis in non-human primates. *Circulation.* 1991; 83: IV-41–55.

37. Kelly AB, Marzec UM, Krupski W, et al. Hirudin interruption of heparin-resistant arterial thrombus formation in baboons. *Blood.* 1991; 77: 1006–1012.

38. Taylor K, Pitcher L, Nichol J, et al. Inappropriate elevation and loss of feedback regulation of thrombopoietin in essential thrombocythaemia (ET). *Blood.* 1995; 86: 49a (abstract no 183).

39. Nichol J, Hornkohl A, Selesi D, Wyres M, Hunt P. TPO levels in plasma of patients with thrombocytopenia or thrombocytosis. *Blood.* 1995; 86: 371a (abstract no 1474).

40. Harker LA, Hunt P, Marzec UM, et al. Regulation of platelet production and function by megakaryocyte growth and development factor (MGDF) in nonhuman primates. *Blood.* 1996; 87: 1833–1844

41. Toombs CF, Lott FD, Nelson AG, Stead RB. Megakaryocyte growth and development factor (MGDF) promotes platelet production in vivo without affecting in vivo thrombosis. *Blood.* 1995; 86: 369a (abstract no 1466).

42. Harker LA, Finch CA. Thrombokinetics in man. *J Clin Invest.* 1969; 48: 963–974.

43. Slichter SJ, Harker LA. Thrombocytopenia mechanisms and management of defects in platelet production. *Clin Haematol.* 1978; 7: 523–539.

44. Tomer A, Scharf RE, McMillan R, Ruggeri ZM, Harker LA. Bernard-Soulier syndrome: quantitative characterization of megakaryocytes and platelets by flow cytometric and platelet kinetic measurements. *Eur J Haematol.* 1994; 52: 193–200.

45. Miyakawa Y, Oda A, Druker BJ, et al. Thrombopoietin induces tyrosine phosphorylation of Stat3 and Stat5 in human blood platelets. *Blood.* 1996; 87: 439–446.

46. Peng J, Friese P, Wolf RF, et al. Relative reactivity of platelets from thrombopoietin- and interleukin-6-treated dogs. *Blood.* 1995; 86: 370a (abstract no 1467).

47. Torii Y, Nishiyama U, Yuki C, Akahori H, Kato T, Miyazaki H. Effects of thrombopoietin on platelet aggregation in normal thrombocythemic and thrombocytopenic mouse model in vitro and ex vivo. *Blood.* 1995; 86: 370a (abstract no 1469).

48. Wun T, Paglieroni T, Kaushansky K, Hammond W, Foster D. Thrombopoietin (Tpo) is synergistic with other hematopoietic growth factors (HGF) for platelet activation. *Blood.* 1995; 86: 912a (abstract no 3635).

49. Ault KA, Mitchell J, Knowles C. Recombinant human thrombopoietin augments spontaneous and ADP-induced platelet activation both in vitro and in vivo. *Blood.* 1995; 86: 367a (abstract no 1456).

50. Oda A, Miyakawa Y, Druker BJ, et al. Thrombopoietin primed human platelet aggregation induced by shear stress, ADP, epinephrine, serotonin, vasopressin or collagen in a manner independent of thromboxane production. *Blood.* 1995; 86: 21a (abstract no 72).

51. Chen JL, Herceg-Harjacek L, Groopman JE, Grabarek J. Regulation of platelet activation in vitro by the c-mpl ligand, thrombopoietin. *Blood.* 1995; 86: 285a (abstract no 1126).

52. Grabarek J, Groopman JE, Lyles YR, et al. Human kit ligand (stem cell factor) modulates platelet activation in vitro. *J Biol Chem.* 1994; 269: 21,718–21,724.

53. Shimoda K, Okamura S, Harada N, Kondo S, Okamura T, Niho Y. Identification of a functional receptor for granulocyte colony-stimulating factor on platelets. *J Clin Invest.* 1993; 91: 1310–1313.

54. Decoster G, Rich W, Brown SL. Safety profile of filgrastim (r-metHuG-CSF). In: Morstyn G, Dexter TM (eds). *Filgrastim (r-metHuG-CSF) in Clinical Practice.* New York: Marcel-Dekker; 1994: 267–290.

55. Sheridan WP, McNiece I. Stem cell factor. In: Armitage JO, Antman KH (eds). *High-Dose Cancer Therapy: Pharmacology, Hematopoietins, Stem Cells.* Baltimore, MD: Williams & Wilkins; 1995: 429–441.

56. Hanson SR, Pareti FI, Ruggeri ZM, et al. Effects of monoclonal antibodies against the platelet glycoprotein IIb/IIIa complex on thrombosis and hemostasis in the baboon. *J Clin Invest.* 1988; 81: 149–158.

57. Harker LA, Hanson SR. Platelet factors predisposing to arterial thrombosis. In: Meade TW (ed). *Bailliere's Clinical Haematology*, vol. 7. London: Bailliere Tindall; 1994: 499–522.

58. Harker LA, Slichter SJ. The bleeding time as a screening test for evaluation of platelet function. *N Engl J Med.* 1972; 287: 155–159.

59. Shulman NR, Watkins SP Jr, Itscoitz SB, Students AB. Evidence that the spleen retains the youngest and hemostatically most effective platelets. *Trans Assoc Am Physician.* 1968; 81: 302–313.

60. Blajchman MA, Senyi AF, Hirsh J, Genton E, George JN. Hemostatic function, survival, and membrane glycoprotein changes in young versus old rabbit platelets. *J Clin Invest.* 1981; 68: 1289–1294.

61. Thiagaraian P, Tait JF. Binding of annexin V/placental anticoagulant protein I to platelets. Evidence for phosphatidylserine. *J Biol Chem.* 1990; 265: 17,420–17,423.

62. Dachary-Prigent J, Freyssinet J-M, Pasquet J-M, Carron J-C, Nurden AT. Annexin V as a probe of aminophospholipid exposure and platelet membrane vesiculation: a flow cytometry study showing a role for free sulfhydryl groups. *Blood.* 1993; 81: 2554–2565.

63. Folts JD, Crowell EB, Rowe GG. Platelet aggregation in partially obstructed vessels and its elimination with aspirin. *Circulation* 1976; 54: 365–370.

64. Demrow HS, Slane PR, Folts JD. Administration of wine and grape juice inhibits in vivo platelet activity and thrombosis in stenosed canine coronary arteries. *Circulation.* 1995; 91: 1182–1188.

65. Ramjit DR, Lynch JJ Jr, Sitko GR, et al. Antithrombotic effects of MK-0852, a platelet fibrinogen receptor antagonist, in canine models of thrombosis. *J Pharmacol Exp Ther.* 1993; 266: 1501–1511.

66. Olbrich C, Aepfelbacher M, Siess W. Epinephrine potentiates calcium mobilization and activation of protein kinases in platelets stimulated by ADP through a mechanism unrelated to phospholipase C. *Cell Signal.* 1989; 1: 483–492.

67. Bondy GS, Gentry PA. Characterization of the normal bovine platelet aggregation response. *Comp Biochem Physiol.* 1989; 92: 67–72.

68. Ardlie NG, Cameron HA, Garrett J. Platelet activation by circulating levels of hormones: a possible link in coronary heart disease. *Thromb Res.* 1984; 36: 315–322.

69. Lin H, Young DB. Opposing effects of plasma epinephrine and norepinephrine on coronary thrombosis in vivo. *Circulation.* 1995; 91: 1135–1142.

20

Efficacy of Mpl Ligands and Other Thrombopoietic Cytokines in Animal Models

Ann M. Farese and Thomas J. MacVittie

1. Introduction

Despite the therapeutic utility shown with hemopoietic growth factors, thrombocytopenia and neutropenia remain as dose-limiting sequelae after radiation or chemotherapy. Dose intensification and/or schedule compression will likely extend the obligate periods of cytopenia, particularly those of neutropenia and thrombocytopenia, and therefore, new cytokines or combinations of cytokines will be required to eliminate or further decrease these periods, thus lessening the likelihood of hemorrhage and infectious episodes.

In the last decade, potent stimulation of granulopoiesis has been realized through the use of the cytokines, granulocyte colony-stimulating factor (G-CSF) and granulocyte macrophage colony-stimulating factor (GM-CSF). Preclinical studies using these cytokines in myelosuppressed rodents, canines, and nonhuman primates have demonstrated significant reductions in the duration of neutropenia and time to recovery of neutrophils to normal values *(1–15)*. In large animal models of severe radiation-induced myelosuppression, G-CSF or GM-CSF administered concurrently with clinical support (antibiotics and fresh, irradiated platelets or whole blood) resulted in a marked reduction in infectious complications, which translated to reduced morbidity and mortality *(1,2,5,7,8,11)*. The clinical risk of infection and subsequent complications are directly related to the depth and duration of neutropenia *(16)*, and the longer severe neutropenia persists, the greater the risk of secondary infections *(17,18)*. An additional benefit of G-CSF and GM-CSF is their respective ability to increase the functional capacity of the neutrophil and monocyte/macrophage, and thereby, enhance cellular host defense *(19–26)*. These properties, in addition to a large in vitro data base suggesting *(27–32)* hemopoietic synergy with other growth factors, support the use of either G-CSF or GM-CSF in combination with other multilineage growth factors or thrombopoietic cytokines to promote the dual lineage recovery of neutrophils and platelets in myelosuppressed hosts.

In contrast to the therapeutic efficacy of G-CSF or GM-CSF, the ability to stimulate platelet production in a consistently efficient manner has remained elusive. Interleukin (IL)-3 has been characterized as a multilineage hemopoietic growth factor with an im-

From: *Thrombopoiesis and Thrombopoietins: Molecular, Cellular, Preclinical, and Clinical Biology*
Edited by: D. J. Kuter, P. Hunt, W. Sheridan, and D. Zucker-Franklin Humana Press Inc., Totowa, NJ

portant role in the viability, proliferation, and differentiation of hemopoietic progenitor cells, but its benefits in both preclinical *(1,33–35)* and clinical *(36–43)* studies of platelet and neutrophil recovery after radiation- or chemotherapy-induced myelosuppression were ambiguous. However, several other thrombopoietic factors have shown unequivocal preclinical efficacy and are currently in various phases of clinical trials. These include the GM-CSF/IL-3 fusion protein (PIXY-321), IL-6, IL-11, and Synthokine (a synthetic IL-3 receptor agonist) *(34,35,44–50)*. Although earlier studies document the utility of PIXY-321 for chemotherapy-induced myelosuppression *(51)*, recent clinical trials in bone marrow transplantation have shown equivocal results. IL-6 and IL-11 are in clinical trials, and current data suggest that they both accelerate hemopoietic recovery from chemotherapy-induced thrombocytopenia, although associated toxicities may limit the therapeutic index of IL-6 (*see* Chapter 10). Synthokine clinical trials are in progress. Currently, however, platelet transfusions remain the only treatment for severe thrombocytopenia, and transfusion recipients remain at risk for alloimmune and allergic reactions, as well as for transmission of infectious diseases (*see* Chapter 4). The need for a thrombopoietic agent is evident.

The purification, identification, and cloning of the previously elusive, primary physiological regulator of megakaryocytopoiesis and thrombopoiesis, Mpl ligand, have stimulated great interest in defining the clinical potential of this cytokine for ameliorating thrombocytopenia *(52–56)*. Recent investigation of the biological properties and preclinical efficacy of Mpl ligand has shown that it has potent effects on megakaryocytopoiesis and platelet production *(52–64)*. It has lessened or abolished the degree and duration of thrombocytopenia in several animal models of radiation- or drug-induced thrombocytopenia, and it may have direct and/or indirect effects on stimulating granulopoiesis and erythropoiesis *(61,63,65–74)*. The data suggest that Mpl ligand is a lineage-dominant, physiologic regulator of platelet production.

This chapter will review the body of comparative preclinical knowledge supporting the therapeutic efficacy of the current and new-generation cytokines used as single or multiple agents either to enhance the production of platelets in thrombocytopenic hosts or to effect the production of both platelets and neutrophils in myelosuppressed hosts.

2. Single Cytokines

The production of platelets requires a regulated series of steps proceeding from the proliferation and differentiation of the earliest progenitor cell to the megakaryocyte, followed by extensive endoreduplication and maturation to form platelets *(75–77)* (*see* Chapter 1). IL-3 *(78-83)*, GM-CSF *(23,84–85)*, IL-6 *(44,86–91)*, IL-11 *(92)*, leukemia-inhibitory factor (LIF) *(93,94)*, and oncostatin-M (OSM) *(95)* have in vivo effects on megakaryocytopoiesis and platelet production.

2.1. Interleukin-3

IL-3, despite significant in vitro data suggesting a pivotal role in the stimulation of megakaryocytopoiesis, has variable efficacy as a single cytokine in the stimulation of platelet production in myelosuppressed primates. Gillio et al. *(33)* showed that whereas IL-3 enhanced myeloid recovery and abrogated the duration of neutropenia in cyclophosphamide- or 5-fluorouracil (FU)-treated cynomolgus monkeys, improved platelet recovery was not seen. In contrast, in a nonhuman primate model of radiation-induced

Table 1
Duration and Mean Days of Cytopenias in Sublethally Irradiated and Cytokine-Treated Rhesus Monkeys[a]

Treatment	Neutropenia		Thrombocytopenia	
	Days	Days of duration	Days	Days of duration
HSA	16	(5–20)	10	(10–19)
GM-CSF	11[b]	(6–16)	8	(10–17)
IL-3	15	(6–20)	6[b]	(10–15)
IL-3 and GM-CSF (days 1–21)	6[b,c]	(6–9, 12–13)	3[b,c]	(11–13)
IL-3 (days 1–7), GM-CSF (days 7–21)	12	(6–17)	8	(10–17)
HSA (days 1–7), GM-CSF (days 7–21)	12	(6–17)	8	(10–17)

[a]Rhesus monkeys given whole-body irradiation with 450 cGy of mixed fission neutron γ-radiation were treated with control protein (HSA) or the indicated cytokines. Neutropenia is an absolute neutrophil count $< 1.0 \times 10^9$/L and thrombocytopenia is a platelet count $< 30 \times 10^9$/L.

[b]$p < 0.05$ versus HSA-treated protocols.

[c]$p < 0.05$ versus sequential protocols and GM-CSF protocol.

HSA = human serum albumin; IL = interleukin; GM-CSF = granulocyte-macrophage colony-stimulating factor.

myelosuppression, IL-3 enhanced the regeneration of platelets and reduced the duration of thrombocytopenia, but had no effect on the recovery of neutrophils (Table 1) *(1,34)*. The variable effectiveness of this cytokine was further emphasized using a nonhuman primate model of hepsulfam-induced pancytopenia. IL-3 therapy administered once daily had no demonstrable effect on recovery of either platelets or neutrophils *(35)*. IL-3 is also characterized by a relatively narrow therapeutic index (*see* Chapter 10). The intrinsic inflammatory activity of this cytokine adds an additional question to the inconclusive preclinical data.

2.2. Synthokine

Synthokine, an IL-3 analog, was developed in response to the relatively narrow therapeutic index characterized by the modest and variable activity of endogenous IL-3 in stimulating hemopoietic recovery relative to its intrinsic inflammatory activity. Synthokine, a high-affinity IL-3 receptor ligand, demonstrated greater in vitro multilineage hemopoietic activity on human bone marrow-progenitor cells compared with IL-3 without increased inflammatory activity *(96)*.

The therapeutic efficacy of Synthokine relative to administration schedule and dose was evaluated in a nonhuman primate model of radiation-induced myelosuppression *(46)*. Synthokine was found to reduce significantly the duration of thrombocytopenia versus vehicle-treated control animals regardless of dosage or protocol length. Although the duration of neutropenia was not significantly altered, the neutrophil nadir was improved in all protocols evaluated. Two doses (25 and 100 μg/kg/day) of Synthokine were evaluated based on the modest hemopoietic activity of endogenous IL-3 and Synthokine in rhesus monkey bone marrow-derived cell cultures (attributable to 81

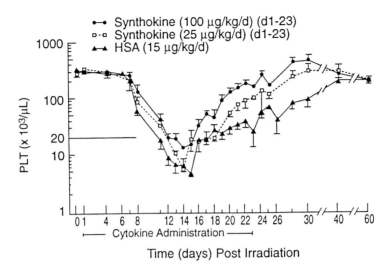

Fig. 1. Recovery of circulating platelets (mean ± SEM) after sublethal irradiation of rhesus monkeys treated with HSA or Synthokine. Synthokine was administered at dosages of 100 μg/kg/day for 23 days (●), or 25 μg/kg/day for 23 days (□), and HSA was administered for 23 days (▲) *(46)*.

and 64% homology, respectively, to rhesus IL-3) (McKearn JP, unpublished results; *97*). The greatest improvement in the duration of thrombocytopenia (3.5 days versus 12.5 days in control animals) was in animals administered 100 μg/kg/day for 23 days (Fig. 1). These data suggest that Synthokine may be clinically useful in the treatment of myelosuppression, and if used in combination with G-CSF or GM-CSF, may provide the desired dual-lineage stimulus for production of both platelets and neutrophils.

2.3. The GP130 Cytokines, IL-6, IL-11, LIF, and OSM

The structurally related pleiotropic cytokines IL-6, IL-11, LIF, and OSM use a common receptor subunit, the 130-kDa glycoprotein (GP130), and share several biological properties, including stimulation of platelet production *(98,99)*. Within this family of cytokines, it appears that IL-11 and IL-6 are most effective in stimulating thrombopoiesis in myelosuppressed animal models.

2.3.1. Interleukin-11

IL-11, administered to mice after exposure to a combined modality regimen of sublethal irradiation and carboplatin, accelerated the recovery of both platelets and hematocrit (Fig. 2) *(48)*. Leukocyte recovery remained unaffected, although marrow and splenic-derived granulocyte-macrophage colony-forming cells (GM-CFC) were stimulated, suggesting the potential for combined cytokine studies with G-CSF or GM-CSF. IL-11 administration was also efficacious in mice subjected to cyclophosphamide-induced marrow aplasia. Recovery of both platelets and leukocytes, as well as bone marrow-derived myeloid progenitors, was enhanced relative to control animals *(100)*. In a canine model of sublethal irradiation, IL-11 showed more modest activity than would have been predicted from the results in rodent models *(101)*. In this study, IL-11 administration did not significantly reduce the duration of thrombocytopenia, although

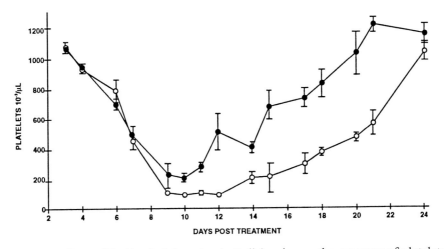

Fig. 2. The effect of IL-11 administration in Balb/c mice on the recovery of platelets after myelosuppression. The data represent the mean ± SE from three separate experiments, with a minimum of 10 mice for each time-point. Mice were bled from the retro-orbital sinus on various days after carboplatin and irradiation. In comparison to vehicle-treated controls (○), platelet counts in the IL-11-treated mice (●) were significantly higher ($p < 0.01$) on days 10–21 (unpaired Student's t-test) *(48)*.

there was a modest trend toward faster initial platelet recovery. There was no effect on neutrophil recovery.

Several studies have demonstrated the efficacy of IL-11 on marrow reconstitution after syngeneic bone marrow transplantation in mice. Administration of IL-11 accelerated recovery of both platelets and neutrophils relative to vehicle-treated controls *(45)*. Two additional models studied transplantation of marrow cells transfected with human IL-11 cDNA *(102,103)*. Animals receiving the IL-11-expressing marrow cells had accelerated, early, and sustained (17 weeks) reconstitution of both platelet counts and the hematocrit. Furthermore, there were no apparent pathological consequences of the sustained, high IL-11 plasma levels *(103)*.

2.3.2. Interleukin-6

IL-6 has numerous biologic activities and appears to play important roles in regulation of the immune response, the acute phase reaction, and hemopoiesis *(86,104,105)*. In animal models of either chemotherapy- or radiation-induced marrow aplasia, IL-6 has consistently reduced both depth and duration of thrombocytopenia *(34,35,44,47,106–111)*. Several rodent models of radiation- or drug-induced myelosuppression have demonstrated a multilineage stimulatory effect of IL-6 on hemopoietic recovery, including an accelerated platelet recovery *(106,108,109)*. In a noted exception, Carrington et al. *(89)* demonstrated that neither IL-3 nor IL-6 alone could shorten the 5-FU-induced period of thrombocytopenia. Inoue et al. *(108)* recently showed increased therapeutic efficacy of a poly[ethylene glycol] (PEG) form of IL-6 relative to unmodified IL-6 in rodent models of myelosuppression induced by radiation or 5-FU. No mention was made of the leukocyte-recovery kinetics. Platelet-recovery time was decreased and platelet nadir abolished.

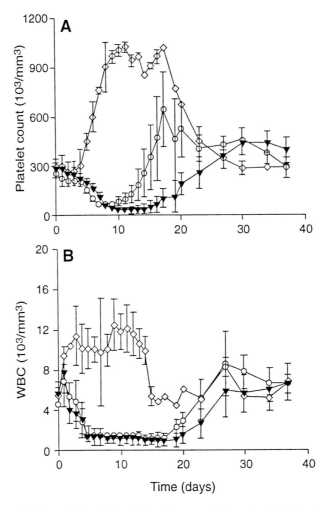

Fig. 3. Changes in the peripheral blood platelet count **(A)** and white blood cell (WBC) count **(B)** in irradiated baboons treated subcutaneously (sc) from days 1–13 postirradiation with IL-6 10 μg/kg/day (○, mean of six subjects ± SD) or the vehicle only (▼, mean of six subjects ± SD). Two sham-irradiated baboons treated with 10 μg/kg/day from days 1–13 (◇, mean ±SD) served as controls *(47)*.

Burstein et al. *(44)* and Selig et al. *(107)* used sublethally irradiated dogs to demonstrate that IL-6 significantly accelerated recovery from thrombocytopenia, although the radiation-induced anemia was exacerbated and granulocyte recovery was not enhanced in either model relative to the vehicle-treated control animals.

Studies in nonhuman primates confirmed the efficacy of IL-6 in enhancing platelet recovery from radiation- or drug-induced thrombocytopenia, although IL-6 dose, radiation exposure level, injection protocol, and form of IL-6 all varied *(34,35,47,110,111)*. Evaluation of IL-6 in more severe nonhuman primate models of radiation-induced myelosuppression confirmed the significantly accelerated recovery of platelets relative to vehicle-treated control animals (Fig. 3) *(34,47,111)*. Treatment with IL-6 reduced the platelet nadir, duration of thrombocytopenia, and time to recovery of platelets to preirradiation levels. Lineage specificity was apparent, in that neither the duration of neutropenia nor neutrophil recovery time were modified by IL-6 therapy. The neutrophil nadir was improved, although this effect did not translate into sustained neutrophil production. Similarly, IL-6 administered for 3 weeks after hepsulfam-induced thrombocytopenia significantly improved both the platelet nadir and duration of thrombocytopenia *(35)*. In this model, IL-6 therapy also significantly lessened the period of neutropenia.

2.3.3. Leukemia-Inhibitory Factor

The therapeutic potential of LIF was suggested by in vitro studies using rodent and human bone marrow-derived colony-forming cells *(112,113)*. LIF, when used alone, was found to promote megakaryocyte maturation, but not megakaryocyte colony formation. However, when combined with IL-3, LIF stimulated increased growth of both megakaryocyte colonies and blast-cell colonies *(112,114)*.

In normal mice, LIF increased the number of both marrow and splenic megakaryocyte progenitor cells, as well as increased the circulating platelet count twofold *(93)*. Normal nonhuman primates administered LIF showed a significant, but somewhat delayed, increase in platelet counts, evident at the end of a 14-day injection period *(94)*. Leukocyte production was unaffected in both normal mice and nonhuman primates.

Farese et al. *(115)* assessed the therapeutic efficacy of LIF in a nonhuman primate model of radiation-induced myelosuppression and noted that LIF administration significantly decreased the duration of thrombocytopenia. Although neither the duration of neutropenia (absolute neutrophil count [ANC] $<1.0 \times 10^9$/L) nor neutrophil-recovery kinetics in the LIF-treated animals differed from the vehicle-treated control animals, absolute neutropenia (ANC = 0) was abolished by LIF administration.

2.3.4. Oncostatin

In vitro, OSM alone cannot stimulate megakaryocyte colony formation. However, it potentiates colony formation induced by IL-3 in a manner analogous to IL-6 and IL-11. In addition, OSM may also promote maturation of megakaryocytes, as has been noted for IL-11, IL-6, and LIF *(113,116)*.

Administration of OSM accelerated platelet recovery and decreased the duration of thrombocytopenia *(95)* in irradiated mice experiencing modest levels of myelosuppression. In contrast, OSM administered to nonhuman primates after a severe dose of irradiation was found to be ineffective in modulating either the duration and the degree of thrombocytopenia, or the recovery of circulating platelets (MacVittie TJ, unpublished). It may be argued that the therapeutic efficacy of OSM may be dose- and protocol-dependent, and further evaluation of these relationships may reconcile the results of these two studies.

3. Mpl Ligand

Recent efforts by several research groups culminated in the discovery of the elusive growth factor thrombobopoietin (TPO), the endogenous Mpl ligand, also described as megakaryocyte growth and development factor (MGDF), or megapoietin *(52–56)* (*see* Chapters 8 and 9). This cytokine appears to be lineage dominant for the production of megakaryocytes and platelets *(52–58)*, although recent data suggest a costimulatory/potentiating effect with erythropoietin (EPO) on the production of erythroid progenitors and red blood cells *(60,63,117)* (reviewed in Chapter 7). In addition, results from some in vivo experiments in normal and myelosuppressed animals demonstrate multilineage hemopoietic stimulation *(59,60,63,66,72)*.

3.1. Normal Animal Models

The thrombopoietic potential of recombinant forms of Mpl ligand was underscored early by the magnitude of the response obtained in normal nonhuman primates relative to that observed for IL-3, IL-6, LIF, and IL-11. Platelet counts in normal nonhuman primates administered IL-6 and IL-11 increase within the first week of administration, with peak responses (approximately twofold or greater) occurring within the second week of the treatment schedule *(47,83,86–88,110,118,119)*. Administration of LIF in normal nonhuman primates elicited a dose-dependent, but somewhat delayed, increase in platelet numbers during the second week of a 14-day administration period *(94)*. Recombinant IL-3 administration to nonhuman primates, although capable of eliciting a significant twofold increase in platelet numbers, did not do so consistently *(78,79,81–83,120)*. IL-3 may very well have stimulatory effects on megakaryocyte colony-forming cells (MK-CFC) and megakaryocyte formation, but has no effect on the further maturation and endoreduplication necessary to produce large numbers of platelets.

In response to daily injections of recombinant human (rHu) MGDF for 10 days, platelet levels increased to approximately 600% of baseline within several days and reached maximal levels between days 12 and 14 (Fig. 4) *(59)*. In addition, there were significant increases in the concentration of bone marrow MK-CFC and mixed colony-forming cells (GEMM-CFC), whereas the concentration of GM-CFC or erythroid burst-forming cells (E-BFC) remained unchanged.

Recombinant forms of Mpl ligand have been evaluated in both an N-terminal nonglycosylated, truncated form rHuMGDF and a derivitized form in which poly[ethylene glycol] is attached to the parent molecule (PEG-rHuMGDF). Derivitization with poly[ethylene glycol] increased the in vivo potency of the molecule in normal rodents approximately 20-fold *(66)* and in normal nonhuman primates approximately 10-fold (Farese AM, unpublished). Pegylation of the molecule also significantly increased the circulating half-life approximately 10-fold that of the native full-length, glycosylated rHuMGDF (Cheung E, unpublished observation) *(see* Chapter 13).

3.2. Myelosuppressed Animal Models

The preclinical therapeutic efficacy of Mpl ligand has recently been demonstrated in both rodent and nonhuman primate models of chemotherapy- or radiation-induced moderate and severe myelosuppression *(61,63,65–73)*. Indeed, its demonstrated lineage-dominant biological activity, as opposed to the more pleiotropic activities of IL-3, IL-6, IL-11, LIF, and OSM, make it an ideal candidate for a clinically effective thrombopoietic agent.

Although rHuMGDF completely prevented thrombocytopenia in a murine model of moderate myelosuppression induced by carboplatin alone *(61)*, the unambiguous efficacy of PEG-rHuMGDF was further demonstrated in a severe, lethal model of carboplatin plus radiation-induced marrow aplasia in Balb/c mice (Fig. 5) *(66)*. A similar, although nonlethal, model in a different mouse strain was used to demonstrate the therapeutic efficacy of IL-11 (Fig. 2) *(48)*. PEG-rHuMGDF was effective at reducing

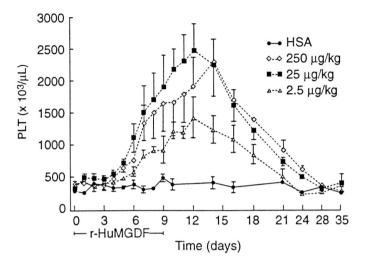

Fig. 4. Effects of r-HuMGDF administration on peripheral blood platelet counts in normal primates. The platelet counts (PLT) (mean ± SEM) observed in normal rhesus monkeys after r-HuMGDF or HSA (●) administrated subcutaneously (sc), every day for 10 consecutive days at dosages of 2.5 (△) (*n* = 3), 25 (■) (*n* = 3), or 250 (◇) (*n* = 2) μg/kg/day. Control animals (*n* = 3) received 25 μg/kg/day of HSA sc, every day for 10 days *(59)*.

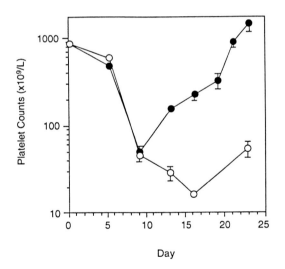

Fig. 5. The effects of PEG-rHuMGDF on platelet counts in myelosuppressed Balb/c mice. Mice were treated on day 0 with carboplatin, irradiation, and received PEG-rHuMGDF at 50 μg/kg/day (●; *n* = 17) or excipient (○; *n* = 25) from days 1–23. Platelet counts (10^9/L) were determined on the indicated days. Data are combined from five separate experiments *(66)*.

the 100% mortality to 14%, while alleviating severe thrombocytopenia, lessening the duration of leukopenia, and improving the anemia caused by combined carboplatin and irradiation *(66)*. More recent studies have confirmed the powerful therapeutic efficacy of this molecule in rodent models of radiation- and/or drug-induced marrow aplasia

Table 2
Thrombocytopenia in Sublethally Irradiated
and Cytokine-Treated Rhesus Monkeys:
Duration, Nadir, and Time to Recovery to Baseline Values[a]

	Duration, d		Nadir, μL^{-1}		Time to recovery, d	
	THROM	NEUT	Platelet	ANC	Platelet	ANC
HSA	12.2	15.5	4,000	2	40	24
r-metHuG-CSF	6.7[b]	12.3[b]	9,000	11	28	18
rHuMGDF	0.25[b]	12.3[b]	28,000[b]	40[b]	22	18
PEG-rHuMGDF	0.0[b]	13.3	43,000[b]	38[b]	19	17
PEG-rHuMGDF and r-metHuG-CSF	0.5[b]	9.0[b,c]	30,000[b]	58[b,c]	21	27

[a]Monkeys were given whole-body irradiation with 700 cGy with ^{60}Co γ-radiation and treated with HSA or the indicated cytokines. Neutropenia (NEUT) is an absolute neutrophil count (ANC) $< 0.5 \times 10^9/$L and thrombocytopenia (THROM) is a platelet count $< 20 \times 10^9/$L.
[b]$p < 0.05$ versus HSA-treated controls.
[c]$p < 0.05$ versus r-metHuG-CSF-treated animals.
Days to recovery to baseline values were not evaluated statistically.
HSA = human serum albumin; r-metHuG-CSF = recombinant methionyl human granulocyte colony-stimulating factor (Filgrastim); ANC = absolute neutrophil count.

(63,67,71). A multilineage response was noted in the model of severe myelosuppression, but the platelet-lineage effect was stronger in the model of moderate marrow aplasia *(61)*. The direct or indirect nature of the MGDF-induced effects on myeloid and erythroid recovery remains to be determined.

Several groups are currently evaluating the therapeutic efficacy of various recombinant forms of Mpl ligand in nonhuman primate models of drug- or radiation-induced myelosuppression *(68–70,72–74)*. Farese et al. *(72)* first reported the efficacy of PEG-rHuMGDF alone and in combination with recombinant human methionyl (r-metHu) G-CSF in a rhesus monkey model of severe radiation-induced thrombocytopenia. Several key variables are used to assess lineage-specific therapeutic efficacy in nonhuman primate models of myelosuppression, including the duration of severe thrombocytopenia (platelets $< 20 \times 10^9/$L), platelet nadir, time of recovery to preirradiation platelet levels, and number of platelet or whole blood transfusions required. Although PEG-rHuMGDF is more biologically active than rHuMGDF *(66,72)*, the effectiveness of both rHuMGDF or PEG-rHuMGDF therapy is exemplified in the significant improvement of all four variables. Severe thrombocytopenia (12.2 days in control animals) was completely abrogated, and the severity of the platelet nadir was improved from $4 \times 10^9/$L in human serum albumin (HSA)-treated control animals to $28 \times 10^9/$L and $43 \times 10^9/$L in irradiated rhesus monkeys treated with rHuMGDF and PEG-rHuMGDF, respectively (Fig. 6A and Fig. 6B; Table 2). This stimulation of platelet production translated into complete independence from platelet transfusion. In addition, the recovery time was reduced by several weeks, from 40 days in control animals to approximately 21 days after treatment with either form of MGDF. Results of colony assays using bone marrow from irradiated animals treated with MGDF support the concept that MGDF stimu-

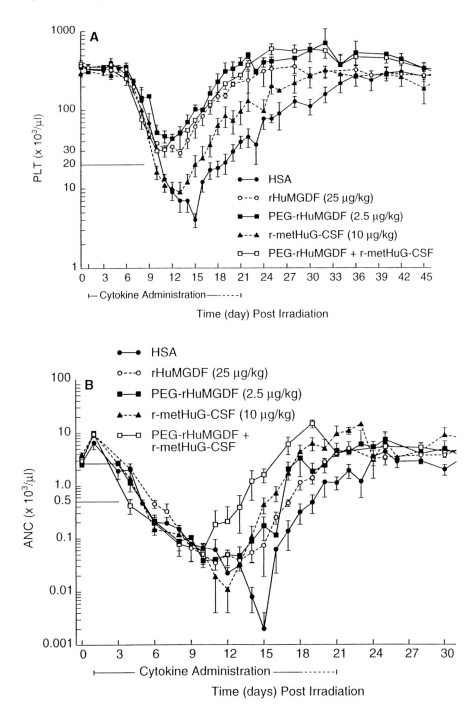

Fig. 6. Effects of cytokine administration on platelet counts (PLT) (mean ± SEM) **(A)** and absolute neutrophil counts (ANC) (mean ± SEM) **(B)** in irradiated rhesus monkeys. Animals were treated subcutaneously with rHuMGDF (25 µg/kg/day) (○) ($n = 4$) and PEG-rHuMGDF (2.5 µg/kg/day) (■) ($n = 4$) from days 1–18, r-metHuG-CSF (10 µg/kg/day) (▲) ($n = 4$) from days 1–21, PEG-rHuMGDF + r-metHuG-CSF (2.5 and 10 µg/kg/day, respectively) (□) ($n = 4$) from days 1–18, or HSA (15 µg/kg/day) (●) ($n = 13$) from days 1–21 *(72)*.

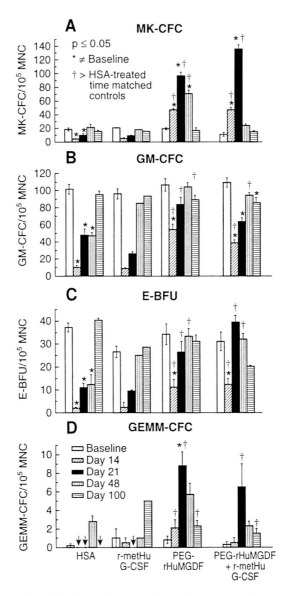

Fig. 7. Bone marrow (BM) MK-CFC **(A)**, GM-CFC **(B)**, E-BFC **(C)**, and GEMM-CFC **(D)** observed in irradiated rhesus primates before (☐) and on days 14 (▨), 21 (■), 48 (▥), and 100 (▤) following total body irradiation (TBI) and administration with PEG-rHuMGDF, r-metHuG-CSF, PEG-rHuMGDF combined with r-metHuG-CSF, or HSA (mean ± SEM of the CFC/10⁵ bone marrow mononuclear cell [MNC] cells) *(72)*.

lates megakaryocytopoiesis in addition to platelet production. PEG-rHuMGDF treatment significantly improved the concentration of marrow-derived MK-CFC within 14 days after total body irradiation. A significant improvement in recovery kinetics was noted for GM-CFC, E-BFC, and GEMM-CFC versus the HSA-treated time-matched control animals, with their frequency enhanced after irradiation versus the HSA-treated,

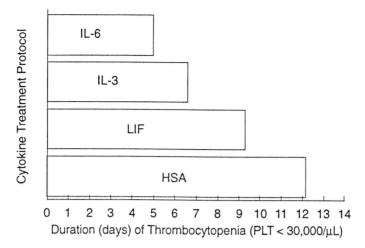

Fig. 8. Duration of thrombocytopenia (platelets < 30 × 10^9/L) versus cytokine treatment protocol in a nonhuman primate model of radiation-induced myelosuppression. IL-3, IL-6, and LIF, or HSA were administered on days 1–23 postirradiation *(34,115)*.

time-matched control group (Fig. 7). The increase in GM-CFC is consistent with the observed improvements in neutrophil nadir, duration of neutropenia, and neutrophil recovery time.

Recombinant forms of IL-6, IL-3, LIF, and OSM have been evaluated in similar nonhuman primate models of radiation-induced myelosuppression (Fig. 8) *(34,115*; MacVittie TJ, unpublished). With the exception of OSM, these cytokines significantly reduced the duration and degree of thrombocytopenia in addition to enhancing the recovery of platelets to within normal levels, although not in a manner as effective as that observed with PEG-rHuMGDF. Comparison of their relative therapeutic efficacy confirms the superior effectiveness of MGDF, either unpegylated or pegylated.

Mpl ligand has been evaluated in three other nonhuman primate models of myelosuppression. One model used X-irradiation and the others, hepsulfam or 5-FU, to induce thrombocytopenia, and each showed thrombopoietic effectiveness *(68–70)*. Administration of Mpl ligand over a 3-week period after X-irradiation abolished the platelet nadir and the duration of thrombocytopenia, and accelerated regeneration and recovery of platelet levels to normal values *(70)*. Reticulocyte recovery was also stimulated, resulting in higher hemoglobin levels. IL-3 and IL-6, also evaluated in this model, were each less effective than Mpl ligand in preventing thrombocytopenia *(70)*. Hepsulfam-induced thrombocytopenia has somewhat different kinetics than that noted with irradiation models. The degree of thrombocytopenia is less severe, occurs later, and its duration is more prolonged *(35,68)*. In this model, administration of PEG-rHuMGDF maintained platelet counts at or above prechemotherapy values without significantly affecting the associated neutropenia and anemia *(68)*. Therapeutic efficacy of Mpl ligand was also evaluated in a baboon model of modest 5-FU-induced myelosuppression where administration of PEG-rHuMGDF led to a more rapid recovery of platelet production *(69)*.

Mpl ligand is the most effective thrombopoietic factor evaluated in both rodent and nonhuman primate preclinical models of radiation- or chemotherapy-induced myelosuppression. Its potential therapeutic benefits include a variable effect on the recovery of neutrophils and the stimulation of erythropoiesis. It is an obvious candidate for use in combination cytokine protocols with either G-CSF, GM-CSF, IL-3, or the synthetic IL-3 receptor agonist, Synthokine.

4. Schedule Variation and Cytokine Efficacy

Cytokine therapy is usually initiated within 24 hours of the myelosuppressive insult. Early availability of growth factors may lead to increased viability and function of critical target cells throughout respective lineages, as well as induce self-renewal and proliferation to rapidly regenerate hemopoiesis. Several studies have evaluated schedule variation by delaying the administration of cytokines for several days after irradiation or cytotoxic drug treatment *(1,5,6,11,48)*. The delayed administration of GM-CSF or G-CSF has led to disparate results. When initiated 7 days after radiation- or drug-induced myelosuppression, GM-CSF did not alter the hemopoietic responses compared with a schedule initiated within 24 hours (Table 1) *(1,6)*. In contrast, G-CSF delayed for 7 days following irradiation of canines or nonhuman primates proved ineffective relative to the initiation of G-CSF treatment within 24 hours of exposure and continued through an equivalent schedule duration of 21–23 days (*11*; Farese AM, unpublished; MacVittie TJ, unpublished).

IL-6, also evaluated in a 7-day delayed schedule, lost all efficacy for platelet production in irradiated nonhuman primates relative to the full schedule initiated within 1 day after irradiation and continued for 21 days (MacVittie TJ, unpublished). There was no significant difference in platelet production between HSA-treated control animals and the delayed IL-6-treated cohorts. In a similar fashion, although using a rodent model of myelosuppression, delaying treatment with IL-11 until day 3 after combined irradiation and carboplatin chemotherapy, reduced the ability of IL-11 to accelerate platelet recovery; delaying treatment until day 7 completely abrogated IL-11's platelet-enhancing activity *(48)*.

In contrast, delaying the addition of PEG-rHuMGDF until 5 days after exposure, although not as effective as the early, day 1 initiation of the 18-day treatment schedule, gave significantly better results than those seen with the HSA-treated controls in reducing the duration of thrombocytopenia *(74)*. The full-schedule protocol (day 1–18) remained significantly better than the delayed schedule, improving both duration of thrombocytopenia and platelet nadir *(74)*.

Comparison of the thrombopoietic stimulus provided by the delayed PEG-rHuMGDF schedule to that provided by either IL-6 or IL-3 administered in a full schedule from day 1–21 after irradiation in the same animal model of radiation-induced myelosuppression underscores the superiority of PEG-rHuMGDF relative to those cytokines. PEG-rHuMGDF administered in the delayed schedule improved all platelet variables, including duration of thrombocytopenia, nadir, and recovery, in a manner comparable to that noted for IL-6 or IL-3 administered in a full-treatment schedule *(34,74)*.

Table 3
Cytokine-Based Therapy (Potential and Present) for Treatment of Bone Marrow Myelosuppression or Myeloablation with Stem Cell Transplantation

Current Generation	
Single cytokines	Combined cytokines
IL-1, IL-3, IL6 IL-11, G-CSF, GMCSF LIF	IL-1 + G-CSF IL-3 + IL-6, IL-3 + GM-CSF, IL-3 + GM-CSF IL-11 + IL-3, IL-11 + G-CSF, IL-11 + GM-CSF IL-6 + GM-CSF, IL-6 + G-CSF PIXY-321 (IL-3/GM-CSF)
New Generation	
Single cytokines	Combined cytokines
Mpl Ligand (MGDF, TPO) Synthokine (IL-3 receptor agonist)	MGDF + G-CSF TPO + GM-CSF, Synthokine + G-CSF Myelopoietin (IL-3/G-CSF) Promegapoietin (IL-3/Mpl ligand) Flt-3 ligand/PIXY-321

Synthokine = synthetic cytokine IL-3; Myelopoietin = IL-3/G-CSF; Promegapoietin = IL-3/Mpl ligand; PIXY-321 = IL-3/GM-CSF fusion protein (PIXY-321); LIF = leukemia-inhibitory factor; Mpl ligand = thrombopoietin (TPO), megakaryocyte growth and development factor (MGDF); flt-3/flk-2 = fetal liver tyrosine kinase ligand.

5. Combined Cytokine Protocols

Despite effective monotherapy with the hemopoietic growth factors after radiation or chemotherapy exposure, multilineage recovery has been elusive. The regeneration of both critical cell lineages, neutrophils and platelets, will undoubtedly require a combination of existing and/or new-generation cytokines (Table 3) that may include a multilineage growth factor, or addition of two lineage-restricted cytokines, such as Mpl ligand and G-CSF. In vitro analyses of cytokine interactions on primitive colony-forming cells promote the concept that early multipotent stem cells require multiple cytokines to stimulate proliferation and differentiation into more committed progenitors requiring single lineage cytokine interaction. Indeed, the cytokines IL-1, IL-6, IL-7, IL-11, IL-12, IL-13, stem cell factor (SCF), and the flt-3 ligand all appear to enhance synergistically the colony-forming ability of growth factors, such as G-CSF, GM-CSF, and IL-3 *(30,121–130)*. SCF, flt-3 ligand, and IL-3 increase survival of primitive hemopoietic colony-forming cells and may have important roles in dampening radiation- or cytotoxic drug-induced apoptosis *(131–133)*. A number of combination cytokine protocols have been evaluated in nonhuman primate models of radiation- or drug-induced myelosuppression, including IL-3/GM-CSF, IL-3/IL-6, IL-6/G-CSF, IL-6/GM-CSF, and PIXY-321 *(1,34,35,49,50,111,134–136)*. To these we can now add the recently evaluated combinations of Synthokine/G-CSF and PEG-rHuMGDF/r-metHuG-CSF *(72,137)*.

5.1. Interleukin-3/Granulocyte Macrophage Colony-Stimulating Factor

IL-3 is conceptually a key cytokine for use in combination studies, based on its capacity for stimulation of multilineage hemopoietic colony-forming cells, marked in vitro synergy with lineage-specific colony-stimulating factors for production of neutrophils and megakaryocytes, and survival-promoting ability for primitive hemopoietic progenitors. However, the proinflammatory effects of IL-3 have reduced its therapeutic index and clinical utility.

Combined IL-3/GM-CSF protocols have been evaluated in two nonhuman primate models, radiation- and hepsulfam-induced myelosuppression *(1,136)* (Table 1 and Fig. 9). Our laboratory assessed the therapeutic efficacy of two administration protocols, sequential administration (IL-3, days 1–7; GM-CSF, days 7–21 after exposure) and coadministration (IL-3, days 1–21; GM-CSF, days 1–21 after exposure) of IL-3/GM-CSF relative to monotherapy in mixed neutron-γ irradiated nonhuman primates *(1)*. The coadministered IL-3/GM-CSF schedule was more effective in enhancing recovery of both platelets and neutrophils relative to the sequential schedule or of monotherapy with either cytokine. Coadministered IL-3/GM-CSF reduced both platelet and neutrophil nadirs, durations of thrombocytopenia and neutropenia, and recovery time of both cell lineages relative to either respective control groups or the sequential IL-3/GM-CSF schedule. The sequential IL-3/GM-CSF schedule had no greater effect on neutrophil production than GM-CSF alone and was less effective than IL-3 alone in reducing thrombocytopenia. The therapeutic efficacy of coadministered IL-3/GM-CSF was confirmed in a rhesus monkey model of hepsulfam-induced myelosuppression *(134)*. IL-3 coadministered with GM-CSF from days 1–22 after drug exposure completely abrogated the significant decline in platelet counts noted in either control animals or animals treated with IL-3 or GM-CSF as monotherapy *(136)* (Fig. 9).

5.2. Interleukin-3/Interleukin-6

The therapeutic efficacy of IL-3 plus IL-6 has also been evaluated in nonhuman primate models of radiation- or chemotherapy-induced myelosuppression *(34,35)*. With regard to the treatment of radiation-induced marrow aplasia, the sequential use of IL-3/IL-6 significantly increased production of platelets compared with both HSA-treated controls and nonhuman primates receiving the coadministered IL-3/IL-6, but did not significantly enhance platelet recovery over either IL-6 or IL-3 as monotherapy, although the duration of thrombocytopenia was improved *(34)*. The IL-3/IL-6 protocols did not significantly enhance recovery of neutrophils. Winton et al. *(35)* showed that although coadministered IL-3/IL-6 was no better than IL-6 alone in reducing the duration of a modest hepsulfam-induced thrombocytopenia, both protocols significantly reduced the duration of severe neutropenia relative to the vehicle-treated control animals.

5.3. Interleukin-6/Granulocyte Colony-Stimulating Factor and Interleukin-6/Granulocyte Macrophage Colony-Stimulating Factor

Coadministered IL-6/GM-CSF rendered hepsulfam-induced thrombocytopenic animals transfusion independent and had the most sustained effect on platelet regeneration relative to IL-6 or GM-CSF alone *(135)*. Neutrophil recovery in this model was significantly enhanced by coadministration of either IL-6 plus G-CSF or IL-6 plus GM-CSF. In addition, coadministered IL-6/G-CSF or IL-6/GM-CSF can accelerate recovery of both platelets and neutrophils relative to HSA-treated control animals after radiation-induced marrow aplasia. The coadministered IL-6/G-CSF or IL-6/GM-CSF, however, did not accelerate platelet or neutrophil recovery relative to IL-6, G-CSF, or GM-CSF monotherapy *(111)*. The cumulative preclinical data suggest that IL-6 com-

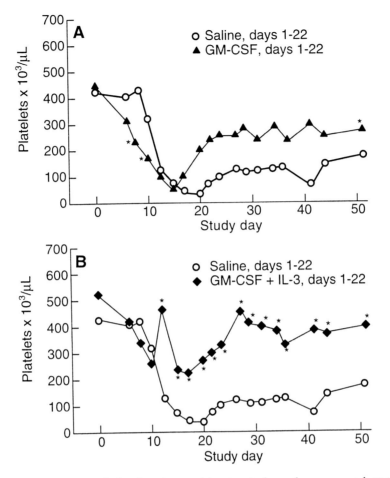

Fig. 9. Regeneration of platelets in cytokine-treated myelosuppressed nonhuman primates. Cytokines were administered subcutaneously, consecutively from days 1–22 posthepsulfam (1500 mg/m²) treatment. Rhesus monkeys were administered either **(A)** GM-CSF alone (5 µg/kg/day, qd) (▲) (*n* = 3) or **(B)** GM-CSF (5 µg/kg/day, qd) plus IL-3 (15 µg/kg/day, bid) (◆) (*n* = 3), or saline (A,B) (○) (*n* = 4) *(136)*.

bined with G-CSF or GM-CSF is effective in enhancing recovery of neutrophils and platelets, although no synergies were shown.

5.4. Interleukin-11/Interleukin-3 and Interleukin-11/ Granulocyte Macrophage Colony-Stimulating Factor

The combination of IL-11 with either IL-3 or GM-CSF has only been evaluated in normal nonhuman primates *(119)*. The use of IL-11/IL-3 or IL-11/GM-CSF resulted in an increase of circulating platelets and neutrophils relative to the single cytokine controls. These protocols have yet to be evaluated in models of myelosuppression.

5.5. Synthokine/Granulocyte Colony-Stimulating Factor

In a nonhuman primate model of radiation-induced myelosuppression, Synthokine/ G-CSF significantly decreased the duration and nadir of thrombocytopenia relative to either HSA-control or G-CSF monotherapy cohorts (Table 4) *(137)*. The combination

Table 4
**Thrombocytopenia and Neutropenia in Sublethally Irradiated
and Cytokine-Treated Rhesus Monkeys:
Duration, Nadir, and Time to Recovery to Baseline Values**[a]

	Duration, days		Nadir, μL^{-1}		Time to recovery, days	
	THROM	NEUT	Platelet	ANC	Platelet	ANC
HSA	11.9	14.8	5000	0	40	24
Synthokine	$3.5^{b,c}$	14.1	$14,000^{b,c}$	$90^{b,c}$	24	21
rHuG-CSF	7.2^{b}	12.4	7000^{b}	4	25	18
Synthokine + rHuG-CSF	$1.9^{b,c}$	10.6	$22,000^{b,c}$	$100^{b,c}$	26	17

[a]Monkeys were given whole-body irradiation with 700 cGy with ^{60}Co γ-radiation and treated with HSA or the indicated cytokines. Neutropenia (NEUT) is an absolute neutrophil count (ANC) $< 0.5 \times 10^9$/L and thrombocytopenia (THROM) is a platelet count $< 20 \times 10^9$/L.
[b]$p < 0.05$ vs HSA-treated controls.
[c]$p < 0.05$ vs r-metHuG-CSF-treated animals.

also improved the duration of thrombocytopenia and platelet nadir noted for Synthokine alone and further reduced, although not significantly, the duration of neutropenia consequent to G-CSF administration. This therapeutic combination also significantly improved the neutrophil nadir relative to G-CSF- or HSA-treated animals.

5.6. Megakaryocyte Growth and Differentiation Factor/ Granulocyte Colony-Stimulating Factor

Monotherapy with rHuMGDF, using either the unmodified or pegylated form of the protein, in a nonhuman primate model of severe radiation-induced myelosuppression significantly improves all key variables of platelet recovery *(72,73)*. Monotherapy with PEG-rHuMGDF was also found to influence granulopoiesis as evidenced by the significantly improved neutrophil nadir, diminished duration of neutropenia, and decreased time to recovery of neutrophils to preirradiation levels (Fig. 6). It was, therefore, of interest to evaluate further the efficacy of combined G-CSF and MGDF administration to enhance maximally the production of both platelets and neutrophils. An additional issue to be evaluated in this model was whether lineage competition would occur within a limited pool of surviving target cells.

PEG-rHuMGDF coadministered with r-metHuG-CSF significantly improved all platelet-recovery variables to the same degree as PEG-rHuMGDF monotherapy (Fig. 6 and Table 2) *(72)*. In addition, PEG-rHuMGDF/r-metHuG-CSF significantly enhanced granulopoiesis as evidenced by a further, significant decrease in the duration of neutropenia versus r-metHuG-CSF alone, as well as a significant improvement in neutrophil nadir versus r-metHuG-CSF alone. The recovery time of neutrophils to normal levels remained similar to that of either r-metHuG-CSF or PEG-rHuMGDF alone. The accelerated recovery of neutrophils and platelets was accompanied by a significant increase in bone marrow colony-forming cells. The frequencies of MK-CFC, GM-CFC, E-BFC, and GEMM-CFC were enhanced after irradiation versus their HSA-treated time-matched controls (Fig. 7). Similar efficacy on multilineage recovery has been noted in rodent models of carboplatin and irradiation-induced myelosuppression *(66,67)*. PEG-rHuMGDF combined with r-metHuG-CSF eliminated the already significantly improved mortality (14%) noted with PEG-rHuMGDF alone in a severe 100% lethal carboplatin/radiation model *(66)*. The PEG-rHuMGDF/r-metHuG-CSF combination significantly shortened the duration of leukopenia as well. In a nonlethal model of chemotherapy with 5-FU, the combination of TPO and G-CSF enhanced neutrophil recovery relative to G-CSF alone and did not compromise the platelet or erythroid response to TPO *(67)*. Combination Mpl ligand and G-CSF therapy clearly offered the best stimulation of both platelets and neutrophils in these models of clinically relevant myelosuppression.

6. Novel Cytokines

The cloning and production of Mpl ligand, as well as the engineering of a new group of agonists of the human IL-3 and G-CSF receptors (Myelopoietin) or Mpl receptor (Promegapoietin), have allowed study of another group of molecules with the potential to ameliorate radiation- or drug-induced myelosuppression *(138–141)*. The ability to form multifunctional receptor agonists may allow for manipulation of relative cytokine binding and signal transduction in hemopoietic target cells and enhancement of the therapeutic index. Myelopoietin stimulates recovery of both neutrophils and platelets in myelosuppressed nonhuman primates in a manner comparable to that noted for combined Mpl ligand and G-CSF *(72,141)*.

7. Summary

The therapeutic efficacy of Mpl ligand in myelosuppressed animals was forecast by the intense thrombopoietic response of normal animals to this cytokine. MGDF is much more effective in stimulating an earlier and larger megakaryocytopoietic and thrombopoietic response in normal nonhuman primates than any other single thrombopoietic cytokine, including IL-3, IL-6, IL-11, or LIF, *(8,47,59,79,82,83,86–88,110,120)*, or even certain cytokine combinations, such as IL-1/IL-3, IL-3/GM-CSF, and IL-3/IL-6 *(79,81–83,120,142)*.

As the major physiologic regulator of platelet production, Mpl ligand significantly improves all key variables associated with restoration of megakaryocyte and platelet production in myelosuppressed nonhuman primates or rodents. Platelet nadir, duration of thrombocytopenia, time to recovery, number of platelet transfusions, and bone marrow MK-CFC clonogenic activity are all substantially improved.

The combined administration of G-CSF with Mpl ligand to myelosuppressed nonhuman primates or rodents enhances recovery of both critical cell lineages, platelets, and neutrophils. An important result is the lack of any demonstrable competition between these potent lineage-dominant cytokines.

It is also of interest that treatment of myelosuppressed nonhuman primates or rodents with Mpl ligand monotherapy stimulates a multilineage response, with increased production of myeloid and/or erythroid cells in addition to platelets *(60,63,66,72)*. In cultures containing EPO, the addition of Mpl ligand enhances the development of early erythroid progenitors *(117)*. The effect on granulopoiesis may be more indirect, considering the recent evidence that megakaryocytes synthesize multiple cytokines both constitutively and in response to other cytokines. Synthesis and release of IL-1, IL-6, GM-CSF, or IL-3 by megakaryocytes within the regenerating marrow microenvironment may augment production of neutrophils, as well as platelets and erythrocytes

(143,144). These results provide a clear rationale for the clinical evaluation of Mpl ligand alone or in combination with G-CSF or GM-CSF.

References

1. Farese AM, Williams DE, Seiler FR, MacVittie TJ. Combination protocols of cytokine therapy with interleukin-3 and granulocyte-macrophage colony-stimulating factor in a primate model of radiation-induced marrow aplasia. *Blood*. 1993; 82: 3012–3018.
2. Farese AM, Kirschner KF, Patchen ML, Zsebo KM, MacVittie TJ. The effect of recombinant canine stem cell factor and/or recombinant canine granulocyte colony stimulating factor on marrow aplasia recovery in lethally irradiated canines. *Exp Hematol*. 1993; 21: 1169 (abstract no 585).
3. Fushiki M, Ono K, Sasai K, et al. Effect of recombinant human granulocyte colony stimulating factor on granulocytopenia in mice induced by irradiation. *Int J Radiat Oncol Biol Phys*. 1990; 18: 353–357.
4. Kobayashi Y, Okabe T, Urabe A, Suzuki N, Takaku F. Human granulocyte colony stimulating factor produced by *Escherichia coli* shortens the period of granulocytopenia induced by irradiation in mice. *J Can Res*. 1987; 78: 763–768.
5. MacVittie TJ, Monroy RL, Patchen ML, Souza LM. Therapeutic use of recombinant human G-CSF in a canine model of sublethal and lethal whole-body irradiation. *Int J Radiat Biol*. 1990; 57: 723–736.
6. Meisenberg BR, Davis TA, Melaragno AJ, Stead R, Monroy R L. A comparison of therapeutic schedules for administering granulocyte colony-stimulating factor to nonhuman primates after high-dose chemotherapy. *Blood*. 1992; 79: 2267–2272.
7. Monroy RL, Skelly RR, MacVittie TJ, et al. The effect of recombinant GM-CSF on the recovery of monkeys transplanted with autologous bone marrow. *Blood*. 1987; 70: 1696–1699.
8. Monroy RL, Skelly RR, Taylor P, Dubois A, Donahue RE, MacVittie TJ. Recovery from severe hemopoietic suppression using recombinant human granulocyte-macrophage colony stimulating factor. *Exp Hematol*. 1988; 16: 334–338.
9. Moore MAS, Warren DJ. Interleukin-1 and G-CSF synergism: in vivo stimulation of stem cell recovery and hematopoietic regeneration following 5-fluorouracil treatment in mice. *Proc Natl Acad Sci USA*. 1987; 84: 7134–7138.
10. Patchen ML, MacVittie TJ, Solberg BD, Souza LM. Therapeutic administration of recombinant human granulocyte colony stimulating factor accelerated hemopoietic regeneration and enhances survival in a murine model of radiation-inducd myelosuppression. *Int J Cell Cloning*. 1990; 8: 107–122.
11. Schuening FG, Storb R, Goehle S, et al. Effect of recombinant human granulocyte colony-stimulating factor on hematopoiesis of normal dogs and on hematopoietic recovery after otherwise lethal total body irradiation. *Blood*. 1989; 74: 1308–1313.
12. Shimamura M, Kobayashi T, Yuo A, et al. Effect of human recombinant granulocyte-colony-stimulating factor on hemopoietic injury in mice induced by 5-fluorouracil. *Blood*. 1987; 69: 353–355.
13. Tanikawa S, Nakao I, Tsuneska K, Nobio N. Effects of recombinant granuloctye colony-stimulating factor (rG-CSF) and recombinant granuloctye-macrophage colony-stimulating factor (rGM-CSF) on acute radiation hematopoietic injury in mice. *Exp Hematol*. 1989; 17: 883–888.
14. Tanikawa S, Nose M, Yoshiro A, Tsuneoka K, Shikita M, Nara N. Effects of recombinant human granuloctye colony-stimulating factor on the hematologic recovery and survival of irradiated mice. *Blood*. 1990; 76: 445–449.
15. Welte K, Bonilla MA, Gillio AP, et al. Recombinant human granulocyte colony-stimulating factor. Effects on hematopoiesis in normal and cyclophosphamide-treated primates. *J Exp Med*. 1987; 165: 941–948.

16. Bodey GP, Buckley M, Sathe YS, Freireich EJ. Quantitative relationships between circulating leukocytes and infection in patients with acute leukemia. *Ann Intern Med*. 1966; 64: 328–340.

17. Schimpff SC. Infection prevention in patients with cancer and granulocytopenia. In: Grieco MH (ed). *Infections in the Abnormal Host*. New York: New York Medical Books; 1980: 926–950.

18. Shenep JL. Empiric antimicrobial treatment in febrile neutropenic cancer patients. *Infect Med*. 1992; April: 39–47.

19. Weisbart RH, Golde DW, Clark SC, Wong GG, Gasson JC. Human granulocyte-macrophage colony-stimulating factor is a neutrophil activator. *Nature*. 1985; 314: 361–363.

20. Weisbart RH, Gasson JC, Golde DW. Colony-stimulating factors and neutrophils. Colony-stimulating factors and host defense. *Ann Intern Med*. 1989; 110: 297–303.

21. Arnaout MA, Wang EA, Clark SC, Sieff CA. Human recombinant granulocyte-macrophage colony-stimulating factor increases cell-to-cell adhesion and surface expression of adhesion-promoting surface glycoproteins on mature granulocytes. *J Clin Invest*. 1986; 78: 597–601.

22. Gasson JC, Weisbart RH, Kaufman SE, et al. Purified human granulocyte-macrophage colony-stimulating factor: direct action on neutrophils. *Science*. 1984; 226: 1339–1342.

23. Mayer P, Lam C, Obenaus H, Liehl E, Besemer J. Recombinant human GM-CSF induces leukocytosis and activates peripheral blood polymorphonuclear neutrophils in nonhuman primates. *Blood*. 1987; 70: 206–213.

24. Mayer P, Schütze E, Lam C, Kricek F, Liehl E. Recombinant murine granulocyte-macrophage colony-stimulating factor augments recovery and enhances resistance to infections in myelosuppressed mice. *J Infect Dis*. 1991; 163: 584–590.

25. Cohen AM, Hines DK, Korach ES, Ratzkin BJ. In vivo activation of neutrophil function in hamsters by recombinant human granulocyte-macrophage colony-stimulating factor. *Infect Immun*. 1988; 56: 2861–2865.

26. Kitagawa S, Yuo A, Suoza LM, Saito M, Miura Y, Takaku F. Recombinant human granulocyte-macrophage colony-stimulating factor enhances superoxide release in human granulocytes stimulated by chemotactic peptide. *Biochem Biophys Res Commun*. 1987; 144: 1143–1146.

27. McNiece IK, McGrath HE. Granulocyte colony-stimulating factor augments in vitro megakaryocyte colony formation by interleukin-3. *Exp Hematol*. 1988; 16: 807.

28. Paquette RL, Zhou JY, Yang YC, Clark SC, Koeffler HP. Recombinant gibbon interleukin-3 acts synergistically with recombinant human G-CSF and GM-CSF in vitro. *Blood*. 1988; 71: 1596.

29. McNiece IK, Andrew R, Stewart M, Clark S, Boone T, Quesenberry P. Action of interleukin-3, G-CSF, and GM-CSF on highly enriched human hematopoietic progenitor cells: Synergistic interaction of GM-CSF plus G-CSF. *Blood*. 1989; 74: 110.

30. Ikebuchi K, Ihle JN, Hirai Y, Wong GG, Clark SC, Ogawa M. Synergistic factors for stem cell proliferation: further studies of the target stem cells and the mechanism of stimulation by interleukin-1, interleukin-6, and granulocyte colony-stimulating factor. *Blood*. 1988; 72: 2007–2014.

31. Robinson BE, McGrath HE, Quesenberry, PJ. Recombininat murine granulocyte macrophage colony stimulating factor has megakaryocyte stimulating action and augments megakaryocyte colony stimulation by IL-3. *J Clin Invest*. 1987; 79: 1648.

32. Bruno E, Miller ME, Hoffman R. Interacting cytokines regulate in vitro human megakaryocytopoiesis. *Blood*. 1989; 73: 671–677.

33. Gillio AP, Gasparetto C, Laver J, et al. Effects of interleukin-3 on hematopoietic recovery after 5-fluorouracil or cyclophosphamide treatment of cynomolgus primates. *J Clin Invest*. 1990; 85: 1560–1565.

34. MacVittie TJ, Farese AM, Patchen ML, Myers L A. Therapeutic efficacy of recombinant interleukin-6 (IL-6) alone and combined with recombinant human IL-3 in a nonhuman primate model of high-dose, sublethal radiation-induced marrow aplasia. *Blood*. 1994; 84: 2515–2522.

35. Winton EF, Srinivasiah J, Kim BK, et al. Effect of recombinant human interleukin-6 (rhIL-6) and rhIL-3 on hematopoietic regeneration as demonstrated in a nonhuman primate chemotherapy model. *Blood*. 1994; 84: 65–73.

36. Ganser A, Seipelt G, Lindemann A, et al. Effects of recombinant human interleukin-3 in patients with myelodysplastic syndromes. *Blood*. 1990; 76: 455–462.

37. Kurzrock R, Talpaz M, Estrov Z, Rosenblum MG, Gutterman JU. Phase I study of recombinant human interleukin-3 in patients with bone marrow failure. *J Clin Oncol*. 1991; 9: 1241–1250.

38. Biesma B, Willemse PH, Mulder NH, et al. Effects of interleukin-3 after chemotherapy for advanced ovarian cancer. *Blood*. 1992; 80: 1141–1148.

39. Ganser A, Lindemann A, Ottmann OG, et al. Sequential in vivo treatment with two recombinant human hematopoietic growth factors (interleukin-3 and granulocyte-macrophage colony-stimulating factor) as a new therapeutic modality to stimulate hematopoiesis: results of a phase I study. *Blood*. 1992; 79: 2583.

40. Nemunaitis J, Buckner CD, Appelbaum FR, et al. Phase I trial with recombinant human interleukin-3 (rhIL-3) in patients with lymphoid cancer undergoing autologous bone marrow transplantation (ABMT). *Blood*. 1992; 80: 85a (abstract no 331)

41. Gerhartz HH, Walther J, Bunica O, et al. Clinical, hematological, and cytokine response to interleukin-3 (IL-3) supported chemotherapy in resistant lymphomas: a phase II study. *Proc ASCO*. 1992; 11: 329 (abstract no 1123).

42. Fibbe WE, Raemaekers J, Verdonck LF, et al. Human recombinant interleukin-3 after autologous bone marrow transplantation for malignant lymphoma. *Ann Oncol*. 1992; 3: 163 (abstract).

43. Vellenga E, de Vries EF. Effects of interleukin-3 after chemotherapy for advanced ovarian cancer. *Blood*. 1992; 80: 1141.

44. Burstein SA, Downs T, Friese P, et al. Thrombocytopoiesis in normal and sublethally irradiated dogs: response to human interleukin-6. *Blood*. 1992; 80: 420–428.

45. Du XX, Neben T, Goldman S, Williams DA. Effects of recombinant interleukin-11 on hematopoietic reconstitution in transplant mice: acceleration of recovery of peripheral blood neutrophils and platelets. *Blood*. 1993; 81: 27–34.

46. Farese AM, Herodin F, Baum C, Burton E, McKearn JP, MacVittie TJ. Acceration of hematopoietic reconstitution with a synthokine (SC-55494) after radiation-induced bone marrow aplasia. *Blood*. 1996; 87: 581–591.

47. Herodin F, Mestries JC, Janodet D, et al. Recombinant glycosylated human interleukin-6 accelerates peripheral blood platelet count recovery in radiation-induced bone marrow depression in baboons. *Blood*. 1992; 80: 688–695.

48. Leonard JP, Quinto CM, Goldman SJ, Kozitza MK, Neben TY. Recombinant human interleukin-11(rhIL-11) multilineage hematopoietic recovery in mice after a myelosuppressive regiman of sublethal irradiation and carboplatin. *Blood*. 1994; 83: 1499–1506.

49. Williams DE, Dunn JT, Park LS, et al. A GM-CSF/ IL-3 fusion protein promotes neutrophil and platelet recovery in sublethally irradiated rhesus monkeys. *Biotechnol Ther*. 1993; 4: 17–29.

50. Williams DE, Farese A, MacVittie TJ. PIXY321, but not GM-CSF plus IL-3, promotes hematopoietic reconstitution following lethal irradiation. *Blood*. 1993; 82: 366a (abstract no 1448).

51. Vadhan-Raj S, Papadopoulos NE, Burgess MA, et al. Effects of PIXY321, a granulocyte-macrophage colony-stimulating factor/interleukin-3 fusion protein, on chemotherapy-induced multilineage myelosuppression in patients with sarcoma. *J Clin Oncol*. 1994; 12: 715–724.

52. Bartley T, Bogenberger J, Hunt P, et al. Identification and cloning of a megakaryocyte growth and development factor that is a ligand for the cytokine receptor Mpl. *Cell*. 1994; 77: 1117–1124.

53. Kato T, Ogami K, Shimada Y, et al. Purification and characterization of thrombopoietin. *J Biochem*. 1995; 118: 229–236.

54. de Sauvage FJ, Hass PE, Spencer SD, et al. Stimulation of megakaryocytopoiesis and thrombopoiesis by the c-Mpl ligand. *Nature*. 1994; 369: 533–538.

55. Lok S, Kaushansky K, Holly RD, et al. Cloning and expression of murine thrombopoietin cDNA and stimulation of platelet production in vivo. *Nature*. 1994; 369: 565–568.

56. Kuter DJ, Beeler DL, Rosenberg RD. The purification of megapoietin: a physiological regulator of megakaryocyte growth and development production. *Proc Natl Acad Sci USA*. 1994; 91: 11,104–11,108.

57. Kaushansky K, Lok S, Holly RD, et al. Promotion of megakaryocyte progenitor expansion and differentiation by the c-Mpl ligand thrombopoietin. *Nature.* 1994; 369: 568–571.

58. Wendling F, Maraskovsky E, Debill N, et al. c-Mpl ligand is a humoral regulator of megakaryocytopoiesis. *Nature.* 1994; 369: 571–74.

59. Farese AM, Hunt P, Boone TC, MacVittie TJ. Recombinant human megakaryocyte growth and development factor (r-HuMGDF) simulates megakaryocytopoiesis in normal primates. *Blood.* 1995; 86: 54–59.

60. Kaushansky K, Broudy VC, Grossmann A, et al. Thrombopoietin expands erythroid progenitors, increases red cell production, and enhances erythroid recovery after myelosuppressive therapy. *J Clin Invest.* 1995; 96: 1683–1687.

61. Ulich TR, del Castillo J, Yin S, et al. Megakaryocyte growth and development factor ameliorates carboplatin-induced thrombocytopenia in mice. *Blood.* 1995; 86: 971–976.

62. Winton EF, Thomas GR, Marian ML, Bucur SZ, McClure HM, Eaton DL. Prediction of a threshold and optimally effective thrombocytopoietic dose of recombinant human thrombopoietin (rhTPO) in nonhuman primates based on murine pharmacokinetic data. *Exp Hematol.* 1995; 23: 879a.

63. Kaushansky K, Lin NL, Grossmann A, Humes J, Sprugel KH, Broudy VC. Thrombopoietin expands erythroid, granulocyte-macrophage, and megakaryocyte progenitor cells in normal and myelosuppressed mice. *Exp Hemato*l. 1996; 24: 265–269.

64. Harker LA, Hunt P, Marzec UM, et at. Regulation of platelet production and function by megakaryocyte growth and development factor in nonhuman primates. *Blood.* 1996; 87: 1833–1844.

65. Sprugel KH, Humes JM, Grossmann A, Ren HP, Kaushansky K. Recombinant thrombopoietin stimulates rapid platelet recovery in thrombocytopenic mice. *Blood* 1994; 84: 242a (Abstract 952).

66. Hokom MM, Lacey D, Kinsler O, et al. Megakaryocyte growth and development factor abrogates the lethal thrombocytopenia associated with carboplatin and irradiation in mice. *Blood.* 1995; 86: 4486–4492.

67. Grossmann A, Lenox JS, Humes JM, Ren HP, Kaushansky K, Sprugel KH. Effects of the combined administration of TPO and G-CSF on recovery from myelosuppression in mice. *Blood.* 1995; 86: 371a (abstract no 1473).

68. Harker LA, Marzec UM, Kelly AB, Hanson SR. Enhanced hematopoietic regeneration in primate model of myelosuppressive chemotherapy by pegylated recombinant human megakaryocyte growth and development factor (PEG-rHuMGDF) in combination with granulocyte colony stimulating factor (G-CSF). *Blood.* 1995; 86: 497a (abstract no 1976).

69. Andrews RG, Winkler A, Woogerd P, et al. Recombinant human megakaryocyte growth and development factor (rHuMGDF) stimulates thrombopoiesis in normal baboons and accelerates platelet recovery after chemotherapy. *Blood.* 1995; 86: 371a (abstract no 1471).

70. Neelis KJ, Wognum AW, Eaton D, Thomas R, Wagemaker G. Preclinical evaluation of thrombopoietin in rhesus monkeys. *Blood.* 1995; 86: 256a (abstract no 1011)

71. Thibodeaux H, Mathias J, Eaton DL, Thomas GR. Evaluation of thrombopoietin (TPO) in murine models of thrombocytopenia induced by whole body irradiation and cancer chemotherapeutic agents. *Blood.* 1995; 86: 497a (abstract no 1977).

72. Farese AM, Hunt P, Grab LB, MacVittie TJ. Combined administration of recombinant human megakaryocyte growth and development factor and granulocyte colony stimulating factor enhances multi-lineage hematopoietic reconstitution in nonhuman primates following radiation-induced marrow aplasia. *J Clin Invest.* 1996; 97: 2145–2151.

73. Farese AM, Hunt P, Grab LB, MacVittie TJ. Evaluation of administration protocols of pegylated megakaryocyte growth and development factor (PEG-rHuMGDF) on platelet recovery in a primate model of radiation-induced bone marrow aplasia. *Blood.* 1995; 86: 497a (abstract no 1975).

74. MacVittie TL, Farese AM, Grab LB, Hunt P. Effect of delayed administration of recombinant human megakaryocyte growth and development factor on hematopoietic reconstitution in non-

human primates following radiation-induced marrow aplasia. *Exp Hematol.* 1995; 23: 311a (abstract no 830).

75   Hoffman R. Regulation of megakaryocytooiesis. *Blood.* 1989; 74: 1196–1212.

76.   Gordon MS, Hoffman R. Growth factors affecting human thrombocytopoiesis: potential agents for treatment of thrombocytopenia. *Blood.* 1992; 80: 302–307.

77.   Kaushansky K. Thrombopoietin: the primary regulator of platelet production. *Blood.* 1995; 86: 419–431.

78.   Mayer P, Valent P, Schmidt G, Liehl E, Bettelheim P. The in vivo effects of recombinant human interleukin-3: Demonstration of basophil differentiation factor, histamine-producing activity, and priming of GM-CSF responsive progenitors in nonhuman primates. *Blood.* 1989; 74: 613–621.

79.   Krumwieh D, Weinmann E, Seiler FR. Different effects of interleukin-3 (IL3) on the hematopoiesis of subhuman primates due to various combinations with granulocyte-macrophage colony-stimulating factor (GM-CSF) and granulocyte colony stimulating factor (G-CSF). *Int J Cell Cloning.* 1990; 8: 229–248.

80.   Wagemaker G, van Gils FCJM, Burger H, et al. Highly increased production of bone marrow derived blood cells by administration of homologous interleukin-3 to rhesus monkeys. *Blood.* 1990; 76: 2235–2241.

81.   Monroy RL, Davis TA, Donahue RE, MacVittie TJ. In vivo stimulation of platelet production in a primate model using IL-1 and IL-3. *Exp Hematol.* 1991; 19: 629–635.

82.   Stahl CP, Winton EF, Monroe MC, et al. Differential effects of sequential, simultaneous, and single agent interleukin-3 and granulocyte macrophage colony-stimulating factor on megakaryocyte maturation and platelet response in primates. *Blood.* 1992; 80: 2479–2485.

83.   Geissler K, Valent P, Bettelheim P, et al. In vivo synergism of recombinant human interleukin-3 and recombinant human interleukin-6 on thrombopoiesis in primates. *Blood.* 1992; 79: 1155–1160.

84.   Emerson SG, Yang YC, Clark SC, Long MW. Human recombinant granulocyte-macrophage colony stimulating factor and interleukin-3 have overlapping but distinct hematopoietic activities. *J Clin Invest.* 1988; 82: 1282–1287.

85.   Stahl CP, Winton EF, Monroe MC, et al. Recombinant human granulocyte-macrophage colony stimulating factor promotes megakaryocyte maturation in nonhuman primates. *Exp Hematol.* 1991; 19: 810–816.

86.   Mayer P, Geissler K, Valent P, Ceska M, Bettelheim P, Liehl E. Recombinant human interleukin 6 is a potent inducer of the acute phase response and elevates the blood platelets in nonhuman primates. *Exp Hematol.* 1991; 19: 688–696.

87.   Asano S, Okano A, Ozawa K, et al. In vivo effects of recombinant human interleukin-6 in primates: stimulated production of platelets. *Blood.* 1990; 75: 1602–1605.

88.   Stahl CP, Zucker-Franklin D, Evatt BL, Winton EF. Effects of human interleukin-6 on megakaryocyte development and thrombocytopoiesis in primates. *Blood.* 1991; 78: 1467–1475.

89.   Carrington PA, Hill RJ, Levin J, Verotta D. Effects of interleukin 3 and interleukin 6 on platelet recovery in mice treated with 5-fluorouracil. *Exp Hematol.* 1992; 20: 462–469.

90.   Hill RJ, Warren MK, Stenberg P, et al. Stimulation megakaryocytopoiesis in mice by human recombinant interleukin-6. *Blood.* 1991; 77: 42–48.

91.   Ishibashi T, Kimura H, Shikama Y, et al. Interleukin-6 is a potent thrombopoietic factor in vivo in mice. *Blood.* 1989; 74: 1241–1244.

92.   Neben TY, Loebelenz J, Hayes L, et al. Recombinant human interleukin-11 stimulates megakayocytopoiesis and increases peripheral platelets in normal and splenectomized mice. *Blood.* 1993; 81: 901–908.

93.   Metcalf D, Nicola N A, Gearing DP. Effects of injected leukemia inhibitory factor (LIF) on hemopoietic and other tissues in mice. *Blood.* 1990; 76: 50–56.

94.   Mayer P, Geissler K, Ward M, Metcalf D. Recombinant human leukemia inhibitory factor in-

duces acute phase proteins and raises the blood platelet counts in nonhuman primates. *Blood*. 1993; 81: 3226–3233.

95. Wallace PM, MacMaster JF, Rillema JR, Peng J, Burstein SA, Shoyab M. Thrombopoietic properties of Oncostatin M. *Blood*. 1995; 86: 1310–1315.

96. Thomas JW, Baum CM, Hood WF, et al. Potent interleukin-3 receptor agonist with selectively enhanced hematopoietic activity relative to recombinant human interleukin 3. *Proc Natl Acad Sci USA*. 1995; 92: 3779–3783.

97. van Gils FCJM, Budel LM, Burger H, van Leen RW, Lowenberg B, Wagemaker G. Interleukin-3 receptors on rhesus monkey bone marrow cells: species specificity of human IL-3, binding characteristics, and lack of competition with GM-CSF. *Exp Hematol*. 1994; 22: 248–255.

98. Taga T, Kishimoto T. Cytokine receptors and signal transduction. *FASEB J*. 1992; 6: 3387–3396.

99. Rose TM, Bruce AG. Oncostatin M is a member of a cytokine family that includes leukemia-inhibitory factor, granulocyte colony-stimulating factor, and interleukin 6. *Proc Natl Acad Sci USA*. 1991; 88: 8641–8645.

100. Hangoc G, Yin J, Copper S, Schendel P, Yang YC, Broxmeyer H. In vivo effects of recombinant interleukin-11 on myelopoiesis in mice. *Blood*. 1993; 81: 965–972.

101. Nash RA, Seidel K, Storb R, et al. Effects of rhIL-11 on normal dogs and after sublethal radiation. *Exp Hematol*. 1995; 23: 389–396.

102. Paul SR, Hayes LL, Palmer R, et al. Interleukin-11 expression in donor bone marrow cells improves hematological reconstitution in lethally irradiated recipient mice. *Exp Hematol*. 1994; 22: 295–301.

103. Hawley TS. Progenitor cell hyperplasia with rare development of myeloid leukemia in interleukin-11 bone marrow chimeras. *J Exp Med*. 1993; 178: 1175–1188.

104. Hirano T, Akira S, Taga T, Kishimoto T. Biological and clinical aspects of interleukin-6. *Immunol Today*. 1990; 11: 443–449.

105. Kishimoto T. The biology of interleukin-6. *Blood*. 1989; 74: 1–10.

106. Patchen ML, MacVittie TJ, Williams JL, Schwartz GN, Souza LM. Administration of interleukin-6 stimulates multilineage hematopoiesis and accelerates recovery from radiation-induced hematopoietic depression. *Blood*. 1991; 77: 472–480.

107. Selig C, Kreja L, Müller H, Seifried E, Nothdurft W. Hematologic effects of recombinant human interleukin-6 in dogs exposed to a total-body radiation dose of 2.4 Gy. *Exp Hematol*. 1994; 22: 551–558.

108. Inoue H, Kadoya T, Kabaya K, et al. A highly enhanced thrombopoietic activity by monmethoxy polyethylene glycol-modified recombinant human interleukin-6. *J Lab Clin Med*. 1994; 124: 529–536.

109. Takatsuki F, Okano A, Suzuki C, et al. Interleukin 6 perfusion stimulates reconstitution of the immune and hematopoietic systems after 5-fluorouracil treatment. *Cancer Res*. 1990; 50: 2885–2890.

110. Zeidler C, Kanz L, Hurkuck F, et al. In vivo effects of interleukin-6 on thrombopoiesis in healthy and irradiated primates. *Blood*. 1992; 80: 2740–2745.

111. Farese AM, Myers LA, MacVittie TJ. Therapeutic efficacy of the combined administration of either recombinant human interleukin-6 and rh-granulocyte colony stimulating factor or rh-granulocyte-macrophage colony stimulating factor in a primate model of radiation-induced marrow aplasia. *Exp Hematol*. 1994; 22: 684 (abstract no 27).

112. Metcalf D, Hilton D, Nicola N A. Leukemia inhibitory factor can potentiate murine megakaryocytopoiesis in vitro. *Blood*. 1991; 77: 2150–2153.

113. Burstein SA, Mei R, Henthron J, Friese P, Turner K. Leukemia inhibitory factor and interleukin-11 promote the maturation of murine and human megakaryocytes *in vitro*. *J Cell Physiol*. 1992; 153: 305–312.

114. Leary AG, Wong GG, Clark SC, Smith AG, Ogawa M. Leukemia inhibitory factor differentiation-inhibiting activity/human interleukin for DA cells augments proliferation of human hematopoietic stem cells. *Blood*. 1990; 75: 1960–1964.

115. Farese AM, Myers LA, MacVittie TJ. Therapeutic efficacy of recombinant human leukemia inhibitory factor in a primate model of radiation-induced marrow aplasia. *Blood.* 1994; 84: 3675–3678.

116. Ishibashi T, Kimura H, Uchida T, Karyone S, Friese P, Burstein SA. Human interleukin 6 is a direct promoter of maturation of megakaryocytes in vitro. *Proc Natl Acad Sci USA.* 1989; 86: 5953–5957.

117. Broudy VC, Lin NL, Kaushansky K. Thrombopoietin (*c*-mpl ligand) acts synergistically with erythropoietin, stem cell factor, and interleukin-11 to enhance murine megakaryocyte colony growth and increases megakaryocyte ploidy in vitro. *Blood.* 1995; 85: 119–126.

118. Bree A, Schlerman F, Timony G, McCarthy K, Stoudemire J, Garnick M. Pharmacokinetics and thrombopoietic effects of recombinant human interleukin 11 (rhIL-11) in nonhuman primates and rodents. *Blood.* 1991; 78: 132a (abstract no 519).

119. Schlerman F, Bree A, Schaub R. Effects of subcutaneous administration of recombinant human interleukin 11 (rhIL-11) or recombinant human granulocyte macrophage colony stimulating factor (rhGM-CSF) alone, or rhIL-11 in combination with recombinant human interleukin 3 (rhIL-3) or rhGM-CSF in nonhuman primates. *Blood.* 1992; 80: 64a (abstract no 245).

120. Donahue RE, Seehra J, Metzger M, et al. Human IL-3 and GM-CSF act synergistically in stimulating hematopoiesis in primates. *Science.* 1988; 241: 1820–1822.

121. Stanley ER, Martocci, A Patinkin D, Rosendaal M, Bradley TR. Regulation of very primitive, multipotent, hemopoietic cells by hemopoietin-1. *Cell.* 1986; 45: 667–674.

122. Ikebuchi K, Wong GG, Clark SC, Ihle JN, Hirai T, Ogawa M. Interleukin 6 enhancement of interleukin 3-dependent proliferation of multipotential hemopoietic progenitors. *Proc Natl Acad Sci USA.* 1987; 84: 9035–3909.

123. Jacobsen SE, Veiby OP, Smeland EB. Cytotoxic lymphocyte maturation factor (interleukin 12) is a synergistic growth factor for hemtopoietic stem cells. *J Exp Med.* 1993; 178: 413–418.

124. Jacobsen SE, Okkenhaug C, Veibly OP, Caput D, Ferrara P, Minty A. Interleukin 13: novel role in direct regulation of proliferation and differentiation of primitive hematopoietic progenitor cells. *J Exp Med.* 1994; 180: 75–82.

125. Jacobsen FW, Veiby OP, Skjønsberg C, Jacobsen SE. Novel role of interleukin 7 in myelopoiesis: stimulation of primitive murine hematopoietic progenitor cells. *J Exp Med.* 1993; 178: 1777–1782.

126. Musashi M, Yang Y-C, Paul SR, Clark SC, Sudo T, Ogawa M. Direct and synergistic effects of interleukin 11 on murine hemopoiesis in culture. *Proc Natl Acad Sci U S A.* 1991; 88: 765–769.

127. Metcalf D, Nicola N.A. Direct proliferative actions of stem cell factor on murine bone marrow cells in vitro: effects of combination with colony-stimulating factors. *Proc Natl Acad Sci USA.* 1991; 88: 6239–6243.

128. Williams N, Bertoncello I, Kavnoudias H, Zsebo K, McNiece I. Recombinant rat stem cell factor stimulates the amplification and differentiation of fractionated mouse stem cell populations. *Blood.* 1992; 79: 58–64.

129. Li CL, Johnson GR. Rhodamine 123 reveals heterogeneity within murine Lin⁻ Sca1⁺ hemopoietic stem cells. *J Exp Med.* 1992; 175: 1443–1447.

130. Zsebo KM, Wypych J, McNiece IK. Identification, purification, and biological characterization of hematopoietic stem cell factor from buffalo rat liver-conditioned medium. *Cell.* 1990; 63: 195–201.

131. Aglietta M, Sanavio F, Stacchini A, et al. Interleukin-3 in vivo: kinetic of response of target cells. *Blood.* 1993; 82: 2054–2061.

132. Brandt JE, Bhalla K, Hoffman R. Effects of interleukin-3 and *c-kit* ligand on the survival of various classes of human hematopoietic progenitor cells. *Blood.* 1994; 83: 1507–1514.

133. Muench MO, Roncarolo MG, Menon S, et al. FLK-2/FLT-3 ligand regulates the growth of early myeloid progenitors isolated from human fetal liver. *Blood.* 1995; 85: 963–972.

134. Gonter PW, Hillyer CD, Strobert EA, et al. The effect of varying ratios of administered rhIL-3 and rhGM-CSF on post-chemotherapy marrow regeneration in a nonhuman primate model. *Blood.* 1993; 82: 365a (abstract no 1443).

135. Gonter PW, Hillyer CD, Strobert EA, et al. Enhanced post-chemotherapy platelet and neutrophil recovery using combination rhIL-6 and rhGM-CSF in a nonhuman primate model. *Blood.* 1993; 82: 365a (abstract no 1444).

136. Winton EF, Hillyer CD, Srinivasiah J, et al. A nonhuman primate model for study of effects of cytokines on hematopoietic reconstitution following chemotherapy-induced marrow damage. In: MacVittie TJ, Weiss J, Browne D (eds). *Advances in the Treatment of Radiation Injuries.* Oxford England: Elsevier Science Limited; 1996: 75–81.

137. MacVittie TJ, Farese AM, Heodin F, Grab LB, Baum C, McKearn JP. Combination therapy for radiation-induced bone marrow aplasia in nonhuman primates using Synthokine SC55494 and recombinant human granulocyte colony stimulating factor. *Blood.* 1996; 87: 4129–4135.

138. Monahan JB, Hood WF, Joy WD, et al. Functional characterization of SC68420- A multifunctional agonist which activates both IL-3 and G-CSF receptors. *Blood.* 1995; 86: 154a (abstract no 604).

139. McKearn JP, Hood WF, Monahan JB, et al. Myelopoietin- A multifunctional agonist of human IL-3 and G-CSF receptors. *Blood.* 1995; 86: 259a (abstract no 1022).

140. Beckmann MP, Heimfeld S, Fei R, et al. Novel hematopoietic factors enhance ex vivo expansion of human CD34+ hematopoietic progenitor cells. *Blood.* 1995; 86: 492a (abstract no 1957).

141. MacVittie TJ, Farese AM, Grab LB, Baum C, McKearn JP. Stimulation of multilineage hematopoietic recovery in a nonhuman primate, bone marrow aplasia model by a multifunctional agonist of human IL-3 and G-CSF receptors. *Blood.* 1995; 86: 499a (abstract no 1987)

142. Geissler K, Valent P, Mayer P, et al. Recombinant human IL-3 expands the pool of circulating hematopoietic progenitor cells in primates. Synergism with recombinant human granulocyte-macrophage colony stimulating factor. *Blood.* 1990; 75: 2305–2310.

143. Jiang S, Levine JD, Fu Y, et al. Cytokine production by primary bone marrow megakaryocytes. *Blood.* 1994; 84: 4151–4156.

144. Wickenhauser C, Lorenzen J, Thiele J, et al. Secretion of cytokines (Interleukins-1α, -3, and -6 and granulocyte macrophage colony-stimulating factor) by normal human bone marrow megakaryocytes. *Blood.* 1995; 85: 685–691.

21

Genetic Manipulation of Mpl Ligand and Thrombopoietin In Vivo

Frederic J. de Sauvage and Mark W. Moore

1. Introduction

The molecular cloning of the Mpl ligand, thrombopoietin (TPO), has revealed a system specifically involved in the regulation of platelet homeostasis. A number of preclinical studies of TPO in animal models indicate the potential usefulness of this molecule in the treatment of thrombocytopenia associated with myelosuppressive therapy. A powerful approach to determine the physiological role of this receptor/ligand system in the regulation of megakaryocyte and platelet production, and its potential effect on other hemopoietic lineages, is the disruption of the *c-mpl* and TPO genes in mice by homologous recombination.

2. *c-mpl* Knockout Models

The proto-oncogene *c-mpl* (*see* Chapter 6) is a member of the cytokine receptor superfamily with sequence similarity to the erythropoietin (EPO) receptor and the granulocyte colony-stimulating (G-CSF) receptor *(1,2)*. Expression of *c-mpl* in normal mice appears to be restricted primarily to hemopoietic tissue, primitive hemopoietic stem cells, megakaryocytes, and platelets *(2,3)*. *c-mpl* antisense oligodeoxynucleotides selectively inhibit in vitro megakaryocytic-colony formation without affecting the growth of erythroid or granulocyte-macrophage colonies, suggesting that c-*mpl* and its ligand may function in regulating megakaryocytopoiesis *(3)* (*see* Chapter 7). The involvement of *c-mpl* and TPO in the control of platelet production was powerfully demonstrated by the generation of mice deficient in *c-mpl* *(4)*. A targeting vector containing a *neoʳ* cassette inserted into the third exon of a 6.6-kb *c-mpl* mouse genomic clone was linearized and electroporated into embryonic stem (ES) cells *(5)*. After homologous recombination, the mutated allele (Fig. 1A) was detected by the addition of a HindIII site from the *neoʳ* gene using an oligonucleotide probe designed from sequences 5' to those present on the targeting vector. Gene targeting was detected in 8 of 288 ES colonies screened, and 5 colonies were selected for microinjection into the blastocoel cavity of 3.5-d C57BL/6J blastocysts *(6)*. Chimeric male mice were mated with C57BL/6J female mice, and agouti-color offspring were screened for germ-line transmission by polymerase chain reaction (PCR) analysis for the *neoʳ* gene and confirmed by Southern

From: *Thrombopoiesis and Thrombopoietins: Molecular, Cellular, Preclinical, and Clinical Biology*
Edited by: D. J. Kuter, P. Hunt, W. Sheridan, and D. Zucker-Franklin Humana Press Inc., Totowa, NJ

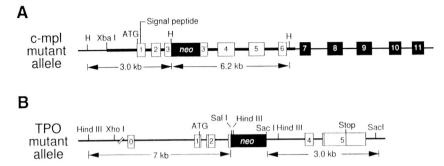

Fig. 1. Structure of mutated alleles after homologous recombination. (**A**) *c-mpl* mutant allele. (**B**) *TPO* mutant allele. In both cases, the intron-exon organization corresponds to the human gene, but restriction enzyme sites correspond to the murine gene.

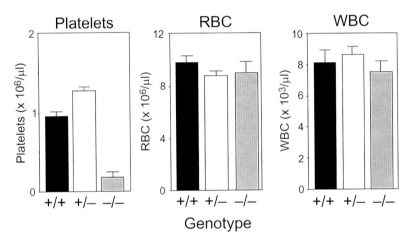

Fig. 2. Total blood cell count in *c-mpl*-deficient mice. Blood was collected by retro-orbital venous puncture and analyzed in a Serono-Baker Diagnostics System 9000 Diff Model Hematology Analyzer to determine platelet counts (*n* = 5 mice/group).

blot analysis of tail DNA. Heterozygous mice were interbred to obtain homozygous animals. Four clones gave germ-line transmission and were interbred to generate homozygous gene-targeted mice (*c-mpl –/–*).

Complete blood cell counts performed on *c-mpl –/–* and *c-mpl +/+* mice revealed a dramatic 85% decrease in platelet counts in the gene-targeted animals with 100% penetration (Fig. 2). Platelet volume was significantly increased, with a mean of 6.25 fL for *c-mpl –/–* mice compared with 4.7 fL for *c-mpl +/+* mice. Histopathology studies demonstrated a similar loss of megakaryocytes in spleen and bone marrow (Fig. 3). There was no significant alteration in either platelet counts, platelet volume, or megakaryocyte numbers in heterozygous animals (*c-mpl +/–*). There was no significant difference in red blood cells, total white blood cells, neutrophils, bands, or eosinophils between *c-mpl +/+* and *c-mpl –/–* mice, as determined by differential cell counts. Size

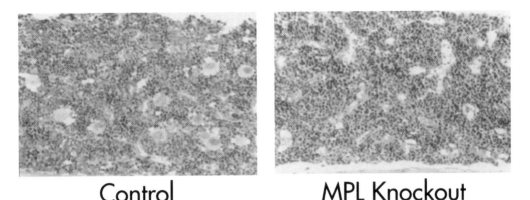

Control   MPL Knockout

Fig. 3. Histology of transverse sections through the sternum of control mice or *c-mpl*-deficient MPL knockout mice.

and cellularity of lymphoid organs were normal, and no alterations were detected in bone marrow, spleen, or thymus when cell-type ratios or cell-maturation markers of B- and T-cells were measured. Thus, disruption of the *c-mpl* gene results in a highly specific loss of both megakaryocytes and platelets, leaving other cell lineages unaffected.

3. TPO Knockout Models

TPO is a cytokine with a unique structure; the amino terminal domain is homologous to EPO, whereas the C-terminal glycosylated domain is unrelated to any known protein *(7–9)*. It was shown that only the EPO-like domain is required for binding and activation of the receptor *(9)* (*see* Chapter 13). The structure of the *TPO* gene reveals that this domain is encoded by the splicing of four exons, whereas the C-terminal tail is present on a single exon *(10)* (*see* Chapter 11). To disrupt the TPO gene efficiently, a 10-kb mouse genomic clone containing the entire *TPO* gene was subcloned, and most of the third coding exon was replaced with a *neo*^{r} cassette *(11)*. Deletion of this fragment located in the EPO-like domain ensured that homologous recombination would lead to a nonfunctional *TPO* gene. After linearization, this construct was electroporated into ES cells, and G418-resistant clones were selected and screened for homologous recombination by Southern blot analysis using an oligonucleotide probe designed from sequences 5' to those present on the targeting vector. After homologous recombination, the mutated allele (Fig. 1B) was detected by the addition of a HindIII site from the *neo*^{r} gene. Gene targeting was detected in one of 80 ES colonies screened, and five colonies were selected for microinjection into blastocysts. All five clones gave germline transmission, and three separate lines were interbred to generate homozygous gene-targeted mice (*TPO* –/–). The litters generated had a normal Mendelian distribution of the three genotypes, indicating viable fetal development of the *TPO* –/– mice. The adult mice were healthy and displayed no overt abnormalities. When expression of TPO mRNA in the liver and kidney was assessed by Northern blot analysis, a 1.8-kb band could be detected in both tissues in *TPO* +/+ mice. The band intensity was reduced by half in the heterozygotes and was undetectable in the liver and kidney of *TPO* –/– animals.

As observed in the *c-mpl* –/– mice, complete blood cell counts done on *TPO* –/– and *TPO* +/+ mice revealed an 88% decrease in platelet counts in the gene-targeted animals with 100% penetration in mice derived from three independent ES cell clones. There was also a significant increase in volume of the remaining platelets, with a mean of 8.7 fL for *TPO* –/– mice compared with 4.7 fL for *TPO* +/+ mice. However, in contrast to *c-mpl* heterozygous mice, a significant alteration in platelet counts (67% of wild-type) was observed in TPO heterozygous mice. Histopathologic examinations of the sternums also demonstrated a major loss of megakaryocytes in the bone marrow of *TPO* –/– mice (<10% of control) and heterozygous mice (60% of control). Again, the phenotype was highly specific for megakaryocyte and platelets, since no significant difference in red blood cells, total white blood cells, neutrophils, bands, or eosinophils in peripheral blood between *TPO* +/+ and *TPO* –/– mice could be detected by differential cell counts. The size and cellularity of lymphoid organs were also unaffected.

4. TPO Stimulates Both Megakaryocyte Proliferation and Maturation

Micrometric measurement of megakaryocytes from *c-mpl* knockout mice indicates that these cells are approx 20% smaller in overall size as well as in nuclear size than the wild-type megakaryocytes. This lower maturation stage of the megakaryocytes was confirmed by ploidy analysis of bone marrow megakaryocytes from *c-mpl*-knockout and *TPO*-knockout mice compared with normal mice (Fig. 4). There was not only a reduction in the absolute number of megakaryocytes, but also a reduction in their maturity as measured by DNA content. These results show that TPO is a physiological regulator of megakaryocyte maturation.

Analysis of megakaryocyte colony-forming cells (MK-CFC) from *TPO* –/–, *c-mpl* –/–, and normal mice in serum-containing semisolid media in the presence of TPO, stem cell factor (SCF), or interleukin (IL)-3 indicates that the number of megakaryocyte progenitors is dramatically reduced in both knockout mice strains compared with control mice *(11,12)*. This indicates that TPO not only affects megakaryocyte maturation as originally thought, but that TPO may act on progenitors earlier than the MK-CFC. It also means that IL-3 does not stimulate a population of progenitor cells completely independently of TPO. This lower number of megakaryocyte progenitors is also reflected in the long duration of TPO treatment required for *TPO* –/– mice to reach normal platelet levels. The platelet levels in *TPO* –/– mice eventually surpass untreated wild-type levels, indicating that protein replacement therapy using TPO can correct the engineered genetic defect, which is not the case in *c-mpl* –/– mice *(11,12)*.

Furthermore, analysis of progenitor cells from the other hemopoietic lineages in *TPO* –/– and *c-mpl* –/– mice also revealed a dramatic decrease in granulocyte-macrophage colony-forming cells (GM-CFC), erythroid burst-forming cells (E-BFC), and mixed colony-forming cells (GEMM-CFC) *(12)*. This decrease in progenitors from all lineages indicates that TPO probably acts on a very early progenitor cell common to all lineages. This is consistent with the observation that TPO enhances not only megakaryocyte colony growth *(8,13,14)*, but also the proliferation of erythroid progenitors *(15,16)*. The involvement of TPO and *c-mpl* at an early stage correlates with the detection of *c-mpl* expression in an AA4[+] Sca[+] murine stem cell population *(17)*. The effect of TPO on this most primitive stem cell population still remains to be investigated.

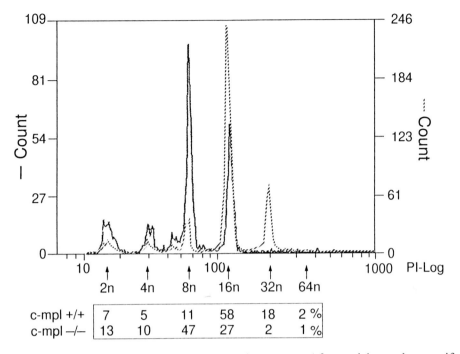

	2n	4n	8n	16n	32n	64n	
c-mpl +/+	7	5	11	58	18	2	%
c-mpl –/–	13	10	47	27	2	1	%

Fig. 4. Ploidy analysis of *c-mpl*-deficient megakaryocytes. After enrichment by centrifugation on a Percoll gradient, megakaryocytes were stained with anti-CD61 (Integrin b_3 chain)-FITC (Pharmingen) and the DNA intercalating fluorescent dye propidium iodide. The DNA content in CD-61-positive cells was quantified by two-color flow cytometry on a Coulter Epics Elite. Solid line, *c-mpl –/–*; dotted line, *c-mpl +/+*.

5. Regulation of TPO Levels

Analysis of *TPO* and *c-mpl* knockout mice has provided some major insights into the understanding of how endogenous TPO levels are regulated as a function of platelet mass. These mice provided animal models of chronic thrombocytopenia and have been used to analyze the mechanisms leading to increased concentrations of circulating TPO in response to low platelet levels *(18)*. As is the case in other thrombocytopenic animals, circulating TPO levels are dramatically elevated in *c-mpl –/–* mice (Fig. 5) *(4)*. However, as described earlier, *TPO +/-* mice have a platelet level intermediate between the wild type and the *TPO –/–* mice *(11)*. This gene dosage effect indicates that TPO production is not regulated as a function of the platelet mass. As a consequence, no increase in TPO mRNA level could be detected in any organs of *c-mpl –/–* mice. In an alternative model originally suggested by De Gabrielle and Penington *(19)*, and further developed by Kuter and Rosenberg *(20,21)*, TPO could be released into the circulation at a constant rate and the level of circulating TPO regulated by the platelets themselves. When platelet levels are normal, they will bind and remove from the circulation TPO released by the liver and/or the kidney. In thrombocytopenic situations, the circulating concentration of TPO will increase and stimulate platelet production by bone marrow megakaryocytes. This regulatory mechanism, whereby the ligand availability is controlled by the mature cell, could be similar to the way G-CSF level is regulated in

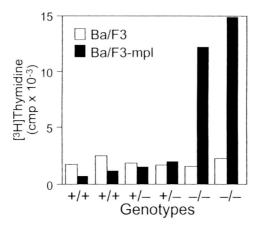

Fig. 5. Effect of *c-mpl* –/– mouse sera on Ba/F3-Mpl⁺ cell proliferation. Proliferation assays were conducted essentially as previously described. (Reprinted with permission from ref. *9*.)

neutropenic states *(22)*. A gene dosage effect is also observed in *G-CSF* +/– mice, where the neutrophil counts of the *G-CSF* +/– mice are 67% of the *G-CSF* +/+ *(23)*. In contrast to platelets from the *c-mpl* –/– mice, normal platelets are capable of binding, internalizing, and degrading TPO. The absence of receptors for TPO in the *c-mpl* –/– mice dramatically affects ^{125}I-rmTPO clearance from the plasma with an initial volume of distribution 40% larger and a steady-state volume of distribution nearly sixfold larger in the *c-mpl* +/+ mice compared with the *c-mpl* –/–mice. This suggests that distribution of ^{125}I-rmTPO is normally cleared from the plasma into platelets or other tissue sites *(18)*. Owing to the specific removal of TPO from the plasma by platelets, the initial half-life ($t_{1/2 \text{ initial}}$) was shorter and the overall systemic clearance (CL/W) of ^{125}I-rmTPO was much faster in the *c-mpl* +/+ mice than in the *c-mpl* –/–mice. Finally, reconstitution of normal platelet levels in *c-mpl* –/– mice, by injection of platelets purified from normal mice, rapidly decreases the plasma levels of TPO to those found in normal mice (Fig. 6). This demonstrates that TPO receptors present on platelets are sufficient to remove quickly the high amount of TPO circulating in the *c-mpl* –/– mice. This model, in which the circulating level of TPO is regulated directly by the platelet mass, is presented in more detail in Chapter 23 of this book.

6. Conclusion

Gene targeting has proven very useful in characterizing the role of *TPO* and *c-mpl* in megakaryocyte and platelet formation. The similar phenotype of the two knockout models demonstrates that there is a nonredundant, one-ligand, one-receptor system. However, even with such a low number of megakaryocytes and platelets, these mice still have some platelets that are normal both structurally and functionally, which prevent them from bleeding. The genes and factors involved in the production of this basal level of platelets and megakaryocytes still remain to be identified. The transcription factor NF-E2 has been identified recently as a key component to the process of platelet formation *(24)*. Other cytokines with megakaryocytopoietic activity, such as IL-3, IL-6, IL-11, or granulocyte-macrophage colony-stimulating factor (GM-CSF), could have

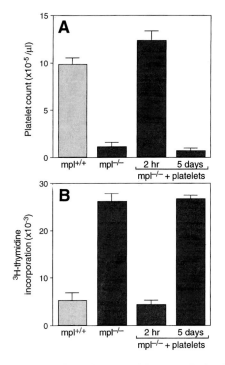

Fig. 6. TPO levels in *c-mpl –/–* mice injected with normal platelets. *c-mpl –/–* mice were injected with 1.5×10^9 normal washed platelets. **(A)** Platelet count and **(B)** circulating TPO levels were determined 2 h and 5 d after injection. TPO levels were measured using the Ba/F3-Mpl proliferation assay.

a role in this process, but are not upregulated in *c-mpl –/–* mice. Breeding of mice deficient in these genes with *TPO*-knockout and *c-mpl*-knockout mice should help to define their role in the process of platelet formation.

References

1. Souyri M, Vigon I, Penciolelli JF, Heard JM, Tamboruin P, Wendling F. A putative truncated cytokine receptor gene tranduced by the myeloproliferative leukemia virus immortalizes hematopoietic progenitors. *Cell.* 1990; 63: 1137–1147.
2. Vigon I, Mornon JP, Cocault L, et al. Molecular cloning and characterization of MPL, the human homolog of the v-mpl oncogene: identification of a member of the hematopoietic growth factor receptor superfamily. *Proc Natl Acad Sci USA.* 1992; 89: 5640–5644.
3. Methia N, Louache F, Vainchenker W, Wendling F. Oligodeoxynucleotides antisense to the proto-oncogene c-mpl specifically inhibit in vitro megakaryocytopoiesis. *Blood.* 1993; 82: 1395–1401.
4. Gurney AL, Carver-Moore K, de Sauvage FJ, Moore MW. Thrombocytopenia in *c-mpl*-deficient mice. *Science.* 1994; 265: 1445–1447.
5. Gossler A, Doetschman T, Korn R, Serfling E, Kemler R. Transgenesis by means of blastocyst-derived embryonic stem cell lines. *Proc Natl Acad Sci USA.* 1986; 83: 9065–9069.
6. Bradley A. Production and analysis of chimaeric mice. In: Robertson E (ed). *Teratocarcinomas and Embryonic Stem Cells: A Practical Approach.* Oxford: IRL; 1987: 113–152.
7. Lok S, Kaushansky K, Holly RD, et al. Cloning and expression of murine thrombopoietin cDNA and stimulation of platelet production in vivo. *Nature.* 1994; 369: 565–568.

8. Kaushansky K, Lok S, Holly RD, et al. Promotion of megakaryocyte progenitor expansion and differentiation by the c-Mpl ligand thrombopoietin. *Nature*. 1994; 369: 568–571.

9. de Sauvage FJ, Hass PE, Spencer SD, et al. Stimulation of megakaryocytopoiesis and thrombopoiesis by the c-Mpl ligand. *Nature*. 1994; 369: 533–538.

10. Gurney AL, Kuang WJ, Xie MH, Malloy BE, Eaton DL, de Sauvage FJ. Genomic structure, chromosomal localization, and conserved alternative splice forms of thrombopoietin. *Blood*. 1995; 85: 981–988.

11. de Sauvage F, Carver-Moore K, Luoh S, et al. Physiological regulation of early and late stages of megakaryocytopoiesis by thrombopoietin. *J Exp Med*. 1996; 183: 2367–2372.

12. Carver-Moore K, Broxmeyer H, Luoh S-M, Cooper S, Peng J, Burstein S, Moore MW, de Sauvage FJ. Low levels of erythroid and myeloid progenitors in TPO- and in c-mpl-deficient mice. *Blood*. 1996.

13. Wendling F, Maraskovsky E, Debili N, et al. cMpl ligand is a humoral regulator of megakaryocytopoiesis. *Nature*. 1994; 369: 571–574.

14. Broudy VC, Lin NL, Kaushansky K. Thrombopoietin (c-mpl ligand) acts synergistically with erythropoietin, stem cell factor, and interleukin-11 to enhance murine megakaryocyte colony growth and increases megakaryocyte ploidy in vitro. *Blood*. 1995; 85: 1719–1726.

15. Kobayashi M, Laver JH, Kato T, Miyazaki H, Ogawa M. Recombinant human thrombopoietin (mpl ligand) enhances proliferation of erythroid progenitors. *Blood*. 1995; 86: 2494–2499.

16. Kaushansky K, Broudy VC, Grossmann A, et al. Thrombopoietin expands erythroid progenitors, increases red cell production, and enhances erythroid recovery after myelosuppressive therapy. *J Clin Invest*. 1995; 96: 1683–1687.

17. Zeigler FC, de Sauvage F, Widmer HR, et al. In vitro megakaryocytopoietic and thrombopoietic activity of c-mpl ligand (TPO) on purified murine hematopoietic stem cells. *Blood*. 1994; 84: 4045–4052.

18. Fielder P, Gurney A, Stefanich E, Marian M, Moore MW, Carver-Moore K, de Sauvage FJ. Regulation of thrombopoietin levels by c-mpl-mediated binding to platelets. *Blood*. 1996; 87: 2154–2161.

19. De Gabriele U, Penington DG. Regulation of platelet production: "hypersplenism" in the experimental animal. *Br J Haematol*. 1967; 13: 384–393.

20. Kuter DJ, Beeler DL, Rosenberg RD. The purification of megapoietin: a physiological regulator of megakaryocyte growth and platelet production. *Proc Natl Acad Sci USA*. 1994; 91: 11,104–11,108.

21. Kuter D, Rosenberg RD. The reciprocal relationship of thrombopoietin (c-Mpl ligand) to changes in the platelet mass during busulfan-induced thrombocytopenia in the rabbit. *Blood*. 1995; 85: 2720–2730.

22. Layton JE, Hockman H, Sheridan WP, Morstyn G. Evidence for a novel in vivo control mechanism of granulopoiesis: mature cell-related control of a regulatory growth factor. *Blood*. 1989; 74: 1303–1307.

23. Lieschke GJ, Grail D, Hodgson G, et al. Mice lacking granulocyte colony-stimulating factor have chronic neutropenia, granulocyte and macrophage progenitor cell deficiency, and impaired neutrophil mobilization. *Blood*. 1994; 84: 1737–1746.

24. Shivdasani R, Rosenblatt M, Zucker-Franklin D, et al. Transcription factor NF-E2 is required for platelet formation independent of the actions of thrombopoietin/MGDF in megakarocyte development. *Cell*. 1995; 81: 695–704.

VI

CLINICAL BIOLOGY

22

Serum Levels of Thrombopoietin in Health and Disease

Janet Lee Nichol

1. Introduction: Cytokine Levels in Plasma and Serum

1.1. Background

The history of measuring cytokine levels in serum and plasma is relatively short (10–15 years). There is increasing interest in being able to quantitate circulating cytokine levels, since these molecules may play a role in the pathogenesis of various diseases. Additionally, the rapid increase in the use of cytokines and cytokine antagonists as therapeutic agents has increased the need for sensitive and reliable assay systems. The development of cytokines for therapeutic use may require not only monitoring of the agent administered, but also monitoring of the levels of induced endogenous cytokines. The interdependence of cytokines in the immune and inflammatory systems and their ability to stimulate secretion of other members of specific cytokine cascades suggest the possibilities of similar interactions between the hemopoietic cytokines *(1,2)*.

Methods used for measuring cytokine levels are predominately of two types. Bioassays rely on a ligand–receptor interaction that results in a measurable biological response (i.e., cell growth), and immunoassays, such as the enzyme-linked immunosorbent assay (ELISA), are based on antibody–antigen interactions for the capture and labeling of specific proteins from a complex solution. It is critical to remember that bioassays measure activity and immunoassays measure immunoreactivity *(3)*. Using both types of assays gives a more complete picture of cytokine levels, and activity in body fluids and culture media.

1.2. Bioassays

Bioassays tend to detect functional forms of cytokines; however, many bioassays are not specific for a particular cytokine and require confirmation by a neutralization control in the form of antibodies or competition by soluble forms of the cytokine receptor. The presence of nonspecific stimulators present in plasma or serum makes confirmation mandatory when patient samples are assayed. Bioassays may involve days of tissue culture and can result in difficulties in fitting standard curves *(4)*. Although bioassays have an advantage of measuring functional properties of proteins, they lack the robust reliability of immunoassays.

From: *Thrombopoiesis and Thrombopoietins: Molecular, Cellular, Preclinical, and Clinical Biology*
Edited by: D. J. Kuter, P. Hunt, W. Sheridan, and D. Zucker-Franklin Humana Press Inc., Totowa, NJ

1.3. Immunoassays

Immunoassays tend to be specific, convenient, and rapid. One problem, however, is that immunoassays may detect inactive forms of the protein. Immunoassays usually use a combination of antibodies. A "capture" antibody (or receptor) stabilized on a plastic microtiter plate specifically binds to the protein of interest. Bound protein levels are then detected using a "signal" antibody tagged with a detection system. These detection systems can be indirect, such as biotin, or direct, such as horseradish peroxidase (HRP). The sensitivity of an ELISA is directly related to the quality of the antibodies used. ELISA assays generally use a chromogenic system quantitated by measuring optical density.

1.4. Steady-State Cytokine Levels in Normal Donors

The bulk of the literature measuring cytokine levels in normal individuals or patients has been generated using ELISA methodology. In some cases, it has not been possible to measure steady-state levels of a cytokine in plasma or serum owing to limited assay sensitivity. Cytokines, such as interleukin (IL)-3, granulocyte-macrophage colony-stimulating factor (GM-CSF), IL-11, and leukemia-inhibitory factor (LIF), have not been detected in normal plasma or serum to date *(5–7)*. It is impossible to distinguish whether this is because of lack of circulating cytokine or lack of sensitivity in the current assays. Another concern is that the antibodies used, particularly monoclonal antibodies (MAb), may be unable to detect cytokines bound to plasma proteins or soluble receptors owing to masking of the relevant epitope(s). This problem has been reported when measuring IL-6 levels *(8)*.

Normal endogenous steady state levels have been measured for several cytokines *(9–25)*. When interpreting these values, unanticipated bimodal effects may be seen. For example, granulocyte colony-stimulating factor (G-CSF) has been found by ELISA to be undetectable in 90% of normal individuals, and its presence at levels of 30–160 pg/mL has been found in the other 10% of the normal population *(9)*. In the case of IL-6, circadian changes in circulating levels have been demonstrated *(10)*. This discussion will be limited to the hemopoietic growth factors and those cytokines suspected of having a role in thrombopoiesis. Table 1 summarizes their reported normal circulating levels, as well as some of the disease states in which changes in these values were observed.

1.5. Cytokine Levels in Disease

Abnormal cytokine levels in some cases have been shown to correlate with increased risk of disease progression and have been implicated in pathogenetic mechanisms for disease *(6,7,12,13,15,16,18,23–25,27,29,30,32–39)*. Correlation of cytokine levels with disease states has also assisted in a better understanding of normal physiological processes. One example of this type of correlation is the inverse relationship between red cell mass (and thus oxygen content) and endogenous erythropoietin (EPO) levels *(20)*.

2. Cytokines and Thrombopoiesis

2.1. Nonspecific Regulators of Megakaryocytopoiesis and Thrombopoiesis

Several cytokines have been reported to affect processes in megakaryocytopoiesis or thrombopoiesis, including stem cell factor (SCF), IL-3, GM-CSF, EPO, IL-6, IL-11,

Table 1
Cytokines Thought to Affect Thrombopoiesis Listed
with Their Reported Normal Circulating Levels[a]

Cytokine	Normal circulating levels	Changes with disease	References
SCF	3300 pg/mL	↑HIV	*(17,18)*
IL-3	<10 pg/mL	↑HD; ↑ssSCA	*(5,26,27)*
	assay detection limit		
GM-CSF	<25 pg/mL	↑MDS; ↑HD;	*(7)*
	assay detection limit	↑ssSCA	
G-CSF	25 pg/mL	↑BMT; ↑PBPC	*(9,15,19,28)*
	(2.5 U/mL)		
M-CSF	3100 pg/mL	↑GVHD	*(29)*
EPO	30–83 pg/mL	↑2°P; ↑dH;	*(20,21,24,25)*
	(15–30 miU/mL)	↑anemia	
IL-6	<0.5 pg/mL	↑GVH; ↓ATP;	*(7,11–13*
	assay detection limit	nc-aplasia BMT;	*27,29,30)*
		↑HD; ↑CLD	
IL-11	<31 pg/mL	nc-BMT	*(26)*
	assay detection limit		
LIF	<3 pg/mL	nc-BMT	*(26,31)*
	assay detection limit		
IL-1b	500 pg/mL	↑HD; FHF	*(7,27)*

[a]Included are some results from measurements in various disease states.

nc, no change; ↑, increased; ↓, decreased; HD, Hodgkin's Disease; BMT, bone marrow transplant; PBPC, peripheral blood progenitor cell transplantation; CLD, chronic liver disease; ATP, acute thrombocytopenic purpura; 2°P, secondary polycythemia; dH, daytime hypoxia; ssSCA, steady-state sickle cell anemia; MDS, myelodysplastic syndromes; HIV, human immunodeficiency virus; GVHD, graft vs host disease; FHF, fulminant hepatic failure; M-CSF, macrophage colony-stimulating factor.

LIF, and IL-1β *(40–64)*. However, their actions are neither specific nor potent for the megakaryocyte lineage and, therefore, they do not appear to be the primary regulators of megakaryocyte development or platelet production.

2.2. The Mpl Ligand: A Primary Regulator of Megakaryocytopoiesis and Thrombopoiesis

Recently, a protein that is a ligand for the cytokine receptor Mpl has been identified, cloned, and expressed (Mpl ligand) *(65–69)* (*see* Chapters 8 and 9). There is now compelling evidence, reviewed throughout this volume, that this protein is a primary regulator of megakaryocytopoiesis and thrombopoiesis. Mpl ligand stimulates growth of megakaryocyte progenitors, and the maturation of developing megakaryocytes, which ultimately results in platelet production in vitro and in vivo *(64,66,67,69–80)*. Mpl ligand is known in its native form as thrombopoietin (TPO).

3. Assays to Detect Endogenous Thrombopoietin

3.1. Bioassays Used During the Discovery and Cloning

During the purification and cloning of TPO, various bioassays were used to detect active protein. These assays used either selected rat bone marrow cells, purified CD34$^+$ human progenitor cells, or murine cell lines (32D and Ba/F3) engineered to express Mpl receptor *(65–68,70,74)*. Rat bone marrow cells depleted of recognizable megakaryocytes were used by Kuter and Rosenberg to monitor ploidy changes by flow cytometry as an indicator of TPO activity *(70)*. An alternate method used platelet GPIIb- and GPIIIa-enriched cells in liquid culture assay pulsed with ^{14}C-5-hydroxytryptamine creatine sulfate (^{14}C-^{5}HT) *(68)*. This assay detected increases in radioactivity as an indirect measure of megakaryocyte activity by exploiting the ability of mature megakaryocytes to take up serotonin (^{5}HT). Bartley et al. used purified human CD34$^+$ cells in liquid culture to monitor megakaryocyte proliferation and maturation *(65)*. Cell proliferation of Mpl-expressing cell lines quantitated by colormetric or radioactive detection systems was also used by this group and others *(65–67)*.

3.2. TPO Immunoassays Currently in Use

An immunoassay was needed to quantitate the amounts of TPO found in sera or plasma from normal donors and from patients with abnormal platelet counts. Scientists from several biotechnology companies have recently developed TPO assays *(26,80–86)*. The assays from Kirin (Tokyo, Japan) *(81)*, a TPO-deleted MAb-based ELISA, and Amgen Inc. (Thousand Oaks, CA), a polyclonal antibody-based ELISA, were sensitive enough to detect TPO in normal plasma with detection limits of 0.5 fmol/mL (approximately 18 pg/mL) and 50 pg/mL, respectively *(81,84)*.

The polyclonal antibody TPO ELISA is able to detect TPO concentrations in samples of normal patient serum or plasma. A polyclonal antibody directed against the receptor binding portion of TPO was selected as the capture antibody based on its good signal-to-noise ratio. The ability of the capture antibody to detect various forms of TPO in human plasma was confirmed by Western blot analysis. The significance and relative bioactivity of these various forms of TPO are uncertain (Chapter 13). A separate polyclonal anti-TPO antibody was conjugated to HRP and selected for use as the signal antibody. The assay is run in a total of 50% serum or plasma, and to maintain this serum concentration, patient samples were diluted in fetal bovine serum (FBS). The FBS was prescreened for undetectable endogenous TPO and negligible background optical density.

Specificity and reproducibility of the assay were confirmed. Serial dilutions of a patient sample with high levels of TPO were run in the same assay with similar dilutions of the prescreened FBS spiked with recombinant human TPO (rHuTPO). The specificity of the TPO ELISA was confirmed by the parallel dose-response curves of these samples. Additionally, a selected patient sample with high TPO levels (approximately 800 pg/mL) and a normal plasma sample (approximately 100 pg/mL) were aliquoted and frozen for use in all assays as benchmarks to assess assay reproducibility. When TPO levels in these samples were compared over a 4-month period, the coefficient of variation was 2.9% for the patient plasma and 7.7% for the normal plasma.

The TPO-depleted MAb ELISA uses a monoclonal mouse anti-rHuTPO as the capture antibody and a biotin-labeled polyclonal rabbit anti-rHuTPO as the signal anti-

Table 2
Platelet Counts and Endogenous TPO Values from Matched Plasma and Serum Collected from Hematologically Normal Individuals[a]

	Platelets, 10^9/L	Plasma TPO, pg/mL	Serum TPO, pg/mL
Range	148–421	22–256	20–313
Mean ± SEM	282 ± 5	81 ± 5	95 ± 6
Median	280	64	78
SD	53	53	57
n	97	97	97

[a]The platelet counts were taken at the time of draw and endogenous TPO values were determined using a polyclonal antibody-based TPO ELISA as the assay system *(84)*.

body. A notable difference in this assay is the diluent used, which contains normal human serum depleted of endogenous TPO by affinity purification over an anti-TPO antibody column. Signal generation was based on a streptavidin alkaline phosphatase detection system. This assay had a sensitivity of 0.5 fmol/mL (approximately 18 pg/mL) and was able to detect TPO in normal donor plasma.

The assay developed by Genentech (South San Francisco, CA) *(82)* uses a chimeric molecule consisting of human Mpl fused to the Fc portion of human IgG as the capture reagent with a biotinylated rabbit anti-TPO for the signal antibody. The limit of detection in this assay is 200 pg/mL, a value above that reported in normal serum and plasma by the other groups.

3.3. Immunoassays: Levels of TPO in Normal Steady State

To determine normal endogenous TPO levels, matched plasma and serum samples from 97 donors with platelet counts >100 × 10^9/L but <450 × 10^9/L were evaluated (Table 2). Values (mean ± SEM) for serum samples averaged 17% higher than the values for plasma samples (95 ± 6 versus 81 ± 5 pg/mL, respectively). The discrepancy between serum and plasma may be owing to the elaboration of immunoreactive TPO into the serum after platelet activation and clotting. The presence of various forms of TPO in lysates of murine and human platelets has been demonstrated *(87,88)*.

The only other report of endogenous TPO levels in normal plasma measured values lower than those reported here *(81)*. Determinations of TPO in samples from 28 normal men and 21 normal women were 0.81 ± 0.3 fmol/mL (29 pg/mL) and 0.70 ± 0.26 fmol/mL (25 pg/mL), respectively. The values in male and female donors were not significantly different.

The reasons for the discrepancies between these two studies remain to be determined. Use of polyclonal capture antibodies could allow for detection of various forms of endogenous TPO that might not be detected by a more restricted MAb if the recognition site for the epitope was cleaved. Detection of all forms may be of questionable advantage, since some of the cleaved forms of the molecule may not be biologically active *(89)*. The use of international standards may assist in rectifying these discrepancies.

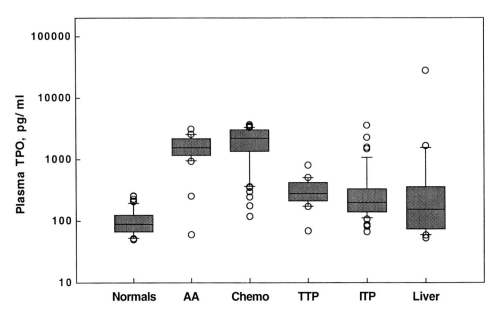

Fig. 1. Plasma levels of endogenous TPO in normal donors and thrombocytopenic patients. Figure includes all patients with listed diagnosis. (AA, aplastic anemia; TTP, thrombotic thrombocytopenia purpura; ITP, immune thrombocytopenia purpura; chemo, chemotherapy; liver, chronic liver disease). TPO was measured by a polyclonal antibody-based ELISA.

3.4. Immunoassays: Levels of Endogenous TPO in Thrombocytopenia and Thrombocytotic States

3.4.1. Thrombocytopenia

The relationship between circulating platelet numbers and endogenous TPO levels was studied in normal donor plasma and thrombocytopenic patient plasma (Fig. 1). The patients were rendered thrombocytopenic by chemotherapy, aplastic anemia, thrombotic thrombocytopenia purpura (TTP), chronic liver disease, or immune thrombocytopenia purpura (ITP).

As a group, these thrombocytopenic patients had a mean platelet count of $29 \pm 2 \times 10^9/L$ (range, $3–96 \times 10^9/L$; median, $22 \times 10^9/L$), and had a mean endogenous TPO level of 3606 pg/mL (range, 37–161,538 pg/mL; $n = 96$). The median value of 1002 pg/mL is greater than 10 times the TPO values found in plasma and serum from normal donors.

The values listed in Table 3 reflect data only from those patients with platelet counts below the normal range ($<100 \times 10^9/L$). The mean values for each diagnostic category were higher than the mean values for normal donor plasma. All the conditions tested had a wide range of endogenous TPO levels that varied from below normal to greater than normal.

The patients with chronic liver disease were evaluated further. These patients appeared to be of two distinct groups: those with high TPO levels (range, 1200–161,500 pg/mL) and those with normal to low values (range, 37–165 pg/mL). The patients with high TPO levels ($n = 4$) had a mean platelet count of $85 \pm 24 \times 10^9/L$ (median, $78 \times 10^9/L$), and a mean TPO value of $48,168 \pm 18,059$ pg/mL (median, 14,992 pg/mL). The

Table 3
Platelet Values from Selected Patients with Thrombocytopenia
Shown in Fig. 1[a]

	Chemotherapy-induced thrombocytopenia		Aplastic anemia		Immune thrombocytopenic purpura		Chronic liver disease	
	Platelet count, 10^9/L	TPO, pg/mL	Platelet count, 10^9/L	TPO, pg/mL	Platelet count, 10^9/L	TPO, pg/mL	Platelet count, 10^9/L	TPO,[b] pg/mL
Range	2–43	71–1829	5–77	45–60,388	18–89	17–313	4–96	37–161,538
Mean ± SEM	15 ± 2	870 ± 69	28 ± 3	4552 ± 2090	41 ± 7	121 ± 24	61 ± 6	11,823 ± 11,517
Median	11	999	26	1452	32	98	62	97
SD	11	430	19	11,825	25	83	23	43,094
n	39	39	32	32	12	12	14	14

[a]Table includes all patients with platelet counts <100 × 10^9/L. TPO levels were measured as in Table 2.
[b]For further breakdown of chronic liver disease, *see text.*

remaining 13 patients evaluated had a mean platelet count of 79 ± 12 × 10^9/L (median, 71 × 10^9/L) with mean TPO values of 97 ± 14 pg/mL (median, 81 pg/mL), which is in the normal range. Since the major site of TPO production is thought to be the liver, it would be useful to examine more patients with liver disease to see if the extreme levels of TPO observed in a few patients might be owing to a dysregulation of TPO production *(66,67)*.

The endogenous TPO levels in plasma from patients with ITP (121 ± 24 pg/mL) were not as high as those found in other patient groups with similarly low platelet numbers. The complex pathology associated with these conditions necessitates careful consideration of potential sources of increased platelet mass, which may be overlooked if only platelet number is measured. Such sources could be platelet stores in the spleen or rapid production in the presence of rapid destruction. These platelets could bind TPO, resulting in low TPO levels even though the platelet counts are also low.

Other researchers found that patients with aplastic anemia (mean platelet count, 56 ± 58 × 10^9/L) had endogenous TPO levels that were significantly higher than normal ($p < 0.01$) at 18.53 ± 112.4 fmol/mL (approximately 676 pg/mL) *(81)*.

Patients ($n = 43$) with chronic ITP (mean platelet count, 52.6 ± 41.2 × 10^9/L) were reported to have endogenous TPO values of 1.78 ± 1.18 fmol/mL (approximately 63 pg/mL) using the TPO-depleted MAb TPO ELISA *(90)*. Thirty percent of these patients had TPO levels in the normal range. Two of these patients were monitored after splenectomy, and in each, the platelet count increased 5- to 10-fold, accompanied by an initial increase in TPO levels (two to three times) followed by a decrease to values similar to the presplenectomy levels.

Endogenous TPO levels in thrombocytopenic children were estimated in 19 cancer patients (7 received submyeloablative therapy, 12 received myeloablative therapy for bone marrow transplantation [BMT]) *(86)*, and 15 patients with ITP. The measurements were done using the chimeric-Mpl IgFc ELISA. All samples were collected dur-

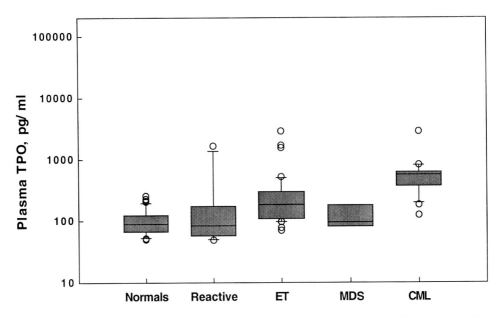

Fig. 2. Plasma levels of endogenous TPO in normal donors and thrombocytotic patients. Figure includes all patients with listed diagnosis, regardless of platelet count at time of blood draw. (ET, essential thrombocytosis; MDS, myelodysplastic syndromes; CML, chronic myelogenous leukemia). TPO levels measured as in Fig. 1.

ing thrombocytopenic episodes and through platelet recovery. The findings were consistent with those reported by other groups: patients with bone marrow failure and subsequent thrombocytopenia because of a lack of platelet production have high levels of TPO. The amount of endogenous TPO measured reflected the severity of the ablative therapy. The TPO data from the patients with ITP are difficult to interpret since measurements were below the sensitivity of the assay (200 pg/mL). They did, however, report analysis of six bone-marrow aspirates from the ITP patients. All of these patients had increased megakaryocyte numbers, typical of this disease.

Another study using this assay confirmed the high levels of activity in aplastic anemia (a stem cell disorder with defective platelet production) *(82)*. This study also reported that endogenous TPO levels from samples of patients with ITP were not measurable.

3.4.2. Thrombocytosis

Plasma and serum from 38 patients with chronic myelogenous leukemia (CML), essential thrombocythemia (ET), myelodysplastic syndrome (MDS), or reactive thrombocytosis were evaluated (Fig. 2).

Patients who were thrombocytotic (platelet counts >500 × 10^9/L) at the time of sampling were analyzed further. Taken together, these patients had a mean platelet count of 903 ± 49 × 10^9/L (range, 547–1303 × 10^9/L; median, 865 × 10^9/L). The mean endogenous TPO level was 400 ± 98 pg/mL (range, 42–1552 pg/mL; median, 235 pg/mL) (*n* = 20).

The mean, range, and median for each thrombocytotic condition (platelet counts >500 × 10^9/L) are shown in Table 4. Low sampling numbers make it difficult to draw

Table 4
Platelet Counts and TPO Values from Selected Patients with Thrombocytosis Shown in Fig. 2

	Chronic myelogenous leukemia		Essential thrombocythemia		Myelodysplastic syndrome		Reactive thrombocytosis	
	Platelet count, 10^9/L	TPO, pg/mL	Platelet count, 10^9/L	TPO, pg/mL	Platelet count, 10^9/L	TPO, pg/mL	Platelet count, 10^9/L	TPO,[b] pg/mL
Range	547–1260	39–836	507–1630	71–1552	698–890	101–259	610–1056	49–1660
Mean ± SEM	934 ± 159	718 ± 122	833 ± 80	236 ± 90	794 ± 96	180 ± 79	847 ± 82	363 ± 260
Median	965	726	785	114	794	180	884	113
SD	318	245	320	360	136	112	202	638
n	4	4	16	16	2	2	6	6

[a]Table includes only patients with platelet counts >500 × 10^9/L. Endogenous TPO levels were measured as for Table 2.

conclusions from these data. In addition, the wide spread of TPO levels may indicate subpopulations within these larger diagnostic categories.

A more detailed description of some of these patients was reported, and it was suggested that the low values for TPO seen in the patients with ET may represent a dysregulation of the feedback loop involved in platelet production *(83)*. Alternatively, the low levels may simply reflect stimulation of thrombopoiesis by cytokines other than TPO. A small number of patients with ET (*n* = 6) were found to have a modest increase in endogenous TPO (approx four times normal values) *(81)*.

3.5. Relationship Between Platelet Number and TPO Levels

Kuter and Rosenberg have suggested that platelet mass may play a role in post-translational regulation of circulating endogenous TPO levels (*see* Chapter 23) *(91)*. Plasma samples from rabbits taken during busulfan-induced thrombocytopenia were found to have TPO levels inversely related to platelet number (mass). The endogenous TPO levels decreased when fresh platelets were transfused into the thrombocytopenic rabbits.

Similar results have been reported in patients who become thrombocytopenic after myeloablation and stem cell transplantation *(73)*. Serum samples from 11 patients were evaluated in a TPO-dependent proliferation bioassay. The changes in endogenous TPO levels were compared with the changes in baseline platelet counts (Fig. 3). This was the first published report of the relationship of endogenous TPO levels to platelet number in any patient population and confirmed the observations of Kuter and Rosenberg.

This relationship has also been examined using the polyclonal antibody TPO ELISA *(26)*. Plasma samples from nine patients who underwent myeloablative therapy before BMT were collected before therapy and through to the recovery of platelets. Levels of TPO, IL-6, IL-11, IL-3, and LIF were monitored by immuno-

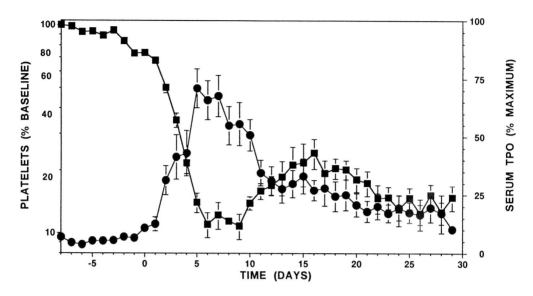

Fig. 3. Platelet counts and endogenous TPO serum levels in patients receiving myeloablative chemotherapy. TPO levels (●) correlate inversely with platelet number (■) in these patients receiving myeloablative therapy and stem cell transplantation ($n = 11$). TPO levels were measured in a 32D-HuMpl+ cell bioassay *(73)*.

assay. Again, there was an inverse correlation between platelet number and endogenous TPO levels; IL-6 levels were also seen to increase as platelet number decreased, but these levels did not decline during the recovery period and, therefore, were determined to be independent of platelet number. There was no change seen in levels of IL-11, IL-3, or LIF throughout the study period. A similar pattern was observed by another group for patients receiving submyeloablative and myeloablative therapy *(86)*.

Cyclic changes have been seen in both neutrophil and platelet numbers in certain pathologic conditions *(85,92)*. To examine further the platelet–TPO relationship using the TPO ELISA, plasma from two patients with cyclic thrombocytopenia was assayed. When plasma from these patients was evaluated over the course of one thrombocytopenic cycle (approximately 25–30 days), the inverse relationship between endogenous TPO levels and platelet number was observed in both (Fig. 4A and Fig. 4B).

3.6. Summary

There are several assays with which to measure endogenous TPO levels in sera or plasma (Table 5). Although these assays suggest different absolute values for the circulating TPO levels, there are several observations that are consistent in all the studies reported:

- TPO is present in serum and plasma from normal individuals at a concentration of <200 pg/mL.
- TPO levels in conditions in which there is marrow ablation and/or lack of platelet production are generally high (10–25 times normal levels).

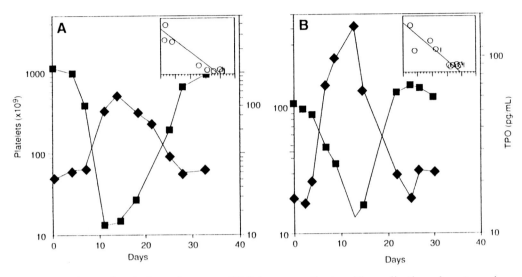

Fig. 4. Plasma levels in endogenous TPO in two patients with cyclic thrombocytopenia. TPO levels (◆) correlate inversely with platelet number (■). Insets show the correlation of platelet count (y-axis) versus TPO levels (x-axis) for each patient. **(A)** $r^2 = 0.915$; **(B)** $r^2 = 0.782$. Thrombopoietin measurements were made by ELISA in heparinized plasma from two patients (A, B) with cyclic thrombocytopenia.

- TPO levels in patients with ITP (increased platelet production and consumption) are generally in the normal to high-normal range. An explanation for this apparent discrepancy is that when platelet production is increased, the passage of platelets through the circulation is enhanced, and TPO clearance may be increased, resulting in little or no accumulation of TPO.
- TPO levels in patients with primary marrow disorders, such as ET, in which platelet production is excessive, are normal to slightly greater than normal. This may suggest a dysregulation of the ability of these platelets to function properly as the posttranslational regulator of TPO.
- TPO levels are inversely related to platelet number (mass) in plasma or serum from animals or patients in whom platelet production is altered. This phenomenon has been observed in patients or animals with myeolsuppressive insult and in patients with cyclic thrombocytopenia.

The findings of near-normal TPO levels in patients with ITP seem to be an exception to the general association of high serum TPO with low platelet count; however, platelets in circulation may not represent all compartments of platelet mass able to remove TPO from the circulation. Potential additional platelet compartments may include the platelets in the spleen *(90)* and perhaps the megakaryocytes in the marrow *(86)*. These potential sources of platelet mass may explain the discrepency between the TPO levels and the platelet counts in this setting. The findings in ET *(80,83)* and in chronic liver disease *(80)* suggest that dysfunctional elements in the production and/or clearance of TPO may exist.

Analysis of circulating concentrations of endogenous TPO in normal and pathologic states may help to resolve many of the questions surrounding TPO production and

Table 5
Comparison of TPO ELISA Methods[a]

Assay group	Capture/ detection sensitivity	Normal levels	Changes with disease
Kirin (81,90)	Monoclonal/biotin 0.5–3.3 fmol/mL (~18–117 pg/mL)	0.8 fmol/mL (~27 pg/mL) serum	lg↑AA; s↑ET hn-ITP
Genetech (82,88)	Chimera-HuMpl-IgG/ biotin 200 pg/mL	Undetectable	lg↑Chemo/BMT; s↑ITP; ↑AA
Amgen (26,80,83–85)	Polyclonal/HRP 20–50 pg/mL	81 ± 5 pg/mL plasma 95 ± 6 pg/mL serum	g↑AA; lg↑Chemo; s↑TTP; n-lg↑CLD, hn-ITP; hn-ET; hn-MDS; n-CML; n-RA; ↑CTw < Plt

[a]Normal levels and levels in various disease states reflect results from preliminary studies with each assay.

n, normal; hn, high normal; lg↑, large increase; ↑, increase; s↑, slight increase; ↓, decrease; BMT, bone marrow transplantation; CLD, chronic liver disease; ITP, immune thrombocytopenia purpura; MDS, myelodysplastic syndromes; ET, essential thrombocythemia; CML, chronic myelogenous leukemia; CTw < Plt, cyclic thrombocytopenic with ↓platelets; TTP, thrombotic thrombocytopenia purpura; Chemo, chemotherapy; AA, aplastic anemia.

regulation, and may also provide new insights into mechanisms underlying thrombocytopenia and thrombocytosis. In addition, these assays can be easily adapted to measure recombinant Mpl ligand forms and should prove useful for clinical pharmacokinetic studies.

References

1. Miyajima A, Miyatake S, Schrears J, et al. Coordinate regulation of immune and inflammatory responses by T cell-derived lymphokines. *FASEB J.* 1988; 2: 2462–2473.
2. Tracey KJ, Beutler B, Lowry SF, et al. Shock and tissue injury induced by recombinant human cachectin. *Science.* 1986; 234: 470–474.
3. Whiteside TL. Cytokine measurements and interpretation of cytokine assays in human disease. *J Clin Immunol.* 1994; 14: 327–339.
4. Rossio JL, Rager HC, Goudry CS, Crisp EA. Cytokine testing in clinical trial monitoring. In: Rose NR, DeMarcario EC, Fahey JL, Friedman H, Penn GM (eds). *Manual of Clinical Laboratory Immunology.* Washington, DC: ASM; 1992: 942–947.
5. Zwierzina H, Schollenberger S, Herold M, Schmalzl F, Besemer J. Endogenous serum levels and surface receptor expression of GM-CSF and IL-3 in patients with myelodysplastic syndromes. *Leukemia Research.* 1992; 16: 1181–1186.
6. Sekiyama KD. Circulating proinflammatory cytokines (IL-1β, TNF-α, and IL-6) and IL-1 receptor antagonist (IL-1Ra) in fulminant hepatic failure and acute hepatitis. *Clin Exp Immunol.* 1994; 98: 71–77.
7. Barak V, Levi-Schaffer F, Nisman B, Nagler A. Cytokine dysregulation in chronic graft versus host disease. *Leukemia Lymphom.* 1995; 17: 169–173.

8. May LT, Viguet H, Kenney JS, Ida N, Allison AC, Sehgal PB. High levels of "complexed" interleukin-6 in human blood. *J Biol Chem.* 1992; 267: 19,698–19,704.

9. Watari K, Asano S, Shirafuji N, et al. Serum granulocyte colony-stimulating factor levels in healthy volunteers and patients with various disorders as estimated by enzyme immunoassay. *Blood.* 1989; 73: 117–122.

10. Sothern RB. Circadian characteristics of circulating interleukin-6 in men. *J Allergy Clin Immunol.* 1995; 95: 1029–1035.

11. Pechumer H, Wilhelm M, Ziegler-Heitbrock HWL. Interleukin-6 (IL-6) levels in febrile children during maximal aplasia after bone marrow transplantation (BMT) are similar to those in children with normal hematopoiesis. *Ann Hematol.* 1995; 70: 309–312.

12. Seymour JF, Talpaz M, Cobanillas F, Wetzler M, Kurzrock R. Serum interleukin-6 levels correlate with prognosis in diffuse large-cell lymphoma. *J Clin Oncol.* 1995; 13: 575–582.

13. Bartalena L, Brogioni S, Grasso L, Martino E. Increased serum interleukin-6 concentration in patients with subacute thyroiditis: relationship with concomitant changes in serum T4-binding globulin concentrations. *J Endocrinol Invest.* 1993; 16: 213–218.

14. Kucharzik T, Stoll R, Lugering N, Domschke W. Circulating antiinflammatory cytokine IL-10 in patients with inflammatory bowel disease (IBD). *Clin Exp Immunol.* 1995; 100: 452–456.

15. Greendyke RM, Sharma K, Gifford R. Serum cytokine levels in patients with neutropenia. *Arch Pathol Lab Med.* 1994; 118: 1193–1195.

16. Blay JY, Farcet JP, Lavaud A, Radoux D, Chouaib S. Serum concentrations of cytokines in patients with Hodgkin's disease. *Eur J Cancer.* 1994; 30A: 321–324.

17. Bowen D, Yaneik S, Bennett L, Culligan D, Resser K. Serum stem cell factor concentration in patients with myelodysplastic syndromes. *Br J Haematol.* 1993; 85: 63–66.

18. Manegold C, Joblonowski H, Armbrecht C, Strohmeyer G, Pietsch T. Serum levels of stem cell factor are increased in asymptomatic human immunodeficiency virus-infected patients and are associated with prolonged survival. *Blood.* 1995; 86: 243–249.

19. Haas R, Gericke G, Witt B, Cayeux S, Hunstein W. Increased serum levels of granulocyte colony-stimulating factor after autologous bone marrow or blood stem cell transplantation. *Exp Hematol.* 1993; 21: 109–113.

20. Erslev AJ. Erythropoietin titers in health and disease. *Semin Hematol.* 1991; 28: 2–8.

21. Oster W, Mertelsmann R. The role of erythropoietin in patients with anemia and normal renal function. In: Schaefer RM, Heidland A, Horl WH (eds). *Erythropoietin in the 90's.* Basel: Karger, 1990: 26–35.

22. Ida N, Sakurai S, Hosoi K, Kunitomo T. A highly sensitive enzyme-linked immunosorbent assay for the measurement of interleukin-8 in biological fluids. *J Immunol Methods.* 1992; 156: 27–38.

23. Gangarossa S, Romano V, Munda SE, Sciotto A, Schiliro G. Low serum levels in interleukin-6 in children with post-infective acute thrombocytopenic purpura. *Eur J Haematol.* 1995; 55: 117–120.

24. Fitzpatrick MF, Mackay T, Whyte KF, et al. Nocturnal desaturation and serum erythropoietin: a study in patients with chronic obstructive pulmonary disease and in normal subjects. *Clin Sci.* 1993; 84: 319–324.

25. Garcia JF, Ebbe SN, Hollander L, Cutting HO, Miller ME, Cronkite EP. Radioimmunoassay of erythropoietin: circulating levels in normal and polycythemic human beings. *J Lab Clin Med.* 1982; 99: 624–635.

26. Bernstein SH, Baer MR, Lawrence D, Herzig GP, Bloomfield CD, Wetzler M. Serial determination of thrombopoietin (TPO), IL-3, IL-6, IL-11 and leukemia inhibitory factor (LIF) levels in patients undergoing chemotherapy for acute myeloid leukemia (AML). *Blood.* 1995; 86: 46a (abstract no 170).

27. Gause A, Jung W, Keymis S, et al. The clinical significance of cytokines and soluble forms of membrane-derived activation antigens in the serum of patients with Hodgkin's disease. *Leukemia Lymphoma.* 1992; 7: 439–447.

28. Cairo MS, Suen Y, Sender L, et al. Circulating granulocyte colony-stimulating factor (G-CSF) levels after allogeneic and autologous bone marrow transplantation: endogenous G-CSF production correlates with myeloid engraftment. *Blood*. 1992; 79: 1869–1873.

29. Imamura M, Hashino S, Kobayashi H, et al. Serum cytokine levels in bone marrow transplantation: synergistic interaction of interleukin-6, interferon-γ, and tumor necrosis factor-a in graft-versus-host disease. *Bone Marrow Transplant*. 1994; 13: 745–751.

30. Tilg H, Wilmer A, Vogel W, et al. Serum levels of cytokines in chronic liver diseases. *Gastroenterology*. 1992; 103: 264–274.

31. Usuki, K. Serum concentrations of thrombopoietin, erythropoietin, IL-6, and IL-11 in various hematological disorders. *FASEB J*. 1995; 2: 2462–2473.

32. Koeffler HP, Goldwasser E. Erythropoietin radioimmunoassay in evaluating patients with polycythemia. *Ann Intern Med*. 1981; 94: 44–47.

33. Barosi G. Inadequate erythropoietin response to anemia: definition and clinical relevance. *Ann Hematol*. 1994; 68: 215–223.

34. Madson KL, Moore TL, Lawrence JM, Osborn TG. Cytokine levels in serum and synovial fluid of patients with juvenile rheumatoid arthritis. *J Rheumatol*. 1994; 21: 2359–2363.

35. Stasi R, Zinzani PL, Galieni P, et al. Clinical implications of cytokine and soluble receptor measurements in patients with newly-diagnosed aggressive non-Hodgkin's lymphoma. *Eur J Haematol*. 1995; 54: 9–17.

36. Yamamoto T, Yoneda K, Ueta E, Osaki T. Serum cytokines, interleukin-2 receptor, and soluble intercellular adhesion molecule-1 in oral disorders. *Oral Surg Oral Med Oral Pathol*. 1994; 78: 727–735.

37. Chang DM. The role of cytokines in heat stroke. *Immunol Invest*. 1993; 22: 553–561.

38. Riche F, Dosquet C, Panis Y, et al. Levels of portal and systemic blood cytokines after colectomy in patients with carcinoma or Crohn's disease. *J Am Coll Surg*. 1995; 180: 718–724.

39. Deehan DJ, Heys SD, Simpson WG, Broom J, Franks C, Eremin O. In vivo cytokine production and recombinant interleukin 2 immunotherapy: an insight into the possible mechanisms underlying clinical responses. *Br J Cancer*. 1994; 69: 1130–1135.

40. Berridge MV, Fraser JK, Carter JM, Lin FK. Effects of recombinant human erythropoietin on megakaryocytes and on platelet production in the rat. *Blood*. 1988; 72: 970–977.

41. Bruno E, Briddell R, Hoffman R. Effect of recombinant and purified hematopoietic growth factors on human megakaryocyte colony formation. *Exp Hematol*. 1988; 16: 371–377.

42. Bruno E, Cooper RJ, Briddell RA, Hoffman R. Further examination of the effects of recombinant cytokines on the proliferation of human megakaryocyte progenitor cells. *Blood*. 1991; 77: 2339–2346.

43. Bruno E, Hoffman R. Effect of interleukin 6 on in vitro human megakaryopoiesis: its interaction with other cytokines. *Exp Hematol*. 1989; 17: 1038–1043.

44. Bruno E, Miller ME, Hoffman R. Interacting cytokines regulate in vitro human megakaryopoiesis. *Blood*. 1989; 73: 671–677.

45. Chang M, Sven Y, Lee SM, et al. Transforming growth factor-β1, macrophage inflammatory protein-1a, and interleukin-8 gene expression is lower in stimulated human neonatal compared with adult mononuclear cells. *Blood*. 1994; 84: 118–125.

46. Dan K, Gomi S, Inokochi K, et al. Effects of interleukin-1 and tumor necrosis factor on megakaryocytopoiesis: mechanism of reactive thrombocytosis. *Acta Heamatol*. 1995; 93: 67–72.

47. Debili N, Hegyi E, Navarro S, et al. In vitro effects of hematopoietic growth factors on the proliferation, endoreplication and maturation of human megakaryocytes. *Blood*. 1991; 77: 2326–2338.

48. Debili N, Issad C, Masse J-M, et al. Expression of CD34 and platelet glycoproteins during human megakaryocytic differentiation. *Blood*. 1992; 80: 3022–3035.

49. Debili N, Masse JM, Katz A, Guichard J, Breton-Gorius J, Vainchenker W. Effects of the recombinant hematopoietic growth factors interleukin-3, interleukin 6, stem cell factor, and leu-

kemia inhibitory factor on the megakaryocytic differentiation of CD34+ cells. *Blood.* 1993; 82: 84–95.

50. Hegyi E, Navarro S, Debili N, et al. Regulation of human megakaryocytopoiesis: analysis of proliferation, ploidy and maturation in liquid cultures. *Int J Cell Cloning.* 1990; 8: 236–244.

51. Ishibashi T, Koziol JA, Burstein SA. Human recombinant erythropoietin promotes differentiation of murine megakaryocytes in vitro. *J Clin Invest.* 1987; 79: 286–289.

52. Lemoli RM, Fogli M, Fortuna A, et al. Interleukin-11 stimulates the proliferation of human hematopoietic CD34+ and CD34+CD33-DR- cells and synergizes with stem cell factor, interleukin-3 and granulocyte-macrophage colony-stimulating factor. *Exp Hematol.* 1993; 21: 1668–1672.

53. Long MW, Hutchinson RJ, Gragowski LL, Heffner CH, Emerson SG. Synergistic regulation of human megakaryocyte development. *J Clin Invest.* 1988; 82: 1779–1786.

54. Martin DI, Zon LI, Mutter G, Orkin SH. Expression of an erythroid transcription factor in megakaryocytic and mast cell lineages. *Nature.* 1990; 344: 444–447.

55. Mazur, E.M., Cohen, J.L., Newton, J., et al. Human serum megakaryocyte colony-stimulating activity appears to be distinct from interleukin-3, granulocyte-macrophage colony-stimulating factor, and lymphocyte-conditioned medium. *Blood.* 1990; 76: 290–297.

56. Mazur EM, Cohen JL, Wong GG, Clark SC. Modest stimulatory effect of recombinant human GM-CSF on colony growth from peripheral blood human megakaryocyte progenitor cells. *Exp Hematol.* 1987; 15: 1128–1133.

57. McDonald TP, Clift RE, Cottrell MB. Large, chronic doses of erythropoeitin cause thrombocytopenia in mice. *Blood.* 1992; 80: 352–358.

58. Straneva J, Bruno E, Beyer G, Floyd A, Hoffman R. Detection of serum factors that regulate cytoplasmic and nuclear maturation of human megakaryocytes. *Clin Res.* 1984; 32: 324.

59. Straneva JE, Yang H, Bruno E, Hoffman R. Separate factors control terminal cytoplasmic maturation of human megakaryocytes. In: Levin RF, Williams N, Leven J, Evatt BC (eds). *Progess in Clinical and Biological Research: Megakaryocyte Development and Function.* New York: Alan R Liss; 1986: 253–257.

60. Straneva JE, Goheen MP, Hui SL, Bruno E, Hoffman R. Terminal cytoplasmic maturation of human megakaryocytes in vitro. *Exp Hematol.* 1986; 14: 919–929.

61. Straneva JE, Yang HH, Hui SL, Bruno E, Hoffman R. Effects of megakaryocyte colony-stimulating factor on terminal cytoplasmic maturation of human megakaryocytes. *Exp Hematol.* 1987; 15: 657–663.

62. Teramura M, Katahira J, Hoshino S, Motoji T, Oshimi K, Mizoguchi H. Effect of recombinant hemopoietic growth factors on human megakaryocyte colony formation in serum-free cultures. *Exp Hematol.* 1989; 17: 1011–1016.

63. Warren MK, Guertin M, Rudzinski I, Seidman MM. A new culture and quantitation system for megakaryocyte growth using cord blood CD34+ cells and the GPIIb/IIIa marker. *Exp Hematol.* 1993; 21: 1473–1479.

64. Broudy VC, Lin NL, Kaushansky K. Thrombopoietin (*c-mpl* ligand) acts synergistically with erythropoietin, stem cell factor and interleukin-11 to enchance murine megakaryocyte colony growth and increases megakaryocyte ploidy in vitro. *Blood.* 1995; 85: 1719–1726.

65. Bartley TD, Bogenberger J, Hunt P, et al. Identification and cloning of a megakaryocyte growth and development factor that is a ligand for the cytokine receptor Mpl. *Cell.* 1994; 77: 1117–1124.

66. de Sauvage FJ, Hass PE, Spencer SD, et al. Stimulation of megakaryocytopoiesis and thrombopoiesis by the *c-Mpl* ligand. *Nature.* 1994; 369: 533–538.

67. Lok S, Kaushansky K, Holly RD, et al. Cloning and expression of murine thrombopoietin cDNA and stimulation of platelet production in vivo. *Nature.* 1994; 369: 565–568.

68. Kato T, Ogami K, Shimada Y, et al. Purification, and characterization of thrombopoietin. *J Biochem.* 1995; 118: 229–236.

69. Kuter DJ, Beeler DL, Rosenberg RD. The purification of megapoietin: aphysiological regulator

of megakaryocyte growth and platelet production. *Proc Natl Acad Sci USA*. 1994; 91: 11,104–11,108.

70. Kuter DJ, Rosenberg RD. Appearance of a megakaryocyte growth promoting activty, megapoietin during acute thrombocytopenia in the rabbit. *Blood*. 1994; 84: 1464–1472.

71. Wendling F, Maraskovsky E, Debili N, et al. *c-Mpl* ligand is a humoral regulator of megakaryo-poiesis. *Nature*. 1994; 369: 571–574.

72. Kaushansky K, Lok S, Holly RD, et al. Promotion of megakaryocyte progenitor expansion and differentiation by the *c-Mpl* ligand thrombopoietin. *Nature*. 1994; 369: 568–571.

73. Nichol JL, Hokom MM,Hornkohl AC, et al. Megakaryocyte growth and development factor: Analyses of in vitro effects on human megakaryopoiesis and endogenous serum levels during chemotherapy-induced thrombocytopenia. *J Clin Invest*. 1995; 95: 2973–2978.

74. Hunt P, Li Y-S, Nichol JL, et al. Purification and biologic characterization of plasma-derived megakaryocyte growth and development factor. *Blood*. 1995; 86: 540–547.

75. Ulich TR, Del Castillo J, Yin S, et al. Megakaryocyte growth and development factor amelio-rates carboplatin-induced thrombocytopenia in mice. *Blood*. 1995; 86: 971–976.

76. Eaton DL, de Sauvage FJ. Thrombopoietin and the humoral regulation of thrombocytopoiesis. *Curr Opinion Hematol*. 1995; 2: 167–171.

77. Dale DC, Bolyard AA, Hammond WP. Cyclic neutropenia: natural history and effects of long-term treatment with recombinant human granulocyte colony-stimulating factor. *Cancer Invest*. 1993; 11: 219–223.

78. Choi ES, Hokom M, Bartley T, et al. Recombinant human megakaryocyte growth and develop-ment factor (rHuMGDF), a ligand for *c-mpl*, produces functional human platelets in vitro. *Stem Cells*. 1995; 13: 317–322.

79. Farese AM, Hunt P, Boone T, McVittie TJ. Recombinant human megakaryocyte growth and development factor stimulates thrombocytopoiesis in normal nonhuman primates. *Blood*. 1995; 86: 54–59.

80. Nichol JL, Hornkohl A, Selesi D, Wyers M, Hunt P. TPO levels in plasma of patients with thrombocytopenia or thrombocytosis. *Blood*. 1995; 86: 371a (abstract no 1474).

81. Tahara T, Usuki K, Sato H, et al. Serum thrombopoietin levels in healthy volunteers and pa-tients with various disorders measured by enzyme immunoassay. *Blood*. 1995; 86: 910a (ab-stract no 3627).

82. Emmons RVB, Shulman NR, Reid DM, et al. Thrombocytopenic patients with aplastic anemia have high TPO levels whereas those with immune thrombocytopenia are much lower. *Blood*. 1995; 86: 372a (abstract no 1475).

83. Taylor K, Pitcher L, Nichol J, et al. Inappropriate elevation and loss of feedback regulation of thrombopoietin in essential thrombocythemia (ET). *Blood*. 1995; 86: 49a (abstract no 183).

84. Hornkohl A, Selesi D, Bennett L, Hockman H, Nichol J, Hunt P. Thrombopoietin and sMpl concentrations in the serum and plasma of normal donors. Blood. 1995; 86: 898a (abstract no 3580).

85. Zent CS, Hornkohl A, Arepally G, et al. Cyclic thrombocytopenia: thrombopoietin response to spontaneous changes in platelet counts. *Blood*. 1995; 86: 370a (abstract no 1470).

86. Chang M, Suen Y, Meng G, et al. Regulation of TPO mRNA expression and protein production: TPO gene regulation appears post transcriptional, and endogenous levels are inversely correlated to megakaryocyte mass and circulating platelet count. *Blood*. 1995; 86: 368a (abstract no 1460).

87. deSauvage FJ, Louh S-M, Carver-Moore K, et al. Deficiencies in early and late stages of megakaryocytopoiesis in TPO-KO mice. *Blood*. 1995; 86: 255a (abstract no 1007).

88. Fielder PJ, Hass P, Nagel M, et al. Human platelets as a model for the binding, internalization, and degradation of thrombopoietin (TPO). *Blood*. 1995; 86: 365a (abstract no 1450).

89. Kato T, Ozawa T, Muto T, et al. Essential structure for biological activity of thrombopoietin. *Blood*. 1995; 86: 365a (abstract no 1448).

90. Kosugi S, Kurata Y, Tomiyama Y, et al. Circulating thrombopoietin level in chronic immune thrombocytopenic purpura. 1996: (manuscript in preparation).

91. Kuter DJ, Rosenberg D. The reciprocal relationship of thrombopoietin (c-Mpl ligand) to changes in the platelet mass during busulfan-induced thrombocytopenia in the rabbit. *Blood*. 1995; 85: 2720–2730.

92. Dale DC, Bolyard AA, Hammond WP. Cyclic neutropenia: natural history and effects of long term treatment with recombinant human granulocyte-colony stimulating factor. *Cancer Invest*. 1993; 11: 219–223.

23

The Regulation of Platelet Production In Vivo

David J. Kuter

1. Introduction

Until recently our understanding of the regulation of platelet production in vivo was very poor. For other circulating blood cells, such as erythrocytes and neutrophils, physiological observations quickly led to the identification of their regulatory hemopoietic factors. The response to anemia and high altitude showed the importance of erythropoietin (EPO) and the leukocytosis associated with infection identified granulocyte colony-stimulating factor (G-CSF) and granulocyte-macrophage colony-stimulating factor (GM-CSF). Ultimately these growth factors were purified, cloned, and introduced into clinical practice, but no such clinical conditions or environmental exposures had clearly shown the existence of the "thrombopoietin" (TPO) *(1)*, which regulated platelet production.

Although initially believed to be particles of dust or microorganisms on the stained blood smear *(2)*, Wright demonstrated in 1906 *(3)* that platelets were enucleate cells that entered the circulation and were responsible for hemostasis. Furthermore, he demonstrated that these cells somehow "budded off" from bone marrow megakaryocytes *(3,4)*. As summarized below, many subsequent studies described a putative feedback loop between bone marrow megakaryocytes and their platelet progeny. The regulation of this feedback loop was usually compared with the well-understood control of red cell production by EPO. For the red cell, anemia decreases oxygen delivery to the kidney peritubular fibroblasts *(5)* and is "sensed" by an efficient cytochrome system *(6,7)*, which in turn rapidly increases the transcription rate of the EPO gene. With no storage pool, EPO is released into the circulation where it acts on bone marrow precursor cells to increase red cell production. By analogy, a decreased platelet count is somehow assumed to be detected by a "sensor" that in turn increases production of endogenous TPO, which then enters the circulation and stimulates the production of platelets by the bone marrow megakaryocytes. Unfortunately for this model (Fig. 1), no sensing organ had ever been identified nor was any platelet substance or function a clear candidate to be "sensed." Many investigators even doubted the existence of TPO (*see* Chapter 5).

The recent identification and purification of TPO have led to a clearer understanding of the normal physiology of platelet production. This knowledge provides an explana-

From: *Thrombopoiesis and Thrombopoietins: Molecular, Cellular, Preclinical, and Clinical Biology*
Edited by: D. J. Kuter, P. Hunt, W. Sheridan, and D. Zucker-Franklin Humana Press Inc., Totowa, NJ

ERYTHROPOIETIN   THROMBOPOIETIN
MODEL   MODEL

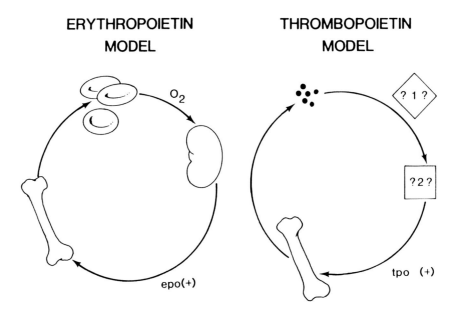

Fig. 1. Comparison of EPO and TPO "sensor" models. As described in the text, the kidney has an efficient oxygen sensor that can measure changes in the red cell mass and alter the EPO transcription rate. By analogy, a platelet sensor (?2?) has long been assumed to be present that measured some platelet attribute (?1?) and changed the TPO transcription rate.

tion of much of the physiological data published previously and a framework for understanding pathological changes of the normal physiology.

2. Lessons Learned About Platelet Physiology from Early Animal and Human Experimentation

Before the purification of TPO, extensive animal and human experimentation had demonstrated a number of principles of platelet physiology. First, the platelet count is stable in any one individual for years, probably for life *(8)*, unless altered by disease or other physiological conditions, such as pregnancy. There is no diurnal variation, but there are modest changes with the menstrual cycle.

Second, although constant for any one individual, there is a wide range of platelet counts that are considered normal for any one species or ethnic group. The average normal human platelet count is 219×10^9/L with a range of 150×10^9/L–450×10^9/L *(9)*. For Mediterranean populations *(10)*, the average normal values are lower, 161×10^9/L with a range of 89×10^9/L–290×10^9/L. When other species are analyzed *(11)*, the average normal platelet count varies enormously, as does the range within any one species; mice have an average platelet count of 1040×10^9/L, whereas porcupines have an average normal platelet count of 20×10^9/L. In contrast, for all of these species, the red blood cell mass is virtually constant.

Third, this wide difference in normal platelet counts within a species or between species is offset by differences in the platelet size *(11,12)*. Within the distribution of normal human platelet counts, those with counts of 430×10^9/L have a mean platelet volume (MPV) of 8.6 fL, whereas those with counts of 160×10^9/L have MPV of 10.6

fL *(9)*. The Mediterranean population *(10)* has an average platelet count of 161×10^9/L and an MPV of 17.8 fL (range, 10.8–29.2 fL) compared with the Northern European population, which has an average platelet count of 219×10^9/L and a MPV of 12.4 fL (range, 9.9–15.6 fL). When extended to other species *(11)*, mice have an MPV of 2.1 fL, whereas porcupines have an MPV of 105 fL. Although the simple product of MPV and platelet count usually fails to describe a constant number, with some corrections for the surface area/cytoplasmic mass in calculating the MPV, the product does describe a constant platelet mass per kilogram body weight for all humans and probably for all higher vertebrates *(11,13)*.

Fourth, the concept that the body defends the mean platelet mass, not the platelet count, has been well demonstrated by Aster *(14,15)*, Penny et al. *(16)*, and de Gabriele and Penington *(17)* in humans and in animals. Approximately one-third of the total human platelet mass is normally concentrated in the spleen *(14)*. In patients with splenomegaly, thrombocytopenia develops as the circulating pool of platelets is redistributed to the spleen in these patients; this exchangeable splenic pool may contain as much as 90% of the body's platelets *(14,16)*. However, the total body platelet mass (circulating plus splenic pools) is normal, as is the platelet life-span. Further proof of this concept comes from experiments in which normal and splenectomized rats *(15,17)* were compared with rats that had their spleens variably increased in size by the injection of methylcellulose. The platelet counts of the splenectomized mice were elevated above normal, whereas the counts of the mice with splenomegaly were decreased proportionally to the extent of organomegaly. However, when the animals were euthanized and the platelet content of the circulating and splenic pools compared, all mice had the same total body mass of platelets.

Fifth, there is a well-characterized relationship between bone marrow megakaryocytes and the circulating platelet mass. After an acute decline in the platelet mass, megakaryocyte size and ploidy start to increase after 2 days and reach a maximum after 4 days *(18)*. The number of megakaryocytes also increases and is maximal after 10 days *(18)*. Platelet production starts increasing after 2 days *(19)* and reaches a maximum rate of approximately 10-fold *(20)* over the next week. Overall, the maximum increase in megakaryocyte mass is 7- to 10-fold *(18,20)*. When the platelet count is experimentally increased by platelet transfusion, the bone marrow megakaryocytes show the opposite effects. An inverse and proportional relationship therefore exists between platelet mass and the bone marrow megakaryocyte ploidy (Fig. 2), a fact that has been used clinically for many years in the evaluation of patients with idiopathic thrombocytopenic purpura (ITP).

These clinical and experimental observations were strong evidence for the existence of a feedback loop between bone marrow megakaryocytes and the circulating platelet mass, which was presumably mediated by the then-uncharacterized endogenous hemopoietic growth factor, TPO. Several experimental models helped further to describe the nature of this regulatory system and predicted some of the attributes of TPO. First, this feedback loop between platelet and megakaryocyte had a definite lag time of up to 24 hours before it was fully activated *(19)*. When mice were made severely thrombocytopenic (Fig. 3) and then transfused to a normal platelet count at various times thereafter, the expected megakaryocyte ploidy increase could be abrogated after 3 hours but not after 24 hours (Fig. 4) of thrombocytopenia. This suggested that either it took 24

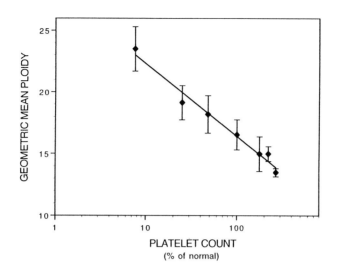

Fig. 2. Relationship of megakaryocyte ploidy to the platelet mass in vivo. Rats (7–11 in each group) were made variably thrombocytopenic or thrombocytotic by the administration of antiplatelet antibody or by the transfusion of platelets *(19)*, respectively, and 2 days later compared with 356 normal rats. Bone marrow was harvested from all rats, labeled with antiplatelet antibody and propidium iodide, and the ploidy distribution determined by flow cytometry *(19)*. For each ploidy distribution, the geometric mean ploidy was determined *(19)* and compared with the platelet mass (here quantified by the platelet count). $r^2 = 0.978$.

hours for endogenous TPO to be maximally generated or that it took 24 hours for TPO to stimulate its receptor maximally.

In an important experiment, Jackson and Edwards *(21)* showed that platelets played some direct role in this feedback loop. When transfused into normal recipient rats, platelets from normal rats, but not from vincristine-treated rats, were effective in regulating megakaryocyte ploidy. Since there was no effect on platelet life-span, these studies showed that there was some vincristine-sensitive platelet attribute that was important in this feedback system.

This platelet attribute did not appear to be any classical platelet function. In a large study of patients with abnormal platelet counts, Harker and Finch found no relationship between disorders in which platelet function was deficient and changes in the megakaryocyte mass *(20)*. Disorders usually associated with diminished platelet function, such as essential thrombocythemia, had the same bone marrow megakaryocyte characteristics as did disorders, such as ITP, in which platelet function was increased. This reinforced the concept that no platelet hemostatic attribute mediated the feedback loop between platelet and megakaryocyte.

The Harker and Finch study *(20)* also demonstrated a discordance between effects of the putative feedback loop on megakaryocyte number versus megakaryocyte ploidy. In reactive thrombocytosis, megakaryocyte number was increased, but the ploidy was decreased below normal, whereas in essential thrombocythemia and in ITP both were increased above normal. These results suggested that megakaryocyte ploidy, not number, was the more accurate indicator of the extent of activation of the feedback loop.

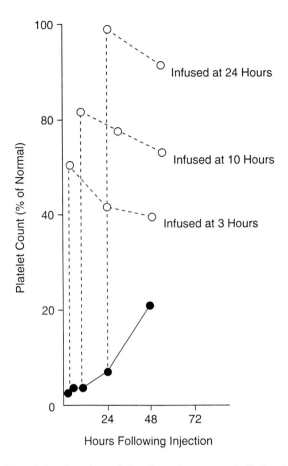

Fig. 3. Relationship of the duration of the thrombocytopenic "stimulus" to the extent of platelet count change. Twenty rats were made severely thrombocytopenic (platelet count 7–9% of normal) by the infusion of antiplatelet antibody *(19)*. Three, 10, and 24 hours later, a constant amount of rat platelets was transfused into some of these rats (three rats for each time-point) (broken lines). The other 11 thrombocytopenic rats were not transfused (solid line). All the rats (transfused and nontransfused) plus 17 normal littermates were harvested for flow cytometry 48 hours after the antibody injection.

3. Lessons from Recent In Vivo Physiological Experiments with TPO

TPO is not an absolute requirement for either megakaryocyte differentiation, endomitosis, or platelet shedding. The elegant studies by de Sauvage et al. have shown that megakaryocyte differentiation, endomitosis, and maturation occur, although at a rate 15% of normal, when the TPO receptor (Mpl) *(22)*, or TPO (Mpl ligand) *(23)* was eliminated by homologous recombination in mice (*see* Chapter 21). Furthermore, Choi et al. have demonstrated that rHuTPO does not regulate and is not required for the shedding of platelets from megakaryocytes *(24)* (*see* Chapter 17). Indeed, elevated amounts of TPO actually inhibit platelet shedding in vitro and in vivo *(24)*. The role of TPO, therefore, is to amplify the wide range of differentiated megakaryocytes ranging from the megakaryocyte colony-forming

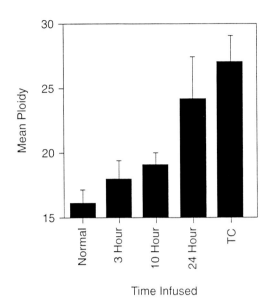

Fig. 4. The duration of the thrombocytopenic "stimulus" determines the extent of mega-karyocyte ploidy "response." As described in Fig. 3, bone marrow megakaryocyte ploidy distributions were determined by flow cytometry in 17 normal rats, 11 severely thrombocytopenic (TC) rats, and a total of 9 thrombocytopenic rats that had been reinfused with platelets 3, 10, or 24 hours after the injection of antiplatelet antibody *(19)*. All ploidy distributions were expressed as the geometric mean ploidy (±SD). Statistical significance was determined by the Mann-Whitney U test (among normal, 3-, and 10-hour values, NS; between 24-hour and TC values, NS; 24-h and TC values versus all others, $p < 0.01$.)

cell (MK-CFC) to the mature megakaryocyte. As such, its known actions are to increase the number and cycling of MK-CFC, the rate of endomitosis, the size of megakaryocytes (and presumably their specific protein contents), shorten the time to maturation, and, thereby, increase platelet production.

The circulating level of endogenous TPO is, therefore, the crucial variable in determining the extent of activation of the feedback loop between platelet and megakaryocyte. It controls the extent of stimulation of the bone marrow megakaryocytes and, ultimately, platelet production. Much has recently become known about the circulating levels of TPO using both immunoassays and bioassays in a number of animal models and in some human disorders (*see* Chapter 22).

After the acute onset of thrombocytopenia in rabbits *(25)*, endogenous TPO levels are not elevated after 3 hours, become half-maximal by 10 hours, and are at their peak by 24 hours (Fig. 5). With prolongation of thrombocytopenia for several more days *(25)*, the levels remain elevated to the same extent (Fig. 6). With recovery from thrombocytopenia (Figs. 5 and 6), the platelet count increases and TPO levels rapidly decrease.

TPO levels vary inversely and proportionally to the platelet count in most experimental models of thrombocytopenia. Following the generation of thrombocytopenia in animals by antiplatelet antibody *(26,27)*, irradiation *(27)*, or chemotherapeutic agents

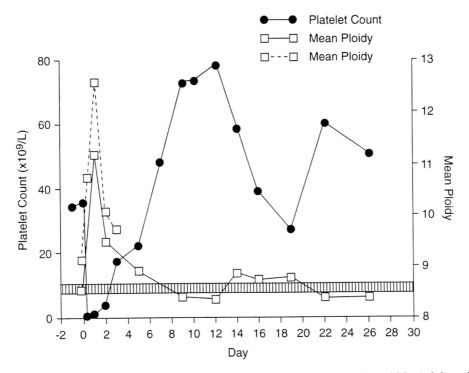

Fig. 5. The appearance of TPO during acute thrombocytopenia in the rabbit. A 3-kg rabbit was injected on day 0 with 3 mL of goat antibody against rabbit platelets, and the platelet count determined daily. Blood samples were obtained from day 2 to day 26, and their TPO content determined by a sensitive bioassay *(25)*, which measured the extent of megakaryocyte ploidization (mean ploidy) in vitro. The hatched area indicates the normal (preinjection) values (±2 SD). All TPO (mean ploidy) values falling outside the hatched area are statistically different ($p < 0.05$) from the normal (preinjection) values. Also shown (dashed line) are the results of a second assay of samples from this animal showing additional data for time-points between 0.3 and 3 d *(25)*. (Reprinted with permission from ref. *25*.)

(26–28), endogenous TPO levels are inversely proportional (Fig. 7) to the platelet count (or more precisely to the platelet mass). TPO levels increase exponentially as the platelet count decreases linearly (Fig. 8). In humans undergoing myeloablative chemotherapy, normal levels of 85 ± 7 pg/mL increase exponentially as the platelet count decreases and reach a peak level of 1268 ± 139 pg/mL at the platelet nadir *(29,30)*. A similar relationship exists between linear changes in the red blood cell mass in iron deficiency and the exponential response of the concentration of endogenous EPO in the circulation *(31)*.

Transfusion of platelets into thrombocytopenic animals *(26)* rapidly decreases the circulating concentration of TPO. When thrombocytopenic rabbits were given platelet transfusions, the circulating concentration of endogenous TPO rapidly declined, with an apparent half-life of <45 minutes (Fig. 9). When the effects of the platelet transfusion had abated, TPO levels again increased, only to decrease once the endogenous platelet production finally increased.

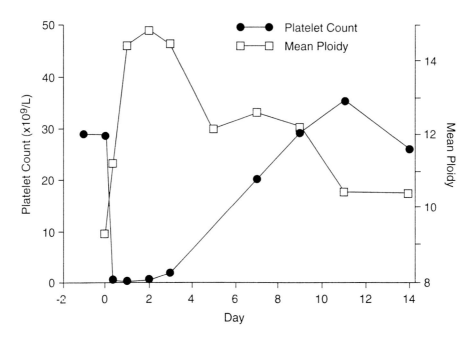

Fig. 6. The relationship of TPO levels to the platelet count after the induction of acute, prolonged thrombocytopenia. A 3-kg rabbit was made thrombocytopenic by the injection at day 0 of a large volume of goat antibody reacting against rabbit platelets, and serial plasma samples were assayed for TPO content (mean ploidy) as described in Fig. 5 *(25)*. (Reprinted with permission from ref. *25.*)

Circulating levels of endogenous TPO are regulated primarily by direct clearance of TPO from the circulation. The reduction of TPO concentration after platelet transfusion described above could be explained by one of two mechanisms. Either the transfused platelets turned off the "platelet sensor," or the platelets directly bound and removed TPO from the circulation. To explore the latter mechanism, platelets were added to thrombocytopenic plasma in an amount that would approximate a normal platelet count in vitro. As shown in Fig. 10, addition of platelets (but not any other blood cell) to thrombocytopenic plasma removed virtually all the endogenous TPO *(32)*. Similar results had long ago been described by de Gabriele and Penington *(33)* and Mazur et al. *(34)*, who found that the stimulatory activity of thrombocytopenic plasma in their assays was reduced when incubated in vitro with viable platelets. When thrombocytopenic plasma was then reconstituted in vitro with variable amounts of platelets to approximate decreasing degrees of thrombocytopenia, the residual level of TPO in the samples was approximately the same as that found in vivo in animals with the same degree of thrombocytopenia (Fig. 11). Since platelets have been shown to contain large amounts of *c-mpl* mRNA *(35)*, the number of TPO receptors on platelets was determined by Scatchard analysis (Table 1). Rat platelets contain 189 ± 20 receptors/cell, whereas sheep platelets have approximately 367 ± 50 receptors/cell, both with approximately equivalent binding constants. This difference in receptor number may account for some of the difference between the platelet counts of rats ($1147 \pm 55 \times 10^9$/L) and sheep ($508 \pm 86 \times 10^9$/L). Although the details of how platelets bind and

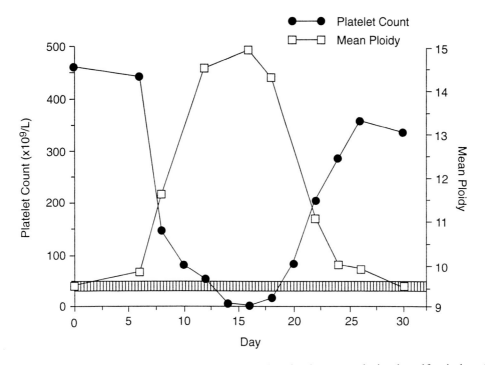

Fig. 7. TPO levels are inversely proportional to the platelet count during busulfan-induced thrombocytopenia in the rabbit. A 3-kg rabbit was injected with busulfan on day 0 and day 3 *(26)*, and blood samples were obtained from day 0 to day 100 after the first injection of busulfan. The level of TPO was quantified by measuring the stimulation of megakaryocyte mean ploidy in a sensitive bioassay *(26)* and is shown for the first 30 days as a function of the platelet count. The pretreatment (day 0) TPO (± 2 SD) level is indicated by the hatched bar. All values outside this region show a statistically significant difference (*p* < 0.05) from pretreatment values. When compared with a purified TPO standard, the pretreatment TPO concentration was 0.25 p*M*, whereas that during the platelet nadir was approximately 50 p*M (26)*. (Reprinted with permission from ref. *26*.)

clear TPO from the circulation have not yet been worked out, it is probably through receptor-mediated endocytosis. Once bound, platelets degrade endogenous TPO *(23)*.

As illustrated by Fig. 12, these observations have led to a new model *(32)* of the regulation of platelet production by TPO. TPO is constitutively produced by the liver and enters the circulation. In the absence of platelets during thrombocytopenia (or in the presence of a normal number of platelets with a defective TPO-clearance mechanism), there is little clearance of TPO, levels increase, bone marrow megakaryocytes are stimulated, and platelet production increases. In contrast, in the presence of platelets (or in the presence of a reduced number of platelets with an enhanced clearance mechanism), TPO clearance by the platelets increases, levels are low, megakaryocytes are not stimulated, and basal platelet production ensues.

Considerable recent experimental evidence supports this model. TPO mRNA seems to be constitutively produced in the liver. When measured by Northern blot analysis *(23,28,29)* and quantitative polymerase chain reaction (PCR) *(36)*, there is no increase

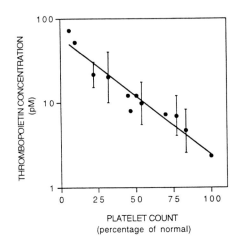

Fig. 8. TPO levels are inversely and exponentially proportional to the platelet count during busulfan-induced thrombocytopenia in sheep. Six sheep were injected with busulfan on day 0 and day 3 *(32)* and blood samples taken daily thereafter. TPO levels were quantified using a sensitive bioassay standardized with purified TPO *(26)* and are plotted (±SD) vs the platelet count. $r^2 = 0.933$.

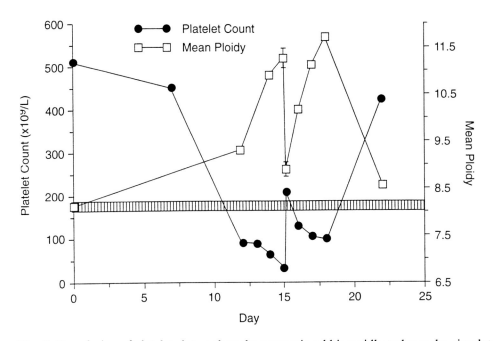

Fig. 9. Transfusion of platelets into a thrombocytopenic rabbit rapidly reduces the circulating concentration of TPO. A 3-kg rabbit was injected on day 0 and day 3 with busulfan *(26)*, and the platelet count and TPO level measured at frequent intervals. TPO was measured using a sensitive bioassay that quantified the stimulation of megakaryocyte ploidy (mean ploidy) in vitro *(26)*. At the beginning of the platelet nadir, purified rabbit platelets were infused over 3 minutes. The hatched bar indicates the TPO content of pretreatment rabbit plasma (±2 SD). All assay values outside this range differ from normal with a statistical significance of $p < 0.05$ *(26)*. (Reprinted with permission from ref. *26*.)

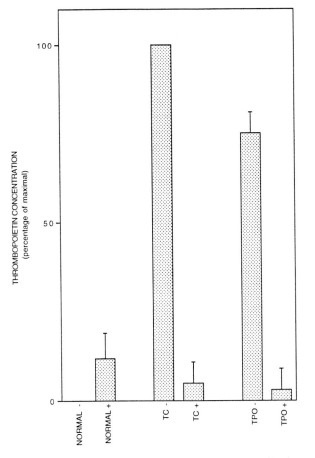

Fig. 10. Platelets bind TPO. Sheep platelets (+) prepared as described *(32)* or an equivalent amount of buffer (–) were incubated for 1 hour at room temperature with normal and thrombocytopenic (TC) sheep plasma, as well as with normal sheep plasma into which purified sheep TPO had been added (to a level approximately 75% that found in the TC plasma). After centrifugation twice at 3000*g* for 15 minutes, the supernatants were assayed for residual TPO activity (±SD) using a sensitive bioassay with 100% being defined as that in TC plasma and 0% that in normal plasma. (Reprinted with permission from ref. *32.*)

in the amount of TPO mRNA in the liver, the major site of TPO production during thrombocytopenia, despite a 100-fold increase in the circulating level of TPO. Although these experiments did not assess possible translational control of TPO production, analysis of the recent TPO "knockout" mice *(23)* seems to exclude such a mechanism. The platelet count and circulating levels of TPO in the heterozygote TPO "knockout" mice were half of that of normal mice, a gene dosage effect that can only be explained by constitutive, unregulated expression of TPO (*see* Chapter 21).

In general, the circulating levels of late-acting hemopoietic growth factors (e.g., EPO, G-CSF, monocyte colony-stimulating factor [M-CSF]) are determined by two different, but not exclusive, mechanisms: change in the production of the factor or change in the clearance of the factor from the circulation by the terminally differ-

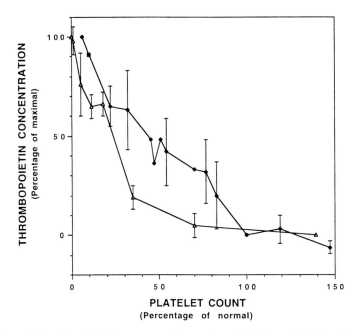

Fig. 11. The relationship of TPO concentration to the platelet count in vitro is comparable to the relationship found in vivo. A pool of thrombocytopenic plasma was made from several severely thrombocytopenic sheep (platelet count 1% of normal), and various amounts of sheep platelets were added, incubated for 1 hour at room temperature, centrifuged, and the residual TPO activity in the supernatant determined as described in Fig. 10. The TPO concentration in the pooled plasma was assigned a value of 100% and that in pretreatment plasma (platelet count 100% of normal) was assigned a value of 0%. The in vitro TPO concentrations (±SD, open triangles) are compared with the TPO concentrations (±SD, filled squares) obtained in six sheep in vivo (*see* Fig. 8). (Reprinted with permission from ref. *32*.)

entiated cell it regulates. EPO levels are primarily determined by the first mechanism *(7)*. An efficient renal "sensor" measures oxygen delivery to the kidney and, in turn, modulates transcription of EPO mRNA. Mature red blood cells do not contain receptors for EPO and do not clear it from the circulation. In contrast, M-CSF levels are almost entirely accounted for by the second mechanism. Monocytes contain receptors for M-CSF and actively clear it from the circulation. A decline in the monocyte count reduces clearance of M-CSF from the circulation, and levels increase, as does monocyte production *(37–39)*. There is little or no change in the transcription of M-CSF mRNA. G-CSF levels are determined by both mechanisms. The terminally differentiated neutrophil binds and clears G-CSF from the circulation and, during neutropenia, levels of G-CSF increase *(40–44)*. In addition, other cytokines can alter the transcription of G-CSF mRNA and thereby alter G-CSF levels (45,46).

There is no "sensor" of the platelet mass; instead, similar to the mechanism for neutrophils and monocytes, the platelet itself determines the circulating level of TPO. Whether other stimuli can affect the constitutive synthesis of TPO is not clear at present. It is conceivable that other physiological stimuli, drugs, or cytokines may alter the rate

Table 1
Characterization of the Receptors on Rat
Platelets[a]

	K_d, pM	Receptors/platelet
20°C ($n = 3$)	38 ± 10	223 ± 32
4°C ($n = 3$)	127 ± 45	189 ± 20

[a]Scatchard analysis was performed as previously described
(32) using purified TPO and rat platelets. Rat platelets were harvested from platelet-rich plasma by centrifugation and the pelleted
platelets (400×10^6 platelets) resuspended in 1 mL of platelet-
poor plasma-derived serum (previously dialyzed versus Hank's
Balanced Salt Solution) containing varying amounts of TPO,
incubated for 1 hour at either 20°C or 4°C, and the platelets then
removed by centrifugation. Linear Scatchard plots were obtained
in all experiments. The data are the average ($\pm$SD) of three separate experiments using platelets from different rats.

of production of TPO by the liver; what is clear is that the circulating level of platelets
is not one of these.

4. Implications and Predictions of this Model of the Regulation
of Platelet Production

There are a number of predictions that emanate from this model of platelet production. First, it provides a unified explanation for pathological disorders of platelet production. It predicts that some disorders of platelet production will result from defects in
the ability of platelets to clear TPO from the circulation. Changes in the amount of
TPO-receptor or binding characteristics of the TPO receptor or the endocytosis mechanism could increase or decrease the platelet's clearance of TPO, and thereby, alter
platelet production. For example, the elevated platelet mass seen in essential thrombocythemia (ET) may be owing to a qualitative defect in ET platelets. The platelets in this
clonal disorder may simply have a defect in their ability to bind and clear TPO from the
circulation. This would result in a transient increase in TPO concentration, an increase
in platelet production accompanied by an increase in TPO clearance, and ultimately, a
new, higher steady-state platelet count and only a slightly elevated TPO level. TPO
levels in patients with ET have recently been measured *(30,47)* and are only slightly
greater (339 ± 101 pg/mL) than those in normal individuals (85 ± 7 pg/mL) (*see* Chapter 22). Although this may not initially appear to be a significant increase in TPO level
compared with the much higher TPO levels seen in patients with aplastic anemia or
after chemotherapy, at the new steady state, this concentration of TPO may be more
than adequate to maintain an increased rate of platelet production.

In another clinical example, chronic ITP, TPO levels *(29,48)* were found to be normal. Compared with the several hundred-fold rise described in animal models of acute
ITP or in animals or humans with thrombocytopenia as a result of chemotherapy, this
appears to present a paradox. Although some *(29,48)* have sought to attribute these
normal levels in ITP to increased binding of TPO to the increased number of bone
marrow megakaryocytes, a simpler explanation comes from considering the differences

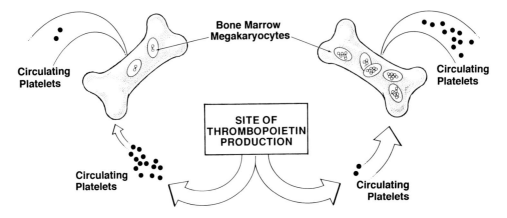

Fig. 12. The proposed mechanism by which TPO regulates megakaryocyte growth and platelet production. (Reprinted with permission from ref. *32*.)

in the platelet production rate between the models of acute ITP in animals, chronic ITP in humans, and after chemotherapy in animals or humans. In the acute ITP models (lasting only 1–3 days), the platelet production rate has barely increased above normal and, in the chemotherapy models, is actually reduced. In marked contrast, in chronic ITP, the production rate is increased five- to eightfold *(20)*. Given the constitutive rate of TPO production, in thrombocytopenic patients with a normal or reduced production rate of platelets, overall TPO clearance is reduced and levels increase greatly. However, in the presence of an increased production rate (of younger, larger, more metabolically active platelets) in chronic ITP, there is increased clearance of TPO and normal levels of TPO. This situation is identical to many other physiological regulatory systems in which the regulatory cytokine is only marginally increased at any elevated steady-state value of cell production. For example, during severe, compensated hemolytic anemias, EPO levels are only slightly elevated above baseline.

A second prediction of this model is that small molecules may be able to alter the platelets' ability to clear TPO and subsequently increase or decrease the rate of platelet production. The unexplained decrease in the platelet count of patients treated with the drug anagrelide *(49)* may be because anagrelide alters cAMP levels in platelets and thereby increases the platelets' ability to endocytose TPO. Conversely, after the administration of low doses of vincristine, there is a brief burst of platelet production *(21,50,51)*. This effect may be owing to the paralysis of platelet endocytosis of TPO, transient elevation of TPO levels, and an associated increase in platelet production until a new steady state is reached. Molecules similar to anagrelide or vincristine might be synthesized to alter the platelets' ability to clear TPO, and thereby decrease or increase platelet production.

The third implication of this model is that other physiological effects, cytokines, or drugs may alter the basal hepatic production rate of TPO. Although it is clear that the circulating platelet mass does not have any effect on endogenous TPO production, there is every reason to expect other physiological regulators may alter *TPO* gene expression just as interleukin (IL)-1, tumor necrosis factor (TNF)-α, and infection all suppress *EPO* gene expression *(52–54)*.

This model also explains all of the older platelet physiology data. For example, platelet mass, not number, is regulated because it is the total mass of platelets (circulating and splenic pool) that is responsible for the clearance of TPO from the circulation.

The ability of platelet transfusions after 3, but not after 24 hours, to abort the expected megakaryocyte ploidy increase after the onset of acute thrombocytopenia *(19)* is now clearly seen as being because of the 24 hours it requires for TPO levels to increase. With the acute destruction of almost all platelets after antiplatelet antibody administration, most TPO clearance (but probably not the small contribution of clearance by TPO receptors on endothelial cells) is eliminated, and TPO levels increase at a rate reflecting the constitutive hepatic TPO production.

In all of these acute, antibody-induced thrombocytopenia models, there has been one curious finding: the TPO curves are assymetrical *(26,27)*, i.e., at the same reduced platelet count, the TPO levels are invariably lower when the counts are increasing than when the counts are decreasing. This most likely reflects differences in the average age of platelets and their relative ability to clear TPO; older platelets would be expected to clear TPO less well than the younger, larger cohorts.

The discrepancy between platelet number and ploidy in chronic inflammatory states *(20)* is now also easily understood. The inflammatory stimulus probably mediated by IL-6 increases the number of MK-CFC and ultimately the platelet mass. However, the increased mass of functionally normal platelets causes increased clearance of TPO, decreased levels, and a decrease in megakaryocyte ploidy. In contrast to ITP and essential thrombocythemia in which normal or slightly elevated levels of TPO are predicted, those in chronic inflammatory states should be reduced. Recent studies *(29,30)* have confirmed the normal or slightly elevated levels in the former two conditions, but no studies have reported TPO levels in thrombocytosis associated with inflammatory conditions.

Yan et al. *(55)* have described a fivefold increase in the platelet count of mice transplanted with bone marrow overexpressing the murine *TPO* gene. As the platelet count increased, the initially markedly elevated levels of TPO declined to normal values despite constant production of the cytokine. Presumably as the platelet mass increased, it increased the total clearance of TPO and decreased the level almost to normal, despite a greatly increased platelet production rate.

Finally, Shivdasani et al. *(56)* have recently disrupted the hemopoietic transcription factor NF-E2 in mice and found that homozygous (NF-E2$^{-/-}$) mice were absolutely thrombocytopenic, and yet 5–10% survived into adulthood. Bone marrow megakaryocytes in these mice were increased two- to fourfold and had an appropriately increased ploidy, but lacked granules and demarcation membranes (*see* Chapter 12). Surprisingly, TPO levels were not elevated despite normal amounts of transcripts for both TPO and its receptor. Although not yet supported by experimental evidence, the most likely explanation for the low TPO levels is that megakaryocyte membrane fragments, not intact platelets, are being shed into the circulation and remove TPO. Alternatively, megakaryocytes may bind and metabolize TPO. Such membrane fragments are hemostatically active *(57)* and could also explain the striking lack of bleeding in some of the animals.

5. Is TPO the Only Regulator of Platelet Production?

The data reviewed above conclusively prove that TPO is the only physiologically relevant regulator of platelet production described to date. However, other molecules, such as IL-6, IL-11, and IL-3, can also increase megakaryocyte growth in vitro as well as increase the platelet count when administered in vivo *(58)* (*see* Chapter 10). As has been demonstrated in vitro, these molecules may synergize with TPO and enhance its effects in vivo. They may also contribute to the basal platelet production rate seen in the *c-mpl* and TPO "knockout" mice. They may also be responsible for the thrombocytosis associated with inflammation. For example, in Castleman's disease, platelet counts are usually markedly elevated and IL-6 levels are strikingly increased. When antibody to IL-6 was administered to a patient with this disorder, the platelet count rapidly declined *(59)*.

6. Conclusions and Future Directions

Since the identification and purification of TPO, much has been learned about the regulation of platelet production. The process appears to be a simple one of constant hepatic TPO production by the liver and direct clearance of TPO from the circulation by the platelets. Although some patients will be found to have defects in the production of TPO or its receptor, most of the known normal and pathological states of platelet production will probably be caused by variations in the ability of platelets to bind and metabolize TPO. It is not just the quantity of platelets, but their quality that is important in this interaction. Variations in the ability of platelets to metabolize TPO are probably just as important as variations in the mass of platelets with regard to how the circulating level of TPO is determined in normal and pathological states. Pharmacological agents might be developed to alter the ability of platelets to metabolize TPO, and thereby modify platelet production and possibly platelet function.

Although much remains to be learned about the direct interaction of normal and pathological platelets with TPO, other areas of potential physiological regulation remain almost totally unexplored. These areas are not dependent on TPO, and include the basic process by which cells commit to the megakaryocytic lineage and the mechanism of platelet formation from mature megakaryocytes. Other as yet uncharacterized, physiologically relevant hemopoietic factors may be responsible for these events.

Acknowledgment

This work was supported by NIH Grant HL54838.

References

1. Kelemen E, Cserhati I, Tanos B. Demonstration and some properties of human thrombopoietin in thrombocythaemic sera. *Acta Haematol.* 1958; 20: 350–355.
2. Bizzozero G. Uber einen neuen formbestandtheil des blutes und dessen rolle bei der thrombose-und der blutgerinnung. *Virchows Arch.* 1882; 90: 261–332.
3. Wright JH. The origin and nature of blood plates. *Boston Med Surg J.* 1906; 154 : 643–645.
4. Wright JH. The histogenesis of the blood platelets. *J Morphol.* 1910; 21: 263–278.
5. Bachmann S, LeHir M, Eckardt KU. Co-localization of erythropoietin mRNA and ecto-5'-nucleotidase immunoreactivity in peritubular cells of rat renal cortex indicates that fibroblasts produce erythropoietin. *J Histochem Cytochem.* 1993; 41: 335–341.
6. Lacombe C, Bruneval P, Da Silva J-L, et al. Production of erythropoietin in the kidney. *Semin Hematol.* 1991; 28: 14–19.
7. Goldberg MA, Imagawa S, Strair RK, Bunn HF. Regulation of the erythropoietin gene in Hep 3B cells. *Semin Hematol.* 1991; 28: 35–39.

8. Brecher G, Schneiderman M, Cronkite EP. The reproducibility and constancy of the platelet count. *Am J Clin Pathol.* 1953; 23: 15–26.

9. Giles C. The platelet count and mean platelet volume. *Br J Haematol.* 1981; 48: 31–37.

10. vonBehrens WE. Mediterranean macrothrombocytopenia. *Blood.* 1975; 46: 199–208.

11. vonBehrens WE. Evidence of phylogenetic canalisation of the circulating platelet mass in man. *Thrombosis Diathesis Haemorrhagica.* 1972; 27: 159–172.

12. Levin J, Bessman DJ. The inverse relation between platelet volume and platelet number. Abnormalities in hematologic diseases and evidence that platelet size does not correlate with platelet age. *J Lab Clin Med.* 1983; 101: 295–307.

13. Thompson CB. From precursor to product: How do megakaryocytes produce platelets? In: Levine RF, Williams N, Levin J, Evatt BL (eds). *Megakaryocyte Development and Function.* New York: Liss; 1986: 361–371.

14. Aster RH. Pooling of platelets in the spleen: Role in the pathogenesis of "hypersplenic" thrombocytopenia. *J Clin Invest.* 1966; 45: 645–657.

15. Aster RH. Studies of the mechanism of "hypersplenic" thrombocytopenia in rats. *J Lab Clin Med.* 1967; 70: 736–751.

16. Penny R, Rozenberg MC, Firkin BG. The splenic platelet pool. *Blood.* 1966; 27: 1–16.

17. de Gabriele G, Penington DG. Regulation of platelet production: "hypersplenism" in the experimental animal. *Br J Haematol.* 1967; 13: 384–393.

18. Harker LA. Kinetics of thrombopoiesis. *J Clin Invest.* 1968; 47: 458–465.

19. Kuter DJ, Rosenberg RD. Regulation of megakaryocyte ploidy in vivo in the rat. *Blood.* 1990; 75: 74–81.

20. Harker LA, Finch CA. Thrombokinetics in man. *J Clin Invest.* 1969; 48: 963–974.

21. Jackson CW, Edwards CC. Evidence that stimulation of megakaryocytopoiesis by low dose vincristine results from an effect on platelets. *Br J Haematol.* 1977; 36: 97–105.

22. Gurney AL, Carver-Moore K, de Sauvage FJ, Moore MW. Thrombocytopenia in *c-mpl*-deficient mice. *Science.* 1994; 265: 1445–1447.

23. de Sauvage FJ, Luoh SM, Carver-Moore K, et al. Deficiencies in early and late stages of megakaryocytopoiesis in TPO-KO mice. *Blood.* 1995; 86: 255 (abstract).

24. Choi ES, Hokom MM, Chen JL, et al. The role of megakaryocyte growth and development factor (MGDF) in terminal stages of thrombopoiesis. *Blood.* 1995; 86: 285 (abstract).

25. Kuter DJ, Rosenberg RD. The appearance of a megakaryocyte growth-promoting activity, megapoietin, during acute thrombocytopenia in the rabbit. *Blood.* 1994; 84: 1464–1472.

26. Kuter DJ, Rosenberg RD. The reciprocal relationship of thrombopoietin (c-Mpl ligand) to changes in the platelet mass during busulfan-induced thrombocytopenia in the rabbit. *Blood.* 1995; 85: 2720–2730.

27. Hunt P, Li L, Nichol JL, Hokom MM, et al. Purification and biologic characterization of plasma-derived megakaryocyte growth and development factor (MGDF). *Blood.* 1995; 86: 540–547.

28. Ulich TR, del Castillo J, Yin S, et al. Megakaryocyte growth and development factor ameliorates carboplatin-induced thrombocytopenia in mice. *Blood.* 1995; 86: 971–976.

29. Chang M, Suen Y, Meng G, et al. Regulation of TPO mRNA expression and protein production: TPO gene regulation appears post transcriptional, and endogenous levels are inversely correlated to megakaryocyte mass and circulating platelet count. *Blood.* 1995; 86: 368a.

30. Nichol J, Hornkohl A, Selesi D, Wyres M, Hunt P. TPO levels in plasma of patients with thrombocytopenia or thrombocytosis. *Blood.* 1995; 86: 371a.

31. Erslev AJ. Erythropoietin titers in health and disease. *Semin Hematol.* 1991; 28: 2–8.

32. Kuter DJ, Beeler DL, Rosenberg RD. The purification of megapoietin: a physiological regulator of megakaryocyte growth and platelet production. *Proc Natl Acad Sci USA.* 1994; 91: 11,104–11,108.

33. de Gabriele G, Penington DG. Regulation of platelet production: "thrombopoietin." *Br J Haematol.* 1967; 13: 210–215.

34. Mazur EM, de Alarcon P, South K, Miceli L. Evidence that human megakaryocytopoiesis is controlled in vivo by a humoral feedback regulatory system. In: Golde DW, Marks PA (eds). *Normal and Neoplastic Hematopoiesis.* New York: Liss; 1983: 533–543.

35. Debili N, Wendling F, Cosman D, et al. The Mpl receptor is expressed in the megakaryocyte lineage from late progenitors to platelets. *Blood.* 1995; 85: 391–401.

36. Stoffel R, Wiestner A, Skoda RC. Thrombopoietin in thrombocytopenic mice: Evidence against regulation at the mRNA level and for a direct regulatory role of platelets. *Blood.* 1996; 87: 567–573.

37. Bartocci A, Mastrogiannis DS, Migliorati G, Stockert RJ, Wolkoff AW, Stanley ER. Macrophages specifically regulate the concentration of their own growth factor in the circulation. *Proc Natl Acad Sci USA.* 1987; 84: 6179–6183.

38. Guilbert LJ, Stanley ER. The interaction of ^{125}I-colony-stimulating factor-1 with bone marrow-derived macrophages. *J Biol Chem.* 1986; 261: 4024–4032.

39. Tushinski RJ, Oliver IT, Guilbert LJ, Tynan PW, Warner JR, Stanley ER. Survival of mononuclear phagocytes depends on a lineage-specific growth factor that the differentiated cells selectively destroy. *Cell.* 1982; 28: 71–81.

40. Watari K, Asano S, Shirafuji N, et al. Serum granulocyte colony-stimulating factor levels in healthy volunteers and patients with various disorders as estimated by enzyme immunoassay. *Blood.* 1989; 73: 117–122.

41. Cebon J, Layton JE, Maher D, Morstyn G. Endogenous haemopoietic growth factors in neutropenia and infection. *Br J Haematol.* 1994; 86: 265–274.

42. Layton JE, Hockman H, Sheridan WP, Morstyn G. Evidence for a novel in vivo control mechanism of granulopoiesis: mature cell-related control of a regulatory growth factor. *Blood.* 1989; 74: 1303–1307.

43. Haas R, Gericke G, Witt B, Cayeux S, Hunstein W. Increased serum levels of granulocyte colony-stimulating factor after autologous bone marrow or blood stem cell transplantation. *Exp Hematol.* 1993; 21: 109–113.

44. Shimazaka C, Uchiyama H, Fujita N, et al. Serum levels of endogenous and exogenous granulocyte-stimulating factor after autologous blood stem cell transplantation. *Exp Hematol.* 1995; 23: 1497–1502.

45. Zsebo KM, Yuschenkoff VN, Schiffer S, et al. Vascular endothelial cells and granulopoiesis: Interleukin-1 stimulates release of G-CSF and GM-CSF. *Blood.* 1988; 71: 99–103.

46. Koeffler HP, Gasson J, Ranyard J, Souza L, Shepard M, Munker R. Recombinant human TNF-α stimulates production of granulocyte colony-stimulating factor. *Blood.* 1987; 70: 55–59.

47. Taylor K, Pitcher L, Nichol J, et al. Inappropriate elevation and loss of feedback regulation of thrombopoietin in essential thrombocythemia (ET). *Blood.* 1995; 86: 49a.

48. Emmons RVB, Shulman NR, Reid DM, et al. Thrombocytopenic patients with aplastic anemia have high TPO levels whereas those with immune thrombocytopenia are much lower. *Blood.* 1995; 86: 372a.

49. Anagrelide Study Group. Anagrelide, a therapy for thrombocythemic states: experience in 577 patients. *Am J Med.* 1992; 92: 69–76.

50. Carbone PP. Clinical studies with vincristine. *Blood.* 1963; 21: 640–647.

51. Harris RA, Penington DG. The effects of low-dose vincristine on megakaryocyte colony-forming cells and megakaryocyte ploidy. *Br J Haematol.* 1984; 57: 37–48.

52. Jelkmann W, Pagel H, Wolff M, Fandrey J. Monokines inhibiting erythropoietin production in human hepatoma cultures and in isolated perfused rat kidneys. *Life Sci.* 1992; 50: 301–308.

53. Faquin WC, Schneider TJ, Goldberg MA. Effect of inflammatory cytokines on hypoxia-induced erythropoietin production. *Blood.* 1992; 79: 1987–1994.

54. Vannucchi AM, Grossi A, Bosi A, et al. Effects of cyclosporin A on erythropoietin production by the human Hep3B hepatoma cell line. *Blood.* 1993; 82: 978–984.

55. Yan X-Q, Lacey D, Fletcher F, et al. Chronic exposure to retroviral vector encoded MGDF (mpl-ligand) induces lineage-specific growth and differentiation of megakaryocytes in mice. *Blood.* 1995; 86: 4025–4033.

56. Shivdasani RA, Rosenblatt MF, Zucker-Franklin D, et al. Transcription factor NF-E2 is required for platelet formation independent of the actions of thrombopoietin/MGDF in megakaryocyte development. *Cell.* 1995; 81: 695–704.
57. Scigliano E, Fruchtman S, Isola L, et al. Infusible platelet membrane (IPM) for control of bleeding in thrombocytopenic patients. *Blood.* 1995; 86: 446a.
58. Broudy VC, Lin NL, Kaushansky K. Thrombopoietin (*c-mpl* ligand) acts synergistically with erythropoietin, stem cell factor, and interleukin-11 to enhance murine megakaryocyte colony growth and increases megakaryocyte ploidy in vitro. *Blood.* 1995; 85: 1719–1726.
59. Beck JT, Hsu SM, Wijdenes J, et al. Brief report: alleviation of systemic manifestations of Castleman's disease by monoclonal anti-interleukin-6 antibody. *N Engl J Med.* 1994; 330: 602–605.

Glossary*

5-FU	5-flurouracil		*c-mpl*	cellular homologue of viral oncogene (*v-mpl*)
ABMT	autologous bone marrow transplantation		CPM	counts per minute
AChE	acetylcholinesterase		CN1	collagen type 1
ADP	adenosine diphosphate		CNTF	ciliary neurotrophic factor
AIDS	acquired immune deficiency syndrome		CR	complete remission/response
			CSF	colony-stimulating factor
ALL	acute lymphocytic leukemia		DEAE	diethylaminoethyl
AML	acute myeloid leukemia		DMEM	Dulbecco's minimal essential medium
ANC	absolute neutrophil count			
ATP	adenosine triphosphate		DMS	demarcation membrane system
AUC	area-under-the-curve		DMSO	dimethyl sulfoxide
AZT	zidovudine (azidothymidine)		DTS	dense tubular system
Baso-CFC	basophil colony-forming cell		EA	endarterectomized
BDNF	brain-derived neurotrophic factor		*E coli*	*Escherichia coli*
bFGF	basic fibroblast growth factor		E-BFC	erythroid burst-forming cell
BHK	baby hamster kidney		EBV	Epstein-Barr virus
Bl-CFC	blast colony-forming cell		E-CFC	erythroid colony-forming cell
BMT	bone marrow transplantation		EDTA	ethylenediamine tetraacetic acid
bp	base pair		EGF	epidermal growth factor
BSA	bovine serum albumin		ELISA	enzyme-linked immunosorbent assay
CAFC	cobblestone area-forming cell			
CALGB	Cancer and Leukemia Group B		EMS	2-ethylmethanesulphonate
cDNA	complementary DNA		Eo	eosinophil (s)
CFC	colony-forming cell		Eo-CFC	eosinophil colony-forming cell
CFR	cyclic flow reduction		EPO	erythropoietin
CHAPS	(3-[3-cholamindopropyl) dimethyl-ammonio]-1-propanosulfonate		ES cell	embryonic stem cell
			EST	expressed sequence tags
CHO	Chinese hamster ovary		ET	essential thrombocythenia
CM	carboxymethyl		FAB	French-American-British
CML	chronic myeloid/myelogenous leukemia		FACS	fluorescence-activated cell sorting
			FBS	fetal bovine serum

*Note on Terminology of Mpl Ligand, MGDF, and TPO: We have adopted the following standard of terms: Mpl ligand is a generic term for both the endogenous and the recombinant-derived molecular species; native thrombopoietin (TPO) is the preferred term for the endogenous Mpl ligand. Various recombinant-derived forms of thrombopoietin are specified where necessary using their species name, e.g., rHuTPO. The term "megakaryocyte growth and development factor" (MGDF) refers to a recombinant, *E coli*-derived form of Mpl ligand, encompassing the EPO-like domain, developed for clinical use.

Note on Terminology of Progenitor Cells: We have adopted a standard terminology for hemopoietic progenitor cells of the following style: granulocyte-macrophage colony-forming cell = GM-CFC (= CFU-GM, granulocyte-macrophage colony-forming unit).

FDA	US Food and Drug Administration	MAPK	mitogen activation protein kinase
FISH	fluorescence *in situ* hybridization	MDS	myelodysplastic syndromes
FITC	fluorescein isothiocynate	Meg-CSA	megakaryocyte colony-stimulating activity
fL	femtoliter	MGDF	megakaryocyte growth and development factor
fmol	femtomole		
F-MuLV	Friend murine leukemia virus	MIP	macrophage-inhibiting factor
G-CFC	granulocyte colony-forming cell	MK	megakaryocyte
G-CSF	granulocyte colony-stimulating factor	MK-BFC	megakaryocyte burst-forming cell
GEMM -CFC	granulocyte-erythroid-macrophage-megakaryocyte colony-forming cell	MK-CFC	megakaryocyte colony-forming cell
		MK-Pot	megakaryocyte potentiating factor
GM-CFC	granulocyte-macrophage colony-forming cell	MNC	mononuclear cell
		Mpl	receptor for endogenous thrombopoietin
GM-CSF	granulocyte-macrophage colony-stimulating factor	Mpl-L	ligand for Mpl
GP	platelet glycoprotein	MPLV	myeloproliferative leukemia virus
GPA	glycophorin	MPV	mean platelet volume
GTP	guanosine triphosphate	M_r	relative molecular weight
GVHD	graft-versus-host disease	MSF	megakaryocyte stimulating factor
Gy	Gray	Multi-CSF	multi-colony-stimulating factor, IL-3
HEK	human embryonic kidney		
HGF	hemopoietic growth factor	NAP	neutrophil-activating product
HIV	human immunodeficiency virus	NHL	non-Hodgkin's lymphoma
HPLC	high-performance liquid chromatography	OCS	open canalicular system
		OSM	oncostatin
HPPC	high proliferative-potential cell	PAF	platelet-activating factor
HPP-CFC	high proliferative-potential colony-forming cell	PAGE	polyacrylamide gel electrophoresis
		PBPC	peripheral blood progenitor cell
HRD	hematopoietin receptor domain	PBSC	peripheral blood stem cell
HRP	horseradish peroxidase	PCR	polymerase chain reaction
Hu	human	PCV	packed cell volume
IFN	interferon	PDGF	platelet-derived growth factor
IL	interleukin	PE	phycoerythrin
IMP	intramembranous particle	PEG	poly [ethylene glycol]
ITP	immune thrombocytopenic purpura	PF	platelet factor
IU	International unit	PF-MK	proplatelet-displaying megakaryocyte
kb	kilobase		
kd	kilodalton	PHA- LCM	phytohemagglutinin antigen–lectin-stimulated lymphocytes
LCK	lymphocyte-specific protein tyrosine kinase		
LCR	locus control region	PMN	polymorphonuclear neutrophil
LIBS	ligand-induced binding site	PPP	platelet-poor plasma
LIF	leukemia-inhibitory factor	PRP	platelet-rich plasma
LTBMC	long-term bone-marrow culture	PSCT	peripheral stem cell transplantation
LTC-IC	long-term culture-initiating cell	r	recombinant
LTR	long terminal repeats	RGDS	arginine-glycine-aspartate-serine
M-CFC	macrophage/monocyte colony-forming cell	RBC	red blood cell
		rc	recombinant canine
M-CSF	monocyte colony-stimulating factor	RER	rough endoplasmic reticulum

rHu	recombinant human	TBI	total body irradiation
rMu	recombinant murine	TGF	transforming growth factor
ro	recombinant ovine	TMS	trimethyl silane
rp	recombinant porcine	TNF	tumor necrosis factor
RPM	rat promegakaryoblast	TPO	thrombopoietin
rRh	recombinant rhesus monkey	TRP	thrombocytopenic rat plasma
RT-PCR	reverse transcriptase polymerase chain reaction	TSF	thrombocytosis-stimulating factor
S-CFC	spleen colony-forming cell	TTP	thrombotic thrombocytopenic purpura
SCCS	surface-connected canalicular system	UV	ultraviolet
SCF	stem cell factor	VG	vascular graft
SCLC	small-cell lung cancer	*v-mpl*	oncogene of myeloproliferative leukemia virus
SDS/PAGE	sodium dodecyl sulfate polyacryl-amide gel electrophoresis	VN	vitronectin
s-Mpl	soluble form of TPO receptor	vWF	von Willebrand factor
TAR syndrome	thrombocytopenia absent radius syndrome	WBC	white blood cell
		WF	Wistar-Furth
		WGA	wheat germ aggultinin

Index